Microbiology in the Era of Artificial Intelligence

Written by leading experts, the book covers a broad range of topics pertaining to the myriad uses of artificial intelligence in microbiology. The book explores how AI and computation can play a key role in understanding and uncovering microscopic mysteries that defy other means of microbiological study. Like other fields of life science, the impact of next generation sequencing and bioinformatics are revolutionizing microbiology. In addition, the emerging role of quantum and nanotechnology in understanding the nature of microbial life is also explored. A special feature of the book is fascinating discussion of the transformation currently underway from classic microbiology to next generation microbiology. This is a must-read book for microbiology students and researchers who want to be at the forefront of this exciting field.

Key Features:

- Provides an overview and perspectives on the future of microbiology
- Documents recent advances in microbiology
- Contributions from an international team of leading researchers
- Reviews the emerging role of applications from other fields like nanotechnology, artificial intelligence, and genomics
- Stimulates academics and researchers to pursue multidisciplinary research

Microbiology in the Era of Artificial Intelligence

Nanotechnology, Quantum, and Next Generation Sequencing

Edited by
Muhammad Ali Syed, Shahzad Ali, and Tanveer Hussain

CRC Press
Taylor & Francis Group
Boca Raton London New York

CRC Press is an imprint of the
Taylor & Francis Group, an informa business

Designed cover image: Shutterstock

First edition published 2025
by CRC Press
2385 Executive Center Drive, Suite 320, Boca Raton, FL 33431

and by CRC Press
4 Park Square, Milton Park, Abingdon, Oxon, OX14 4RN

CRC Press is an imprint of Taylor & Francis Group, LLC

ISBN: 9781032712376 (hbk)
ISBN: 9781032727233 (pbk)
ISBN: 9781003410164 (ebk)

DOI: 10.1201/9781003410164

Typeset in Times
by Newgen Publishing UK

Dedication

Dedicated to our teachers, family and students.

Contents

About the Editors

Muhammad Ali Syed

Muhammad Ali Syed is serving as Chairman and Associate Professor in Microbiology at The University of Haripur, Pakistan. He has vast experience of teaching and research in microbiology. His areas of expertise are infectious diseases, vaccine efficacy, and nanomedicine.

Shahzad Ali

Shahzad Ali is Associate Professor in Wildlife and Ecology at the University of Veterinary and Animal Sciences, Lahore, Pakistan. He has long experience of teaching and research in zoonotic infectious diseases, microbial diversity and ecology, and the One Health concept.

Tanveer Hussain

Tanveer Hussain is currently employed at the Department of Biological Sciences of the Virtual University of Pakistan, Lahore. His areas of expertise are molecular biology and genomics. He has vast teaching and research experience in the relevant field.

Contributors

Anza Abbas
Department of Microbiology, The University of Haripur, Pakistan

Sidra Abbas
Microbiology and Biotechnology Research Laboratory, Department of Biotechnology, Fatima Jinnah Women University, Rawalpindi, Pakistan

Syed Muhammad Afzal
College of Basic Medicine, Jilin University, China

Rafiq Ahmad
Department of Microbiology, The University of Haripur, Pakistan

Gholamreza Ahmadian
Department of Industrial and Environmental Biotechnology, National Institute of Genetic Engineering and Biotechnology (NIGEB), Tehran, Iran

Mohsen Bidar Ajorloo
Biology Research Center, Faculty of Basic Sciences, Imam Hossein University, Tehran, Iran

Amjad Ali
Integrative Biology Laboratory, Atta-ur-Rahman School of Applied Biosciences (ASAB), National University of Sciences and Technology (NUST), Islamabad, Pakistan

Muhammad Muddassir Ali
Institute of Biochemistry and Biotechnology, University of Veterinary and Animal Sciences, Lahore, Pakistan

Shahzad Ali
Department of Wildlife and Ecology, University of Veterinary and Animal Sciences, Lahore, Pakistan

Muhammad Arif
Department of Bone Joint and Sports Medicine, Qingdao Hospital, University of Health and Rehabilitation Sciences, Qingdao, China

Furqan Awan
Department of Epidemiology and Public Health, University of Veterinary and Animal Sciences, Lahore, Pakistan

Howra Bahrulolum
Department of Industrial and Environmental Biotechnology, National Institute of Genetic Engineering and Biotechnology (NIGEB), Tehran, Iran

Fiza Faiz
Integrative Biology Laboratory, Atta-ur-Rahman School of Applied Biosciences (ASAB), National University of Sciences and Technology (NUST), Islamabad, Pakistan

Farah Alam Gill
Department of Biosciences, Shifa Tameer-e-Millat University, Islamabad, Pakistan

Uzair Haider
Microbial Genomics and Bioinformatics Research Group, Atta-ur-Rahman School of Applied Biosciences (ASAB), National University of Sciences and Technology (NUST), Islamabad, Pakistan

Abbas Hajizade
Biology Research Center, Faculty of Basic Sciences, Imam Hossein University, Tehran, Iran

Muhammad Hamza
Department of Bioinformatics, Government Postgraduate College Mandian Abbottabad, Pakistan

Laiba Hassan
Department of Microbiology, The University of Haripur, Pakistan

Muhammad Hayat
State Key Laboratory of Microbial Technology, Institute of Microbial Technology, Shandong University, Qingdao, China

Zakieh Sadat Hoseini
Department of Medical Nanotechnology, School of Advanced Technologies in Medicine, Tehran University of Medical Sciences, Iran

Muhammad Imran
Institute of Biochemistry and Biotechnology, University of Veterinary and Animal Science, Lahore, Pakistan

Usama Imtiaz
Department of Microbiology, The University of Haripur, Pakistan

Moein Iranmanesh
Biology Research Center, Faculty of Basic Sciences, Imam Hossein University, Tehran, Iran

Zaara Ishaq
Integrative Biology Laboratory, Atta-ur-Rahman School of Applied Biosciences (ASAB), National University of Sciences and Technology (NUST), Islamabad, Pakistan

Camellia Katalani
Biology Research Center, Faculty of Basic Sciences, Imam Hossein University, Tehran, Iran

Misbah Khalil
Department of Microbiology, The University of Haripur, Pakistan

Kishmain Ahmad Khan
Department of Microbiology, The University of Haripur, Pakistan

Sohail Khan
College of Resources and Environment, University of Sciences, Beijing, China

Sohrab Khan
National Marine Fisheries Research Institute, Gdynia, Poland

Ejaz Mohammad
Department of Microbiology, Government Postgraduate College Mandian, Abbottabad, Pakistan

Hafsah Muhammad
Institute of Basic Medical Sciences, Khyber Medical University, Peshawar, Pakistan

Javed Muhammad
Department of Microbiology, The University of Haripur, Pakistan

Hafiz Muhammad Waqas Munir
Department of Oncology, University of Turin, Italy

Saghi Nooraei
Department of Industrial and Environmental Biotechnology, National Institute of Genetic Engineering and Biotechnology (NIGEB), Tehran, Iran

Maria Rukan
Department of Microbiology, The University of Haripur, Pakistan

Maimoona Sabir
Department of Microbiology, The University of Haripur, Pakistan

Ayesha Saeed
Clinical and Biomedical Research Centre, Foundation University, Islamabad, Pakistan

Afshan Saleem
Department of Microbiology, The University of Haripur, Pakistan

Nayyer Siddique
Institute of Basic Medical Sciences, Khyber Medical University, Peshawar, Pakistan

Fariha Masood Siddiqui
Department of Biosciences, Shifa Tameer-e-Millat University, Islamabad, Pakistan

Anabia Sohail
Khalifa University of Science and technology, Abu Dhabi, United Arab Emirates

Muhammad Ali Syed
Department of Microbiology, The University of Haripur, Pakistan

Faouzia Tanveer
Department of Biosciences, Shifa Tameer-e-Millat University, Islamabad, Pakistan

Javeria Tanvir
Department of Physics, The University of Haripur, Pakistan

Yousof Tarverdizadeh
Biology Research Center, Faculty of Basic Sciences, Imam Hossein University, Tehran, Iran

Arif Ullah
Department of Microbiology, The University of Haripur, Pakistan

Mohsen Yaghoubi
Biology Research Center, Faculty of Basic Sciences, Imam Hossein University, Tehran, Iran

Mostafa Zare
Biology Research Center, Faculty of Basic Sciences, Imam Hossein University, Tehran, Iran

Syeda Fatima Zehra
Department of Computer Science, COMSATS University Islamabad, Wah Cantt, Pakistan

Foreword

The microbial world lies at the heart of the bioeconomy, driving innovations in sustainable agriculture, environmental management, and industrial biotechnology where microbes silently power industries, from agriculture to energy, by acting as nature's original recyclers, breaking down waste, producing biofuels, and enabling eco-friendly manufacturing processes. Furthermore, microbial fermentation is crucial in the production of pharmaceuticals, enzymes, and food products, making the microbial world a cornerstone for advancing a sustainable and circular bioeconomy. In the biotechnology industry, the integration of smart technologies—such as artificial intelligence, machine learning, and advanced data analytics—has revolutionized the approach to microbes and diseases, significantly impacting revenue generation. Smart technologies enhance the precision and efficiency of microbial diagnostics, enabling faster identification and characterization of pathogens, which in turn accelerates the development of targeted therapies and vaccines. By leveraging these technologies to improve healthcare outcomes and operational efficiencies, companies in the multi-billion-dollar biotechnology sector, such as diagnostics, nanodiagnostics, smart chip-based devices, microfluidic devices, drugs, therapeutics, monoclonals, biologics, vaccines, and probiotics, can enhance their competitive edge and achieve substantial revenue growth. The integration of smart technologies—such as artificial intelligence, machine learning, and advanced data analytics—has revolutionized how we approach microbes and diseases. Dr. Muhammad Ali Syed and his team have done a great job in exploring these avenues in an easy-to-comprehend manner for diverse readers—a valuable book for microbiologists and all biologists.

Prof. Habib Ali Bokhari (PhD)
Vice Chancellor, Kohsar University, Murree, Pakistan

Preface

Microbiology stands at the threshold of an evolutionary era where convergence of artificial intelligence, nanoscience and technology, next generation sequencing, and quantum is transforming human understanding of the world of microorganisms. The boundaries of scientific investigation are continuously expanding and the tools available for research are increasing and becoming more sophisticated. Classical microbiology is greatly influenced by the massive interdisciplinary research data available today and the level of automation and integrated systems to explore the world of microbes.

In this book, the authors embark on an exciting journey to explore the interesting intersection of microbiology with these cutting-edge fields that have already revolutionized many other areas of science and technology. In the modern age, microbiology may benefit from innovative research with these emerging technologies and tools. We delve into the possibilities and challenges of applying computation and bioinformatics for molecular biology data analysis of microbial identification, and the power of nanotechnology and quantum to explore the secrets of the microbial world, which still remains largely unexplored.

We attempt to inspire new generations of microbiologists to embrace the interdisciplinary approach to shape the future of microbiology. Join us in this exciting journey that uncovers wonders of the microbial world and unlocks the secrets of this planet's tiny mysterious creatures.

Muhammad Ali Syed (PhD)

Acknowledgements

Firstly, we would like to acknowledge the publisher CRC Taylor & Francis for offering us an opportunity to publish with them. We are grateful to Prof. Chuck Crumly and Mr. Kara Roberts for their continued support and guidance throughout the entire process.

We are grateful to our institutions, colleagues, families, and students for encouraging us in the task and providing necessary technical and moral support.

We would like to offer our sincerest gratitude to all the authors of this book for their valuable contribution, insight, and timely response.

Thanks to you all for your contribution, support, and dedication. This book would not have been possible without you.

Abbreviations

CHAPTER 1

AI = artificial intelligence
MALDI-TOF = matrix assisted laser desorption/ionization time-of-flight
MEMS = microelectromechanical systems
ML = machine learning
NGS = next-generation sequencing
VLPs = virus-like particles
WGS = whole genome sequencing

CHAPTER 2

ESI = electrospray ionization
GC-MS = gas chromatography-mass spectrometry
LC-MS = liquid chromatography with mass spectrometry
MALDI-TOF MS = matrix assisted laser desorption/ionization time-of-flight mass spectrometry
MS = mass spectrometry
NAATs = nucleic acid amplification tests
NGS = next-generation sequencing
OTU = operational taxonomic unit
PCR = polymerase chain reaction
qPCR = quantitative polymerase chain reaction

CHAPTER 3

Ag NPs = silver nanoparticles
Au NPs = gold nanoparticles
CVD = chemical vapor deposition
E. coli = *Escherichia coli*
ETEC = enterotoxigenic *E. coli*
PLGA = poly(lactic-co-glycolic acid)
SARS-CoV-2 = severe acute respiratory syndrome coronavirus
S. aureus = *Staphylococcus aureus*
SCFAs = short-chain fatty acids
SPIONs = superparamagnetic nanoparticles
SWCNs = single-walled carbon nanotubes
TiO_2 NPs = titanium dioxide nanoparticles
ZnO NPs = zinc oxide nanoparticles

CHAPTER 4

ATP = adenosine triphosphate
Au NPs = gold nanoparticles
AuNUs = gold nano-urchins
BA = biotin-streptavidin
CFU ml^{-1} = colony-forming units per milliliter
CNTs = carbon nanotubes

DPV = differential pulse voltammetric
ELISA = enzyme-linked immunosorbent assay
FRET = fluorescence resonance energy transfer
GO = graphene oxide
GQD = graphene quantum dots
HBsAb = hepatitis B surface antibody
HBsAg = hepatitis B surface antigen
HBV = hepatitis B virus
HCR = hybridization chain reaction
ICP-MS = inductively coupled plasma mass spectrometry
LOD = limit of detection
MDR = multiple drug-resistant
PCR = polymerase chain reaction
QDs = quantum dots
rGO = reduced graphene oxide
SEB = staphylococcal enterotoxin B
SELEX = systematic evolution of ligands by exponential enrichment
SPR = surface plasmon resonance

CHAPTER 5

AF4-MALS = asymmetric field flow coupled with multi-angle light scattering detectors
AFM = atomic force microscopy
AIPs = auto-inducing peptides
APCs = antigen-presenting cells
BHK = baby hamster kidney
CHO = Chinese hamster ovary
CSP = circumsporozoite protein
DENV = Dengue virus
DLS = dynamic light scattering
ES-DMA = electrospray-differential mobility analysis
eVLPs = enveloped virus-like particles
FD = freeze-drying
FDA = food and drug administration
GAS = group A *Streptococcus*
HBV = hepatitis B virus
HEK = human embryonic kidney
HEV = hepatitis E virus
HPV = human papillomavirus
MERS-CoV = middle east respiratory syndrome coronavirus
PEI = polyethyleneimine
PTMs = post-translational modifications
RSV = respiratory syncytial virus
RT-PCR = reverse transcription polymerase chain reaction
RV = rotavirus
SIV = simian immunodeficiency virus
SDS-PAGE = sodium dodecyl sulfate-polyacrylamide gel electrophoresis
ssRNA = single-stranded RNA
SSTIs = skin and soft tissue infections
TEM = transmission electron microscopy

TLRs = Toll-like receptors
VLPs = virus-like particles
VLP-J8 = modular carrier peptide J8

CHAPTER 6

AI = artificial intelligence
Apt1 = aptamer 1
BioMEMS = biomedical microelectromechanical systems
CA15-3 = carbohydrate antigen 15-3
CEA = carcinoembryonic antigen
CFC = carbon fiber cloth
CNTs = carbon nanotubes
ECL = electrochemiluminescence
EPS = extracellular polymeric substances
GCE = glassy carbon electrode
GMR = giant magnetoresistive sensor
ICs = integrated circuits
IoE = Internet of Everything
IoT = Internet of Things
LOC = lab-on-a-chip
MEMS = microelectromechanical systems
MOF = metal-organic framework
NPs = nanoparticles
OCS = oxygen storage capacity
PCOS = polycystic ovarian syndrome
PDA@Gr = polydopamine@graphene oxide
PRL = prolactin
PtCo NDs = platinum and cobalt alloyed nanodendrites
QDs = quantum dots
SPR = surface plasmon resonance

CHAPTER 7

AFM = atomic force microscopy
Ag NPs = silver nanoparticles
CARD-FISH = catalyzed reporter deposition FISH
CLASI-FISH = combinatorial labeling and spectral imaging FISH
CLEM = correlative light and electron microscopy
CLSEM = correlative light and scanning electron microscopy
CLSM = confocal laser scanning microscopy
Cryo-EM = cryo-electron microscopy
Cryo-ET = cryo-electron tomography
Cryo-SEM = cryo-scanning electron microscopy
DESI = desorption electrospray ionization
diSPIM = dual-view inverted selective plane illumination microscope
DOPE-FISH = double labeling of oligonucleotide probes for FISH
DVC = direct viable bacterial cells
EM = electron microscope/microscopy
EMDB = Electron microscopy data bank

f3D-SIM = Fast 3D-structured illumination microscopy
FIB-SEM = Focused ion beam scanning electron microscopy
FISH = fluorescence in situ hybridization
FP = fluorescent protein
FPALM = fluorescence photoactivation localization microscopy
FRE = fine root endophytes
HCI = high-content imaging
IMS = imaging mass spectrometry
INSTIs = integrase strand-transfer inhibitors
LAIV = live attenuated influenza vaccine
LLSM = lattice light-sheet microscopy
LM = light microscopy
LNA = locked nucleic acid
LSM = laser scanning microscopy
MALDI = matrix-assisted laser desorption ionization
MALDI IMS = MALDI imaging mass spectrometry
MP-IVM = multi-photon excitation intravital microscopy
mSPIM = multidirectional selective plane illumination microscopy
OPM = optical projection microscopy
NA = numerical aperture
NAM = nucleic acid mimics
NAM-FISH = nucleic acid mimics FISH
OPM = oblique plane microscopy
PALM = photoactivated localization microscopy
RBV = antiviral ribavirin
RCM = re-scan confocal microscopy
RING-FISH = recognition of individual fenes FISH
rRNA = ribosomal-RNA
SEM = scanning electron microscope
sfGFP = superfolder GFP
SIM = structured illumination microscopy
SIM-OS = structured illumination microscopy optical-sectioning microscopy
SIMS = secondary-ion mass spectrometry
SMLM = single-molecule localization microscopy
SPIM = Selective plane illumination microscopy
ssOPM = stage-scanning OPM
STED = stimulated emission depletion
STM = scanning tunneling microscope
STORM = random optical reconstruction microscopy
TEM = transmission electron microscope
TOF-SIMS = time-of-flight secondary ion mass spectrometry
TPEF = two-photon excited fluorescence

CHAPTER 8

FMO = Fenna-Matthews-Olson
QKD = quantum key distribution
QM = quantum microbiology

CHAPTER 9

2D-PAGE = two-dimensional polyacrylamide gel electrophoresis
AdhE2 = alcohol dehydrogenase
CE-MS = capillary electrophoresis-MS
ChIP = chromatin immunoprecipitation
FL = Fludioxonil
GC-MS = gas chromatography-MS
LAB = lactic acid bacteria
LC-MS = liquid chromatography-mass spectrometry
LOC = lab-on-a-chip
MALDI = matrix-assisted laser desorption/ionization
MALDI-TOF-MS = MALDI time-of-flight mass spectrometry
MFA = metalaxyl-M-fludioxonil-Azoxystrobin
MS = mass spectrometry
NGS = next-generation sequencing
NMR = nuclear magnetic resonance
RNA Seq = RNA sequencing
WGS = whole genome sequencing

CHAPTER 10

BMA = bio model analyzer
ESI = electrospray ionization
MALDI = matrix-assisted laser desorption ioniaztion
MS = mass spectrometery
NJ = neighbor-joining
Pas = process algebras

CHAPTER 11

ABMs = agent-based models
AE = autoencoder
AI = artificial intelligence
AUROC = area under the receiver operating curve
CNN = convolutional neural network
COPD = chronic obstructive pulmonary disease
CPRD = clinical practice research datalink
DBN = deep belief network
DE = discrete event
DL = deep learning
DRL = deep reinforcement learning
DTL = deep transfer learning
DWT = discrete wavelet transform
EKRRMDA = ensemble of kernel ridge regression-based MiRNA-disease association
FA = factor analysis
GAN = generative adversarial network
GLCM = gray-level co-occurrence matrix
GMDH = group method of data handling
HHO = Harris hawks optimization

ICA = independent component analysis
IoT = Internet of Things
kNN = k-nearest neighbors
LDA = linear discriminant analysis
LLE = locally linear embedding
LPP = locality preserving projection
MDS = multi-dimensional scaling
ML = machine learning
MLP = multi-layer perceptron
NCDs = non-communicable diseases
PBMDA = path-based MiRNA-disease association
PCA = principal component analysis
PSO = particle swarm optimization
RBM = restricted Boltzmann machine
RF = random forests
RNN = recurrent neural network
SEIR = susceptible-exposed-infected-removed
SHAP = SHapley Additive exPlanations
SINDy = sparse identification of nonlinear dynamics
SIR = susceptible-infected-recovered
SOM = self-organizing map
STFT = short-time Fourier transform
SVD = singular value decomposition
SVM = support vector machine

CHAPTER 12

2-DE = two-dimensional gel electrophoresis
AI = artificial intelligence
ARGs = antibiotic resistance genes
ASR = archive for sequence reads
BLAST = basic local alignment search tool
BLOSUM = blocked substitution matrix
CARD = comprehensive antibiotic resistance database
CATH = class, architecture, topography and homology
CDD = conserved domain database
COGs = clusters of orthologous group of proteins
DDBJ = DNA database of Japan
DNA = deoxyribonucleic acid
DTI = drug–target interaction
EBI = European bioinformatics institute
EMBL = European molecular biology laboratory
GDPR = general data protection regulation
GENTRL = generative tensorial reinforcement learning
GINA = genetic information nondiscrimination act
HIPAA = health insurance portability and accountability act
ICAT = isotope-coded affinity tag labelling
IMG = integrated microbial genomes
INSDC = International Nucleotide Sequence Database Collaboration
KEGG = Kyoto encyclopedia of genes and genomes

MaizeGDB = maize genetics and genomics database
MGDB microbial genome database
ML = maximum likelihood
MLST = multi-locus sequencing typing
MMD = molecular modeling database
MP = maximum parsimony
MPSS = massively parallel signature sequencing
MSA = multiple sequence alignment
NCBI = national center for biotechnology information
NDB = nucleic acid database
NGS = next-generation sequencing
NIG = national institute of genetics
NIH = national institute of health
NJ = neighbor-joining
NMR = nuclear magnetic resonance
OMIM = online Mendelian inheritance in man
PAM = point accepted mutation
PDB = protein data bank
PIR = protein information resources
PSA = pairwise sequence alignment
RCSB = research collaborators for structural bioinformatics
RNA = ribonucleic acid
SCOP = Structural classification of proteins database
SDS-PAGE = sodium dodecyl sulphate polyacrylamide gel electrophoresis
TrEMBL = translated European molecular biology laboratory
UniProt = universal protein resource
VAST = vector alignment search tool
VFDB = virulence factor database
WGS = whole genome sequencing
wwPBD = worldwide protein data bank

CHAPTER 13

CNVs = copy number variations
ELSI = ethical, legal, and social implications
NGS = next-generation sequencing
PCR = polymerase chain reaction
SNPs = single nucleotide polymorphisms
SVs = structural variations
WGS = whole genome sequencing

CHAPTER 14

A = adenine
AST = antibiotic susceptibility testing
C = cytosine
ddNTP = dideoxynucleotide triphosphate
G = guanine
HLA = human leukocyte antigen
NGS = next-generation sequencing

ONT = Oxford nanopore technologies
PTP = picotiter plate
SMRT = single-molecule sequencing in real time
SVs = structural variants
T = thymine
TGS = third-generation sequencing
TGSTs = third-generation sequencing techniques
WGS = whole genome sequencing

CHAPTER 15

AI = artificial intelligence
AMR = antimicrobial resistance
BLAST = basic local alignment search tool
CNNs = convolutional neural networks
DCNN = deep convolutional neural networks
DNP-AAP = deep neural pursuit-average activation potential
DQN = deep Q networks
FTIR = Fourier-transformed infrared spectroscopy
GANs = generative adversarial networks
GIS = geographic information system
HMM = hidden Markov models
IBD = inflammatory bowel disease
ML = machine learning
MLP = multilayer perceptron
MRSA = methicillin-resistant *Staphylococcus aureus*
NLP = natural language processing
RNNs = recurrent neural networks
SVM = support vector machines
t-SNE = t-distributed stochastic neighbor embedding

CHAPTER 16

AMR = antimicrobial resistance
ASM = American Society for Microbiology
BiP = Biotechnology Immersion Program
CSM = Canadian Society of Microbiologists
FMO = Fenna-Matthews-Oslon
FSH = follicle stimulating hormone
HTS = high throughput sequencing
iGEM = International Genetically Engineered Machine
LIS = laboratory information system
MP = microbial protein
NGS = next-generation sequencing
PMI = postmortem interval
QIIME = quantitative insights into microbial ecology
SCP = single cell protein
STXM = scanning transmission X-ray microscopy
TSBE = transdisciplinary synthetic biology education
WGS = whole genome sequencing

1 Introduction

Ejaz Mohammad, Shahzad Ali, and Muhammad Ali Syed

Microbiology is the branch of biology that deals with the study of microscopic organisms and can be divided into several subdisciplines including virology, bacteriology, mycology, parasitology, and phycology (1). These subdisciplines further encompass broad areas of several specific fields. These fields include pathogenic microbiology (study of etiology, epidemiology, and disease processes of pathogenic microorganisms); immunology (analysis of the immune system and how it protects the biological system from harmful microbes and substances); microbial physiology (structure–function relationships of microbes); microbial genetics (linked to molecular biology); industrial microbiology (microbes involved in production of substances like antibiotics, vaccines, etc.); food microbiology (microbes causing food spoilage and foodborne diseases/intoxications) and so on (2). Microbiology focuses on studying microbial physiology, microbial diversity, microbial ecology, microbial genetics, interconnectedness with microbial communities, and the relationship of microbes with the host and environment (3).

Microorganisms are key drivers of planetary processes and evolution by performing a wide range of reactions that support life on earth (4). Microbial processes have a significant impact on the biological inhabitants of the earth (5). Microorganisms in the environment play a crucial role in nutrient recycling, maintaining the equilibrium of trophic chains, performing essential physiological functions in plants and animals, and preserving natural ecological environments (6).

Microbiology is a dynamic discipline that began a few centuries ago with Antonie van Leeuwenhoek's discovery of microorganisms, followed by significant contributions from Louis Pasteur and Robert Koch, leading to the advancement of molecular genetics. The current century marks the beginning of a transformative era, characterized by the integration of various interdisciplinary approaches, including the combination of AI, nanotechnology, and quantum innovations, to drive revolutionary change.

The field of microbiology is now distinguished by the unparalleled gathering of biological data using advanced high-throughput intelligent technology. Extending beyond conventional laboratory and field research, there is an increasing demand for expertise in utilizing computational methods to get significant insights from these extensive datasets. The multifaceted field of microbiology is now on the verge of a revolutionary era due to the integration of artificial intelligence (AI). By using a wide range of technologies such as machine learning (ML), computer vision, and object identification, AI is poised to revolutionize this field. The application of this technology extends to computationally intensive activities, including the prediction of drug targets and vaccine candidates, identification of microorganisms responsible for infectious diseases, classification of drug resistance to antimicrobial therapy, prediction of disease outbreaks, investigation of interactions among microorganisms, quality assurance, bacterial identification, and maintenance of hygiene standards (7–9).

AI integration in microbiology extends beyond a single field and encompasses virology, parasitology, mycology, and bacteriology. The combination of AI with sophisticated data-collecting devices, increased computing capabilities, and worldwide networks, exemplified by Big Data, Moore's Law, and the Internet, has propelled the successful implementation of AI in the field of

DOI: 10.1201/9781003410164-1

microbiology (10, 11). This convergence is especially noticeable in the domain of pharmaceutical design, where artificial intelligence-driven methodologies are leading the advancement of innovative medicines to address the formidable challenge of multidrug-resistant organisms, which poses a significant risk to worldwide health (12). Moreover, AI in pharmaceutical microbiology has a significant role in utilizing predictive analytics, thereby reducing the cost and the need for labor-intensive methods of extensive physical testing of every sample or batch (13).

Similarly, its influence on clinical microbiology is revolutionary, simplifying data processing and improving diagnostic precision. Techniques like MALDI-TOF mass spectrometry coupled with AI are revolutionizing microbe identification and antibiotic resistance profiling, giving rapid, reliable, and cost-effective solutions (14). Moreover, AI's aptitude to manage 'big data'—such as genetic data, digital images, DNA, RNA, and protein sequences from new diagnostic technologies is crucial in improving laboratory results and expediting diagnostic processes (8).

AI has a wide range of applications beyond human health and genetics. It can also be utilized in environmental protection, particularly in challenging environments like aquacultures, soil, and extreme environmental conditions. Zhai et al. conducted a study on the bacterial biofilms found on metallic alloys at a depth of 5,700 m undersea, utilizing AI technology. They obtained the sequencing data of the microbial communities in the biofilms and used advanced data analytics techniques to analyze the dataset and compare the microbial composition on various alloy surfaces (15). McElhinney et al. also examined the effectiveness of ML algorithms in analyzing the large microbiological datasets generated due to the emergence of microbial omics (16).

The disciplines of microbiology and nanotechnology have significantly contributed to the field of science and technology, providing innovative solutions for human well-being and environmental balance. Nevertheless, given the concerning rise in antibiotic/multidrug resistance in microorganisms caused by the excessive use of drugs, as well as the potential impact of metal nanoparticles on the food chain, it has become imperative to develop interdisciplinary approaches that merge microbiology and nanotechnology. These approaches aim to address the resulting human health issues and mitigate the environmental and ecological damage. The combination of these fields logically provides creative and environmentally friendly solutions.

Microbial infections represent a significant public health challenge, as numerous bacteria, yeasts, fungi, and other pathogenic microorganisms can develop resistance to medications (17). Moreover, biofilm formation by certain pathogenic microorganisms also poses a threat to human health (18). Therefore, it is essential to create new and effective antimicrobial agents, particularly in light of the current situation. The application of nanotechnology presents a promising strategy. Some nanomaterials have been developed with distinctive features that offer exceptional antimicrobial properties, aiding in the treatment of resistant infectious diseases (19): for instance, applications of nanoparticles in antibacterial coatings for medical implants and devices to inhibit infections and enhance wound healing, systems for delivering antibiotics and detecting bacteria, and antimicrobial vaccines (20).

Nanomaterials are currently being utilized as vaccine delivery vectors and adjuvants to augment vaccine-induced targeted immune responses and antigen immunogenicity. They find extensive application in infectious disease prevention and tumor immunotherapy. Meanwhile, nanomaterials such as liposomes, polymeric nanoparticles, and exosomes can serve as delivery vehicles for precise medication distribution, regulated release, and efficient therapy. Through the precise adjustment of their surface functional groups, nanomaterials such as magnetic nanoparticles and quantum dots can be employed in biomedical imaging, enabling precise identification of particular targets and live tracking of disease advancement.

Recognizing infectious diseases and their root cause is essential for effectively managing and treating the condition. The conventional methods used to determine the etiological agents of infectious diseases lack high sensitivity, specificity, and robustness. Therefore, a need for advanced diagnostic technologies is prerequisite to tackling such drawbacks and limitations of conventional diagnostic approaches (21). The development of nanoscale biosensors with enhanced sensitivity and specificity in detecting and identifying microorganisms has significantly influenced the diagnosis of infectious diseases (22, 23). Furthermore, the advancement of rapid diagnostic tests utilizing nanoparticles has significantly improved speed efficiency (24).

In addition to diagnostics, nanotechnology has been useful in enhancing vaccine development and medicine delivery (25). The nanotechnologies created for application in vaccines include nanocarriers with diverse compositions, sizes, and surface characteristics. A variety of vaccine nanocarriers have been developed and studied for their effectiveness in delivering antigens and adjuvants to immune cells, aiming to enhance a protective immune response (26). Regrettably, while immune cells can uptake antigens, a lack of adequate adjuvant activity may lead to reduced immunogenicity. Some methods involve the design of nanocarriers that can simultaneously deliver an antigen along with an adjuvant (27). Nanocarriers enable the precise targeting and prolonged release of antigens or adjuvants to antigen-presenting cells.

Currently, a diverse array of nanoparticles, especially liposomes, polymeric and inorganic nanoparticles, as well as self-assembled protein nanoparticles and virus-like particles (VLPs), are under investigation as carriers for antigens. The nano-sized characteristics of various biological systems, including proteins and viruses, offer significant advantages (28, 29).

Incorporating nanotechnology into microbiology not only improves our capacity to study and understand microorganisms, but also offers new solutions to important worldwide health problems. We can improve healthcare worldwide by using these state-of-the-art technologies to develop better diagnostic tools, vaccines, and antimicrobial treatments.

Quantum mechanics, a science typically linked to physics, is now being applied to microbiology through the new discipline of quantum biology, resulting in significant advancements. This field of research investigates the impact of quantum principles on biological processes at the molecular level, providing fresh perspectives on the behavior and functioning of microorganisms.

The field of quantum biology emerged in close alignment with the advancement of quantum mechanics. Through the contributions of Bohr, Heisenberg, Pauli, Schrödinger, Dirac, Born, Jordan, Fermi, and other scholars, the mathematical foundation of the new quantum mechanics was established by 1927 (30). Accompanied by their achievement in mastering the atomic realm, and fueled by the hubris of their youth, numerous trailblazers in the field of quantum physics confidently ventured beyond their physics laboratories and beyond their traditional educational settings to explore uncharted scientific territories. Microbiology, together with the developing fields of genetics and the chromosome theory of heredity, remained uncharted domains. However, an increasing number of biophysicists and biochemists started to actively engage in research on quantum biology (31).

DNA may function similarly to a quantum computer, existing in superposition states that ultimately collapse into a single state. The counterargument lies in the fact that DNA struggles to maintain a coherent state for sufficient durations to receive feedback on which mutations or sequences might be advantageous, as coherence times typically range from femto to pico seconds (32). An example from quantum microbiology: green sulfur bacteria achieve nearly complete efficiency in transferring photonic energy through a chlorosome antenna and a pigment-protein complex known as the FMO complex (33). The energy transfer did not adhere to a random trajectory; instead, it was observed in both locations simultaneously, effectively demonstrating a quantum superposition state (34).

It was also hypothesized that the emergence of pre-biotic life was facilitated by quantum information processing. However, as evolutionary processes at the organism level progressed, this initial quantum processing became integrated into the mechanisms of DNA replication, transcription, and translation (35, 36).

STRUCTURE AND SCOPE OF THE BOOK

This book explores the integration of cutting-edge fields such as nanotechnology, artificial intelligence, and quantum science within the realm of microbiology. The book demonstrates how these advanced fields have transformed our understanding of microbial systems. In the 21st century, AI has a profound impact on nearly every field and is currently revolutionizing microbiology through the use of various modeling and data analysis techniques. These advancements can predict a range of prospects, including drug development, the transmission dynamics of microbial pathogens, vaccine design, and epidemic forecasting. In a similar vein, nanotechnology presents significant advancements in microbial detection through the use of nanoscale biosensors. Furthermore, nanoparticles have been engineered to effectively deliver vaccines and drugs. The book also delves into the emerging field of quantum biology, investigating quantum phenomena in tunneling, coherence, and DNA processing in essential microbial processes. It also offers extensive literature on advancements in NGS, emphasizing the significance of this technique in the genomics of microbes. This book offers a thorough exploration of the most recent developments in applied microbiology, organized into 16 chapters, starting with this introduction.

CHAPTER 2: A JOURNEY FROM CLASSICAL MICROBIOLOGY TO NEXT GENERATION MICROBIOLOGY

It explores the chronological developments in the field of microbiology, starting with the conventional approaches used for the identification of microbes followed by the limitations of such methodologies. The chapter further discusses the development of advanced microbiological techniques along with their applications

CHAPTER 3: ADVANCEMENTS IN NANOTECHNOLOGY AND THEIR APPLICATIONS IN MICROBIOLOGY

A concise overview of nanotechnology is presented, followed by an analysis of the progress and innovations in nanomaterials. The use of nanotechnology in clinical microbiology, nanodiagnostics, environmental microbiology, and water and food microbiology are also discussed.

CHAPTER 4: NANOTECHNOLOGY-BASED ASSAYS FOR MICROBIAL DETECTION AND IDENTIFICATION

This chapter focuses on cutting-edge techniques for rapid and accurate microbial detection using nanotechnology. It reviews various nanotechnology-based assays, including nanoparticle-based biosensors as well as examples of applications of nanobiosensors in microbial detection. The chapter emphasizes the advantages of these assays in terms of sensitivity, specificity, and speed, as well as their applications in clinical diagnostics, food safety, and environmental monitoring.

CHAPTER 5: VIRUS-LIKE PARTICLES: APPLICATIONS IN VACCINE DEVELOPMENT AND DRUG DELIVERY FOR INFECTIOUS AGENTS

This chapter delves into the various types, production methods, and uses of virus-like particles in the creation of vaccines targeting microbes such as bacteria, viruses, and certain parasites. It additionally exemplifies successful VLP-based vaccines and the application of VLP in drug delivery.

CHAPTER 6: NANOSENSORS AND MICROELECTROMECHANICAL SYSTEMS FOR MICROBIAL AGENTS

It reviews the principles and designs of various nanosensors and MEMS devices, discussing their applications in detecting pathogens, monitoring environmental microbes, and studying microbial interactions.

CHAPTER 7: RECENT ADVANCES IN MICROBIOLOGY TECHNIQUES FOR IMAGING MICROBES AND THEIR COMPONENTS

This chapter offers a thorough examination of traditional imaging technology utilized in the past, followed by advanced imaging techniques currently employed to visualize microbes. Additionally, the use of these imaging technologies in studies of microbial pathogenesis, as well as in drug development and delivery, and in the field of biotechnology, is explored.

CHAPTER 8: QUANTUM MICROBIOLOGY

This chapter explores the incorporation of quantum phenomena in microbiology, highlighting quantum effects on microbial systems. It highlights potential applications of quantum mechanics and technologies for a better understanding of biological processes and answers to many questions that need to be addressed.

CHAPTER 9: OMICS TECHNOLOGIES IN MICROBIOLOGY

This chapter delves into the diverse applications of omics technologies within the realm of microbiological research, encompassing areas such as environmental, agricultural, medical, food, and industrial microbiology. The pertinent techniques based on omics in the field of microbiology are also explored.

CHAPTER 10: COMPUTATIONAL BIOLOGY AND ITS ROLE IN LIFE SCIENCE RESEARCH

This chapter provide a comprehensive discussion of the computational methodologies utilized for interpreting biological data and their application in life-science-related data and research.

CHAPTER 11: COMPUTATION AND DISEASE PREDICTION MODELS

This chapter explores the advancement and utilization of computational models for the prognosis and management of diseases. The chapter also emphasizes the utilization of machine and deep learning models to forecast disease outbreaks, comprehend disease dynamics, and guide public health treatments.

CHAPTER 12: BIOINFORMATICS APPLIED TO MICROBIOLOGY

This chapter highlights the significance of bioinformatics in the analysis and interpretation of intricate microbiological data. It covers bioinformatics tools and approaches used for the study of genomics, transcriptomics, and metagenomics data. The chapter also explores the incorporation of bioinformatics with other fields, such as clinical microbiology, and antibiotic resistance to promote the progress of microbial research.

CHAPTER 13: WHOLE GENOME SEQUENCING AND MAKING SENSE OF MOLECULAR DATA

This chapter explores the basics of whole genome sequencing and how to make sense of the extensive data generated, including different tools used for data analysis as well as challenges.

CHAPTER 14: TYPES OF WHOLE GENOME SEQUENCING TECHNIQUES AND THEIR USEFULNESS IN MICROBIOLOGY

In this chapter, a comprehensive introduction to different types of WGS is presented. In addition, the application of whole genome sequencing in microbiology is discussed in detail.

CHAPTER 15: ARTIFICIAL INTELLIGENCE AND MACHINE LEARNING TECHNIQUES APPLIED TO MICROBIOLOGY

Use of artificial intelligence has been increasing in all disciplines of science. Microbiology is also expected to fully benefit from enormous opportunities offered by AI/ML tools in clinical and laboratory diagnosis of disease, devising highly sensitive techniques for microbial detection and identification, disease prediction models, as well as control and prevention of diseases along with other applications. This chapter highlights the basics of artificial intelligence and various AI/ML-based approaches with possible applications in microbiology research.

CHAPTER 16: FUTURE MICROBIOLOGY EDUCATION

This chapter highlights the current challenges in microbiology curricula and teaching methodologies. The subject of microbiology is experiencing massive changes since new knowledge is being added and new techniques are replacing the existing ones. Moreover, the chapter also discusses the emerging technologies and allied subjects followed by the incorporation of these allied fields in microbiology education.

REFERENCES

1. Lateef, A., O. Darwesh, and I. Matter, Microbial nanobiotechnology: the melting pot of microbiology, microbial technology and nanotechnology. *Microbial Nanobiotechnology: Principles and Applications*. 2021: Springer. p. 1–19.
2. Siddique, S., How expert guide to microbiology: 101 tips to learn about the history, *applications, research, universities, and careers in microbiology*. 2021: HowExpert.
3. Willey, J.M., L.M. Sherwood, and C.J. Woolverton, *Prescott's microbiology*. 2014: McGraw-Hill.
4. Timmis, K., et al., *The contribution of microbial biotechnology to sustainable development goals*. 2017: Wiley Online Library. p. 984–987.
5. Corvalan, C., S. Hales, and A.J. McMichael, *Ecosystems and human well-being: health synthesis*. 2005: World Health Organization.
6. Hunter-Cevera, J.C., The value of microbial diversity. *Current Opinion in Microbiology*, 1998. **1**(3): p. 278–285.
7. Russo, G., et al., The combination of artificial intelligence and systems biology for intelligent vaccine design. *Expert Opinion on Drug Discovery*, 2020. **15**(11): p. 1267–1281.
8. Mohseni, P. and A. Ghorbani, Exploring the synergy of artificial intelligence in microbiology: advancements, challenges, and future prospects. *Computational and Structural Biotechnology Reports*, 2024. **1**: p. 100005.
9. Lopatkin, A.J. and J.J. Collins, Predictive biology: modelling, understanding and harnessing microbial complexity. *Nature Reviews Microbiology*, 2020. **18**(9): p. 507–520.

10. Shelke, Y.P., A.K. Badge, and N.J. Bankar, Applications of artificial intelligence in microbial diagnosis. *Cureus*, 2023. **15**(11):e49366.
11. Juhas, M. *Brief Lessons in Microbiology: From the Origin of Life to Artificial Intelligence*. 2023: Springer. p. 93–109.
12. Mishra, A., et al., Evolution of diagnostic and forensic microbiology in the era of artificial intelligence. *Cureus*, 2023. **15**(9):e45738.
13. Lopez-Jimenez, F., et al. Artificial intelligence in cardiology: present and future. Mayo Clinic Proceedings. 2020;**95**:1015–39.
14. Wilson, S., S. Steele, and K. Adeli, Innovative technological advancements in laboratory medicine: predicting the lab of the future. *Biotechnology & Biotechnological Equipment*, 2022. **36**(sup1): p. S9–S21.
15. Zhai, X., et al., Study on the bacterial communities of the biofilms on titanium, aluminum, and copper alloys at 5,772 m undersea in Yap Trench. *Frontiers in Microbiology*, 2022. **13**: p. 831984.
16. McElhinney, J.M., et al., Interfacing machine learning and microbial omics: a promising means to address environmental challenges. *Frontiers in Microbiology*, 2022. **13**: p. 851450.
17. Teixeira, M.C., et al., Nanomedicines for the delivery of antimicrobial peptides (AMPs). *Nanomaterials*, 2020. **10**(3): p. 560.
18. Barzegari, A., et al., The battle of probiotics and their derivatives against biofilms. *Infection and Drug Resistance*, 2020. **13**: p. 659–672.
19. Reshma, V.G., et al., Engineered nanoparticles with antimicrobial property. *Current Drug Metabolism*, 2017. **18**(11): p. 1040–1054.
20. Souza, G.M., et al., Green synthesized nanoparticles as a promising strategy for controlling microbial biofilm. In *Nanotechnology for Advances in Medical Microbiology*, ed. N.R. Maddela, S. Chakraborty, and R. Prasad. Springer; 2021: p. 1–28.
21. Thwala, L.N., et al., Nanotechnology-based diagnostics for diseases prevalent in developing countries: current advances in point-of-care tests. *Nanomaterials*, 2023. **13**(7): p. 1247.
22. Wan, Y., et al., New insights for biosensing: lessons from microbial defense systems. *Chemical Reviews*, 2022. **122**(9): p. 8126–8180.
23. Ramesh, M., et al., Nanotechnology-enabled biosensors: a review of fundamentals, design principles, materials, and applications. *Biosensors*, 2022. **13**(1): p. 40.
24. Qasim, M., et al., Nanotechnology for diagnosis and treatment of infectious diseases. *Journal of Nanoscience and Nanotechnology*, 2014. **14**(10): p. 7374–7387.
25. Peek, L.J., C.R. Middaugh, and C. Berkland, Nanotechnology in vaccine delivery. *Advanced Drug Delivery Reviews*, 2008. **60**(8): p. 915–928.
26. Kim, M.-G., et al., Nanotechnology and vaccine development. *Asian Journal of Pharmaceutical Sciences*, 2014. **9**(5): p. 227–235.
27. Garg, A. and H.K. Dewangan, Nanoparticles as adjuvants in vaccine delivery. *Critical Reviews in Therapeutic Drug Carrier Systems*, 2020. **37**(2): p. 183–204.
28. Bezbaruah, R., et al., Nanoparticle-based delivery systems for vaccines. *Vaccines*, 2022. **10**(11): p. 1946.
29. Laval, J.-M., P.-E. Mazeran, and D. Thomas, Nanobiotechnology and its role in the development of new analytical devices presented at SAC 99, Dublin, Ireland, July 25–30, 1999. *Analyst*, 2000. **125**(1): p. 29–33.
30. Mohseni, M., et al., *Quantum effects in biology*. 2014: Cambridge University Press.
31. McFadden, J. and J. Al-Khalili, The origins of quantum biology. *Proceedings of the Royal Society A*, 2018. **474**(2220): p. 20180674.
32. Gariaev, P.P., et al., DNA as basis for quantum biocomputer. *DNA Decipher Journal*, 2011. **1**(1): p. 025–046.
33. Ishizaki, A., et al., Quantum coherence and its interplay with protein environments in photosynthetic electronic energy transfer. *Physical Chemistry Chemical Physics*, 2010. **12**(27): p. 7319–7337.
34. Collini, E., et al., Coherently wired light-harvesting in photosynthetic marine algae at ambient temperature. *Nature*, 2010. **463**(7281): p. 644–647.
35. Trevors, J. and L. Masson, Quantum microbiology. *Current Issues in Molecular Biology*, 2011. **13**(2): p. 43–50.
36. Koruga, D., Classical and quantum information processing in DNA-protein coding. In *Cell and Tissue Engineering*, ed. B. Obradović. Springer; 2012: p. 9–26.

2 A Journey from Classical Microbiology to Next Generation Microbiology

Rafiq Ahmad, Sohrab Khan, Syed Muhammad Afzal, Sohail Khan, Muhammad Arif, and Muhammad Hayat

2.1 INTRODUCTION

The study of microorganisms that live in every corner of the planet has faced several challenges, including the discovery of unknown creatures and knowledge regarding how they interact with their environment and humans. The study of microbial populations has evolved from Leeuwenhoek's initial report on microorganisms and their oral species in 1676 (1) to characterization using modern techniques. Antonie van Leeuwenhoek's invention simplified the examination of microorganisms under a microscope. Pasteur's and Koch's scientific achievements have transformed the microscope into a useful tool for medical practitioners and microbiologists to visualize microorganisms (2, 3). Van Leeuwenhoek and Pasteur contributed to the discovery of microbes, and in the nineteenth century multidisciplinary methodologies were used to examine the planet's various microbiological activities. Early scientists attempted to isolate these "invisible" organisms and quickly became familiar with the cultivation of bacteria on solid media. The technique was applied in Cohn's and Koch's laboratories around 1875 (4). Moreover, Koch introduced the preparation of pure culture by streaking bacteria on solidified gelatin in 1877 (5).

Like other branches of life sciences, microbiology is also rapidly evolving by continuously benefiting from scientific and technological advancements and the emergence of cutting-edge technologies, automation, integrated systems, computation, and the use of artificial intelligence and nanotechnology. This chapter highlights the journey of the science of microbiology from early developments to the present time.

2.2 DEVELOPMENT OF CONVENTIONAL METHODS

Using agar as a solidifying agent substantially improved bacterial isolation and culture since agar is inert to most bacteria and remains solid at 37°C, where most pathogenic bacteria are grown (6, 7). Julius Richard Petri (1852–1921), a German bacteriologist, designed a covered dish known as a Petri plate, in which media solidified and avoided contamination; later, Koch modified the Petri dish (8). In addition, bacterial cultures in the lab account for less than 5%, or probably as little as 1%, of all bacteria, since the majority of bacteria are non-culturable in the laboratory. When all viruses infecting organisms are counted, the number decreases by several orders of magnitude. Traditional microbiological approaches have disadvantages because they are based on a statistically small fraction of all bacteria (9). However, it is appropriate for medically relevant species because traditional microbiological techniques as a "gold standard," particularly those intended for clinical laboratory use, are biased toward human infections. When microbial analysis approaches involve the description and type of environmental organisms rather than microbial pathogens, many laboratory procedures, particularly clinical ones, may be ineffective or provide inaccurate findings.

 DOI: 10.1201/9781003410164-2

Classical microbiology identifies bacteria using both gross and microscopic characteristics. A bacterial colony's shape, size, and surface properties are all considered gross morphology. Microbe morphological structure is important because it helps us identify different species of bacteria and gives us more insight into their physiology, pathogenic mechanisms, and antigenic characteristics (10). In addition, Koch's 1876 demonstration of the anthrax bacillus life cycle for the first time revealed that a particular microorganism was the root cause of a known disease. Tuberculosis was believed to be an inherited disease; however, in 1882, Koch developed a methylene-blue stain from which he identified a rod-shaped bacterium and concluded that tubercle bacilli, caused the disease (11). In 1884, after the disease-causing bacteria were discovered, Hans Christian Gram developed the Gram-staining method to identify pathogens. The stains used in this process are crystal violet, Gram's iodine, and safranin, which are used to identify and differentiate between Gram-positive and Gram-negative bacteria (12, 13). Gram staining is used to identify polymorphonuclear infiltration and prominent microbial morphologies. Generally, morphological techniques used dichotomic keys, with Gram staining as the first test, followed by catalase and oxidase assays. Initial attempts made by scientists were largely focused on their morphological features, biochemical characteristics, such as enzyme profile, pH-based reactions, acid detection carbon source utilization, etc., and serological properties (14, 15).

2.3 LIMITATIONS OF CLASSICAL MICROBIOLOGY TECHNIQUES

Classical microbiological techniques involve cultivating microorganisms in a sample on specific media and using the serial dilution method to count the number of cells in each taxonomic group (16–18). Classical microbiology, usually called conventional microbiology, has been the primary subject of study of microorganisms for many years. As discussed above, traditional microbiology testing has historically used microscopy, antigen detection, serology, and culture. Since the beginning of microbiology, microscopic evaluation of direct clinical specimens has been crucial for rapid diagnosis of bacterial, fungal, and protozoal infections (19, 20). The primary means of traditional testing is culture. Clinical specimens are inoculated on various media and then incubated in the case of bacteria and fungi (21). Conventional, culture-based techniques for microbial identification are regarded as the "gold standard" by which other laboratory procedures are evaluated (22). It is crucial to recognize its limitations even if it has substantially contributed to our understanding of the microbial world.

The primary limitation of traditional microbiology is its dependence on culturing methods. Microorganisms are cultivated in a lab environment using specific media. Nevertheless, not all microbes are amenable to conventional methods of cultivation. It is hard to fulfill the growth requirements of many microbes in the lab, which makes it challenging to research them with standard methods. This constraint limits our comprehension of microbial communities' complete intricacy and diversity.

The traditional techniques for microbiological diagnostics have the key advantage of being well-validated, ensuring the accuracy of the results. These techniques suit hospitals and laboratories for conventional laboratory diagnostic tests and microbial identification (23). However, classical microbiology has been ineffective in accurately reflecting microbial community dynamics and interaction, which is a limitation. Microorganisms rarely survive in isolation; instead, they engage in intricate interactions with both their surroundings and one another. Classical microbiology's main emphasis is the study of individual microorganisms in culture, which frequently overlooks the complex interactions and relationships within microbial communities. This limitation makes it more challenging to comprehend how microbes behave and change in their native environments. Traditional methods cannot be used to cultivate many microorganisms found in the environment, leaving a substantial scientific gap. These uncultivable microbes might be essential to many ecological processes, including the cycle of nutrients and the onset of disease. We would be unable to comprehend their

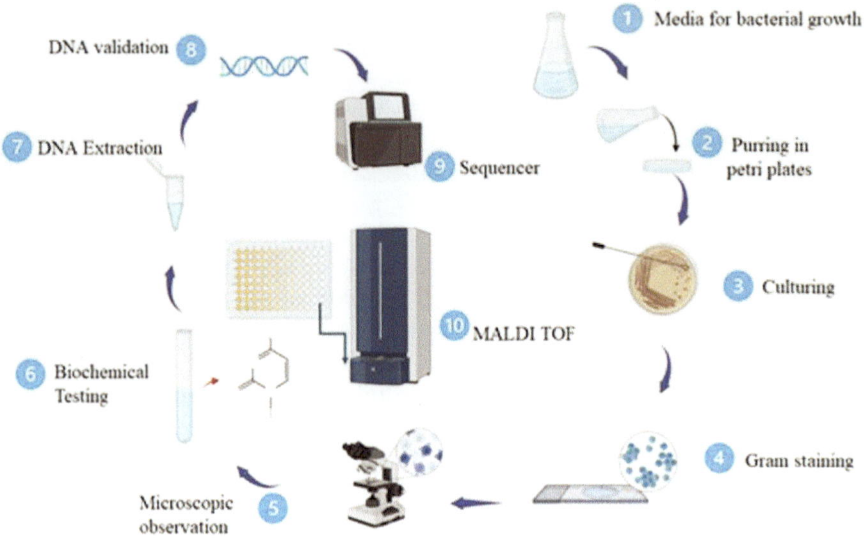

FIGURE 2.1 Journey of microbiology from classical techniques to modern-day techniques.

TABLE 2.1
The Pros and Cons of the Cultivation Methods

Pros of the Cultivation Method	Cons of the Cultivation Method
Isolation and identification	Culturability bias
Pure culture establishment	Slow and labor-intensive
Characterization of growth requirements	Loss of unculturable microorganisms
Study of microbial diversity	Altered growth conditions
Testing antimicrobial agents	Inability to study complex interactions

contributions to the entire microbial ecology if we could not conduct thorough research. In addition, these tests typically take weeks to complete and have a minimum turnaround time of 48 to 72 hours for detecting slow-growing organisms. A lengthy wait for test findings could impact how serious illnesses are dealt with. If the diagnosis relates to a disease that poses a risk to life, these techniques are of less use (24). Moreover, the count of colony-forming units can only account for a tiny percentage of microorganisms (0.1–10%) found in environmental samples (25).

There are also difficulties for classical microbiology in examining the functional potential of microbes. Although it can provide insight into the distribution and abundance of particular microbes, it usually fails to identify their functional capacities. Comprehending the functional characteristics of microorganisms is crucial for multiple applications such as ecological restoration, biotechnology, and the well-being of humans. The functional variation in microbial communities may not be fully understood by classical microbiology alone. Studying microorganisms in real time or in their natural habitat is a limited capability of classical microbiology (Figure 2.1).

In contrast to their native environments, microorganisms often display unique behavior in laboratory conditions. This limitation may cause differences between results obtained in the lab and actual situations. Although it is difficult to achieve this using traditional microbiological methods, studying microorganisms in their natural environment would yield a more accurate depiction of their behavior and interactions (Table 2.1) (26).

Classical microbiology has several restrictions, even though it has contributed significantly to our understanding of microbes. The difficulties in studying microorganisms in their natural environments, the reliance on culture techniques, the inability to capture the dynamics of microbial communities, the difficulty of studying unculturable microorganisms, and the limited competence to identify functional potential are a few of these limitations.

2.4. MOLECULAR METHODS USED IN MICROBIOLOGY

Microbial identification through culture remains a cornerstone in microbiology; however, its efficacy can be enhanced by emerging diagnostic modalities like genotypic techniques or those reliant on species-specific protein profiling, thus mitigating inherent limitations (26). Close collaboration among clinicians, microbiologists, and bioinformaticians, along with utilizing computational tools, is essential for critically assessing complex and advanced diagnostic methods and interpreting their results (27, 28).

2.4.1 Polymerase Chain Reaction

The advent of polymerase chain reaction (PCR) marked a pivotal advancement in biological research, ushering in the contemporary era of molecular diagnostics. As exemplified by PCR, nucleic acid amplification tests (NAATs) offer heightened sensitivity and specificity, revolutionizing clinical laboratory practices and establishing a new "gold standard" for diagnosing numerous infectious diseases (29). This innovation has significantly enhanced patient care capabilities within clinical settings. PCR-based diagnostics are widely used across diverse microbial species due to their exceptional sensitivity, capable of detecting 1–100 genome equivalents of bacterial or fungal DNA. With high specificity and rapid amplification rates, PCR is recommended for identifying organisms that cannot be cultured *in vitro* or when existing culture techniques are insensitive or time-consuming (30). Common PCR modifications utilized in routine diagnostics include multiplex PCR, nested PCR, and quantitative PCR (qPCR) (31). This test has reduced the time required by each assay and proved to be a reliable technique for detecting infectious agents directly from patient samples. An excellent example of PCR's usefulness in diagnostics is in the diagnosis of COVID-19, HIV, and hepatitis B viral infections.

2.4.2 Next-generation Sequencing

Next-generation sequencing (NGS) refers to a set of advanced techniques used for rapid and high-throughput sequencing of DNA or RNA molecules. Sanger sequencing, introduced in 1976, was a groundbreaking method in nucleic acid sequencing. It precisely records nucleotide sequences within analyzed genes (32, 33). Next-generation sequencing (NGS) has since emerged as an alternative, with two primary streams: 16S rRNA and shotgun metagenomics. The 16S rRNA, a component of the bacterial ribosome, consists of 1,500 base pairs divided into nine highly conserved and hypervariable regions (V1–V9). Shotgun metagenomics is a PCR-independent method, eliminating the need for primer design based on gene structure. This approach facilitates whole-organism genome sequencing, enabling the discovery of new species. The 16S rRNA gene NGS method is anticipated to soon become a routine diagnostic tool due to its speed, simplicity, and cost-effectiveness compared to shotgun metagenomics (26).

2.4.3 Metagenomics and Functional Genomics

It has been difficult to identify previously unknown microorganisms or to understand their relationship with the environment, due to several obstacles (34). Most studies on microbes rely on cultivating organisms in cultures, which has certain downsides (35). Typically, the most crucial stage in bacterial research is obtaining these microorganisms in pure culture. Nevertheless, the technique only explains 1% or less of the total diversity found in the environment by the standards of culturing of the majority of environmental samples (36). Uncultured microorganisms account for the majority of Earth's natural biodiversity. As a result, there is a need for culture-independent ways to identify and characterize such microbes, as well as discovering their ecological functions in the biosphere (37). The vast microbial world has been revealed thanks to the rapid advancement of molecular microbiology and next-generation sequencing (NGS) technology. Significant progress has been achieved in increasing the cultivability of previously uncultured microorganisms. The culture-independent approach is the most promising and economical way of recognizing and extracting gene products of organisms in their environment (38, 39). Metagenomics is the study of genetic material (genomes) from a heterogeneous population of species. To put it briefly, metagenomics is a novel approach to studying microbes in a particular setting using sequencing analysis or functional gene screening. Metagenomics research is primarily concerned with microbial diversity, population structure, genetic and evolutionary links, functional activity, cooperative partnerships, and environmental interactions. Metagenomics research is quickly growing in medicine, agriculture, environmental protection, and other disciplines (40). Two primary and distinct outcomes emerge from the metagenomic analysis: diverse genetic determinants for biologically active molecules and processes that have potential medical and biotechnological applications; and outlooks on the complexity of the microbiome, both in species diversity and dynamics on a community scale. With the development of next-generation (ultra-high-throughput) sequencing technologies, DNA sequencing has greatly dropped in cost, signifying a new era for metagenomics. This strategy will help solve formerly unanswerable biological questions and speed up biotechnological discoveries (41).

Whole-genome shotgun (WGS) metagenomics and amplicon or marker gene studies are now the two major approaches used in high-throughput sequencing studies of microbial communities. To assess the biodiversity and functional capacities of the microbial community under study, WGS metagenomics strives to sequence every genome present in an environmental sample. Once all of a sample's genetic material has been retrieved, its gene content and diversity—including that of viruses, plasmids, bacteria, eukaryotes, and archaea—can be fully characterized. Marker gene studies, on the other hand, rely on the sequencing of a gene-specific area to disclose the composition and diversity of certain taxonomic groupings inside an environmental sample. The 16S rRNA gene is the most popular marker gene commonly used to identify the presence of archaea and bacteria in microbial ecology (42). The ITS region and the 18S rRNA genes are employed to analyze the structure of the fungal community and to demonstrate the presence of eukaryotes (43, 44). Zhang et al. (40) made the first attempt at grouping metagenomics into functional and sequencing metagenomics. Research in functional metagenomics aims to discover new bioactive chemicals and functional genes through systematic analysis of genomes. Metagenomics sequencing was used to analyze the diversity of the microbiome. Breitwieser et al. refer to shotgun sequencing data as metagenomics and amplicon sequencing data as metataxonomics (45). Here, we will discuss in detail the two approaches of metagenomics.

2.4.3.1 Types of Metagenomics Approaches

Sequencing every gene in the microbial population is necessary for shotgun metagenomics. For taxonomy investigations, amplicon sequencing concentrates on a substantial portion of the gene, such as the internal transcribed spacer (ITS) region, 16S rRNA, and 18S rRNA (46).

2.4.3.2 Whole-genome Shotgun Sequencing (First Generation DNA Sequencing)

Extracting microbial DNA from an environmental sample is the first step in the metagenomic analysis process. The metagenomic DNA specimen that has been obtained needs to be of the highest purity and quality, precisely representing all of the species present, both in terms of quality and quantity. By directly sequencing the collected metagenomic DNA and then doing relevant bioinformatics analysis, it is possible to understand the functional characteristics of microbes that colonize novel environments (38). It provides a coordinated understanding of community structure, composition, genetic and demographic variability, functional capacity, and probable metabolic pathways for a modest cost, in less time, and with more information. Sequencing extracted genomic DNA directly without amplifying the target gene circumvents the bias caused by PCR (47). The gold standard for sequencing is dideoxy chain termination sequencing, while shotgun sequencing, or first-generation sequencing, is based on the conventional Sanger sequencing methodology. Compared to second- and third-generation sequencing methods, shotgun sequencing has a better accuracy. It can handle repetitive sequences with high precision and determine long reads between 700 and 1,000 bp. The drawback is that it takes a lot of time and effort to identify more than one template at once (40).

2.4.3.3 Next-generation Sequencing (NGS) High-Throughput Sequencing

The emergence of "next-generation sequencing (NGS)" technology during the years 2004 to 2006 had a transformative impact on the field of biomedical research, resulting in a significant increase in the quantity of sequencing data being generated (48). The phrase "next-generation sequencing" is a broad term given to sequencing systems that employ post-Sanger technology to sequence huge quantities of DNA fragments simultaneously (49). The first phase of second-generation techniques, as seen on Illumina and Ion Torrent platforms, is commonly characterized by steps that include DNA fragmentation, surface attachment, adaptor ligation, end repair, and in-situ amplification. In this process of short-read sequencing, millions of small sequencing processes happen simultaneously during the massively parallel sequencing of short reads. However, because short-read methods need the reassembling of sequencing data over lengthy DNA segments, they pose difficulties when dealing with low-complexity or structurally varied areas (50). The 454-sequencing method from Roche, which uses pyrosequencing, is one of the currently widely used second-generation sequencing technologies. Using this method, the sequence is determined by tracking the pyrophosphate release that occurs when nucleotides are added to the DNA template (51). The fundamental drawback of this method is that it is prone to high mistake rates when reading homopolymer sequences, where n nucleotides might occasionally be read as n-1 nucleotides. The first bacterium to be sequenced using this method was the soil-dwelling *Myxococcus xanthus* (52). Another method for determining the sequence is called Ion Torrent sequencing, based on the emission of H+ ions during the synthesis of DNA. Reversible dye terminators serve as the foundation for the sequencing-by-synthesis technique used by the Illumina sequencing platform. SOLiD sequencing (sequencing by oligonucleotide ligation and detection), another emerging method, uses a ligation-based strategy with reversible terminators to determine the DNA sequence (51). An oil-water emulsion is used to amplify PCR reactions involving adaptors flanking template DNA fragments. PCR amplicons fixed on the beads are placed on a stable flat surface and undergo hybridization with a universal PCR primer that is compatible with adaptors. The specific location of each nucleotide is elucidated during every sequencing phase by attaching a fluorescently labeled DNA octamer to the universal primer. A pentamer is left on the DNA template after further chemical breakage. This process repeats itself until the DNA sequence is revealed. This platform has 99.94% accuracy because it uses a two-base coding method, which makes miscalls easier to identify (53).

Third-generation sequencing systems are far more capable than Sanger or short-read sequencing methods, reaching read lengths of up to 10 kb. Short-read problems, such as the identification of structural variants and genome-wide repetitions, can be solved by these "long-read" methods. Technologies have evolved to include fewer steps in library preparation and immediate targeting of unfragmented DNA molecules, where the generation of high-molecular-weight DNA is a constraint as compared to second-generation methods. While it has enhanced progression of software analysis, the precision of the reads was one of the earliest flaws of these third-generation methods in contrast to the second one (50). The initial long-read methodology to be commercially available was the single-molecule-real-time (SMRT®) technology developed by Pacific Biosciences (54). This method involves attaching an adaptor molecule to both ends of the target DNA molecule intended for sequencing, resulting in the formation of a closed circular DNA molecule during the library preparation phase. Subsequently, the circular DNA molecule is placed into a 150,000 zeptolitre well SMRT® cell, each containing a singular immobilized DNA polymerase at the bottom. The initiation of replication occurs when the DNA polymerase binds to hairpin adaptors located on the circular target DNA molecules. Following this, four fluorescently labeled nucleotides are introduced into the reaction wells. A light pulse is generated upon the enzymatic integration of each base, facilitating the identification of the base and allowing for iterative analysis to ultimately produce the DNA sequence (55). The initiation of the MinION Access Programme (MAP) facilitated the integration of MinION (Oxford Nanopore Technology) in the year 2014. Utilization of electrophoresis is imperative for the conveyance of DNA/RNA molecules via a nanopore. This system encompasses the application of a consistent electric field and the utilization of electrolytic solutions. Observations are conducted on the prevailing pattern and size alteration as the nucleic acid traverses through the nanopore. To initiate the library preparation process, double-stranded DNA is fragmented using a Covaris g-TUBE, followed by the repair of fragmented DNA through a PreCR procedure. In preparation for the addition of a poly-A tail to the 3′-OH end, an end repair procedure is employed to form blunt-ended DNA molecules. Integration of a Y adapter, named due to its structure, and a hairpin adaptor are essential steps in the DNA process. At the Y adapter juncture, a motor protein unwinds the double-stranded DNA, allowing the single strand to move through the nanopore. Subsequently, base calling is executed, yielding a read length of several hundred thousand base pairs with an accuracy ranging between 65% and 88%. The base calling system is categorized as 1-dimensional (1D) when data from a single strand is utilized, or as a 2-dimensional (2D) system otherwise (56).

Sample metagenomic DNA shotgun sequencing at random loci can identify target genes, the likely phylogeny of which can be determined by sequencing homologous sequences in non-redundant databases with the help of BLAST analysis. Hence, random sequencing may confirm the presence of the known genes with proven effects in the corresponding newly discovered microorganism and point to its homologs that could yield similar benefits and eventually lead to the development of new effective proteins, enzymes, and catalysts. Moreover, this newly discovered gene could either create a new catalytic protein or be able to function in the extreme conditions that are typical for the industrial process (38).

The species or strain of a microbial population can be determined using metagenomic sequencing. For instance, the research team discovered *Candidatus kryptonia*, a novel, highly thermophilic bacterium that was undetectable by 16S rDNA, using a combination of metagenome sequencing and single-cell sequencing (57). A different research team identified 1.2 million previously unidentified genes while characterizing the community of marine microorganisms of the Sargasso Sea. They also made the first attribution of bacterial species with rhodopsin-like photoreceptors (58). Applying the shotgun approach using metagenomic sequencing, Oh et al. examined the microbial composition and performance of the healthy human skin microbiome (59).

2.4.3.4 Amplicon Sequencing

The most popular high-throughput screening technique for the microbiome at the moment is amplicon sequencing, which involves amplifying one or more gene sequences, such as the ITS region, recA, gryA, and 16S rRNA genes. For phylogenetic purposes, several of these target genes are well-studied molecular clocks prevalent throughout particular domains or clades of species (60). It would be a breakthrough to develop genus-specific PCR primers that can amplify and sequence these hypervariable regions and hence help in identifying the species of the bacteria that are implicated with food samples. An 18S rDNA can help to differentiate the fungi. Operational taxonomic units (OTU) are formed by grouping these sequences according to how similar their nucleotide sequences are. Subsequently, OTUs are cross-referenced with databases to determine which bacteria are part of the microbiome. Identifying a microbial community from Sargasso Sea picoplankton was the first step toward characterizing the microbiome using this method (61). Amplicon sequencing has been widely used to evaluate microbial diversity in environmental materials because it is easily accessible using well-established assays and is inexpensive using free analytical software packages (62). The precise detection of bacterial community is possible at a significantly lower cost and more capacity with the later addition of NGS platforms to 16S rDNA sequencing. Higher coverage per sample can be achieved because only a short amplicon is sequenced (63). However, 16S rDNA sequencing frequently fails to identify bacteria above the genus level because of shorter reads generated by NGS techniques, particularly from Illumina platforms (64).

2.5 BIOINFORMATICS AND DATA ANALYSIS IN MICROBIOLOGY

This past decade has been marked by great improvements in computer technology and memory capacity, leading to possible solutions for grand challenge problems that include complete sequencing of genomes and network management of large integrated datasets. Nowadays, all researchers around the world can easily use the huge volume of genomic and proteomic data since the technology enabling this is computer capacity and miniaturization of biochemical processes and microarray chips and analysis. The boom of genome and proteome analysis brought about by data availability has produced several tools and discoveries that are not achievable in wet lab research (65). Bioinformatics is an information technology application that uses computer-based algorithms to assist in handling and analyzing data generated in biological research and experimentation. It is useful for creating patterns for disease outbreaks and novel biological products, as well as for DNA barcoding. Additionally, proteomics makes it easier to investigate protein structures and locate areas where proteins interact with one another (66).

The ensuing flood of data has mostly been kept up with by microbial bioinformatics, which is now distinctly developing as a unique field propelled forward by a committed global community of microbial bioinformaticians (67). To understand such datasets, a growing number of bioinformatic tools, pipelines, and databases are available. Before testing, these tools are usually chosen based on the scientific topic, kind of sample, tool features, and accessibility to reference sources (68). It is important to note that applying bioinformatics to the study of microbial genes, genomes, and metagenomes presents a unique challenge: instead of considering the finite, relatively easy-to-control genome of a human, or any other animal, or a plant, we have to deal with the genomic information of millions of microbial pathogens and thousands of commensal species, as well as up to billions of environmental microbial species (69). In 1995, the first complete microbial genome of *Haemophilus influenzae* was sequenced, marking a turning point in microbial biology (70) as a result of the collaboration of national laboratories, academic universities, several drug development companies, non-profit organizations, and businesses in the agricultural and bioremediation sectors, as well as federal health agencies (65). As new technologies and instruments are developed, biological databases are becoming essential research collaborators as biological research advances at a faster pace. The most important bioinformatics work has been acknowledged with the establishment and maintenance of major databases such as the GenBank and the Protein Data Bank. The primary biological databases are the bases for developing the secondary databases that gather high-level knowledge and act as the repositories for data received from experiments (71). The Pfam database of protein families and domains is an illustration of a secondary database of this type (72).

2.6 MASS SPECTROMETRY BASED METHODS

Mass spectrometry (MS) based methods have become popular for microbial identification and typing due to their speed, cost-effectiveness, simplicity, and applicability across various microorganisms, including bacteria, archaea, and fungi (73, 74). Various ionization and separation techniques can be combined with mass spectrometry (MS including gas chromatography (G C) (73, 75, 76), matrix-assisted laser desorption ionization time-of-flight (MALDI-TOF) (77, 78, electromigration techniques (79, 80), and electrospray ionization (ESI) (81, 82). Microbial identification research has primarily focused on minimizing the time needed for routine diagnostic procedures. The integration of liquid chromatography with mass spectrometry (LC-MS) has transformed metabolome analysis, enabling microbial identification by analyzing non-volatile or thermally labile high molecular compounds that are not suitable for gas chromatography-mass spectrometry (GC-MS) approaches (83, 84). LC-MS finds primary utility in clinical settings for microbial identification (85). Yet, it has also proven invaluable in detecting commercially available compounds within the in silico metabolome of *Bacillus subtilis* and *Escherichia coli*, as well as achieving comprehensive coverage of the metabolome of *Saccharomyces cerevisiae* (86).

2.6.1 Matrix-Assisted Laser Desorption/Ionization Time-of-Flight Mass Spectrometry (MALDI-TOF MS)

The latest tool in clinical microbiology for identifying microorganisms is mass spectrometry; specifically, MALDI-TOF MS. MALDI-TOF MS stands as the cutting-edge next-generation tool for the swift identification and classification of microorganisms. It operates by ionizing microbial cells with brief laser pulses and accelerating the particles within a vacuum system through an electric field (87, 88). MALDI-TOF analyzes mass spectra to identify and quantify ionized biological molecules within a 100 Da – 100 kDa range. In the matrix-assisted laser desorption and ionization (MALDI) phase, a sample undergoes ionization from a nitrogen laser beam's energy. Excited molecules transfer to the analyte, causing desorption of newly ionized particles. The time-of-flight (TOF) technique measures ions' flight time, leveraging the principle that ions of varying masses move at different speeds (27). MALDI offers distinct perspectives on bacterial biology and chemistry by detecting specific chemicals and measuring chemical alterations in response to environmental, temporal, or other external influences (89). A significant benefit of MALDI-TOF MS is its capability to directly identify microorganisms from bodily fluids like blood, cerebrospinal fluid, or urine. It is significant in critical conditions such as sepsis or meningitis (90). In MALDI-TOF, microbial identification relies on comparing specific mass peaks with patterns in a database. This distinctive protein profile acts as a molecular "fingerprint" (27, 91), ensuring highly specific results. The method is highly sensitive and is guaranteed by requiring only a small amount of material to be tested. MALDI-TOF MS offers rapid results (typically less than one hour), is relatively inexpensive, and supports high throughput (90, 92).

Combining traditional microbiology with cutting-edge methods and technologies like DNA sequencing, single-cell genomics, and metagenomics is necessary to overcome constraints. By integrating these approaches, we can improve our comprehension of the microbial world and its implications for many research areas (Figure 2.1).

2.7 ARTIFICIAL INTELLIGENCE APPLIED TO MICROBIOLOGY

Artificial intelligence is a rapidly growing field of computer science that focuses on producing intelligent machines (93). It includes methods and techniques such as natural language processing (NLP), machine learning (ML), and deep learning (DL). Large language models (LLMs) use deep learning and big datasets to comprehend, make a summary, and create and anticipate text-based content (94, 95). Currently, machine learning is integrated into almost every research area. It plays a significant role and is extensively used to solve numerous microbiological and biotechnological problems. Machine learning plays a significant role in epidemiology (96, 97), diagnosis (98, 99), drug repurposing (100, 101), and vaccine design (102). For instance, it is used in functional and structural genomics (103, 104), proteomics (105, 106), and metabolomics (107). Similarly, AI is also used in drug discovery (108), drug safety (109), pharmacology (110), pharmacogenetics (111), and pharmacogenomics (112). When used in microbiology and related fields, machine learning and artificial intelligence pose significant challenges. Among these, one key challenge is obtaining high-quality data to train machine learning models. Secondly, generating effective models for multi-faceted microbiology data is problematic. Lastly, understanding and validating black-box machine learning algorithms are challenging (113). Regardless of challenges, the future of AI in microbiology blooms with promise. Once these algorithms are established and trusted, they will become part of the workflow available to scientists (113).

2.8 FUTURE PERSPECTIVE

We stand on the brink of a new era in microbiology, and the future holds mammoth potential for groundbreaking research and discoveries that will transform our understanding of microbes. By merging traditional microbiological techniques with state-of-the-art metagenomics technology, layers of microbial communities are starting to emerge, exposing hitherto unthinkable levels of diversity and complexity.

In the near future, we expect several significant developments. For instance, novel cultivation methods for microorganisms will facilitate the growth of presently unculturable microorganisms. This will increase our understanding of the diversity of microorganisms and make it easier to investigate their functions in diverse ecosystems. Similarly, by improving metagenomics analyses, researchers will be able to enhance our understanding of microbial diversity and communities. This will make it easier to monitor the changes in functionality over time. Another important aspect is the combination of bioinformatics and AI to produce increasingly innovative data analysis tools. The huge volumes of data produced by metagenomic research will be handled by these instruments, enabling quicker and more precise predictions of protein structures and microbial interactions. Correspondingly, nanotechnology in microbiology will create new opportunities for the identification and management of microbes. Nanoscale devices can be utilized to precisely modify specific areas or distribute antimicrobials. Another important aspect is to deeply understand the human microbiome, and personalized medicine will become possible. Therapies that are more successful and have fewer adverse effects may result from customized care based on the distinct microbiological makeup of each patient. Similarly, sustainable agriculture and environmental preservation would significantly benefit from the application of metagenomics. We can create plans to save threatened ecosystems and increase agricultural output by understanding the roles that microbes play in different types of environments. More accurate disease outbreak modeling will enhance public health responses. Understanding disease propagation across microbial communities in greater detail can help develop early detection and intervention tactics.

In conclusion, the combination of conventional and contemporary methods in microbiology signifies a paradigm shift in our understanding of the microbial world, not merely an improvement in our scientific toolkit. The subject of microbiology holds immense potential for the upcoming generation of scientists who could discover secrets that transform our understanding of the tiniest life forms on Earth. The transition from traditional to next-generation microbiology is not only a history of scientific advancement, but also a blueprint for a day when this underrepresented group will play a key role in our pursuit of knowledge and the advancement of humankind.

REFERENCES

1. Schierbeek A. *Measuring the invisible world: the life and works of Antoni van Leeuwenhoek FRS*. London: Abelard-Schuman; 1959.
2. Isenberg HD. Clinical microbiology: past, present, and future. *Journal of Clinical Microbiology*. 2003 Mar;41(3):917.

3. Tang YW, Liu D, Sussman M, Poxton IR, Schwartzman J. *Molecular medical microbiology*. 2nd edition. London: Academic Press; 2014.
4. Löffler F. *Vorlesungen über die geschichtliche Entwickelung der Lehre von den Bacterien*. Leipzig: F.C.W. Vogel; 1887.
5. Koch R. Untersuchungen über Bakterien: VI. Verfahren zur Conserviren und Photographiren der Bakterien. *Beiträge zur Biologie der Pflanzen*. 1877;2:399–434.
6. Hesse W, Gröschel DH. Walther and Angelina Hesse—early contributors to bacteriology. *ASM News*. 1992;58(8):425–428.
7. Hoffmann H. Über die Kultur und Färbung von Bakterien. *Botanische Zeitung*. 1869;27:252.
8. Mahajan M. Etymologia: Petri Dish. *Emerging Infectious Diseases*. 2021 Jan;27(1):261.
9. DeLong EF, Pace NR. Environmental diversity of bacteria and archaea. *Systematic Biology*. 2001 Aug 1;50(4):470–478.
10. Li Y, Peng X, Zhou X, Ren B, Xiao L, Li Y, Li M, Guo Q. Basic biology of oral microbes. In *Atlas of Oral Microbiology: From Healthy Microflora to Disease*, ed. X Zhou and Y Li. Singapore: Springer; 2020. p. 1–24.
11. Koch R. The etiology of tuberculosis (1884). In: Carter KC, editor. *Essays of Robert Koch*. Connecticut: Greenwood Press; 1987. p. 129–150.
12. Barnard FA. The germ theory of disease and its relations to hygiene. *Public Health Papers and Reports*. 1873;1:70.
13. Coico R. Gram staining. *Current Protocols in Microbiology*. 2006 Feb 1;Appendix 3:A–3C.
14. Lagier JC, Hugon P, Khelaifia S, Fournier PE, La Scola B, Raoult D. The rebirth of culture in microbiology through the example of culturomics to study human gut microbiota. *Clinical Microbiology Reviews*. 2015;28(1):237–264.
15. Colwell RR, Brayton P, Herrington D, Tall B, Huq A, Levine MM. Viable but non-culturable Vibrio cholerae O1 revert to a cultivable state in the human intestine. *World Journal of Microbiology and Biotechnology*. 1996;12:28–31.
16. Hodges N. Fundamental features of microbiology. *Hugo and Russell's Pharmaceutical Microbiology*, ed. SP Denyer, N Hodges, and SP Gorman. Wiley; 2004. p. 9.
17. Brock TD, Madigan MT, Martinko JM, Parker J. *Brock biology of microorganisms*. Upper Saddle River (NJ): Prentice-Hall; 2003.
18. Osipov GA, Turova ES. Studying species composition of microbial communities with the use of gas chromatography-mass spectrometry: microbial community of kaolin. *FEMS Microbiology Reviews*. 1997;20(3–4):437–446.
19. Austrian R. The Gram stain and the etiology of lobar pneumonia, an historical note. *Bacteriological Reviews*. 1960;24(3):261–275.
20. Gram C. Ueber die isolirte Farbung der Schizomyceten in Schnitt-und Trockenpraparaten. *Fortschritte der Medicin*. 1884;2:185–189.
21. Laupland KB, Valiquette L. The changing culture of the microbiology laboratory. *Canadian Journal of Infectious Diseases and Medical Microbiology*. 2013;24:125–128.
22. Boyles TH, Wasserman S. Diagnosis of bacterial infection. *South African Medical Journal*. 2015;105(5):419.
23. Bunn T, Sikarwar A. Diagnostics: conventional versus modern methods. *Journal of Advances in Medical and Pharmaceutical Sciences*. 2016;8(4):1–7.
24. Kelley SO. New technologies for rapid bacterial identification and antibiotic resistance profiling. *SLAS Technology*. 2017;22(2):113–121.
25. White DC. Validation of quantitative analysis for microbial biomass, community structure, and metabolic activity. *Advances in Limnology*. 1988;31(1):1–18.
26. Zukowska ME. Advanced methods of bacteriological identification in a clinical microbiology laboratory. *Journal of Pre-Clinical and Clinical Research*. 2021;15(2):68–72.
27. Żabicka D, Literacka E. Nowoczesne metody wykrywania i identyfikacji bakterii. *Forum Zakażeń*. 2013;4(1):65–72.
28. Vila J, Gómez MD, Salavert M, Bosch J. Métodos de diagnóstico rápido en microbiología clínica: necesidades clínicas. *Enfermedades Infecciosas Y Microbiología Clínica*. 2017;35(1):41–46.

29. Valones MAA, Guimarães RL, Brandão LAC, Souza PR, Carvalho ADA, Crovela S. Principles and applications of polymerase chain reaction in medical diagnostic fields: a review. *Brazilian Journal of Microbiology*. 2009;40:1–11.
30. Yang S, Rothman RE. PCR-based diagnostics for infectious diseases: uses, limitations, and future applications in acute-care settings. *The Lancet Infectious Diseases*. 2004;4(6):337–348.
31. Buchan BW, Ledeboer NA. Emerging technologies for the clinical microbiology laboratory. *Clinical Microbiology Reviews*. 2014;27(4):783–822.
32. Morganti S, Tarantino P, Ferraro E, D'Amico P, Duso BA, Curigliano G. Next generation sequencing (NGS): a revolutionary technology in pharmacogenomics and personalized medicine in cancer. *Advances in Experimental Medicine and Biology*. 2019;1168:9–30.
33. Boers SA, Jansen R, Hays JP. Understanding and overcoming the pitfalls and biases of next-generation sequencing (NGS) methods for use in the routine clinical microbiological diagnostic laboratory. *European Journal of Clinical Microbiology & Infectious Diseases*. 2019;38:1059–1070.
34. Escobar-Zepeda A, Vera-Ponce de León A, Sanchez-Flores A. The road to metagenomics: from microbiology to DNA sequencing technologies and bioinformatics. *Frontiers in Genetics*. 2015;6:348.
35. Kellenberger E. Exploring the unknown. *EMBO Reports*. 2001;2(1):5–7. https://doi.org/10.1093/embo-reports/kve014
36. Amann RI, Ludwig W, Schleifer KH. Phylogenetic identification and in situ detection of individual microbial cells without cultivation. *Microbiological Reviews*. 1995;59(1):143–169. https://doi.org/10.1128/mr.59.1.143-169.1995
37. Mardanov AV, Kadnikov VV, Ravin NV. Metagenomics: a paradigm shift in microbiology. In: *Metagenomics*, ed. M Nagarajan. London: Academic Press; 2018. p. 1–13. https://doi.org/10.1016/B978-0-08-102268-9.00001-X
38. Coughlan LM, Cotter PD, Hill C, Alvarez-Ordóñez A. Biotechnological applications of functional metagenomics in the food and pharmaceutical industries. *Frontiers in Microbiology*. 2015;6:672. https://doi.org/10.3389/fmicb.2015.00672
39. Degois J, Clerc F, Simon X, Bontemps C, Leblond P, Duquenne P. First metagenomic survey of the microbial diversity in bioaerosols emitted in waste sorting plants. *Annals of Work Exposures and Health*. 2017;61(9):1076–1086.
40. Zhang L, Chen F, Zeng Z, Xu M, Sun F, Yang L, Bi X, Lin Y, Gao Y, Hao H, Yi W, Li M, Xie Y. Advances in metagenomics and its application in environmental microorganisms. *Frontiers in Microbiology*. 2021;12:766364. https://doi.org/10.1093/annweh/wxx075
41. Chistoserdova L. Functional metagenomics: recent advances and future challenges. *Biotechnology and Genetic Engineering Reviews*. 2009;26(1):335–352. https://doi.org/10.5661/bger-26-335
42. Case RJ, Boucher Y, Dahllöf I, Holmström C, Doolittle WF, Kjelleberg S. Use of 16S rRNA and rpoB genes as molecular markers for microbial ecology studies. *Applied and Environmental Microbiology*. 2007;73(1):278–288. https://doi.org/10.1128/AEM.01177-06
43. Schoch CL, Seifert KA, Huhndorf S, Robert V, Spouge JL, Levesque CA, et al. Nuclear ribosomal internal transcribed spacer (ITS) region as a universal DNA barcode marker for fungi. *Proceedings of the National Academy of Sciences of the United States of America*. 2012;109(16):6241–6246. https://doi.org/10.1073/pnas.1117018109
44. Lindahl BD, Nilsson RH, Tedersoo L, Abarenkov K, Carlsen T, Kjøller R, et al. Fungal community analysis by high-throughput sequencing of amplified markers – a user's guide. *The New Phytologist*. 2013;199(1):288–299. https://doi.org/10.1111/nph.12243
45. Breitwieser FP, Lu J, Salzberg SL. A review of methods and databases for metagenomic classification and assembly. *Briefings in Bioinformatics*. 2019;20(4):1125–1136. https://doi.org/10.1093/bib/bbx120
46. Lema NK, Gemeda MT, Woldesemayat AA. Recent advances in metagenomic approaches, applications, and challenges. *Current Microbiology*. 2023;80(11):347. https://doi.org/10.1007/s00284-023-03451-5
47. Castañeda LE, Barbosa O. Metagenomic analysis exploring taxonomic and functional diversity of soil microbial communities in Chilean vineyards and surrounding native forests. *PeerJ*. 2017;5:e3098. https://doi.org/10.7717/peerj.3098
48. Mardis ER. Next-generation sequencing platforms. *Annual Review of Analytical Chemistry (Palo Alto, Calif.)*. 2013;6:287–303. https://doi.org/10.1146/annurev-anchem-062012-092628

49. Schlebusch S, Illing N. Next generation shotgun sequencing and the challenges of de novo genome assembly: review article. *South African Journal of Science*. 2012;108(11):1–8. https://doi.org/10.10520/EJC127515
50. Hu T, Chitnis N, Monos D, Dinh A. Next-generation sequencing technologies: an overview. *Human Immunology*. 2021;82(11):801–811. https://doi.org/10.1016/j.humimm.2021.02.012
51. Satam H, Joshi K, Mangrolia U, Waghoo S, Zaidi G, Rawool S, et al. Next-generation sequencing technology: current trends and advancements. *Biology*. 2023;12(7):Article 7. https://doi.org/10.3390/biology12070997
52. Vos M, Velicer GJ. Genetic population structure of the soil bacterium Myxococcus xanthus at the centimeter scale. *Applied and Environmental Microbiology*. 2006;72(5):3615–3625. https://doi.org/10.1128/AEM.72.5.3615-3625.2006
53. Cao Y, Wang J, Jian F, Xiao T, Song W, Yisimayi A, et al. Omicron escapes the majority of existing SARS-CoV-2 neutralizing antibodies. *Nature*. 2022;602(7898):657–663. https:// doi.org/10.1038/s41586-021-04385-3
54. Eid J, Fehr A, Gray J, Luong K, Lyle J, Otto G, et al. Real-time DNA sequencing from single polymerase molecules. Science. 2009;323(5910):133–138. https://doi.org/10.1126/science.1162986
55. Rhoads A, Au KF. PacBio sequencing and its applications. *Genomics, Proteomics & Bioinformatics*. 2015;13(5):278–289. https://doi.org/10.1016/j.gpb.2015.08.002
56. Lu H, Giordano F, Ning Z. Oxford nanopore MinION sequencing and genome assembly. *Genomics, Proteomics & Bioinformatics*. 2016;14(5):265–279. https://doi.org/10.1016/j.gpb.2016.05.004
57. Eloe-Fadrosh EA, Paez-Espino D, Jarett J, Dunfield PF, Hedlund BP, Dekas AE, et al. Global metagenomic survey reveals a new bacterial candidate phylum in geothermal springs. Nature Communications. 2016;7:10476. https://doi.org/10.1038/ncomms10476.
58. Venter JC, Remington K, Heidelberg JF, Halpern AL, Rusch D, Eisen JA, et al. Environmental genome shotgun sequencing of the Sargasso Sea. *Science*. 2004;304(5667):66–74. https://doi.org/10.1126/science.1093857
59. Oh J, Byrd AL, Deming C, Conlan S, Kong HH, Segre JA. Biogeography and individuality shape function in the human skin metagenome. *Nature*. 2014;514(7520):Article 7520. https://doi.org/10.1038/nature13786
60. Yang B, Wang Y, Qian PY. Sensitivity and correlation of hypervariable regions in 16S rRNA genes in phylogenetic analysis. *BMC Bioinformatics*. 2016;17(1):135. https://doi.org/10.1186/s12859-016-0992-y
61. Giovannoni SJ, Britschgi TB, Moyer CL, Field KG. Genetic diversity in Sargasso Sea bacterioplankton. *Nature*. 1990;345(6270):Article 6270. https://doi.org/10.1038/345060a0
62. Caporaso JG, Lauber CL, Walters WA, Berg-Lyons D, Huntley J, Fierer N, et al. Ultra-high-throughput microbial community analysis on the Illumina HiSeq and MiSeq platforms. *The ISME Journal*. 2012;6(8):1621–1624. https://doi.org/10.1038/ismej.2012
63. Claesson MJ, O'Sullivan O, Wang Q, Nikkilä J, Marchesi JR, Smidt H, et al. Comparative analysis of pyrosequencing and a phylogenetic microarray for exploring microbial community structures in the human distal intestine. *PLOS ONE*. 2009;4(8):e6669. https://doi.org/10.1371/journal.pone.0006669
64. Claesson MJ, Wang Q, O'Sullivan O, Greene-Diniz R, Cole JR, Ross RP, et al. Comparison of two next-generation sequencing technologies for resolving highly complex microbiota composition using tandem variable 16S rRNA gene regions. *Nucleic Acids Research*. 2010;38(22):e200–e200. https://doi.org/10.1093/nar/gkq873
65. Bansal AK. Bioinformatics in microbial biotechnology – a mini review. *Microbial Cell Factories*. 2005;4:19. https://doi.org/10.1186/1475-2859-4-19
66. Rao VS, Srinivas K, Sujini GN, Kumar GN. Protein-protein interaction detection: methods and analysis. *International Journal of Proteomics*. 2014;2014:e147648. https://doi.org/10.1155/2014/147648
67. Loman NJ, Quinlan AR. Poretools: a toolkit for analyzing nanopore sequence data. *Bioinformatics (Oxford, England)*. 2014;30(23):3399–3401. https://doi.org/10.1093/bioinformatics/btu555
68. Walsh LH, Coakley M, Walsh AM, O'Toole PW, Cotter PD. Bioinformatic approaches for studying the microbiome of fermented food. *Critical Reviews in Microbiology*. 2023;49:693–725. https://doi/abs/10.1080/1040841X.2022.2132850

69. Locey KJ, Lennon JT. Scaling laws predict global microbial diversity. *Proceedings of the National Academy of Sciences of the United States of America*. 2016;113(21):5970–5975. https://doi.org/10.1073/pnas.1521291113
70. Fleischmann RD, Adams MD, White O, Clayton RA, Kirkness EF, Kerlavage AR, et al. Whole-genome random sequencing and assembly of Haemophilus influenzae Rd. *Science (New York, N.Y.)*. 1995;269(5223):496–512. https://doi.org/10.1126/science.7542800
71. Likić VA, McConville MJ, Lithgow T, Bacic A. Systems biology: the next frontier for bioinformatics. *Advances in Bioinformatics*. 2010;2010:e268925. https://doi.org/10.1155/2010/268925
72. Finn RD, Mistry J, Tate J, Coggill P, Heger A, Pollington JE, et al. The Pfam protein families database. *Nucleic Acids Research*. 2010;38(suppl_1):D211–D222. https://doi.org/10.1093/nar/gkp98593
73. Fox A. Mass spectrometry: identification and bio detection, lessons learned and future developments. *Identification of Microorganisms by Mass Spectrometry*. 2005;169:23–37.
74. Sauer S, Kliem M. Mass spectrometry tools for the classification and identification of bacteria. *Nature Reviews Microbiology*. 2010;8:74–82.
75. Senes C, Saldan N, Costa W, Svidzinski T, Oliveira C. Identification of Fusarium oxysporum fungus in wheat based on chemical markers and qualitative GC-MS test. *Journal of the Brazilian Chemical Society*. 2018;1:167–178.
76. Fox A. Mass spectrometry for species or strain identification after culture or without culture: past, present, and future. *Journal of Clinical Microbiology*. 2006;44:2677–2680.
77. Jang KS, Kim YH. Rapid and robust MALDI-TOF MS techniques for microbial identification: a brief overview of their diverse applications. *Journal of Microbiology*. 2018;56:209–216.
78. Biswas S, Rolain JM. Use of MALDI-TOF mass spectrometry for identification of bacteria that are difficult to culture. *Journal of Microbiological Methods*. 2013;92:14–24.
79. Buszewski B, Rogowska A, Pomastowski P, Złoch M, Railean-Plugaru V. Identification of microorganisms by modern analytical techniques. *Journal of AOAC International*. 2017;100:1607–1623.
80. Kłodzińska E, Buszewski B. Electrokinetic detection and characterization of intact microorganisms. *Analytical Chemistry*. 2009;81:8–15.
81. Zhang JI, Talaty N, Costa AB, Xia Y, Tao WA, Bell R, Callahan JH, Cooks RG. Rapid direct lipid profiling of bacteria using desorption electrospray ionization mass spectrometry. *International Journal of Mass Spectrometry*. 2011;301:37–44.
82. Vaidyanathan S, Goodacre R. High-throughput microbial characterizations using electrospray ionization mass spectrometry and its role in functional genomics. In: *Identification of microorganisms by mass spectrometry*, ed. CL Wilkins and JO Lay Jr. Hoboken, NJ, USA: John Wiley & Sons, Inc.; 2006. p. 229–256.
83. Lattanzio V, Solfrizzo M, Powers S, Visconti A. Simultaneous determination of aflatoxins, ochratoxin A and Fusarium toxins in maize by liquid chromatography/tandem mass spectrometry after multitoxin immunoaffinity cleanup. *Rapid Communications in Mass Spectrometry*. 2007;21:3253–3261.
84. Warren CR. A liquid chromatography–mass spectrometry method for analysis of intact fatty-acid-based lipids extracted from soil. *European Journal of Soil Science*. 2018;69:791–803.
85. Bakhtiar R, Ramos L, Tse FL. High-throughput mass spectrometric analysis of xenobiotics in biological fluids. *Journal of Liquid Chromatography & Related Technologies*. 2002;25(4):507–540.
86. Van der Werf MJ, Overkamp KM, Muilwijk B, Coulier L, Hankemeier T. Microbial metabolomics: toward a platform with full metabolome coverage. *Analytical Biochemistry*. 2007;370:17–25.
87. Dierig A, Frei R, Egli A. The fast route to microbe identification: matrix assisted laser desorption/ionization—time of flight mass spectrometry (MALDI-TOF MS). *The Pediatric Infectious Disease Journal*. 2015;34(1):97–99.
88. Sandrin TR, Demirev PA. Characterization of microbial mixtures by mass spectrometry. *Mass Spectrometry Reviews*. 2018;37(3):321–349.
89. Lay Jo Jr. MALDI-TOF mass spectrometry of bacteria. *Mass Spectrometry Reviews*. 2001;20(4):172–194.
90. Singhal N, Kumar M, Kanaujia PK, Virdi JS. MALDI-TOF mass spectrometry: an emerging technology for microbial identification and diagnosis. *Frontiers in Microbiology*. 2015;6:791.
91. Hosseini S, Martinez-Chapa SO. Principles and mechanism of MALDI-ToF-MS analysis. In Fundamentals of MALDI-ToF Analysis, ed. S Hosseini and SO Martinez-Chapa. Singapore: Springer; 2017. p. 1–19.
92. Hou TY, Chiang-Ni C, Teng SH. Current status of MALDI-TOF mass spectrometry in clinical microbiology. *Journal of Food and Drug Analysis*. 2019;27(2):404–414.

93. Holzinger A, Kickmeier-Rust M, Müller H. Kandinsky patterns as IQ-test for machine learning. In: *Machine Learning and Knowledge Extraction: Third IFIP TC 5, TC 12, WG 8.4, WG 8.9, WG 12.9 International Cross-Domain Conference, CD-MAKE 2019, Canterbury, UK, August 26–29, 2019, Proceedings*, ed. A Holzinger, P Kieseberg, A Min Tjoa, and E Weippl. Cham: Springer International Publishing; 2019. p. 1–14.
94. Suleimenov IE, Vitulyova YS, Bakirov AS, Gabrielyan OA. Artificial intelligence: what is it? In: *Proceedings of the 2020 6th International Conference on Computer and Technology Applications*; 2020 Apr 22–25. https://doi.org/10.1145/3397125.3397141
95. Davenport T, Kalakota R. The potential for artificial intelligence in healthcare. *Future Healthcare Journal*. 2019;6(2):94. https://doi.org/10.7861/futurehosp.6-2-94
96. Mackey T, Purushothaman V, Li J, Shah N, Nali M, Bardier C, et al. Machine learning to detect self-reporting of symptoms, testing access, and recovery associated with COVID-19 on Twitter: retrospective big data infoveillance study. *JMIR Public Health and Surveillance*. 2020;6:e19509.
97. Santosh KC. AI-driven tools for coronavirus outbreak: need of active learning and cross-population train/test models on multitudinal/multimodal data. *Journal of Medical Systems*. 2020;44(5):93.
98. Albahri AS, Hamid RA, Alwan JK, Al-Qays ZT, Zaidan AA, Zaidan BB, et al. Role of biological data mining and machine learning techniques in detecting and diagnosing the novel coronavirus (COVID-19): a systematic review. *Journal of Medical Systems*. 2020;44:1–11.
99. Elaziz MA, Hosny KM, Salah A, Darwish MM, Lu S, Sahlol AT. New machine learning method for image-based diagnosis of COVID-19. *PLOS ONE*. 2020;15(6):e0235187.
100. Gao K, Nguyen DD, Chen J, Wang R, Wei GW. Repositioning of 8565 existing drugs for COVID-19. *The Journal of Physical Chemistry Letters*. 2020;11(13):5373–5382.
101. Mirabelli C, Wotring JW, Zhang CJ, McCarty SM, Fursmidt R, Pretto CD, et al. Morphological cell profiling of SARS-CoV-2 infection identifies drug repurposing candidates for COVID-19. *Proceedings of the National Academy of Sciences*. 2021;118(36):e2105815118.
102. Ong E, Wong MU, Huffman A, He Y. COVID-19 coronavirus vaccine design using reverse vaccinology and machine learning. *Frontiers in Immunology*. 2020;11:561910.
103. Caudai C, Galizia A, Geraci F, Le Pera L, Morea V, Salerno E, et al. AI applications in functional genomics. *Computational and Structural Biotechnology Journal*. 2021;19:5762–5790. https://doi.org/10.1016/j.csbj.2021.10.009
104. Lin J, Ngiam KY. How data science and AI-based technologies impact genomics. *Singapore Medical Journal*. 2023;64(1):59–66.
105. Xiao Q, Zhang F, Xu L, Yue L, Kon OL, Zhu Y, et al. High-throughput proteomics and AI for cancer biomarker discovery. *Advanced Drug Delivery Reviews*. 2021;176:113844. https://doi.org/10.1016/j.addr.2021.113844
106. Mund A, Coscia F, Hollandi R, Kovács F, Kriston A, Brunner AD, et al. AI-driven deep visual proteomics defines cell identity and heterogeneity. *BioRxiv*. 2021;2021-01. https://doi.org/10.1101/2021.01.25.427969
107. Petrick LM, Shomron N. AI/ML-driven advances in untargeted metabolomics and exposomics for biomedical applications. *Cell Reports Physical Science*. 2022;3(7):100978. https://doi.org/10.1016/j.xcrp.2022.100978
108. David L, Thakkar A, Mercado R, Engkvist O. Molecular representations in AI-driven drug discovery: a review and practical guide. *Journal of Cheminformatics*. 2020;12(1):56. https://doi.org/10.1186/s13321-020-00460-5
109. Diaw MD, Papelier S, Durand-Salmon A, Felblinger J, Oster J. AI-assisted QT measurements for highly automated drug safety studies. *IEEE Transactions on Biomedical Engineering*. 2022;70:1504–15. https://doi.org/10.1109/TBME.2022.3221339
110. Van der Lee M, Swen JJ. Artificial intelligence in pharmacology research and practice. *Clinical and Translational Science*. 2023;16(1):31–36. https://doi.org/10.1111/cts.13431
111. Roche-Lima A, Roman-Santiago A, Feliu-Maldonado R, Rodriguez-Maldonado J, Nieves-Rodriguez BG, et al. Machine learning algorithm for predicting warfarin dose in Caribbean Hispanics using pharmacogenetic data. *Frontiers in Pharmacology*. 2020;10:1550. https://doi.org/10.3389/fphar.2019.01550

112. Lin E, Lin CH, Lane HY. Precision psychiatry applications with pharmacogenomics: artificial intelligence and machine learning approaches. *International Journal of Molecular Sciences*. 2020;21(3):969. https://doi.org/10.3390/ijms21030969
113. Goodswen SJ, Barratt JL, Kennedy PJ, Kaufer A, Calarco L, Ellis JT. Machine learning and applications in microbiology. *FEMS Microbiology Reviews*. 2021;45(5):fuab015.

3 Advancements in Nanotechnology and Their Applications in Microbiology

Zakieh Sadat Hoseini, Moein Iranmanesh, Yousof Tarverdizadeh, Mohsen Yaghoubi, Mostafa Zare, and Abbas Hajizade

3.1 INTRODUCTION

As an emerging tool, nanotechnology has revolutionized almost all scientific and industrial fields. Biological disciplines, including biotechnology, biochemistry, biophysics, cell biology, molecular biology, genetics, and physiology, have experienced significant benefits from the progress in nanotechnology (1–3). Microbiology, an important branch of biology, has an undeniable role in human life. Microorganisms have significantly influenced life on Earth, to the extent that it is inconceivable to envision a world without them. In the realm of biology, including microbiology, nanotechnology has made significant strides. Nanotechnology enhances microbial detection, diagnosis, prevention, treatment of infectious diseases, and microbial control (4, 5). In the present chapter, the applications of nanotechnology in microbiology are discussed. The field of microbiology is diverse, therefore, our main focus is on medical microbiology, though many principles can be applied in other microbiology disciplines, such as environmental microbiology, food microbiology, industrial microbiology, etc.

3.2 MEDICAL MICROBIOLOGY AND ITS CHALLENGES

Medical microbiology is a branch of microbiology that deals with the diagnosis, identification, prevention, and treatment of infectious agents, including bacteria, viruses, parasites, fungi, and prions (6).

Infection control is the first step in combating infectious agents (7). Improved sanitation facilities and access to safe drinking water are two important issues in this regard and governments are responsible for providing these items. At the same time, medical microbiologists try to develop new and effective vaccines to prevent infectious diseases (8). It is one of the main challenges that medical microbiology faces. While significant progress has been made in enhancing vaccine development strategies, there remain numerous infectious agents for which no approved vaccines exist. Indeed, some approved vaccines lack adequate efficiency (9). Here, nanotechnology can play an important role, discussed in the following sections.

Ensuring microbial safety in water resources, wastewater treatment, food products, and hospital surfaces—particularly within surgical units where drug-resistant microbes are increasingly prevalent—is a critical challenge in preventing the spread of infectious agents (10, 11). Detection and identification of infectious agents is another step in combating infectious diseases. Medical microbiologists should be able to identify infectious agents in different samples, including blood, bone marrow, sputum, urine, feces, different tissues, various body fluids, etc. Timely detection and

 DOI: 10.1201/9781003410164-3

identification of pathogens is the main step to fighting against them (12). Despite many advancements in detection and diagnostic tools, there are still many infectious diseases and infectious agents that remain undetectable with available techniques and tools. This is another main challenge in front of medical microbiologists.

Managing infections represents a critical step in combating infectious agents. Despite the rise of multi-drug-resistant bacterial strains (13), antibiotics remain the primary treatment for bacterial diseases. The cellular uptake of antibiotics, particularly in the context of intracellular bacteria like *Brucella* species, pathogenic species of *Shigella*, and *Salmonella* species, poses a significant challenge (14). Furthermore, hydrophobic antibiotics exacerbate the situation, making their effective delivery a formidable task. In the following sections, after a brief introduction to nanotechnology and nanomaterials, we will discuss how nanotechnology can help us overcome some of the challenges that microbiologists face.

3.3 A BRIEF REVIEW OF NANOTECHNOLOGY

The term "nano," originating from the Greek word for "dwarf," denotes a reduction in size or time by a factor of 10^{-9}, which is 1,000 times smaller than a micrometer (15). The nanometer scale (nm) corresponds to one billionth of a meter, or approximately the width of three to five atoms in at least one dimension. Nanomaterials, due to their small size and large surface area, exhibit distinct properties compared to bulk materials. This arises from their high surface-to-volume ratio and quantum effects. The field of nanoscience and nanotechnology holds great promise, particularly as material dimensions shrink below 100 nm, leading to significant alterations in properties such as chemical, physical, and biological behavior (16, 17).

The concept of nanotechnology is commonly associated with the famous physicist and Nobel laureate, Dr. Richard P. Feynman (18, 19). In his classic scientific lecture entitled "There's Plenty of Room Below: An Invitation to a New Field of Physics", Feynman talked about putting the entire Encyclopedia Britannica on the head of a needle, insisting that there was no law preventing such a thing. In 1974, after a decade and a half, Professor Norio Taniguchi from Tokyo University of Science introduced the term "nanotechnology." He employed this expression to characterize the production of precise materials (devices) with dimensions measured in nanometers. The simplest definition of nanotechnology is "*technology at the nanoscale.*" Since then, various definitions of "*atomic precision technology*" or "*atomic precision engineering*" have evolved from nanotechnology (20).

There are also different methods for nanomaterial synthesis, but the two main approaches are bottom-up and top-down methods. Examples of top-down techniques include lithography, mechanical or ball milling, laser ablation, sputtering, electron explosive arc discharge, and thermal decomposition. Examples of bottom-up techniques include chemical vapor deposition (CVD), sol-gel, spinning, pyrolysis, and biosynthesis (21, 22).

3.4 NANOMATERIALS: TYPES AND FUNCTIONS

Nanomaterials have different characteristics such as shape, composition, dimension, porosity, etc., so they are classified in different ways (23, 24). According to their origin, nanomaterials can be classified into (i) natural and (ii) synthetic. Natural nanomaterials are formed during natural processes, like physical fragmentation, photo-oxidation, precipitation reactions, and gas-solid nucleation, while synthetic nanomaterials are designed and fabricated in laboratories (25, 26).

The composition of nanomaterials is another property that is useful for their classification. Based on their composition, they are classified into four different types, including: (i) inorganic nanomaterials, (ii) organic nanomaterials, (iii) carbon nanomaterials, and (iv) composite nanomaterials (Figure 3.1).

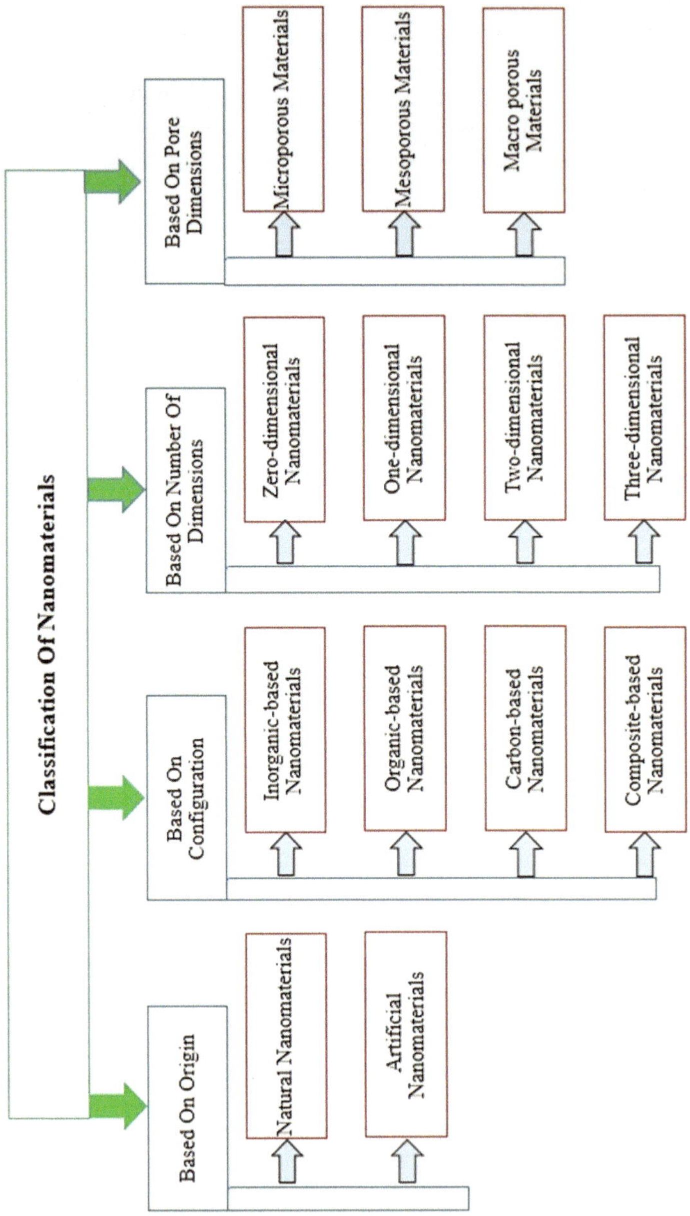

FIGURE 3.1 Diagram depicting classification of nanomaterial on the basis of origin, composition, dimensions, and pore size.

Inorganic nanomaterials are non-degradable nanomaterials that are used for the production of different nanostructures. These nanomaterials are composed of quantum dots (semiconductor nanocrystals with unique optical properties), metal nanoparticles (such as aluminum (Al), cadmium (Cd), cobalt (Co), copper (Cu), gold (Au), iron (Fe), lead (Pb), silver (Ag), and zinc (Zn)), and metal oxide nanoparticles (such as titanium oxide (TiO_2), zinc oxide (ZnO), and aluminum oxide (Al_2O_3)) (27, 28).

Organic nanomaterials are present in different types and shapes. These materials are formed from organic materials consisting of carbon-based molecules, polymers, or biomolecules excluding carbon materials; lipid-based nanomaterials such as liposomes and micelles, polymeric nanomaterials such as chitosan, alginate, poly(lactic-co-glycolic acid) (PLGA) are some examples of organic nanomaterials (29).

Carbon nanomaterials, like fullerenes, graphene, and carbon nanotubes (CNTs), are progressively used in science and industry (30). Carbon-based nanomaterials have many outstanding properties that make them an excellent choice for the development of nanostructures (31, 32).

Composite nanomaterials are constituted by more than one nanoscale compound, with special physical and chemical properties. The performance of composite nanomaterials is significantly enhanced in comparison to individual nanomaterials. Indeed, composite nanomaterials can gain novel features and properties. These nanomaterials are increasingly used in different areas, like regenerative medicine (33, 34), aerospace (35), energy harvesting and storage (36, 37), environmental applications (38), etc.

Nanomaterials can also be classified on the basis of their dimensions. According to this feature, nanomaterials are categorized into (i) zero-dimensional, (ii) one-dimensional, (iii) two-dimensional, and (iv) three-dimensional nanomaterials (24). All dimensions of zero-dimensional nanomaterials are on the nanometer scale. Different NPs, such as metal NPs, polymeric NPs, quantum dots, and lipid NPs are zero-dimensional nanomaterials. In one-dimensional nanomaterials, one of the three dimensions is in nanoscale. Nanowires and nanorods are two examples of one-dimensional nanomaterials. Two dimensions of two-dimensional nanomaterials are in the nanoscale. Nanoplates and nanosheets are examples of this type of nanomaterials. In 3-dimensional nanomaterials, all three dimensions are larger than 100 nm. Nanocages, nanoflowers, and nanocubes are some examples of 3-dimensional nanomaterials (39).

Based on their pores, porous nanomaterials are classified into three groups: (i) microporous materials, (ii) mesoporous materials, and (iii) macroporous materials. Microporous nanomaterials have pore sizes of 2 nm or less (40). Having hundreds of empty channels, these nanomaterials can encapsulate and absorb small molecules or absorbents (41). Metal-organic frameworks and zeolites are two main examples of microporous materials with different applications in industry. For example, zeolites sometimes act as molecular sieves and are used for sterilization purposes (42). Mesoporous nanomaterials have pore sizes between 2 and 50 nm. These materials have uniform nanochannels and, owing to this feature, they have many applications in different fields. For example, mesoporous silica NPs (MSNPs) are widely used for drug and vaccine delivery (43), diagnosis purposes, photodynamic therapy (44), photocatalysis (45), etc. The pore size of macroporous nanomaterials is greater than 50 nm. As an example, fiberlike Fe_2O_3 macroporous nanomaterials are potent photocatalysts (46). Macroporous carbon nanoparticles encapsulating ZnO are used in lithium-ion batteries (47). Figure 3.2, represents the electron micrographs of some nanomaterials.

Due to their unique characteristics, nanomaterials find a number of applications in various fields of science and technology. These nanomaterials are increasingly used in different areas, ranging from regenerative medicine (33, 34, 53) to aerospace (35), energy harvesting and storage (36, 37), environmental applications (38), mechanical engineering (54), food (55, 56), cosmetics (52, 57, 58), etc. (Figure 3.3).

In regenerative medicine, nanomaterials are widely used for different purposes; they are excellent vehicles for the delivery of therapeutics, such as bone morphogenetic proteins (59–61),

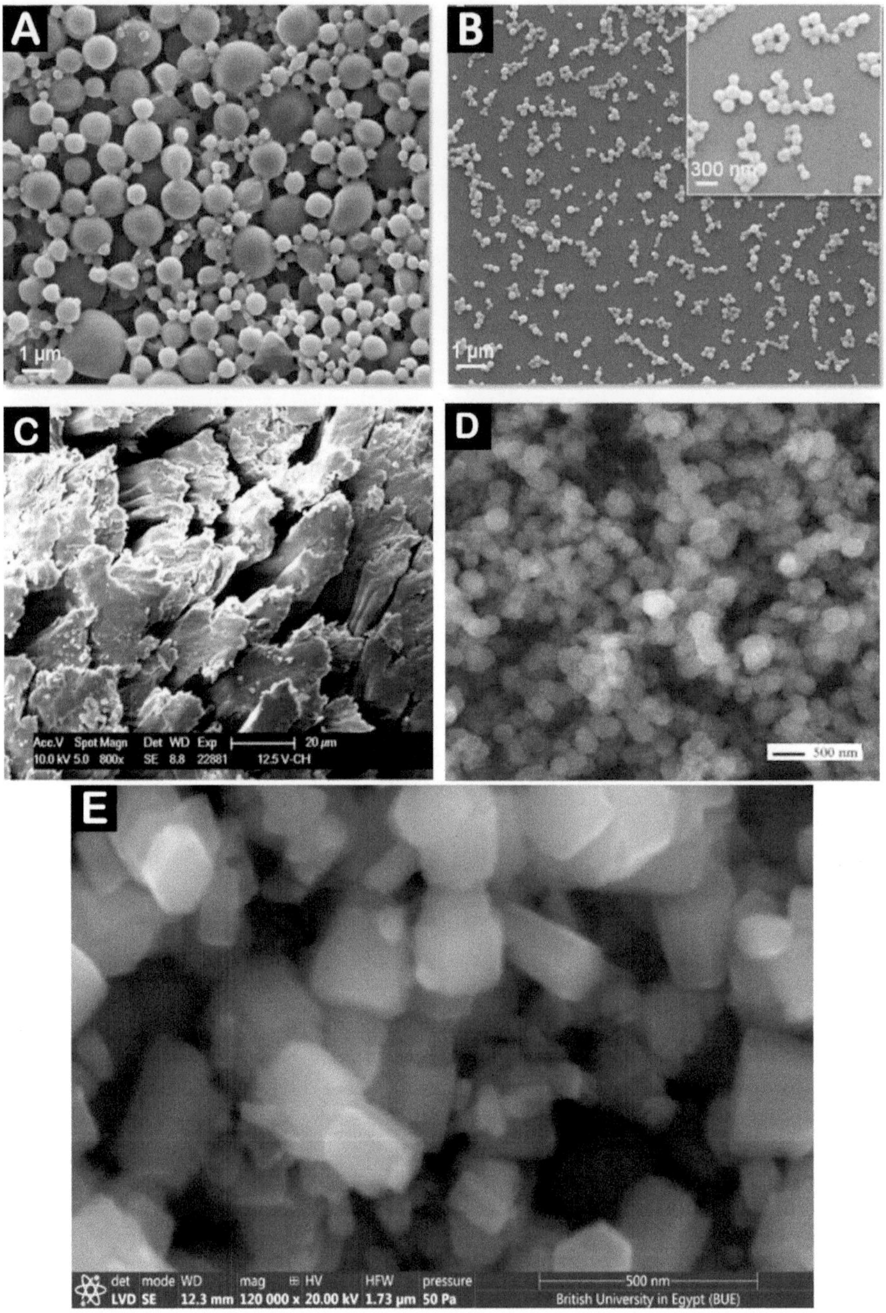

FIGURE 3.2 (A and B) SEM and TEM images of PLGA nanoparticles produced by microchannel emulsification in an interdigital micromixer at different conditions (Adopted from reference (48)). (C) SEM images of chitosan film, and chitosan-viscose rayon biocomposite (Adopted from reference (49)). (D) Particle size distribution of silica nanoparticles synthesized in 2-prophanol (Adopted from reference (50)). (E) SEM image for ZnO assisted CTAB synthesized at concentration of 1% after calcination at 350 °C (Adopted from reference (51)).

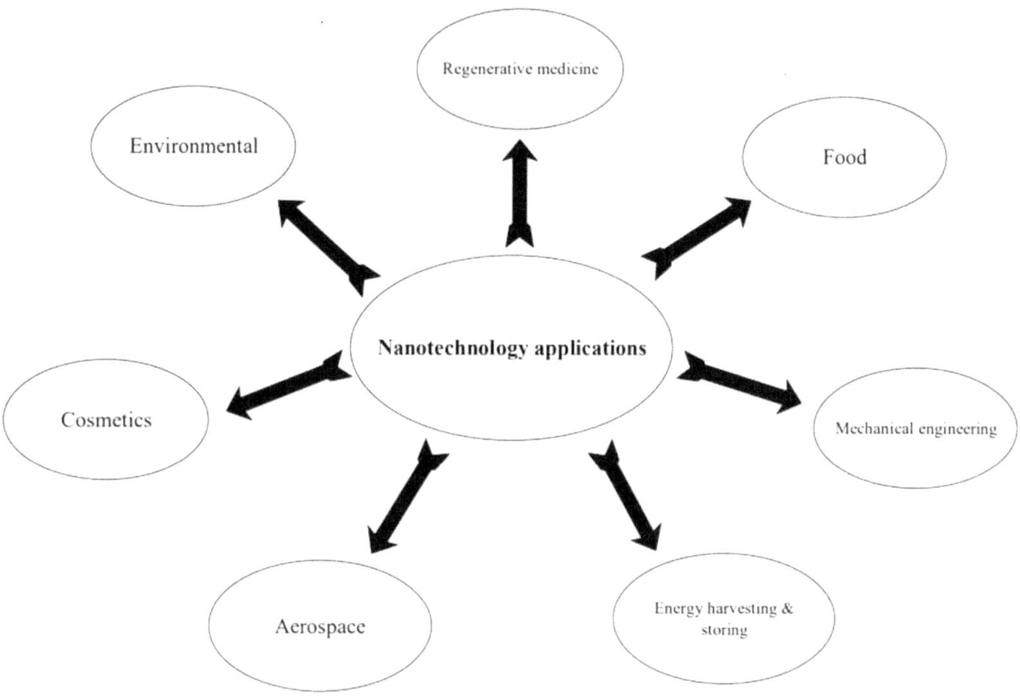

FIGURE 3.3 Applications of nanotechnology in different sectors/industries.

hydroxyapatite (HAP) (62, 63), growth factors (64, 65); they can be used as 3-dimensional scaffolds for tissue engineering (66–68); they can act as therapeutics, for example, metal stents composed of titanium or nitinol nanoparticles, owing to their mechanical properties and durability, can treat cardiovascular diseases like superficial femoral artery stenosis and atherosclerosis (69).

In aerospace, desired features of airplane and satellite components such as weight reduction, power production, radiation protection, mechanical strength, durability, etc., have been considerably improved using nanomaterials (70, 71).

Nanomaterials are a great option to enhance the performance of energy-harvesting devices. They can improve the photovoltaic effect, thermo-electricity, and ferroelectricity of materials (72, 73).

Owing to their smaller size and higher surface-to-volume ratio, nano-based cosmetics have many advantages over micro-based cosmetics. Because of their smaller size, they can penetrate the skin more efficiently and improve the delivery of the ingredients of the product. Due to their high surface-to-volume ratio, they can carry the product's ingredients into the skin more efficiently (74). Currently, nanomaterials are used in different cosmetics, such as shampoos, soaps, sunscreens, toothpastes, skin moisturizers, anti-wrinkle creams, anti-cellulite creams, face powders, perfumes, etc. (75).

3.5 NANOTECHNOLOGY-BASED STRATEGIES TO COMBAT MICROBIAL PATHOGENS

In this section, nanotechnology-based strategies that have been used to fight against microbial pathogens are discussed. Our main focus is on three areas in which nanotechnology has proven its effectiveness and key role: prevention of infectious diseases; detection and diagnosis of infectious agents and diseases; and finally, the treatment of infectious diseases.

3.5.1 Application of Nanotechnology in the Control and Prevention of Infectious Diseases

Control and prevention of infectious diseases is the first step to combat them effectively. Many different programs are implemented locally or globally to control and prevent infectious diseases. These programs range from adopting appropriate health policies, to environmental sanitation, drinking water and food safety, and of course, vaccination (7, 76). The applications of nanomaterials in environmental sanitation as well as water and food safety are discussed in the following sections. In the present section, we will focus on the use of nanomaterials in the era of vaccination.

Undoubtedly, vaccination is one of the best strategies for the control and prevention of diseases, including infectious diseases (77). We have seen the importance of vaccines during the COVID-19 pandemic, where following the introduction of vaccines, disease cases, deaths, hospital admissions, etc., have dramatically decreased (78, 79). The role of vaccination in limiting and management of other infectious diseases has been crucial (80–82). However, the growth of the vaccination field has not been acceptable (83). This field is faced with many challenges. The delivery of vaccines, especially for vaccines that are delivered mucosally, the inability of a vaccine to stimulate both humoral and cellular immune systems, and the stability of vaccines both *in vitro* and *in vivo*, are some of these challenges.

Nanomaterials help to increase the effectiveness of vaccines in many ways. They act as vaccine delivery vehicles. Indeed, they have an intrinsic adjuvanticity (although the adjuvant property varies depending on the nanostructure) (84–88). Since nanomaterials entrap and/or encapsulate the vaccine components, they dramatically increase the stability of vaccines (43, 89, 90). Another interesting feature of nanomaterials in the vaccine field is the ability of these materials to escape from endosomes and lysosomes (91, 92). This is a valuable feature, especially for exogenous antigens that are produced outside the host body or outside the cells. Exogenous antigens mainly activate the immune system via MHC class II pathway and stimulate CD4$_+$ T cells, while the endogenous antigens that are synthesized inside the host cells activate the immune system via MHC class I pathway and stimulate CD8+ T cells. In the earlier pathway (MHC II), humoral immunity is stimulated, which is characterized by the production of antibodies, and can fight extracellular infectious agents (93). It means that these vaccines are not a good choice to combat viruses or intracellular bacteria. Owing to their potential for escaping from endosomes or lysosomes, nanomaterials release their cargoes (vaccine components) into the cytosol (94, 95). By this, antigen (an active component of vaccine) is presented on MHC I molecules, a pathway that is called the cross-presentation pathway (Figure 3.4). Since part of a nanomaterial formulation is escape from lysosomes, delivery of vaccines using these materials can stimulate both humoral and cellular immune systems.

Many investigations have shown the efficiency of nanomaterials in establishing an appropriate immune response, as listed in Table 3.1. Different nanomaterials from different types have been studied for this aim and collectively they show the effectiveness of these materials for vaccine delivery.

3.5.2 Nanodiagnostics: Application of Nanomaterials in the Detection and Diagnosis of Microbes

Timely detection and identification of microbial pathogens is crucial for the control and management of microbial diseases. Traditional methods, such as culturing techniques, are time-consuming so in some cases it takes a week the complete the detection process (108). Nanomaterials have emerged as promising materials for the detection of microbial pathogens. Many useful classifications reveal the applications of nanomaterials for the detection and identification of pathogens (109–111). Because of the importance of nanodiagnostic tools and methods, these methods will be comprehensively discussed in Chapter 4.

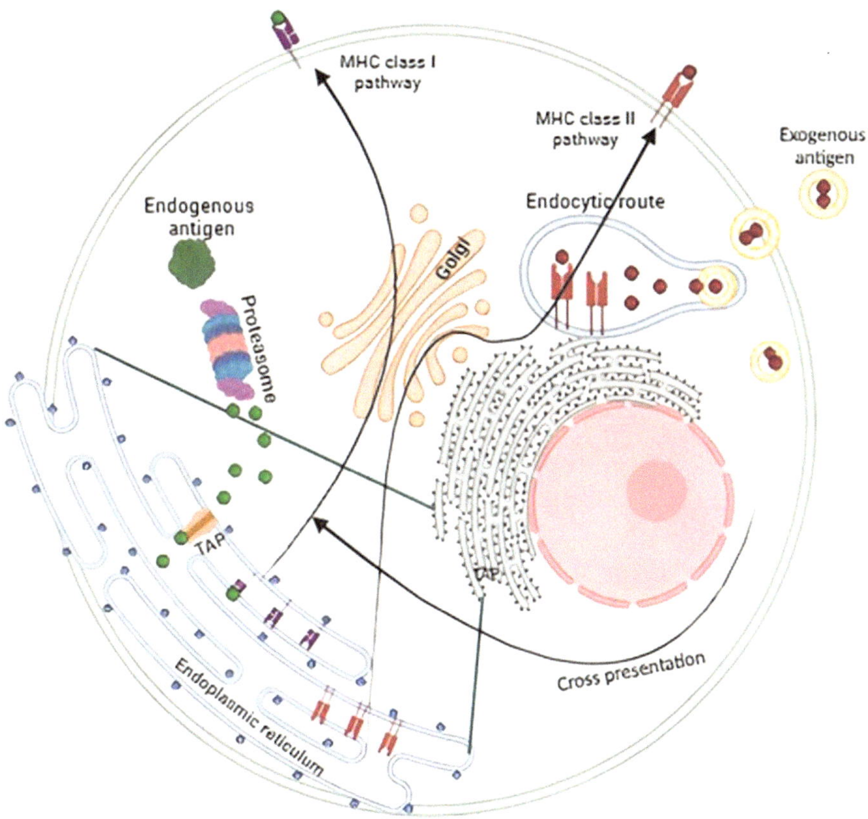

FIGURE 3.4 Depiction of the cross-presentation pathway of the immune system. While typical exogenous antigens are presented on MHC-II molecules and activate humoral immunity, if they can acquire the ability to escape from lysosomes, they can be presented on MHC-I molecules (a pathway that is called the "cross-presentation pathway") and induce the cellular arm of adopted immunity. Nanomaterials can help exogenous antigen to escape from lysosomes.

3.5.3 Nanomaterials in the Treatment of Infectious Diseases

Nanomaterials play their role in the treatment of infectious diseases in two main ways. First, some nanomaterials have antimicrobial activity; and second, they are appropriate carriers for the delivery of antibiotics.

Some nanomaterials, especially metallic NPs, have antimicrobial activity. Gold (Au), silver (Ag), copper (Cu), nickel (Ni), selenium (Se), cadmium (Cd), and tungsten (W) NPs are some examples in which antimicrobial activities have been shown (112–114). Type, size, structure, surface chemistry, surface charge, surface area, and concentration are major characteristics of NPs that determine the extent of antimicrobial activities (115, 116). NPs perform their antimicrobial role with different mechanisms: (i) some NPs are positively charged; hence they can bind to the negatively charged surfaces of bacteria via electrostatic interactions. This interaction helps NPs to disrupt the membrane, and finally the pathogen will be killed (117); (ii) the ability of the metallic NPs to generate reactive oxygen species (ROS) is another mechanism for the antimicrobial activity of NPs (118, 119); (iii) some NPs can interfere with DNA replication and cell division (120); (iv) NPs can induce damage to DNA (121); (v) NPs can be engineered to target the virulence factors of bacteria (122); (vi) engineered NPs can inhibit biofilm formation (123).

TABLE 3.1
Different Studies Investigating the Efficiency of NPs for Vaccine Development

Nanomaterial Type	NP Material	Infectious Agent	Animal Model	Route of Administration	Immunogenicity Results	References
Organic NPs	PLGA	*E. coli* O157:H7	BALB/c Mice	Intraperitoneal	Mice were able to tolerate $200LD_{50}$ of bacteria	(96)
	Alginate	ETEC	Mice/Rabbits	Intraperitoneal	Raised antibodies could neutralize LT toxin in rabbit loop assay	(97)
	Chitosan	*S. aureus*	BALB/c Mice	Nasal	Raised antibodies could neutralize the bacteria	(98)
	Liposome	SARS-CoV-2	BALB/c Mice	Nasal	Systemic neutralizing antibodies, IgA in the lung and nasal compartments were induced; T-cell responses in the lungs of mice were stimulated	(99)
Inorganic NPs	Mesoporous silica NPs	*C. perfringense*	Ross 308 broiler chickens	Oral & Nasal	Immunized chickens tolerate 10 LD50 of bacterium	(100)
	Metallic NPs (Ag NPs and Au NPs)	*Candida albicans*	Rats	Injection	Humoral and cellular immune responses were stimulated efficiently	(101)
	Metallic NPs (Au NPs)	SARS-CoV-2	BALB/c mice	Subcutaneous	The Au NP-adjuvanted protein induced a strong IgG response	(102)
	Metallic NPs (SPIONs)	*Plasmodium yoelii*				
Carbon-based NPs	Carbon nanoparticle	-	BALB/c mice	Oral	Humoral and cellular immune responses were stimulated efficiently	(103)
	SWCNTs	*Aeromonas hydrophila*	Grass carp	Intramuscular	Immunized fish showed a much higher survival rate following the challenge test	(104)
	Graphene oxide	*Chlamydia trachomatis*	BALB/c mice	Subcutaneous	Humoral and cellular immune responses were stimulated efficiently	(105)

TABLE 3.1 (Continued)
Different Studies Investigating the Efficiency of NPs for Vaccine Development

Nanomaterial Type	NP Material	Infectious Agent	Animal Model	Route of Administration	Immunogenicity Results	References
Composite NPs	Selenium NPs encapsulated by alginate-chitosan	*Vibrio cholera*	Caco-2 cells	*In vitro* assay	The formulation significantly increased anti-inflammatory cytokines and reduced the release of pro-inflammatory cytokines.	(106)
	Chitosan-PLGA NPs	*Acinetobacter baumannii*	BALB/c mice	Subcutaneous	CS-PLGA-rOmp22 NPs stimulated both humoral and cellular immunity and protected the immunized mice against acute lethal intratracheal *A. baumannii*	(107)

Besides the antimicrobial activities of NPs, they are a proper carrier for antibiotics, especially hydrophobic ones. Using nanomaterials for antibiotic delivery is important in many directions: (i) nanomaterials protect antibiotics from the harsh conditions in different parts of the body, e.g. the high acidity environment of the stomach lumen (124, 125); (ii) by adding specific ligands to NPs, targeting the delivery of antibiotics to a specific site is possible (126–128); (iii) by encapsulating the antibiotics, their toxicity will be significantly decreased (129, 130); (iv) some NPs, especially lipid-based ones, can efficiently enter the cells and release their cargoes (antibiotics) into the cytoplasm. It is very important for antibiotics that are not able to enter cells efficiently (e.g., hydrophobic antibiotics) (131, 132).

3.6 NANOTECHNOLOGY IN WATER AND FOOD MICROBIOLOGY

In addition to the role of nanotechnology in medical microbiology, in this section, the applications of nanosciences in the field of water and food microbiology are discussed.

One of the basic human needs is access to safe drinking water. Nano-based technologies have helped in detecting and removing pathogenic microbes from water. Electrospun nanofibers and nanobiocides such as metal nanoparticles offer a promising solution to enhance the performance of water filtration membranes. The presence of bacteria in water causes biofouling of membranes, which greatly reduces the quality of drinking water. Nanofibers with functionalized surfaces can effectively prevent the growth of these bacteria. Nanofibers have effectively integrated nanobiocides, showcasing remarkable antimicrobial characteristics (133). Another application of nanomaterials is nanoabsorbents for removing organic and inorganic pollutants from water (134).

Nanotechnology has also made significant progress in the field of the food industry, which is discussed below. Packaging of food products based on nanocomposites is one of the valuable applications that has entered the market. These resources have been created to improve the durability of food items and reduce the reliance on traditional plastics, thus encouraging the adoption of sustainable packaging options. Thermoplastic starch and its derivatives have been extensively studied as biopolymer nanocomposites for packaging purposes. The addition of minute quantities of nanofillers to polymers can significantly enhance their barrier, mechanical, and thermal characteristics. Moreover, this integration can introduce antimicrobial and antioxidant properties, which are especially advantageous for food packaging applications (135). Another application of nanomaterials in food microbiology is the detection of pathogens using nano-biosensors. The high sensitivity of nanomaterials has led to their widespread use in the detection of food pathogens, food quantification, and safety (136).

3.7 OTHER IMPACTS OF NANOTECHNOLOGY AND MICROBIOLOGY: REVIEW OF TWO CASES

In previous sections of the present chapter, the role of nanotechnology in medical, water, and food microbiology was discussed. In the last section of this chapter, we focus on two important areas: the effects of nanomaterials on (i) gut microbiota and (ii) the environment.

3.7.1 Effects of Nanomaterials on Gut Microbiota

The effects of gut microbiota and nanoparticles refer to the interactions between nanomaterials and the complex ecosystem of microorganisms that reside in the gut. These interactions can have various consequences, both positive and negative, and are an area of ongoing research. Many nanoparticles can affect the function of the gut microbiome. Usually, a decrease in the number of beneficial bacteria and microbial diversity is an indicator of gut microbiota dysbiosis. Studies show that, based on their physicochemical properties, different nanoparticles can induce dysbiosis and disturb gut homeostasis (137, 138). Figure 3.5 schematically illustrates the effects of nanomaterials on the gut microbiome.

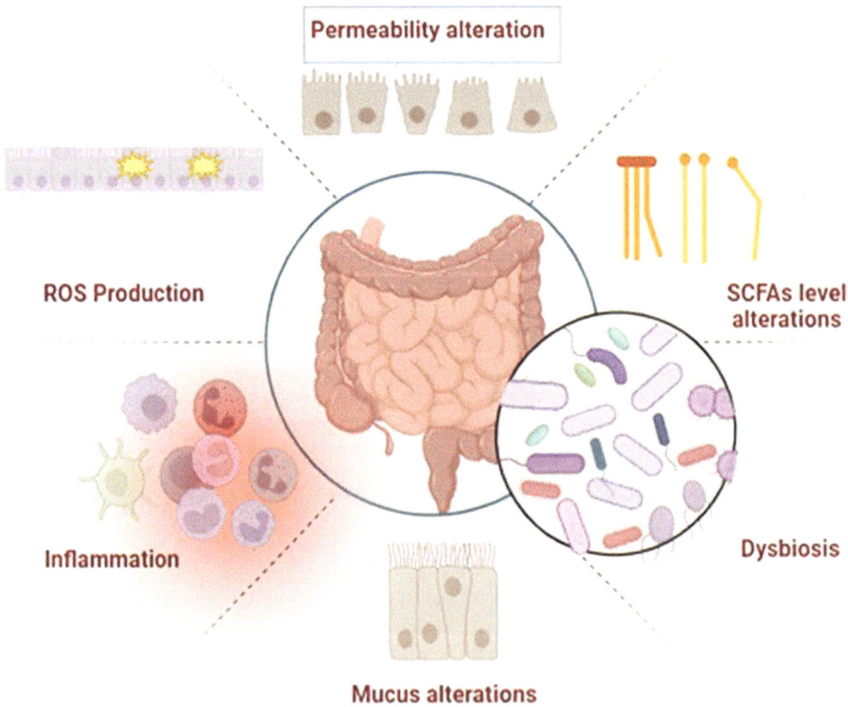

FIGURE 3.5 Schematic illustration of effects of nanoparticles on gut microbiota.

3.7.1.1 Titanium Dioxide Nanoparticles (TiO_2 NPs)

Titanium dioxide nanoparticles (TiO_2 NPs) are commonly used as a food additive for their coloring and opacifying properties. TiO_2 NPs can be found in various food products, including candies, chewing gum, baked goods, and dairy products (139). The consumption of food containing these nanoparticles can lead to concerns about human health due to the interaction with the gut microbiota. One of the effects of TiO_2 nanoparticles is a change in the relative abundance of certain bacteria and, as a result, a malfunction in the digestive system, such as a decrease in the level of short-chain fatty acids (SCFAs), disturbance in gut metabolism and immunity (140). Also, studies on animal models show that the oral consumption of these nanoparticles leads to changes in the structure and composition of the gut microbiota. For example, Li et al. (2018) followed the oral consumption of TiO_2 NPs for one month in an animal model. The results showed that titanium accumulated in the spleen, lung, and kidney, but had no significant effect on the diversity of microbiota, although the structure of villous epithelial cells was changed in a time-dependent manner. *Proteobacteria* were mainly affected. Therefore, chronic consumption of foods containing these type of NPs disrupts microbiota homeostasis (141).

Another effect of nanoparticles on the gut microbiome is oxidative stress induction, the infiltration of inflammatory cells, and immunogenicity in the colon following gut dysbiosis. Mu et al. (2019) showed that the oral administration of TiO_2 NPs for two to three months caused loss weight and intestinal inflammation in animal models. Although the microbial diversity was not affected, the abundance of probiotic strains, especially *Bifidobacterium* and *Lactobacillus*, was greatly reduced. These nanoparticles worsen chronic colitis and change the population of immune cells in lymph nodes (142). Cao et al. (2020) compared the oral consumption of TiO_2 for two months in obese and non-obese animal models. Titanium nanoparticles changed the composition of microbiota; increased abundance of *Firmicutes phylum*, and decreased abundance of *Bacteroidetes phylum* and *Bifidobacterium* and *Lactobacillus* genera. Also, inflammation induction, the increase of immune cells, and cytokines in the colonic mucus were observed; and obese mice were more susceptible to these effects (143).

3.7.1.2 Silver Nanoparticles (Ag NPs)

While Ag NPs are known for their antimicrobial properties, studies show negative consequences of Ag NP exposure on the composition and function of the human gut microbiota. The interaction of these nanoparticles with the gut microbiota can change the abundance and diversity of the microbiome. In particular, it increases the frequency of gram-negative bacteria such as *E. coli* and *Lactobacillus* (144).

According to Han et al. (2014), in the presence of metal nanoparticles (not microparticles) there is a sharp reduction in larval development stages. Although it did not affect the survival of adult Drosophila melanogaster. The diversity of the gut microbiota of larvae treated with silver nanoparticles was greatly reduced and the abundance of *Lactobacillus brevis* increased. Also, the abundance of *Acetobacter* decreased. Therefore, changing the gut microbiome can contribute to the toxicity caused by silver nanoparticles (145).

3.7.1.3 Zinc Oxide Nanoparticles (ZnO NPs)

Zinc oxide nanoparticles show antimicrobial properties by releasing zinc ions; their effects on microbes depend on particle size, shape, concentration, agglomeration, colloidal formulation, and pH of the media (146). Studies show that the green synthesis of these nanoparticles by microbes is more biocompatible with the environment (147). In addition to this advantage, some side effects of zinc oxide nanoparticles are mentioned in this section.

ZnO NPs can destroy the cell membrane and cause cell death by increasing the permeability, which is a possible mechanism involved in the antibacterial properties of these nanoparticles. Liu et al. (2009) studied the antibacterial properties of ZnO NPs on *Escherichia coli* O157:H7. They observed that there was a direct relationship between the increased concentration of nanoparticles and the inhibition of bacterial growth, and a concentration of at least 12 mmol $^{l-}$1 causes the complete inhibition of bacterial growth (148).

Mendes et al. (2022) investigated the effect of ZnO nanoparticles on a number of gram-positive (*Bacillus subtilis*) and gram-negative bacteria (*Escherichia coli*, *Pseudomonas aeruginosa*), which are components of the normal gut microbiota or pathogen (149). The results showed that ZnO NPs inhibit the growth of both gram-positive and negative strains. The destruction of the membrane is probably due to the electrostatic interaction between the negative bacterial membrane and the positive charge of zinc metal ions. Since ZnO nanoparticles are semi-conductors, when they receive more energy than their bandgap, the valence band electron is excited and transferred to the conduction band. Therefore, an electron hole is formed. The hole created by the electron emission is highly oxidizing and causes the oxidation of water molecules or hydroxide anions. This reaction eventually leads to the production of ROS and oxidative stress. Increasing ROS production can lead to protein oxidation, DNA damage and dose-dependent genotoxicity in bacterial cells (150). Zinc ion is one of the co-factors of many enzymes and transcription factors. Therefore, zinc ions released from ZnO nanoparticles can affect many cellular signaling pathways and disrupt cellular metabolic activity (151).

As shown in recent research, the use of these nanoparticles to inhibit the growth of pathogenic bacteria has a dual effect, acting as a double-edged sword due to the simultaneous effect on natural gut microbiota.

3.7.1.4 Carbon-based Nanoparticles

Chen et al. (2018) administered different doses (0.05, 0.5, and 2.5 mg kg-1 body weight/day) of single-walled carbon nanotubes (SWCNTs) orally to mice over seven days. A 2.5 mg kg-1 dose caused intestinal histological damage, increased intestinal permeability, and production of inflammatory cytokines. Also, the oral administration of SWCNT caused a disturbance in the balance of the gut microbiota (decrease of *Firmicutes* and increase of *Bacteroidetes*) and an increase in inflammatory bacteria (152). However, the study of Christophersen et al. (2016) showed that the oral administration of multi-walled carbon nanotubes (MWCNTs) for 10 weeks (4 or 40 μg each week) did not change the composition of the mice gut microbiota (153). Graphene is of great interest to researchers due to its antibacterial properties. Xie et al. (2016) investigated the gut microbiota changes of mice exposed to graphene (for four weeks). The results showed that lower concentrations of graphene show more toxicity than higher concentrations (probably due to the aggregation of high concentrations). In animals treated with graphene, the proportion of gram-negative bacteria increased compared to gram-positive, which is a sign of the greater sensitivity of the gram-positive bacteria to graphene. The results of high-throughput sequencing also showed an increase in the frequency and types of antibiotic resistance genes (ARGs) in the mouse gut (154). Đurašević et al. (2020) studied the long-term effects of fullerene C60 olive/coconut oil solution consumption in rats. The use of fullerene decreased the ratio of bacteria involved in lipid homeostasis, which reduced serum triglyceride concentration (155). Khare et al. (2020) gave rats water or cream containing 1% cellulose nanofibers (CNF) twice a week for five weeks. The consumption of CNF changed the microbial diversity, and as a result, decreased the species producing short-chain fatty acids (156).

3.7.2 Environmental Impact of Nanomaterials

Nanomaterials possess diverse impacts on microorganisms in the air, soil, and water. They can alter the composition and variety of microbial communities, thereby potentially causing ecological consequences. Moreover, nanoparticles present in the environment can influence the behavior of beneficial microorganisms, transforming them into potential hazards. This transformation can occur through various mechanisms such as the disruption of membrane potential, generation of reactive oxygen species (ROS), and impairment of proteins. Exposure to nanomaterials via the environment can transpire through inhalation by humans and other air-breathing species, as well as uptake by aquatic organisms from water or sediments.

Wolny-Koładka et al. (2017) studied the toxicity of silver nanoparticles against antibiotic-resistant airborne strains *Staphylococcus epidermidis* and *Staphylococcus aureus*, which also form biofilms. The study found that the minimal inhibitory concentration (MIC) for *S. epidermidis* strains was 30 μg/mL, while for *S. aureus* strains, it was 60 μg/mL. The high surface-to-volume ratio of silver nanoparticles facilitates interaction with microbial surfaces, leading to their antimicrobial activity. It was also shown that factors such as temperature and exposure time are impressive in the effectiveness of silver nanoparticles (157). Baysal et al. (2017) investigated the interaction of PM2.5 airborne particulates with ZnO and TiO_2 nanoparticles and their effect on gram-positive *Bacillus subtilis*, *Staphylococcus aureus*, and gram-negative *Escherichia coli* and *Pseudomonas aeruginosa* bacteria (158). The results suggest that the toxicity of Ag NPs (silver nanoparticles) in organisms may change after their transformation in the air compartments and the interaction of PM2.5 airborne particulates with ZnO and TiO_2 nanoparticles can increase their toxicity to bacteria.

Nanoparticles are widely used as antimicrobial agents against pathogenic bacteria. The prevalence of nanoparticles in the ecosystem causes beneficial microorganisms to become hazardous to their cells, which prevents the promotion of plant growth and the detoxification of pollutants. For example, the misuse of metallic nanoparticles can lead to toxicity to beneficial soil bacteria. Especially, the toxicity of nanoparticles is much higher than bulk material due to the increase in the ratio of surface area to volume and the strengthening of surface activity. Zhang et al. (2021) studied the effect of silver nanoparticles on denitrification and the microbial community of paddy soil. In this study, concentrations of 0.1 to 50 mg/kg of silver nanoparticles were investigated. The results showed that the concentration of 0.1 to 10 mg/kg of the nanoparticles did not make a significant difference compared to the bulk state, but the concentration of 50 mg/kg caused an increase of more than 60% in the denitrification rate and N2O emission rate and also the microbial community was changed. The most important bacteria involved in the N2O emission of the concentration of 50 mg/kg of Ag NPs were *Firmicutes* and *β-proteobacteria*, such as *Clotridiales*, *Burkholderiales*, and *Anaerolineales* (159). Zhou et al. (2020) investigated the long-term effect (over 70 days) of ZnO and TiO_2 nanoparticles on the soil microbial community: 2% of ZnO reduced 66% of microbial biomass carbon and changed the soil microbial community. It also reduced the accumulation of chlorophyll in *L. perenne.* Titanium nanoparticles showed similar effects with less intensity (160). Nanoparticles are also toxic to aquatic microorganisms. Londono et al. (2017) investigated the effect of TiO_2 at 700 μg/L and ZnO at 70 μg/L in river water samples, and they observed that titanium nanoparticles alone in this concentration did not affect the bacterial communities. In contrast, the addition of ZnO alone or together with TiO_2 induced a significant change in the structure and diversity of the microbial community (161).

3.8 FUTURE PERSPECTIVES

Nanotechnology holds immense promise for a diverse range of applications within the field of microbiology. Almost all fields of microbiology have been influenced by the advancement of nanotechnology. By their roles as adjuvants and delivery vehicles, immunity against infectious diseases has been improved and nanomaterials are finding their place in vaccine development. Pfizer-BioNTech and Moderna mRNA-based vaccines are the most famous examples of the application of NPs in the delivery of vaccines. In microbial detection and identification, NPs have gained an undeniable position. Advancements in the synthesis of NPs with novel features in combination with the development of biological methods are promoting this field. Through the translation of these advancements, *in vivo* detection and diagnosis of microbial agents is not far away. The carrier properties of NPs are not limited to the delivery of vaccines. NPs can also deliver drugs like antibiotics into the host and bacterial cells; therefore, they are a good candidate for efficient fighting against intracellular pathogens. The focus should be on the precise targeting of NPs to deliver drugs in a regulated and controlled manner. NPs are also potent disinfectants. They are good candidates to disinfect critical places, such as surgery rooms, where most drug-resistant pathogens are evolved. NPs are also improving the food industry as well as our environment by these special features. NPs can enhance investigation tools and therefore help microbiologists to understand the microbial world better. Finally, it is worth mentioning that while the potential applications of nanotechnology in microbiology are promising, continuous research and development are indispensable to ensure the secure and efficient utilization of nanomaterials in various microbial applications.

REFERENCES

1. Schaefer HE. *Nanoscience: The Science of the Small in Physics, Engineering, Chemistry, Biology and Medicine*. Springer Science & Business Media; 2010.
2. Tawade PV, Wasewar KL. Nanotechnology in biological science and engineering. In: Wasewar KL, Singh R, editors. *Environmental Applications of Microbial Nanotechnology*. Elsevier; 2023. p. 43–64.
3. Panda PK, Verma SK, Suar M. Nanoparticle–biological interactions: the renaissance of bionomics in the myriad nanomedical technologies. *Future Medicine*. 2021;16(25):2249–54.
4. Hajizade A, Ahmadian G, Fan K, Zhu M, Arpanaei A. Nanotechnology-based detection, prevention and treatment of infectious diseases. *Frontiers in Microbiology*. 2023;14:1224727.
5. Maddela NR, Chakraborty S, Prasad R. *Nanotechnology for Advances in Medical Microbiology*. Springer; 2021.
6. Parija SC. *Textbook of Microbiology and Immunology*. Springer; 2023.
7. Ellwanger JH, Veiga ABGd, Kaminski VdL, Valverde-Villegas JM, Freitas AWQd, Chies JAB. Control and prevention of infectious diseases from a One Health perspective. *Genetics and Molecular Biology*. 2021:44;e20200256.
8. Wagner A, Weinberger B. Vaccines to prevent infectious diseases in the older population: immunological challenges and future perspectives. *Frontiers in Immunology*. 2020;11:717.
9. Hajj Hussein I, Chams N, Chams S, El Sayegh S, Badran R, Raad M, et al. Vaccines through centuries: major cornerstones of global health. *Frontiers in Public Health*. 2015;3:269.
10. Endale H, Mathewos M, Abdeta D. Potential causes of spread of antimicrobial resistance and preventive measures in One Health perspective—a review. *Infection and Drug Resistance*. 2023;16:7515–45.
11. Hayward C, Ross KE, Brown MH, Whiley H. Water as a source of antimicrobial resistance and healthcare-associated infections. *Pathogens*. 2020;9(8):667.
12. Hoorfar J. Rapid detection, characterization, and enumeration of foodborne pathogens. *APMIS*. 2011;119:1–24.
13. Avakh Majalan P, Hajizade A, Nazarian S, Pourmand MR, Amiri Siyavoshani K. Investigating the prevalence of Shigella species and their antibiotic resistance pattern in children with acute diarrhea referred to selected hospitals in Tehran, Iran. *Journal of Applied Biotechnology Reports*. 2018;5(2):70–4.
14. Wang C, Yang Y, Cao Y, Liu K, Shi H, Guo X, et al. Nanocarriers for the delivery of antibiotics into cells against intracellular bacterial infection. *Biomaterials Science*. 2023;11(2):432–44.
15. Hadef F. An introduction to nanomaterials. In: Hadef F, editor. *Environmental Nanotechnology: Volume 1*. Springer; 2018. p. 1–58.
16. Roduner E. Size matters: why nanomaterials are different. *Chemical Society Reviews*. 2006;35(7):583–92.
17. Khan I, Saeed K, Khan I. Nanoparticles: properties, applications and toxicities. *Arabian Journal of Chemistry*. 2019;12(7):908–31.
18. Feynman RP. There's plenty of room at the bottom. *Engineering and Science*. 1959;23(5):22–36.
19. Garg P, Priyadarshi N, Ambule MD, Kaur G, Kaul S, Gupta R, et al. Multiepitope glycan based laser assisted fluorescent nanocomposite with dual functionality for sensing and ablation of Pseudomonas aeruginosa. *Nanoscale*. 2023;15(37):15179–95.
20. National Nanotechnology Initiative. *Leading to the Next Industrial Revolution: A Report by the Interagency Working Group on Nanoscience, Engineering, and Technology*. Committee on Technology, National Science and Technology Council; 2000.

21. Krishnia L, Thakur P, Thakur A. Synthesis of nanoparticles by physical route. In: Thakur P, Thakur A, Krishnia L, editors. *Synthesis and Applications of Nanoparticles*. Springer; 2022. p. 45–59.
22. Khan Y, Sadia H, Ali Shah SZ, Khan MN, Shah AA, Ullah N, et al. Classification, synthetic, and characterization approaches to nanoparticles, and their applications in various fields of nanotechnology: a review. *Catalysts*. 2022;12(11):1386.
23. Barhoum A, García-Betancourt ML, Jeevanandam J, Hussien EA, Mekkawy SA, Mostafa M, et al. Review on natural, incidental, bioinspired, and engineered nanomaterials: history, definitions, classifications, synthesis, properties, market, toxicities, risks, and regulations. *Nanomaterials*. 2022;12(2):177.
24. Mekuye B, Abera B. Nanomaterials: an overview of synthesis, classification, characterization, and applications. *Nano Select*. 2023;4:486–501.
25. Khan FA. Nanomaterials: types, classifications, and sources. In: Khan FA, editor. *Applications of Nanomaterials in Human Health*. Springer; 2020. p. 1–13.
26. Sharma VK, Filip J, Zboril R, Varma RS. Natural inorganic nanoparticles–formation, fate, and toxicity in the environment. *Chemical Society Reviews*. 2015;44(23):8410–23.
27. Wang X, Zhong X, Li J, Liu Z, Cheng L. Inorganic nanomaterials with rapid clearance for biomedical applications. *Chemical Society Reviews*. 2021;50(15):8669–742.
28. Bianchi E, Vigani B, Viseras C, Ferrari F, Rossi S, Sandri G. Inorganic nanomaterials in tissue engineering. *Pharmaceutics*. 2022;14(6):1127.
29. Alshammari BH, Lashin MM, Mahmood MA, Al-Mubaddel FS, Ilyas N, Rahman N, et al. Organic and inorganic nanomaterials: fabrication, properties and applications. *RSC Advances*. 2023;13(20):13735–85.
30. Zaytseva O, Neumann G. Carbon nanomaterials: production, impact on plant development, agricultural and environmental applications. *Chemical and Biological Technologies in Agriculture*. 2016;3(1):1–26.
31. Peng Q, Dearden AK, Crean J, Han L, Liu S, Wen X, et al. New materials graphyne, graphdiyne, graphone, and graphane: review of properties, synthesis, and application in nanotechnology. *Nanotechnology, Science and Applications*. 2014;7:1–29.
32. Kah M, Hofmann T. The challenge: carbon nanomaterials in the environment: new threats or wonder materials? *Environmental Toxicology and Chemistry*. 2015;34(5):954.
33. Chen W, Ming Y, Wang M, Huang M, Liu H, Huang Y, et al. Nanocomposite hydrogels in regenerative medicine: applications and challenges. *Macromolecular Rapid Communications*. 2023;44:2300128.
34. Biglari N, Zare EN. Conjugated polymer-based composite scaffolds for tissue engineering and regenerative medicine. *Alexandria Engineering Journal*. 2024;87:277–99.
35. Kausar A, Ahmad I, Maaza M, Bocchetta P. Self-healing nanocomposites—advancements and aerospace applications. *Journal of Composites Science*. 2023;7(4):148.
36. Chandraraj SS, Xavier JR. Facile synthesis of graphene-based mixed metal sulphide nanocomposite for energy storage applications. *Surfaces and Interfaces*. 2023;36:102515.
37. Al-Muntaser A, Pashameah RA, Saeed A, Alwafi R, Alzahrani E, AlSubhi SA, et al. Boosting the optical, structural, electrical, and dielectric properties of polystyrene using a hybrid GNP/Cu nanofiller: novel nanocomposites for energy storage applications. *Journal of Materials Science: Materials in Electronics*. 2023;34(7):678.
38. Goci MC, Leudjo Taka A, Martin L, Klink MJ. Chitosan-based polymer nanocomposites for environmental remediation of mercury pollution. *Polymers*. 2023;15(3):482.
39. Paras, Yadav K, Kumar P, Teja DR, Chakraborty S, Chakraborty M, et al. A review on low-dimensional nanomaterials: nanofabrication, characterization and applications. *Nanomaterials*. 2022;13(1):160.
40. Wu L, Li Y, Fu Z, Su B-L. Hierarchically structured porous materials: synthesis strategies and applications in energy storage. *National Science Review*. 2020;7(11):1667–701.
41. Saleh TA. *Surface Science of Adsorbents and Nanoadsorbents: Properties and Applications in Environmental Remediation*. Academic Press; 2022.
42. Yilmaz B, Trukhan N, Müller U. Industrial outlook on zeolites and metal organic frameworks. *Chinese Journal of Catalysis*. 2012;33(1):3–10.
43. Hajizade A, Salmanian AH, Amani J, Ebrahimi F, Arpanaei A. EspA-loaded mesoporous silica nanoparticles can efficiently protect animal model against enterohaemorrhagic E. coli O157:H7. *Artificial Cells, Nanomedicine, and Biotechnology*. 2018;46(sup3):S1067–S75.
44. Bayir S, Barras A, Boukherroub R, Szunerits S, Raehm L, Richeter S, et al. Mesoporous silica nanoparticles in recent photodynamic therapy applications. *Photochemical & Photobiological Sciences*. 2018;17:1651–74.

45. Qian X, Fuku K, Kuwahara Y, Kamegawa T, Mori K, Yamashita H. Design and functionalization of photocatalytic systems within mesoporous silica. *ChemSusChem*. 2014;7(6):1528–36.
46. Liu S, Zhang L, Zhou J, Xiang J, Sun J, Guan J. Fiberlike Fe2O3 macroporous nanomaterials fabricated by calcinating regenerate cellulose composite fibers. *Chemistry of Materials*. 2008;20(11):3623–8.
47. Zhang C, Zhang Z, Yin F, Zhang Y, Mentbayeva A, Babaa MR, et al. 3D ordered macroporous carbon encapsulated ZnO nanoparticles as a high-performance anode for lithium-ion batteries. *ChemElectroChem*. 2017;4(9):2359–65.
48. Ortiz de Solorzano I, Uson L, Larrea A, Miana M, Sebastian V, Arruebo M. Continuous synthesis of drug-loaded nanoparticles using microchannel emulsification and numerical modeling: effect of passive mixing. *International Journal of Nanomedicine*. 2016;11:3397–416.
49. Gorade V, Chaudhary B, Parmaj O, Kale R. Preparation and characterization of chitosan/viscose rayon filament biocomposite. *Journal of Natural Fibers*. 2022;19(4):1189–200.
50. Ismail A, Saputri LNMZ, Dwiatmoko AA, Susanto BH, Nasikin M. A facile approach to synthesis of silica nanoparticles from silica sand and their application as superhydrophobic material. *Journal of Asian Ceramic Societies*. 2021;9(2):665–72.
51. Morsy M, Gomaa I, Taha W, Elhamid AEMA, Shoukry H, Mokhtar MM. Optimization and surface functionalization of carbon nanotubes using different hierarchical metal oxide nanoparticles. *HBRC Journal*. 2023;19(1):509–21.
52. Singh V, Yadav P, Mishra V. Recent advances on classification, properties, synthesis, and characterization of nanomaterials. In: Husen A, Siddiqi KS, editors. *Green Synthesis of Nanomaterials for Bioenergy Applications*; Springer; 2020. p. 83–97.
53. Rahimizadeh P, Rezaieyazdi Z, Behzadi F, Hajizade A, Lim SI. Nanotechnology as a promising platform for rheumatoid arthritis management: diagnosis, treatment, and treatment monitoring. *International Journal of Pharmaceutics*. 2021;609:121137.
54. Duan L, Li J, Duan H. Nanomaterials for lubricating oil application: a review. *Friction*. 2023;11(5):647–84.
55. Babu PJ, Tingirikari JMR. A review on polymeric nanomaterials intervention in food industry. *Polymer Bulletin*. 2023;80(1):137–64.
56. Zhang F, Shen R, Li N, Yang X, Lin D. Nanocellulose: an amazing nanomaterial with diverse applications in food science. *Carbohydrate Polymers*. 2023;304:120497.
57. Sharma A, Agarwal P, Sebghatollahi Z, Mahato N. Functional nanostructured materials in the cosmetics industry: a review. *ChemEngineering*. 2023;7(4):66.
58. Kulkarni D, Shelke S, Musale S, Panzade P, Sharma K, Giram P. Nanomaterials in cosmetics and dermatology. In: Husen A, Siddiqi KS, editors. *Advances in Smart Nanomaterials and Their Applications*. Elsevier; 2023. p. 297–323.
59. Bastami F, Paknejad Z, Jafari M, Salehi M, Rad MR, Khojasteh A. Fabrication of a three-dimensional β-tricalcium-phosphate/gelatin containing chitosan-based nanoparticles for sustained release of bone morphogenetic protein-2: implication for bone tissue engineering. *Materials Science and Engineering: C*. 2017;72:481–91.
60. Ki M-R, Kim SH, Rho S, Kim JK, Min KH, Yeo KB, et al. Use of biosilica to improve loading and delivery of bone morphogenic protein 2. *International Journal of Biological Macromolecules*. 2024;254:127876.
61. Xu H, Cui Y, Tian Y, Dou M, Sun S, Wang J, et al. Nanoparticle-based drug delivery systems for enhancing bone regeneration. *ACS Biomaterials Science & Engineering*. 2024;10:1302–22.
62. Abdul Halim NA, Hussein MZ, Kandar MK. Nanomaterials-upconverted hydroxyapatite for bone tissue engineering and a platform for drug delivery. *International Journal of Nanomedicine*. 2021;16:6477–96.
63. Munir MU, Salman S, Ihsan A, Elsaman T. Synthesis, characterization, functionalization and bio-applications of hydroxyapatite nanomaterials: an overview. *International Journal of Nanomedicine*. 2022;17:1903–25.
64. Nandhini J, Karthikeyan E, Rajeshkumar S. Nanomaterials for wound healing: current status and futuristic frontier. *Biomedical Technology*. 2024;6:26–45.
65. Wang Z, Wang Z, Lu WW, Zhen W, Yang D, Peng S. Novel biomaterial strategies for controlled growth factor delivery for biomedical applications. *NPG Asia Materials*. 2017;9(10):e435.
66. Raghav PK, Mann Z, Ahlawat S, Mohanty S. Mesenchymal stem cell-based nanoparticles and scaffolds in regenerative medicine. *European Journal of Pharmacology*. 2022;918:174657.

67. Menaa F, Abdelghani A, Menaa B. Graphene nanomaterials as biocompatible and conductive scaffolds for stem cells: impact for tissue engineering and regenerative medicine. *Journal of Tissue Engineering and Regenerative Medicine*. 2015;9(12):1321–38.
68. Shabbirahmed AM, Sekar R, Gomez LA, Sekhar MR, Hiruthyaswamy SP, Basavegowda N, et al. Recent developments of silk-based scaffolds for tissue engineering and regenerative medicine applications: a special focus on the advancement of 3D printing. *Biomimetics*. 2023;8(1):16.
69. Yang Y, Chawla A, Zhang J, Esa A, Jang HL, Khademhosseini A. Applications of nanotechnology for regenerative medicine; healing tissues at the nanoscale. In: Atala A, editor. *Principles of Regenerative Medicine*. Elsevier; 2019. p. 485–504.
70. Arepalli S, Moloney P. Engineered nanomaterials in aerospace. *MRS Bulletin*. 2015;40(10):804–11.
71. Barako MT, Gambin V, Tice J. Integrated nanomaterials for extreme thermal management: a perspective for aerospace applications. *Nanotechnology*. 2018;29(15):154003.
72. Bunea AC, Dediu V, Laszlo EA, Pistrițu F, Carp M, Iliescu FS, et al. E-skin: the dawn of a new era of on-body monitoring systems. *Micromachines*. 2021;12(9):1091.
73. Dragoman D. Nanomaterials for energy harvesting. *Nanomaterials*. 2023;13:1154.
74. Fytianos G, Rahdar A, Kyzas GZ. Nanomaterials in cosmetics: recent updates. *Nanomaterials*. 2020;10(5):979.
75. Gupta V, Mohapatra S, Mishra H, Farooq U, Kumar K, Ansari MJ, et al. Nanotechnology in cosmetics and cosmeceuticals—a review of latest advancements. *Gels*. 2022;8(3):173.
76. Anthonj C, Githinji S, Höser C, Stein A, Blanford J, Grossi V. Kenyan school book knowledge for water, sanitation, hygiene and health education interventions: disconnect, integration or opportunities? *International Journal of Hygiene and Environmental Health*. 2021;235:113756.
77. Plotkin SA, Orenstein W, Offit PA. *Vaccines E-Book*. Elsevier Health Sciences; 2012.
78. Watson OJ, Barnsley G, Toor J, Hogan AB, Winskill P, Ghani AC. Global impact of the first year of COVID-19 vaccination: a mathematical modelling study. *The Lancet Infectious Diseases*. 2022;22(9):1293–302.
79. Christie A, Henley SJ, Mattocks L, Fernando R, Lansky A, Ahmad FB, et al. Decreases in COVID-19 cases, emergency department visits, hospital admissions, and deaths among older adults following the introduction of COVID-19 vaccine—United States, September 6, 2020–May 1, 2021. *Morbidity and Mortality Weekly Report*. 2021;70(23):858.
80. Excler JL, Saville M, Berkley S, Kim JH. Vaccine development for emerging infectious diseases. *Nature Medicine*. 2021;27(4):591–600.
81. Hinman A. Eradication of vaccine-preventable diseases. *Annual Review of Public Health*. 1999; 20(1):211–29.
82. World Health Organization. *Potential Benefits and Limitations of mRNA Technology for Vaccine Research and Development for Infectious Diseases and Virus-Induced Cancers: Report of the WHO Science Council*. World Health Organization; 2023.
83. Hajizade A, Ebrahimi F, Salmanian AH, Arpanaei A, Amani J. Nanoparticles in vaccine development. *Journal of Applied Biotechnology Reports*. 2014;1(4):125–34.
84. Nooraei S, Sarkar Lotfabadi A, Akbarzadehmoallemkolaei M, Rezaei N. Immunogenicity of different types of adjuvants and nano-adjuvants in veterinary vaccines: a comprehensive review. *Vaccines*. 2023;11(2):453.
85. Sun B, Li M, Yao Z, Yu G, Ma Y. *Advances in Vaccine Adjuvants: Nanomaterials and Small Molecules*. Springer; 2023.
86. Delechiave G, de Oliveira Silva M, de Castro-Amarante MF, Mallaupoma Camarena DE, Venceslau-Carvalho AA, Ferreira LCS, et al. Layer-by-layer assembly of polymeric nanoparticles with heparin-RBD complexes as an adjuvant for SARS-CoV-2 protein-based vaccines. *ACS Applied Nano Materials*. 2024;7:4068–77.
87. Filipić B, Pantelić I, Nikolić I, Majhen D, Stojić-Vukanić Z, Savić S, et al. Nanoparticle-based adjuvants and delivery systems for modern vaccines. *Vaccines*. 2023;11(7):1172.
88. Luzuriaga MA, Shahrivarkevishahi A, Herbert FC, Wijesundara YH, Gassensmith JJ. Biomaterials and nanomaterials for sustained release vaccine delivery. *Wiley Interdisciplinary Reviews: Nanomedicine and Nanobiotechnology*. 2021;13(6):e1735.
89. Pelliccia M, Andreozzi P, Paulose J, D'Alicarnasso M, Cagno V, Donalisio M, et al. Additives for vaccine storage to improve thermal stability of adenoviruses from hours to months. *Nature Communications*. 2016;7(1):13520.

90. Lee ES, Shin HJ, Na K, Bae YH. Poly (l-histidine)–PEG block copolymer micelles and pH-induced destabilization. *Journal of Controlled Release*. 2003;90(3):363–74.
91. Hao P, Wu X, Wang L, Wei S, Xu H, Huang W, et al. One-pot generating subunit vaccine with high encapsulating efficiency and fast lysosome escape for potent cellular immune response. *Bioconjugate Chemistry*. 2020;31(8):1917–27.
92. Du G, Sun X. Engineering nanoparticulate vaccines for enhancing antigen cross-presentation. *Current Opinion in Biotechnology*. 2020;66:113–22.
93. Abbas AK, Lichtman A, Pillai S. *Cellular and Molecular Immunology, 9e*. Elsevier; 2018.
94. Smith SA, Selby LI, Johnston AP, Such GK. The endosomal escape of nanoparticles: toward more efficient cellular delivery. *Bioconjugate Chemistry*. 2018;30(2):263–72.
95. Ahmad A, Khan JM, Haque S. Strategies in the design of endosomolytic agents for facilitating endosomal escape in nanoparticles. *Biochimie*. 2019;160:61–75.
96. Kordbacheh E, Nazarian S, Hajizade A, Farhang A. Entrapment of HETS recombinant protein onto PLGA and alginate NPs improves the immunogenicity of the protein against E. coli O157:H7. *Molecular Immunology*. 2019;114:612–9.
97. Kordbacheh E, Nazarian S, Sadeghi D, Hajizadeh A. An LTB-entrapped protein in PLGA nanoparticles preserves against enterotoxin of enterotoxigenic Escherichia coli. *Iranian Journal of Basic Medical Sciences*. 2018;21(5):517.
98. Hosseini SA, Nazarian S, Ebrahimi F, Hajizade A. Immunogenicity evaluation of recombinant Staphylococcus aureus enterotoxin B (rSEB) and rSEB-loaded chitosan nanoparticles following nasal administration. *Iranian Journal of Allergy, Asthma and Immunology*. 2020;19(2):159–71.
99. An X, Martinez-Paniagua M, Rezvan A, Sefat SR, Fathi M, Singh S, et al. Single-dose intranasal vaccination elicits systemic and mucosal immunity against SARS-CoV-2. *Iscience*. 2021;24(9):103037.
100. Hoseini ZS, Hajizade A, Razmyar J, Ahmadian G, Arpanaei A. Mesoporous silica nanoparticles-based formulations of a chimeric proteinous vaccine candidate against necrotic enteritis disease. *Materials Science and Engineering: C*. 2021;128:112316.
101. Abdelnaby MA, Shoueir KR, Ghazy AA, Abdelhamid SM, El Kemary MA, Mahmoud HE, et al. Synthesis and evaluation of metallic nanoparticles-based vaccines against Candida albicans infections. *Journal of Drug Delivery Science and Technology*. 2022;68:102862.
102. Sekimukai H, Iwata-Yoshikawa N, Fukushi S, Tani H, Kataoka M, Suzuki T, et al. Gold nanoparticle-adjuvanted S protein induces a strong antigen-specific IgG response against severe acute respiratory syndrome-related coronavirus infection, but fails to induce protective antibodies and limit eosinophilic infiltration in lungs. *Microbiology and Immunology*. 2020;64(1):33–51.
103. Wang T, Zou M, Jiang H, Ji Z, Gao P, Cheng G. Synthesis of a novel kind of carbon nanoparticle with large mesopores and macropores and its application as an oral vaccine adjuvant. *European Journal of Pharmaceutical Sciences*. 2011;44(5):653–9.
104. Gong YX, Zhu B, Liu GL, Liu L, Ling F, Wang GX, et al. Single-walled carbon nanotubes as delivery vehicles enhance the immunoprotective effects of a recombinant vaccine against Aeromonas hydrophila. *Fish & Shellfish Immunology*. 2015;42(1):213–20.
105. Zhao L, Shu M, Chen H, Shi K, Li Z. Preparation of graphene oxide–stabilized Pickering emulsion adjuvant for Pgp3 recombinant vaccine and enhanced immunoprotection against Chlamydia trachomatis infection. *Frontiers in Immunology*. 2023;14:1148253.
106. Bagheri-Josheghani S, Bakhshi B. Formulation of selenium nanoparticles encapsulated by alginate-chitosan for controlled delivery of Vibrio cholerae LPS: a novel delivery system candidate for nanovaccine. *International Journal of Biological Macromolecules*. 2022;208:494–508.
107. Du X, Xue J, Jiang M, Lin S, Huang Y, Deng K, et al. A multiepitope peptide, rOmp22, encapsulated in chitosan-PLGA nanoparticles as a candidate vaccine against Acinetobacter baumannii infection. *International Journal of Nanomedicine*. 2021;16:1819–36.
108. MacAulay S, Ellison AR, Kille P, Cable J. Moving towards improved surveillance and earlier diagnosis of aquatic pathogens: from traditional methods to emerging technologies. *Reviews in Aquaculture*. 2022;14(4):1813–29.
109. Syed MA. Advances in nanodiagnostic techniques for microbial agents. *Biosensors and Bioelectronics*. 2014;51:391–400.

110. Zhu X, Radovic-Moreno AF, Wu J, Langer R, Shi J. Nanomedicine in the management of microbial infection–overview and perspectives. *Nano Today*. 2014;9(4):478–98.
111. Alafeef M, Moitra P, Pan D. Nano-enabled sensing approaches for pathogenic bacterial detection. *Biosensors and Bioelectronics*. 2020;165:112276.
112. Syed MA, Manzoor U, Shah I, Bukhari SHA. Antibacterial effects of tungsten nanoparticles on the Escherichia coli strains isolated from catheterized urinary tract infection (UTI) cases and Staphylococcus aureus. *New Microbiologica*. 2010;33(4):329–35.
113. Syed MA, Babar S, Bhatti A, Bokhari H. Antibacterial effects of silver nanoparticles on the bacterial strains isolated from catheterized urinary tract infection cases. *Journal of Biomedical Nanotechnology*. 2009;5(2):209–14.
114. Jamil B, Syed MA. Nano-antimicrobials: a viable approach to tackle multidrug-resistant pathogens. In: Shegokar R, editor. *Nanotechnology Applied to Pharmaceutical Technology*. Springer; 2017. p. 31–54.
115. Niazi JH, Gu MB. Toxicity of metallic nanoparticles in microorganisms—a review. In: Prasad MNV, editor. *Atmospheric and Biological Environmental Monitoring*. Springer; 2009. p. 193–206.
116. Cheng G, Dai M, Ahmed S, Hao H, Wang X, Yuan Z. Antimicrobial drugs in fighting against antimicrobial resistance. *Frontiers in Microbiology*. 2016;7:470.
117. Joshi AS, Singh P, Mijakovic I. Interactions of gold and silver nanoparticles with bacterial biofilms: molecular interactions behind inhibition and resistance. *International Journal of Molecular Sciences*. 2020;21(20):7658.
118. Oberdörster E. Manufactured nanomaterials (fullerenes, C60) induce oxidative stress in the brain of juvenile largemouth bass. *Environmental Health Perspectives*. 2004;112(10):1058–62.
119. Kim S, Ryu DY. Silver nanoparticle-induced oxidative stress, genotoxicity and apoptosis in cultured cells and animal tissues. *Journal of Applied Toxicology*. 2013;33(2):78–89.
120. Li K, Zhao X, Hammer BK, Du S, Chen Y. Nanoparticles inhibit DNA replication by binding to DNA: modeling and experimental validation. *ACS Nano*. 2013;7(11):9664–74.
121. Nejdl L, Kudr J, Moulick A, Hegerova D, Ruttkay-Nedecky B, Gumulec J, et al. Platinum nanoparticles induce damage to DNA and inhibit DNA replication. *PLOS ONE*. 2017;12(7):e0180798.
122. Alavi M, Li L, Nokhodchi A. Metal, metal oxide and polymeric nanoformulations for the inhibition of bacterial quorum sensing. *Drug Discovery Today*. 2023;28(1):103392.
123. Mohamad F, Alzahrani RR, Alsaadi A, Alrfaei BM, Yassin AEB, Alkhulaifi MM, et al. An explorative review on advanced approaches to overcome bacterial resistance by curbing bacterial biofilm formation. *Infection and Drug Resistance*. 2023;16:19–49.
124. Wang Y, Yang Y, Shi Y, Song H, Yu C. Antibiotic-free antibacterial strategies enabled by nanomaterials: progress and perspectives. *Advanced Materials*. 2020;32(18):1904106.
125. Raza S, Matuła K, Karoń S, Paczesny J. Resistance and adaptation of bacteria to non-antibiotic antibacterial agents: physical stressors, nanoparticles, and bacteriophages. *Antibiotics*. 2021;10(4):435.
126. Vassallo A, Silletti MF, Faraone I, Milella L. Nanoparticulate antibiotic systems as antibacterial agents and antibiotic delivery platforms to fight infections. *Journal of Nanomaterials*. 2020;2020:1–31.
127. Yoo J, Park C, Yi G, Lee D, Koo H. Active targeting strategies using biological ligands for nanoparticle drug delivery systems. *Cancers*. 2019;11(5):640.
128. Razei A, Javanbakht M, Hajizade A, Heiat M, Zhao S, Aghamollaei H, et al. Nano and microparticle drug delivery systems for the treatment of Brucella infections. *Biomedicine & Pharmacotherapy*. 2023;169:115875.
129. Alhariri M, Azghani A, Omri A. Liposomal antibiotics for the treatment of infectious diseases. *Expert Opinion on Drug Delivery*. 2013;10(11):1515–32.
130. Fan Y, Lüchow M, Zhang Y, Lin J, Fortuin L, Mohanty S, et al. Nanogel encapsulated hydrogels as advanced wound dressings for the controlled delivery of antibiotics. *Advanced Functional Materials*. 2021;31(7):2006453.
131. Ghosh R, De M. Liposome-based antibacterial delivery: an emergent approach to combat bacterial infections. *ACS Omega*. 2023;8(39):35442–51.
132. Razei A, Cheraghali AM, Saadati M, Ramandi MF, Panahi Y, Hajizade A, et al. Gentamicin-loaded chitosan nanoparticles improve its therapeutic effects on brucella-infected J774A.1 murine cells. *Galen Medical Journal*. 2019;8:e1296.
133. Botes M, Cloete TE. The potential of nanofibers and nanobiocides in water purification. *Critical Reviews in Microbiology*. 2010;36(1):68–81.

134. Kunduru KR, Nazarkovsky M, Farah S, Pawar RP, Basu A, Domb AJ. Nanotechnology for water purification: applications of nanotechnology methods in wastewater treatment. *Water Purification*. Elsevier; 2017. p. 33–74.
135. Sarfraz J, Gulin-Sarfraz T, Nilsen-Nygaard J, Pettersen MK. Nanocomposites for food packaging applications: an overview. *Nanomaterials*. 2021;11(1):10.
136. Singh T, Shukla S, Kumar P, Wahla V, Bajpai VK, Rather IA. Application of nanotechnology in food science: perception and overview. *Frontiers in Microbiology*. 2017;8:1501.
137. Jiang Z, Jacob JA, Li J, Wu X, Wei G, Vimalanathan A, et al. Influence of diet and dietary nanoparticles on gut dysbiosis. *Microbial Pathogenesis*. 2018;118:61–5.
138. Vita AA, Royse EA, Pullen NA. Nanoparticles and danger signals: oral delivery vehicles as potential disruptors of intestinal barrier homeostasis. *Journal of Leukocyte Biology*. 2019;106(1):95–103.
139. Rinninella E, Cintoni M, Raoul P, Mora V, Gasbarrini A, Mele MC. Impact of food additive titanium dioxide on gut microbiota composition, microbiota-associated functions, and gut barrier: a systematic review of in vivo animal studies. *International Journal of Environmental Research and Public Health*. 2021;18(4):2008.
140. Chen Z, Han S, Zhou D, Zhou S, Jia G. Effects of oral exposure to titanium dioxide nanoparticles on gut microbiota and gut-associated metabolism in vivo. *Nanoscale*. 2019;11(46):22398–412.
141. Li J, Yang S, Lei R, Gu W, Qin Y, Ma S, et al. Oral administration of rutile and anatase TiO_2 nanoparticles shifts mouse gut microbiota structure. *Nanoscale*. 2018;10(16):7736–45.
142. Mu W, Wang Y, Huang C, Fu Y, Li J, Wang H, et al. Effect of long-term intake of dietary titanium dioxide nanoparticles on intestine inflammation in mice. *Journal of Agricultural and Food Chemistry*. 2019;67(33):9382–9.
143. Cao X, Han Y, Gu M, Du H, Song M, Zhu X, et al. Foodborne titanium dioxide nanoparticles induce stronger adverse effects in obese mice than non-obese mice: gut microbiota dysbiosis, colonic inflammation, and proteome alterations. *Small*. 2020;16(36):2001858.
144. Williams K, Milner J, Boudreau MD, Gokulan K, Cerniglia CE, Khare S. Effects of subchronic exposure of silver nanoparticles on intestinal microbiota and gut-associated immune responses in the ileum of Sprague-Dawley rats. *Nanotoxicology*. 2015;9(3):279–89.
145. Han X, Geller B, Moniz K, Das P, Chippindale AK, Walker VK. Monitoring the developmental impact of copper and silver nanoparticle exposure in Drosophila and their microbiomes. *Science of the Total Environment*. 2014;487:822–9.
146. Siddiqi KS, ur Rahman A, Tajuddin, Husen A. Properties of zinc oxide nanoparticles and their activity against microbes. *Nanotechnology Reviews*. 2018;7(5):413–41.
147. Murali M, Gowtham H, Shilpa N, Singh SB, Aiyaz M, Sayyed R, et al. Zinc oxide nanoparticles prepared through microbial mediated synthesis for therapeutic applications: a possible alternative for plants. *Materials Today: Proceedings*. 2023;14:354.
148. Liu YJ, He LL, Mustapha A, Li H, Hu Z, Lin MS. Antibacterial activities of zinc oxide nanoparticles against Escherichia coli O157:H7. *Journal of Food Science*. 2009;74(3):M89–93.
149. Mendes CR, Dilarri G, Forsan CF, Sapata VdMR, Lopes PRM, de Moraes PB, et al. Antibacterial action and target mechanisms of zinc oxide nanoparticles against bacterial pathogens. *Scientific Reports*. 2022;12(1):2658.
150. Singh R, Cheng S, Singh S. Oxidative stress-mediated genotoxic effect of zinc oxide nanoparticles on Deinococcus radiodurans. *Scientific Reports*. 2020;10:1–13.
151. Saptarshi SR, Duschl A, Lopata AL. Biological reactivity of zinc oxide nanoparticles with mammalian test systems: an overview. *Nanomedicine*. 2015;10(13):2075–92.
152. Chen H, Zhao R, Wang B, Zheng L, Ouyang H, Wang H, et al. Acute oral administration of single-walled carbon nanotubes increases intestinal permeability and inflammatory responses: association with the changes in gut microbiota in mice. *Advanced Healthcare Materials*. 2018;7(13):1701313.
153. Christophersen DV, Jacobsen NR, Andersen MH, Connell SP, Barfod KK, Thomsen MB, et al. Cardiovascular health effects of oral and pulmonary exposure to multi-walled carbon nanotubes in ApoE-deficient mice. *Toxicology and Applied Pharmacology*. 2016;371:29–40.
154. Xie Y, Wu B, Zhang XX, Yin J, Mao L, Hu M. Influences of graphene on microbial community and antibiotic resistance genes in mouse gut as determined by high-throughput sequencing. *Environmental Science & Technology*. 2016;50(11):5761–72.

155. Đurašević S, Nikolić G, Todorović A, Drakulić D, Pejić S, Martinović V, et al. Effects of fullerene C60 supplementation on gut microbiota and glucose and lipid homeostasis in rats. *Journal of Agricultural and Food Chemistry*. 2020;68(12):3552–60.
156. Khare S, DeLoid GM, Molina RM, Gokulan K, Couvillion SP, Bloodsworth KJ, et al. Effects of ingested nanocellulose on intestinal microbiota and homeostasis in Wistar Han rats. *Nanomedicine: Nanotechnology, Biology, and Medicine*. 2020;29:102216.
157. Wolny-Koładka KA, Malina DK. Silver nanoparticles toxicity against airborne strains of Staphylococcus spp. *Journal of Environmental Science and Health, Part A*. 2017;52(13):1247–56.
158. Baysal A, Saygin H, Ustabasi GS. Interaction of PM2.5 airborne particulates with ZnO and TiO_2 nanoparticles and their effect on bacteria. *Environmental Monitoring and Assessment*. 2018;190(1):1–15.
159. Zhang X, Dang D, Zheng L, Wu L, Wu Y, Li H, et al. Effect of Ag nanoparticles on denitrification and microbial community in a paddy soil. *Frontiers in Microbiology*. 2021;12:785439.
160. Zhou Q, Zhang X, Wu Z. Impact of TiO_2 and ZnO nanoparticles on soil bacteria and the enantioselective transformation of racemic-metalaxyl in agricultural soil with Lolium perenne: a wild greenhouse cultivation. *Journal of Agricultural and Food Chemistry*. 2020;68(40):11242–52.
161. Londono N, Donovan AR, Shi H, Geisler M, Liang Y. Impact of TiO_2 and ZnO nanoparticles on an aquatic microbial community: effect at environmentally relevant concentrations. *Nanotoxicology*. 2017;11(9–10):1140–56.

4 Nanotechnology-based Assays for Microbial Detection and Identification

Zakieh Sadat Hoseini, Moein Iranmanesh, Mohsen Yaghoubi, and Abbas Hajizade

4.1 INTRODUCTION

Microbial detection and identification are very important in different microbial disciplines, including industrial, environmental, and medical microbiology (1–3). Because of the importance of this field, related methods are constantly and regularly developed. Traditional methods, which are based on culturing, morphological characteristics, biochemical and physiological properties, and pathological effects of pathogens, have been used for centuries (4, 5). However, parallel with the growth and improvements in science and technology, these methods are evolving and changing. DNA-based methods, for example, have emerged since the discovery of DNA structure by Watson and Crick in 1953 (6) and following discoveries, such as polymerase chain reaction (PCR) by Kary Banks Mullis in 1983 (7). As another example, following the introduction of aptamers by Ellington and Szostak in 1990 (8), many detection strategies were developed based on this discovery (9–11). On the other hand, discoveries in other scientific fields and the integration of these discoveries with biological methods have opened new ways for more accurate and reliable methods for the detection and identification of microbes. Biophotonics, a multidisciplinary scientific field that employs light-based technologies in biology and medicine (12), is a good example. Among many applications in life sciences and medicine, biophotonics-based methods are also promising in microbial detection (13–15).

Nanotechnology is another important field that has revolutionized almost all disciplines of biology (16–18), including detecting and identifying microorganisms (18–21). Nanotechnology-based methods for microbial detection and identification integrate the advances in nanotechnology and discoveries of biomolecules and microbial properties. In this chapter, following a brief overview of the different detection and identification methods, a comprehensive review will be carried out on nanotechnology-based detection and identification methods for microbial pathogens.

4.2 A BRIEF OVERVIEW OF MICROBIAL IDENTIFICATION METHODS

Since the discovery of microorganisms, detection methods have been regularly developed (22). The field of microbial detection and identification has gained more attention following the discovery of their role as infectious agents. Direct observation of microbes using microscopes was one of the first methods for the detection and identification of microbes. This method is still useful in many cases and is progressively developed by the invention of new microscopes (23). However, microscopy methods are not efficient in microbial identification; low sensitivity, expensive instruments, need for skilled personnel, laboriousness, and time-consuming process are among the main limitations of microscopy-based methods (24, 25). Culture-based methods have also emerged since the discovery era of microorganisms. Although these methods are still the gold standard for the detection of many

DOI: 10.1201/9781003410164-4

pathogens (26), they are time-consuming: in some cases it takes a week to complete the detection process (27). Indeed, culture-based methods usually need additional tests, such as molecular tests, biochemical tests, serological tests, etc. (28, 29). To overcome this limitation, other methods have been introduced in this field. Nucleic acid amplification-based methods such as polymerase chain reaction (PCR) (30), immunological methods such as enzyme-linked immunosorbent assay (ELISA) (31, 32), phage detection methods (33, 34), etc., have been introduced in recent years. In combination with these methods, nanomaterials have emerged as promising materials for the detection of microbial pathogens, and are extensively reviewed in the upcoming sections (35, 36).

4.3 NANO-BIOSENSORS

Nano-biosensors are cutting-edge devices that combine nanotechnology with biosensing principles. These miniaturized sensors play a pivotal role in healthcare, enabling the detection, monitoring, and diagnosis of pathogens such as viruses, bacteria, and fungi. A biosensor comprises three main components:

(1) Receptor: the biological element that recognizes the analyte
(2) Transducer: converts the recognition event into a measurable signal
(3) Readout system: displays or quantifies the detected signal

In nano-biosensors, the receptor is often a biological molecule (e.g., antibodies, enzymes, DNA) that interacts with the target analyte. Nanoparticles, nanowires, carbon nanotubes (CNTs), nanorods, and quantum dots (QDs) are commonly employed.

Nanomaterials may possess distinctive characteristics, including the ability to change color when interacting with analytes, heightened sensitivity due to their extensive surface area, exceptional stability in sensor applications, and excellent electrical and thermal conductivity, which aid in signal transduction. Results of many studies reveal the applications of nanomaterials for the detection and identification of pathogens (22–24). For example, Syed (2014) has categorized nano-diagnostic techniques into microfluidic assays or lab-on-a-chip, nanoparticle-based assays, Förster resonance energy transfer-based biosensors and molecular beacons, and one-dimensional and carbon-derived nanomaterials (22). Another useful classification of nano-diagnostic tools has been presented by Malik et al. (2013) where they have classified nano-biosensors into three categories: nanoparticle-based sensors, nanotube-based sensors, and nanowire-based sensors (25). They also have classified nano-biosensors according to the sensing mechanism into electrochemical, calorimetric, optical, and acoustic biosensors. In this chapter, based on the bioreceptor part (antibody, aptamer, DNA, or bacteriophage), we categorized nano-biosensors into four categories: nano-immunosensors, aptamer-based nano-biosensors, DNA-bases nano-biosensors, and bacteriophage-based sensors. It is also worth mentioning that some of the nano-based detection systems do not fall into these categories.

4.3.1 NANO-IMMUNOSENSORS

Immunological methods that are based on the high affinity and high specificity of the antigen-antibody recognition reaction have been used for a variety of applications, such as medical diagnosis, environmental studies, water and food safety testing, pharmaceutical analyses, etc. These methods include a variety of techniques, for example enzyme-linked immunosorbent assay (ELISA), agglutination tests, immunofluorescence assay (IFA), and Western blotting. Commonly utilized by the general public, rapid tests or lateral flow assays represent one of the most prevalent types of immunosensor kits employing nanoparticles (Figure 4.1).

Accordingly, many analytical methods use the antigen-antibody interaction for microbial detection. In a simple platform, pathogens or their products (toxins, secreted proteins, etc.) can

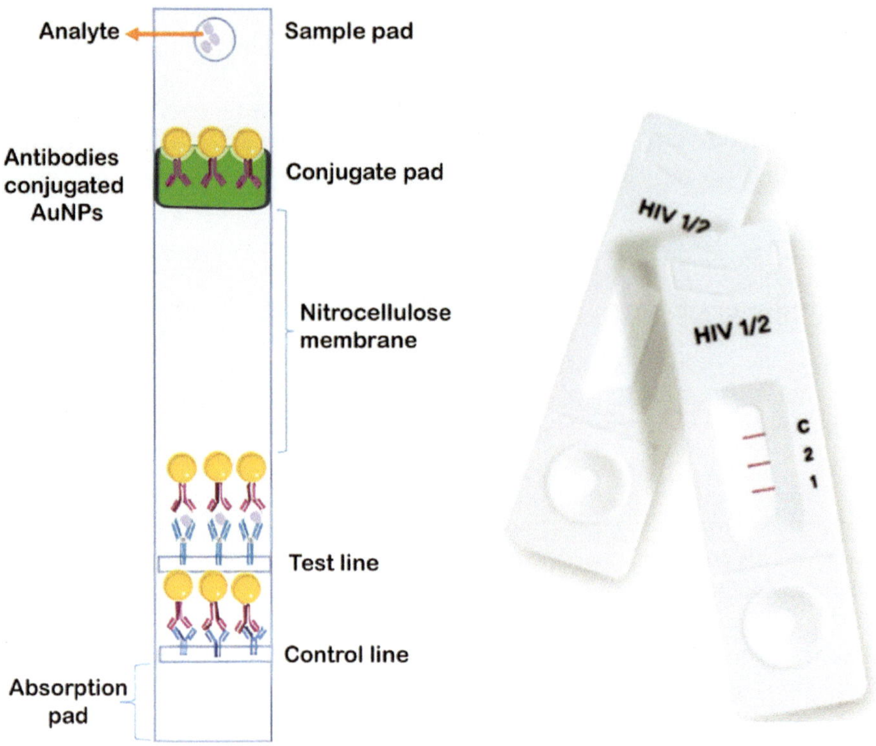

FIGURE 4.1 Schematic representation of different components of immunochromatographic strips or rapid lateral flow assay.

be detected based on immobilized antibodies on NPs. In this platform, specific antibodies (poly or monoclonal) against the desired pathogen or its products are immobilized on NPs. Then, the sample is introduced into the antibody-NP complex. Following the addition of a second antibody against the analyte, the presence of the analyte is investigated. There are also modifications to this method. For example, Cheng and colleagues created a diffusometric nano-immunosensor designed for the swift, straightforward, and quantitative detection of botulinum toxins in biological samples, achieving a detection limit of 10 pg/mL within just 2 minutes (26). Khan et al. developed a nano-biosensor based on monoclonal antibody-conjugated gold NPs for the multiple drug-resistant (MDR) *Salmonella typhimurium* detection. They reached a LOD of 10^3 CFU/mL of bacteria in water and lettuce samples in 5 minutes (27). There is a good review paper about surface plasmon resonance (SPR)-based sensors for the detection of pathogens (28). The application of these sensors is the study of molecular/cellular interactions. Surface modification of nanoparticles with antibodies (and other bio-recognition ligands, such as enzymes and aptamers) allows them to bind the specific portions (receptors) on the target microbes' surface. Following the interaction between ligands and receptors, NPs and microbes aggregate and the SPR/colorimetric response will change. He et al. (2014) developed an inductively coupled plasma mass spectrometry (ICP-MS) sensitive assay for detecting HIV-1 p24 antigen using biotin–streptavidin (BA) interaction and gold nanoparticles (Au NPs). Initially, an immobilized anti-HIV-1 p24 monoclonal antibody captured the p24 antigen. Subsequent immunoreactions involved biotinylated anti-p24 polyclonal antibody and Au NPs-labeled streptavidin. Dissociation of Au NPs using diluted HNO_3 facilitated ICP-MS measurements. The calibration graph was linear for p24 antigen concentrations from 7.5 to 75 pg mL^{-1}, with a detection limit of 1.49 pg mL^{-1}. Importantly, interference from other proteins (e.g.,

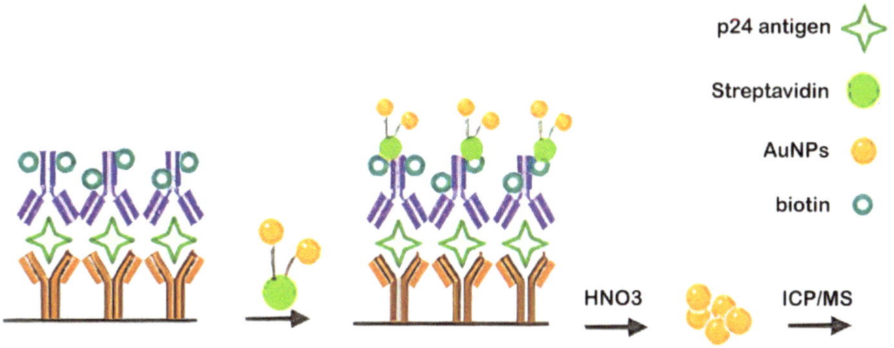

FIGURE 4.2 Schematic representation of the immunosensor based on the biotin–streptavidin system and gold nanoparticles (Au NPs) for p24 detection using inductively coupled plasma mass spectrometry (ICP-MS).

human IgG, human HSA, human CEA, human AFP) was minimal. Compared to traditional enzyme-linked immunosorbent assays (BA-ELISA), this Au NP-based ICP-MS method avoids toxic enzyme substrates and streamlines the experimental process by omitting color rendering steps. Overall, it provides a sensitive approach for HIV-1 p24 antigen determination (29) (Figure 4.2) (Table 4.1).

The use of nanomaterials in label-free immunosensors significantly increases the efficiency of the system due to the improvement of electron transfer (ET) and the facilitation of antibody (Ab) loading. In addition, in sandwich immunosensors, nanomaterials can be used both as carriers and as electroactive or electrocatalytic detectors (non-enzymatic immunosensors). (30). For example, Khashayar et al. (2017) developed stable gold nanoparticles (Au NPs) with an average size of 6 nm with attaching antibodies on their surface. The conjugation process involved both covalent and electrostatic interactions. By using the Au NP-GSH-NHS-ab combination in a 1:1:1 ratio, they successfully preserved the protein's tertiary structure, ensuring the antibody's functionality for subsequent steps. The one-pot method allowed the covalent coupling of antibodies while maintaining their functionality. Additionally, the Au NP content in the solution facilitated electron transfer, leading to an amplified detection signal. This innovative technique holds promise for measuring serum biomarker levels in both single and multiplexed sensor systems (31).

Paper-based electrochemical sensors are also considered a type of immunosensor. They offer advantages such as simplicity, portability, and cost-effectiveness. Researchers have developed a novel biosensor for detecting *Salmonella typhimurium*, a type of bacterium that can cause food poisoning. The biosensor uses a specialized strip with a filter paper pad that serves as a reservoir for the solution being tested. This design allows for easy attachment of antibodies to the surface and collection of samples, making the process more efficient and cost-effective. The strip is divided into two sections, with the first section directly attaching antibodies to the surface and the second section incorporating a specific type of dendrimer molecule that helps to facilitate the immune response. To test the biosensor, researchers used a technique called electrochemical impedance spectroscopy, which measures the electrical properties of a material. They found that as the biosensor detected the presence of antigens, there was a corresponding shift in the electrical signal, which was directly related to the amount of *Salmonella* present in the sample. The biosensor was able to detect *Salmonella* at a concentration of 5 cells per milliliter, and the entire process took less than an hour to complete. This technology has the potential to be used in real-world settings, such as food processing plants or laboratories, to quickly and accurately detect the presence of *Salmonella* and prevent foodborne illness. (32).

The applications of magnetic nanoparticles (MNPs) in the isolation of bacterial pathogens from sample sources is amazing, although they are limited in the identification of parasites. A recent study by Srinivasan et al. (2021) demonstrated a significant improvement in the detection sensitivity of

TABLE 4.1
Example of Some of the Developed Nano-biosensors for the Detection of Pathogens

Nano-biosensor Type	Pathogen	Environment	Limit of Detection (LOD)	Reference
DNA Nano-biosensor	*Nisseria meningitides*	-	1×10^{-15} M	(55)
	Salmonella choleraesuis	-	50 nM	(56)
	Salmonella enteritidis	-	25 CFU/mL	(57)
	Clostridium botulinum	-	0.02–2 ng/mL	(58)
		-		
	Dengue virus (DENV)	-	0.49 fM	(59)
	SARS-CoV-2	-	1 pM	(60)
	Human immunodeficiency virus (HIV)	-	48 fM	(61)
	Hepatitis C virus (HCV)	-	5 pM	(62)
	Human papillomavirus (HPV)	-	6.72 pM	(63)
Immuno Nano Sensor	*S. enteritidis*	Beverage samples	32 *Salmonella* mL^{-1}	(64)
	Hepatitis E virus	Clinical samples	56 RNA copies mL^{-1}	(65)
	Norovirus		69 RNA copies mL^{-1}	
	S. Typhimurium	Chicken meats	43 CFU mL^{-1}	(66)
	S. Typhimurium and V. parahaemolyticus	Aquatic samples	10 CFU mL^{-1} 102 CFU mL^{-1}	(67)
	Salmonella pullorum and Salmonella gallinarum	-	8.8 × 102	(68)
	H1N1 influenza virus	Nasal swab	10^2 PFU/sample	(69)
	Hepatitis B antigen	-	0.1 ng/mL	(70)
	SARS-CoV-2	-	0.01 ag/mL	(71)
	HIV-1	-	37 aM	(72)
Aptamer based nano sensor	*E. coli*	Water	1×10^2 CFU	(73)
	P. aeruginosa	Water	1 cell	(74)
	E. coli	Water	150 CFU mL^{-1}	(75)
	S. aureus	Bacterial suspension and human serum	1.0 CFU mL^{-1}	(76)
	Ochratoxin A	Peanut, soybean, and corn	28.18 pg/mL	(77)
	Salmonella typhimurium	Pork	4 CFU/mL	(78)
	Newcastle disease virus	-	0.4 EID50/m	(79)
	H5N1 avian influenza virus (AIV)	-	5.9 pM	(80)
	Hepatitis B virus	-	0.0014 fg/mL	(81)
	Staphylococcus aureus	Pig skin	8 × 102 CFU/mL; 107 CFU/mL	(82)

parasites by coating Fe_3O_4 MNPs with a conductive polymer, polypyrrole (Ppy). This modification enabled enhanced separation and detection capabilities of the MNPs, allowing for the development of a novel biosensing platform. The researchers employed electrical conductivity measurement as the detection method, which offered high sensitivity and ease of use. By functionalizing the Fe_3O_4@Ppy complex with carboxylic groups and conjugating *E. coli*-specific antibodies, they created a highly sensitive biosensing system capable of detecting bacterial suspensions within a short time frame of 10 minutes. The system demonstrated a detection limit of 10 colony-forming units per

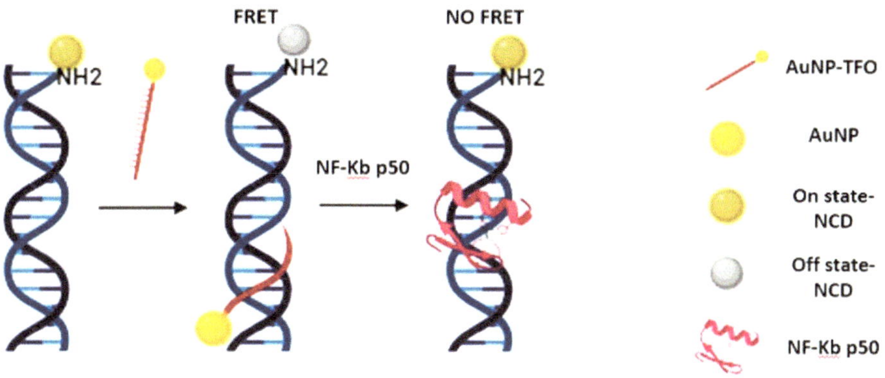

FIGURE 4.3 Diagram of Innovative biosensor combining triplex DNA, NCDs, and Au NPs for rapid NF-kB p50 detection.

milliliter (CFU/mL), with a detection range of 10^1 to 10^6 CFU mL^{-1}, highlighting its potential for rapid and sensitive detection of bacteria (33).

Immunosensors can also be used to detect food poisoning. Existing methods for identifying *S. aureus*, a common foodborne pathogen, are time-consuming and impractical for on-site use. To address this, Magar et al. (2023) developed a label-free impedimetric immunosensor. This sensor, constructed using a self-assembling monolayer of lipoic acid on an AuNPs@Fe_3O_4-screen printed electrode, covalently immobilizes anti-*S. aureus* antibodies. Under optimized conditions, the sensor monitors a specific immune reaction between pathogen cells and immobilized antibodies. The newly developed immunosensor exhibits high sensitivity and selectivity, with a limit of detection of 0.10 cfu/mL and a limit of quantification of 0.34 CFU/mL. It successfully detects *S. aureus* contamination in milk and meat samples (34). Upan *et al.* (2020) developed an innovative label-free electrochemical immunosensor for detecting hepatitis B surface antigen (HBsAg), a crucial marker for hepatitis B virus (HBV). By specifically binding antigen–antibody pairs on the electrode surface, they hindered the oxidation of silver nanoparticles (Ag NPs), which serve as the redox probe. This reduction in Ag NPs' oxidation current was the key. The screen-printed carbon electrode (SPCE) was enhanced with carbon nanotubes decorated with gold nanoparticles (Au NPs-CNT) and silver nanoparticles (Ag NPs). The biocompatible Au NPs provided ample surface area for tightly immobilizing the hepatitis B surface antibody (HBsAb), thereby enhancing the signal. In differential pulse voltammetric (DPV) detection, Ag NPs acted as the sensing probe. Additionally, an automated sequential injection (SI) system streamlined solution-handling while minimizing chemical usage. Notably, the immunosensor achieved a linear detection range of 1–40 ng mL^{-1}, with a low detection limit of 0.86 ng mL^{-1} for HBsAg. This cost-effective and efficient method holds promise for clinical analysis (35).

4.3.2 Aptamer-based Nano-biosensors

Aptamers are single-stranded nucleic acids that can bind to target molecules specifically. Their targets range from small molecules and ions, like vitamin B12, adenosine triphosphate (ATP), Cu_{2+} , Ag^+, etc., to macromolecules and cells. They are selected in an *in vitro* procedure called SELEX (systematic evolution of ligands by exponential enrichment) (36). Their advantages over antibodies, such as the diversity of their targets, higher stability, longer shelf-life, and faster and easier synthesis, make them ideal molecules for different applications, including biosensors (37, 38). Because of these unique features, various aptamer-based nano-biosensors have evolved, which are comprehensively reviewed in various references (39, 40). Wang et al. (2022) developed a novel dual-mode sensing platform for the rapid and sensitive detection of *Clostridium perfringens*, a key foodborne pathogen. This innovative platform integrates fluorescent and electrochemical signals, leveraging an electrochemical aptasensor built using dual-amplification technology based on a DNA walker and hybridization chain reaction (HCR). In the presence of *C. perfringens* genomic DNA, a FAM-labeled aptamer specifically binds to the DNA, triggering the DNA walker on hairpin DNA tracks to initiate the synthesis of double-stranded DNA. The subsequent hybridization chain reaction produces long-chain DNA, leading to enhanced methylene blue absorption and a detectable signal. Additionally, the released FAM-labeled aptamer generates fluorescent signals. The synergistic effects of MB and FAM significantly improve the sensitivity and accuracy of the dual-mode sensor. This ultra-sensitive biosensor shows excellent performance in the detection of *C. perfringens* in the range of 1 to 10^8 CFU g^{-1}. Therefore, it can be highly effective in controlling foodborne diseases (41). Ding et al. (2023) introduced an innovative biosensor designed for highly accurate and sensitive detection of pathogenic bacteria. The biosensor employs a ratiometric dual-signal electrochemical method. It relies on an aptamer recognition-induced rolling circle amplification (RCA)/G-quadruplex strategy. On a gold electrode chip, an aptamer sequence (P1) linked to ferrocene (Fc) captures the target bacterium. This capture releases a different sequence (P2) that triggers the RCA/G-quadruplex cascade. In the presence of potassium ions, the cascade binds the electrochemical indicator methylene blue (MB). By integrating signals from Fc loss and MB appearance (IMB/IFC ratio), the biosensor quantifies the target bacteria concentration. Notably, it achieves excellent detection performance and specificity, with a low detection limit of 10 CFU/mL. Its diagnostic potential for clinical infectious diseases caused by pathogenic bacteria, particularly in personalized medicine applications, is promising (42). Novakovic and colleagues (2024) introduced an aptamer-based sensor for Gram-positive bacteria detection. The biosensor combines vancomycin (an antibiotic) with aptamers. The sensor utilizes a vancomycin-modified screen-printed carbon electrode that selectively captures bacteria susceptible to this antibiotic in a short time frame of 10 minutes, with a remarkably low detection limit of 2 CFU/mL. The identification of bacterial species is achieved through the use of species-specific aptamers. After capturing bacteria, the electrodes are exposed to aptamers, and the resulting changes in current intensity are analyzed using differential pulse voltammetry to provide a detectable signal.The biosensor successfully identifies *Staphylococcus aureus* and *Bacillus cereus* in untreated milk or serum (at 100 CFU/mL) within 45 minutes. This portable biosensor holds promise for clinical diagnostics, food safety analysis, environmental monitoring, and security applications (43). Nodoushan et al. (2019) developed a highly sensitive biosensor for detecting staphylococcal enterotoxin B (SEB). The biosensor uses a screen-printed electrode modified with reduced graphene oxide (rGO) and gold nano-urchins (Au NUs). A single-stranded DNA probe is attached to the Au NU's surface, followed by the attachment of a specific aptamer. When SEB molecules are present, the aptamer detaches from the electrode surface. By recording the peak current in differential pulse voltammetry (DPV) after applying an electrochemical signal generator, the biosensor quantifies SEB concentration. This innovative approach leverages rGO properties and Au NP surface characteristics for enhanced performance (44).

4.3.3 DNA-based Nano-biosensors

The general basis of DNA-based biosensors is the presence of a single-stranded oligonucleotide (as a probe) immobilized on the nanomaterial. If its complementary strand is present, it is hybridized together and as a result of this interaction, a signal is generated and detected by the sensor. (45). This signal can be optical, electrochemical, or piezoelectric. Based on the generated signal, there are different DNA-based nano-biosensors (46) (Figure 4.3).

For example, here we describe an example of optical DNA-based biosensors based on fluorescent nanostructures (45). When a valence electron is excited, it tends to return to its ground state. When the excited electron returns to its ground state, it emits a photon and this is the cause of fluorescence

(46). Biosensors can use fluorescence resonance energy transfer (FRET) technology for the detection of pathogens (47). FRET uses the energy transfer between a donor and an acceptor fluorophore. Photoluminescent nanoparticles, such as graphene quantum dots (GQD), and graphene-like 2D nanomaterials semiconductor quantum dots (QD), have been used as donor fluorophores in FRET assays (48). Because of their strong luminescence, high quantum yield, high photostability, and long fluorescence lifetime, these nanomaterials are appropriate donor fluorophores in FRET assays. Indeed, some nanomaterials, such as graphene oxide (GO) and gold nanoparticles (Au NPs), that have a strong optical absorption ability are used as quenchers (49).

The limitation of the efficiency of nano-optical biosensors is in identifying non-transparent samples such as urine and blood (50). In such cases, it is preferable to use electrochemical biosensors. These biosensors produce an electrical signal. In the case of DNA-based sensors, the electrical signal can be produced from DNA hybridization. (50).

Researchers led by Zhu have developed a multi-analyte detection method for identifying bacteria and detecting nucleic acids. By leveraging the use of nanopores and DNA dumbbell nano-switches, they were able to design a system that can simultaneously detect multiple distinct nucleic acid sequences. Specifically, they created a novel configuration that combines four DNA dumbbell nano switches on a single carrier, enabling the simultaneous detection of four different nucleic acid sequences.The high specificity of these nano-switches allowed the differentiation of single base variants in both DNA and RNA targets. Notably, the researchers were able to develop a strategy for identifying distinct bacterial species even when the genetic sequences of interest exhibited high similarity. By integrating multiple dumbbell nano-switches with barcoded DNA carriers, they enabled accurate identification of different bacterial species, overcoming challenges associated with sequence similarity. Furthermore, the use of strain-specific 16S ribosomal RNA fragments proved to be a valuable tool in facilitating bacterial identification (51).

Mayuri et al. (2023) have recently investigated an innovative method for detecting specific DNA sequences by utilizing curcumin-coated gold nanorods as a novel optical sensing platform. This approach offers a promising avenue for the development of sensitive and specific DNA detection techniques, with potential applications in various fields. In this study, the researchers synthesized gold nanorods by combining curcumin as a reducing agent and silver nitrate in an alkaline base solution.These Au NRs exhibit remarkable sensitivity, detecting specific DNA targets with an impressive low detection limit of 8.6 pM for complementary target (CT) DNA. Overall, this study introduces a straightforward bio-based method to tailor the shape and size of Au NRs, establishing them as a reliable platform for cancer-related DNA detection (52). Spherical nucleic acids (SNAs) are unique nanostructures composed of gold cores enveloped by a densely packed DNA shell. When functionalized with polyadenine (polyA), SNAs enhance the control and reproducibility of DNA assembly on small surfaces (nanointerfaces). Their applications span medical detection and intracellular imaging. However, due to single-site binding physics, SNA-based biosensors typically operate within a fixed 81-fold range of target concentrations. To address this limitation, Gu et al. (2023) proposed a tri-block DNA approach for assembling SNAs. This tri-block structure includes a polyA block (anchors SNAs with tunable surface density), stem block (varies stability using different GC base pairs), and fixed loop block (recognizes specific targets). By adjusting polyA length, temperature, and base pairs, SNAs can detect targets with varying affinity. To expand the dynamic range, they combine two SNAs with different affinities, achieving a 1000-fold range. These tunable SNAs enable sensitive mRNA detection in cells, opening exciting possibilities for biomedical and clinical applications (53).

4.3.4 Bacteriophage-based Nano-biosensors

Bacteriophages are viruses that specifically infect bacteria. Due to their small size, high genetic engineering capability, and high replication speed, these viruses have been considered as a powerful

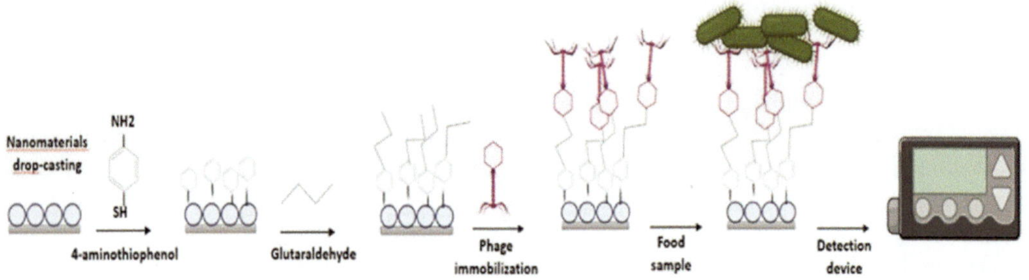

FIGURE 4.4 Schematic representation of the phage-based biosensing method for the swift impedimetric identification of *E. coli.*

tool in making nano-biosensors. By genetically engineering bacteriophages, scientists give them the power to produce specific proteins on their surface. The high speed of reproduction of these viruses has enabled researchers to build very sensitive biosensors based on them. These sensors can identify different types of pathogenic bacteria with high accuracy and speed (55). A research conducted by Abdulhamid and his colleagues (2023) has proven that a bacteriophage-based nano-biosensor can detect *Escherichia coli* O157:H7 bacteria in food samples with high accuracy and extreme sensitivity. The authors of this article made a very sensitive and accurate nanosensor to detect this strain of bacteria by combining gold nanoparticles, carbon nanotubes, and tungsten oxide. This sensor is capable of very accurately identifying this bacterium at a very low concentration of 3.0 CFU/mL using the electrochemical method of impedance measurement and disposable printed electrodes. As a result, this nano-biosensor is suggested for real-time monitoring of food quality (54) (Figure 4.4).

4.4 COMMERCIALIZED NANO-BIOSENSORS FOR THE DETECTION OF MICROBIAL PATHOGENS

There are many commercialized nano-biosensors in the market. Table 4.2 presents some of these biosensors. As can be seen, nano-biosensors are exploited for the detection of different microbial agents, from viruses to bacteria and molds.

4.5 FUTURE PERSPECTIVES

The swift and precise identification of microbial pathogens is essential for inhibiting the transmission of infectious diseases. Nano-biosensors represent an emerging and promising approach, providing exceptional sensitivity and specificity. By combining nanotechnology with biosensing components, we can achieve substantial improvements in the detection of pathogens. An emerging strategy involves the use of metal nanoparticles like gold and silver, known for their distinctive optical characteristics beneficial for identifying pathogens. By attaching antibodies or peptides to these nanoparticles that selectively adhere to microbial pathogens, it's possible to create sensors that operate on colorimetric or fluorescence principles. Utilizing quantum dots as fluorescent markers in biosensors represents another innovative avenue. Quantum dots yield a robust and bright signal that can improve biosensors' sensitivity thresholds. Their ability to emit light at various wavelengths depending on their size enables the concurrent detection of multiple pathogens. Additionally, carbon-based nanostructures such as graphene and carbon nanotubes, with their superior electrical characteristics, are advantageous for electrochemical biosensing. These nanomaterials can be functionalized with bio-recognition elements to sense changes in electrical conductivity caused by pathogens. Research is also delving into the use of magnetic nanoparticles to segregate and concentrate pathogens from intricate biological matrices, which could markedly elevate the sensitivity

TABLE 4.2
Examples of Some of the Commercialized Nano-biosensors for Microbial Detection

Type of Pathogen	Nano-biosensor Type	Pathogen	Environment	Sensitivity	Specificity	Company (Country)	Website
Bacteria	*Immuno nano-bioensor*	*Treponema pallidum*	Serum, plasma, whole blood	99.3%	99.5%	Abbott Laboratories INC. USA	www.globalpointofcare.abbott/ww/en/product-details/bioline-syphilis-test.html
	Immuno nano-biosensor	*Salmonella* TYPHI	human serum, plasma, or whole blood	64.9% (IgG), 94.6% (IgM), 100% (IgG/IgM)	88.3% (IgG), 92.2% (IgM), 85.7% (IgG/IgM)	Abbott Laboratories INC. USA	www.globalpointofcare.abbott/ww/en/product-details/bioline-salmonella-typhi-igg-igm-test.html
	Immuno nano-biosensor	*Mycobacterium tuberculosis complex and other mycobacteria*	Solid cultures (colony, condensation fluid) or liquid cultures	98.6%	100%	Abbott Laboratories INC. USA	www.globalpointofcare.abbott/pk/en/product-details/bioline-tb-ag-mpt64-rapid.html
		Chlamydia/ Gonorrhea/ Trichomoniasis	Vaginal swab	Chlamydia: 97.4% gonorrhea: 97.8% trichomonas: 99.3%	Chlamydia: 97.8 % gonorrhea: 99.1% trichomonas: 96.7%	CLIAwaived, Inc.	https://cliawaived.com/visby-sexual-health-pcr-tests-chlamydia-gonorrhea-trichomoniasis.html
	Immuno nano-biosensor	*Mycoplasma pneumoniae*	Human serum	90%	88%	CLIAwaived, Inc.	https://cliawaived.com/rapid-test-kit-immunocard-ae-elisa-mycoplasma-pneumoniae-igm-antibody-serum-sample.html
	Immuno nano-biosensor	*Clostridium difficile*	Human serum	97%	86%	Trinity Biotech Plc.	www.trinitybiotech.com/products/uni-gold-c-difficile-gdh-control-kit/
		Streptococcus pneumoniae	Urine	88%	92%	Trinity Biotech Plc.	www.trinitybiotech.com/products/uni-gold-s-pneumoniae/
		Legionella pneumophila	Urine	82.1%	99.2%	Trinity Biotech Plc.	https://www.trinitybiotech.com/products/uni-gold-legionella-urinary-antigen-plus/

(continued)

TABLE 4.2 (Continued)
Examples of Some of the Commercialized Nano-biosensors for Microbial Detection

Type of Pathogen	Nano-biosensor Type	Pathogen	Environment	Sensitivity	Specificity	Company (Country)	Website
Virus	*Immuno nano-biosensor*	*SARS-CoV-2*	Whole blood, serum, or plasma	94.4%	98%	InTec PRODUCTS, INC.	www.intecasi.com/rapid-sars-cov-2-antibody-igm-igg-test_p12.html
	Immuno nano-biosensor	*SARS-CoV-2*	Nasopharyngeal/nasal specimen	96.3%	100%	InTec PRODUCTS, INC.	www.intecasi.com/rapid-sars-cov-2-antigen-test-nasopharyngeal-nasal-swab_p54.html
	Immuno nano-biosensor	*SARS-CoV-2*	Whole blood (fingertip/venous), serum, or plasma	97.13%	94.48%	Sugentech	https://sugentech.com/en/sub/product/detail.php?idx=49&cat_name=COVID-19
		SARS-CoV-2	Nasal swab	96.69%	99.48%	VivaChek biotech (Hangzhou) Co., Ltd	www.vivachek.com/en/prods/verinopro-sars-cov-2-ag-rapid-test.html
	Immuno nano-biosensor	*HIV*	Whole blood / serum / plasma	99.91%	99.67%	Abbott Laboratories INC. USA	www.globalpointofcare.abbott/pk/en/product-details/bioline-hiv-syphilis-duo.html
	Immuno nano-biosensor	*Dengue virus serotype DEN-1, 2, 3 and 4*	Serum, plasma	94.6%	96.5%	Abbott Laboratories INC. USA	www.globalpointofcare.abbott/pk/en/product-details/bioline-dengue-igg-igm.html
	Immuno nano-biosensor	*Dengue virus*	human serum, plasma, or whole blood	92.4% (Dengue NS1 Ag), 94.2% (Dengue IgG/IgM)	98.4%(Dengue NS1 Ag), 96.4% (Dengue IgG/IgM)	Abbott Laboratories INC. USA	www.globalpointofcare.abbott/ww/en/product-details/bioline-dengue-duo-ns1-ag-ab-combo.html
	Immuno nano-biosensor	*Influenza virus type A, type B, and A (H1N1)*	Human nasal swab, throat swab, nasopharyngeal swab, or nasal/nasopharyngeal aspirate	Influenza A (H1N1): 77.0%, Influenza A: 62.0%, Influenza B: 84.8%	Influenza A (H1N1): 99.8% InfluenzaA:99.8% InfluenzaB:100.0%	Abbott Laboratories INC. USA	www.globalpointofcare.abbott/pk/en/product-details/bioline-influenza-ag-aba.html

Immuno nano-biosensor	*Hepatitis C virus*	Serum, plasma, whole blood	99.3%	100%	Abbott Laboratories INC. USA	www.globalpointofcare.abbott/pk/en/product-details/bioline-hcv.html
Immuno nano-biosensor	*Hepatitit B virus*	Serum or plasma	100%	100%	Abbott Laboratories INC. USA	www.globalpointofcare.abbott/pk/en/product-details/bioline-hbsag.html
	SARS-CoV-2, influenza A, and influenza B viral RNA	Nasal mucosa	Covid: 97.4% Flu A: 99.3% Flu B: 96.9%	Covid: 98.8% Flu A: 99.3% Flu B: 100%	CLIAwaived, Inc.	https://cliawaived.com/visby-respiratory-health-pcr-tests-covid-19-flu.html
	Hepatitis C virus	Blood/blood-derived body fluids	99%	-	CLIAwaived, Inc.	https://cliawaived.com/oraquick-hcv-rapid-antibody-test.html
Immuno nano-biosensor	*HIV*	Serum, plasma, or whole blood	100%	95.5%	Trinity Biotech Plc.	www.trinitybiotech.com/products/uni-gold-hiv-3/
	SARS-CoV-2	-	88.4%	100%	Access Bio, Inc.	www.prnewswire.com/news-releases/carestart-covid-19-antigen-test-kit-using-nanoact-launched-in-the-us-301187826.html
	Hepatitis C virus	Blood	98%	-	OraSure Technologies Inc., (Bethlehem, PA, USA)	https://orasure.com/products-infectious/OraQuick-HCV.html
	Ebola virus	Blood	94.12%	100%	OraSure Technologies Inc., (Bethlehem, PA, USA)	https://orasure.com/products-infectious/OraQuick-Ebola.html
Immuno nano-biosensor	*Hepatitis E virus*	Whole blood, serum, or plasma	99.2%	99.3%	InTec PRODUCTS, INC.	https://www.intecasi.com/rapid-anti-hev-igm-test_p61.html

(*continued*)

TABLE 4.2 (Continued)
Examples of Some of the Commercialized Nano-biosensors for Microbial Detection

Type of Pathogen	Nano-biosensor Type	Pathogen	Environment	Sensitivity	Specificity	Company (Country)	Website
Protozoa	Immuno nano-biosensor	*Plasmodium falciparum*	Human whole blood	99.7%	99.5%	Abbott Laboratories INC. USA	www.globalpointofcare.abbott/ww/en/product-details/bioline-malaria-ag-pf-pv-hrp-2-pldh.html
	Immuno nano-biosensor	*Plasmodium vivax*	Human whole blood	95.5%	99.5%	Abbott Laboratories INC. USA	www.globalpointofcare.abbott/ww/en/product-details/bioline-malaria-ag-pf-pv-hrp-2-pldh.html

of detection techniques that follow. Moreover, crafting nanocomposites that amalgamate various nanomaterials may result in combined effects that boost biosensor efficacy. Finally, marrying these nano-biosensors with microfluidic technologies might set the stage for compact, on-site diagnostic tools that deliver instantaneous feedback, crucial for prompt medical decision-making and efficacious disease control. With ongoing advancements, it is expected that nano-biosensors will evolve to be more advanced, offering improved proficiency for swift, point-of-use identification of microbial pathogens.

REFERENCES

1. Ferone M, Gowen A, Fanning S, Scannell AG. Microbial detection and identification methods: bench top assays to omics approaches. *Comprehensive Reviews in Food Science and Food Safety*. 2020;19(6):3106–29.
2. Maier RM, Pepper IL, Gerba CP. *Environmental Microbiology*. Academic Press; 2009.
3. Ahmad S, Lohiya S, Taksande A, Meshram RJ, Varma A, Vagha K, et al. A comprehensive review of innovative paradigms in microbial detection and antimicrobial resistance: beyond traditional cultural methods. *Cureus*. 2024;16(6):e61476.
4. Lagier J-C, Edouard S, Pagnier I, Mediannikov O, Drancourt M, Raoult D. Current and past strategies for bacterial culture in clinical microbiology. *Clinical Microbiology Reviews*. 2015;28(1):208–36.
5. Franco-Duarte R, Černáková L, Kadam S, Kaushik SK, Salehi B, Bevilacqua A, et al. Advances in chemical and biological methods to identify microorganisms—from past to present. *Microorganisms*. 2019;7(5):130.
6. Watson JD, Crick F. A structure for deoxyribose nucleic acid. *Nature*. 1953;171:737–38.
7. Faloona F, Mullis K, Scharf S, editors. *Specific Enzymatic Amplification of DNA in Vitro*. Cold Spring Harbor Symp Quant Biol; 1986.
8. Ellington AD, Szostak JW. In vitro selection of RNA molecules that bind specific ligands. *Nature*. 1990;346(6287):818–22.
9. Li D, Liu L, Huang Q, Tong T, Zhou Y, Li Z, et al. Recent advances on aptamer-based biosensors for detection of pathogenic bacteria. *World Journal of Microbiology and Biotechnology*. 2021;37:1–20.
10. Léguillier V, Heddi B, Vidic J. Recent advances in aptamer-based biosensors for bacterial detection. *Biosensors*. 2024;14(5):210.
11. Patra I, Kadhim MM, Mahmood Saleh M, Yasin G, Abdulhussain Fadhil A, Sabah Jabr H, et al. Aptasensor based on microfluidic for foodborne pathogenic bacteria and virus detection: a review. *Critical Reviews in Analytical Chemistry*. 2024;54(4):872–81.
12. Prasad PN. *Introduction to Biophotonics*. John Wiley & Sons; 2004.
13. Dudak FC, Boyacı İH. Rapid and label-free bacteria detection by surface plasmon resonance (SPR) biosensors. *Biotechnology Journal: Healthcare Nutrition Technology*. 2009;4(7):1003–11.
14. Ravindran N, Sandhya K, Mamathi CA, Nirmal TS, Sunil CK. Recent advances in Surface Plasmon Resonance (SPR) biosensors for food analysis: a review. *Critical Reviews in Food Science and Nutrition*. 2023;63(8):1055–77.
15. Gupta A, Singh T, Singh RK, Tiwari A. Numerical analysis of coronavirus detection using photonic crystal fibre–based SPR sensor. *Plasmonics*. 2023;18(2):577–85.
16. Hajizade A, Ebrahimi F, Salmanian A-H, Arpanaei A, Amani J. Nanoparticles in vaccine development. *Journal of Applied Biotechnology Reports*. 2014;1(4):125–34.
17. Rahimizadeh P, Rezaieyazdi Z, Behzadi F, Hajizade A, Lim SI. Nanotechnology as a promising platform for rheumatoid arthritis management: diagnosis, treatment, and treatment monitoring. *International Journal of Pharmaceutics*. 2021;609:121137.
18. Malik S, Muhammad K, Waheed Y. Nanotechnology: a revolution in modern industry. *Molecules*. 2023;28(2):661.
19. Yezdani U, Khan MG, Kushwah N, Verma A, Khan F. Application of nanotechnology in diagnosis and treatment of various diseases and its future advances in medicine. *World Journal of Pharmacy & Pharmaceutical Science*. 2018;7(11):1611–33.
20. Malik S, Muhammad K, Waheed Y. Emerging applications of nanotechnology in healthcare and medicine. *Molecules*. 2023;28(18):6624.

21. Hajizade A, Ahmadian G, Fan K, Zhu M, Arpanaei A. Nanotechnology-based detection, prevention and treatment of infectious diseases. *Frontiers in Microbiology*. 2023;14:1224727.
22. Syed MA. Advances in nanodiagnostic techniques for microbial agents. *Biosensors and Bioelectronics*. 2014;51:391–400.
23. Zhu X, Radovic-Moreno AF, Wu J, Langer R, Shi J. Nanomedicine in the management of microbial infection–overview and perspectives. *Nano Today*. 2014;9(4):478–98.
24. Alafeef M, Moitra P, Pan D. Nano-enabled sensing approaches for pathogenic bacterial detection. *Biosensors and Bioelectronics*. 2020;165:112276.
25. Malik P, Katyal V, Malik V, Asatkar A, Inwati G, Mukherjee TK. Nanobiosensors: concepts and variations. *International Scholarly Research Notices*. 2013;2013:1–9.
26. Cheng H-P, Chuang H-S. Rapid and sensitive nano-immunosensors for botulinum. *ACS Sensors*. 2019;4(7):1754–60.
27. Khan SA, Singh AK, Senapati D, Fan Z, Ray PC. Targeted highly sensitive detection of multi-drug resistant salmonella DT104 using gold nanoparticles. *Chemical Communications*. 2011;47(33):9444–6.
28. Thawany P, Tiwari UK, Deep A. *Surface Plasmon Resonance (SPR)-Based Nanosensors for the Detection of Pathogenic Bacteria. Nanosensors for Point-of-Care Diagnostics of Pathogenic Bacteria*. Springer; 2023. p. 41–57.
29. He Q, Zhu Z, Jin L, Peng L, Guo W, Hu S. Detection of HIV-1 p24 antigen using streptavidin–biotin and gold nanoparticles based immunoassay by inductively coupled plasma mass spectrometry. *Journal of Analytical Atomic Spectrometry*. 2014;29(8):1477–82.
30. Kulkarni MB, Ayachit NH, Aminabhavi TM. Recent advancements in nanobiosensors: current trends, challenges, applications, and future scope. *Biosensors*. 2022;12(10):892.
31. Khashayar P, Amoabediny G, Larijani B, Hosseini M, Vanfleteren J. Fabrication and verification of conjugated AuNP-antibody nanoprobe for sensitivity improvement in electrochemical biosensors. *Scientific Reports*. 2017;7(1):16070.
32. Silva NF, Almeida CM, Magalhães JM, Gonçalves MP, Freire C, Delerue-Matos C. Development of a disposable paper-based potentiometric immunosensor for real-time detection of a foodborne pathogen. *Biosensors and Bioelectronics*. 2019;141:111317.
33. Srinivasan SY, Gajbhiye V, Bodas D. Development of nano-immunosensor with magnetic separation and electrical detection of Escherichia coli using antibody conjugated Fe3O4@ Ppy. *Nanotechnology*. 2020;32(8):085603.
34. Magar HS, Abdelghany H, Abbas MN, Bilitewski U, Hassan RY. Fast analysis of Staphylococcus aureus in food products using disposable label-free nano-electrochemical immunosensor chips. *Microchemical Journal*. 2023;193:109097.
35. Upan J, Banet P, Aubert P-H, Ounnunkad K, Jakmunee J. Sequential injection-differential pulse voltammetric immunosensor for hepatitis B surface antigen using the modified screen-printed carbon electrode. *Electrochimica Acta*. 2020;349:136335.
36. Tuerk C, Gold L. Systematic evolution of ligands by exponential enrichment: RNA ligands to bacteriophage T4 DNA polymerase. *Science*. 1990;249(4968):505–10.
37. Jayasena SD. Aptamers: an emerging class of molecules that rival antibodies in diagnostics. *Clinical Chemistry*. 1999;45(9):1628–50.
38. Syed MA, Pervaiz S. Advances in aptamers. *Oligonucleotides*. 2010;20(5):215–24.
39. Kim YS, Raston NHA, Gu MB. Aptamer-based nanobiosensors. *Biosensors and Bioelectronics*. 2016;76:2–19.
40. Syed MA, Jamil B. Aptamers and aptasensors as novel approach for microbial detection and identification: an appraisal. *Current Drug Targets*. 2018;19(13):1560–72.
41. Wang W, Yuan W, Wang D, Mai X, Wang D, Zhu Y, et al. Dual-mode sensor based on the synergy of magnetic separation and functionalized probes for the ultrasensitive detection of Clostridium perfringens. *RSC Advances*. 2022;12(39):25744–52.
42. Ding Z, Yang S, Wang J, Zhao Z, Xu H, Chen Z, et al. Rolling circle amplification/G-quadruplex-based dual-signal ratiometric electrochemical aptasensor for ultrasensitive detection of pathogenic bacteria. *ChemElectroChem*. 2023;10(17):e202300257.

43. Novakovic Z, Khalife M, Costache V, Camacho MJ, Cardoso S, Martins V, et al. Rapid detection and identification of vancomycin-sensitive bacteria using an electrochemical apta-sensor. *ACS Omega*. 2024;9(2):2841–9.
44. Nodoushan SM, Nasirizadeh N, Amani J, Halabian R, Fooladi AAI. An electrochemical aptasensor for staphylococcal enterotoxin B detection based on reduced graphene oxide and gold nano-urchins. *Biosensors and Bioelectronics*. 2019;127:221–8.
45. Brenneman KL, Poduri S, Stroscio MA, Dutta M. Optical detection of lead (II) ions using DNA-based nanosensor. *IEEE Sensors Journal*. 2013;13(5):1783–6.
46. Schulman SG. *Fluorescence and Phosphorescence Spectroscopy: Physicochemical Principles and Practice*. Elsevier; 2017.
47. Bhupathi P, Elhassan A-Elgadir TM, Mohammed Ali RH, Sanaan Jabbar H, Gulnoza D, Joshi S, et al. Fluorescence resonance energy transfer (FRET)-based sensor for detection of foodborne pathogenic bacteria: a review. *Critical Reviews in Analytical Chemistry*. 2023:1–18.
48. Tian F, Lyu J, Shi J, Yang M. Graphene and graphene-like two-denominational materials based fluorescence resonance energy transfer (FRET) assays for biological applications. *Biosensors and Bioelectronics*. 2017;89:123–35.
49. Li F, Pei H, Wang L, Lu J, Gao J, Jiang B, et al. Nanomaterial-based fluorescent DNA analysis: a comparative study of the quenching effects of graphene oxide, carbon nanotubes, and gold nanoparticles. *Advanced Functional Materials*. 2013;23(33):4140–8.
50. Cagnin S, Caraballo M, Guiducci C, Martini P, Ross M, SantaAna M, et al. Overview of electrochemical DNA biosensors: new approaches to detect the expression of life. *Sensors*. 2009;9(4):3122–48.
51. Zhu J, Tivony R, Bošković F, Pereira-Dias J, Sandler SE, Baker S, et al. Multiplexed nanopore-based nucleic acid sensing and bacterial identification using DNA dumbbell nanoswitches. *Journal of the American Chemical Society*. 2023;145(22):12115–23.
52. Mayuri S, Jha NS, Jha SK. Curcumin-capped gold nanorods as optical sensing platform for sequence specific detection of DNA based on their self-assembly. *International Journal of Biological Macromolecules*. 2023;253:126829.
53. Gu M, Yi X, Xiao Y, Zhang J, Lin M, Xia F. Programming the dynamic range of nanobiosensors with engineering poly-adenine-mediated spherical nucleic acid. *Talanta*. 2023;256:124278.
54. Abdelhamied N, Abdelrahman F, El-Shibiny A, Hassan RY. Bacteriophage-based nano-biosensors for the fast impedimetric determination of pathogens in food samples. *Scientific Reports*. 2023;13(1):3498.
55. Singh R, Verma R, Kaushik A, Sumana G, Sood S, Gupta RK, et al. Chitosan–iron oxide nano-composite platform for mismatch-discriminating DNA hybridization for Neisseria gonorrhoeae detection causing sexually transmitted disease. *Biosensors and Bioelectronics*. 2011;26(6):2967–74.
56. Lee W, Hwang BH. Plasmonic biosensor controlled by DNAzyme for on-site genetic detection of pathogens. *Biotechnology Journal*. 2020;15(5):1900329.
57. Zhang P, Song M, Dou L, Xiao Y, Li K, Shen G, et al. Development of a fluorescent DNA nanomachine for ultrasensitive detection of Salmonella enteritidis without labeling and enzymes. *Microchimica Acta*. 2020;187(7):376.
58. Wang Q, Cheng X, Li H, Yu F, Wang Q, Yu M, et al. A novel DNA quantum dots/aptamer-modified gold nanoparticles probe for detection of Salmonella typhimurium by fluorescent immunoassay. *Materials Today Communications*. 2020;25:101428.
59. Song C, Zhang J, Liu Y, Guo X, Guo Y, Jiang X, et al. Highly sensitive SERS assay of DENV gene via a cascade signal amplification strategy of localized catalytic hairpin assembly and hybridization chain reaction. *Sensors and Actuators B: Chemical*. 2020;325:128970.
60. Park KS, Choi A, Kim HJ, Park I, Eom M-S, Yeo S-G, et al. Ultra-sensitive label-free SERS biosensor with high-throughput screened DNA aptamer for universal detection of SARS-CoV-2 variants from clinical samples. *Biosensors and Bioelectronics*. 2023;228:115202.
61. Diao W, Tang M, Ding S, Li X, Cheng W, Mo F, et al. Highly sensitive surface plasmon resonance biosensor for the detection of HIV-related DNA based on dynamic and structural DNA nanodevices. *Biosensors and Bioelectronics*. 2018;100:228–34.
62. Teengam P, Nisab N, Chuaypen N, Tangkijvanich P, Vilaivan T, Chailapakul O. Fluorescent paper-based DNA sensor using pyrrolidinyl peptide nucleic acids for hepatitis C virus detection. *Biosensors and Bioelectronics*. 2021;189:113381.

63. Su A, Liu Y, Cao X, Xu W, Liang C, Xu S. A universal CRISPR/Cas12a-mediated AuNPs aggregation-based surface-enhanced Raman scattering (CRISPR/Cas-SERS) platform for virus gene detection. *Sensors and Actuators B: Chemical.* 2022;369:132295.
64. Zhao X, Smith G, Javed B, Dee G, Gun'ko YK, Curtin J, et al. Design and development of magnetic iron core gold nanoparticle-based fluorescent multiplex assay to detect Salmonella. *Nanomaterials.* 2022;12(21):3917.
65. Ganganboina AB, Chowdhury AD, Khoris IM, Doong R-a, Li T-C, Hara T, et al. Hollow magnetic-fluorescent nanoparticles for dual-modality virus detection. *Biosensors and Bioelectronics.* 2020;170:112680.
66. Kulkarni MB, Ayachit NH, Aminabhavi TM. Biosensors and microfluidic biosensors: from fabrication to application. *Biosensors.* 2022; 12(7):543.
67. Liu L, Hong J, Wang W, Xiao S, Xie H, Wang Q, et al. Fluorescent aptasensor for detection of live foodborne pathogens based on multicolor perovskite-quantum-dot-encoded DNA probes and dual-stirring-bar-assisted signal amplification. *Journal of Pharmaceutical Analysis.* 2022;12(6):913–22.
68. Sun Q, Zhao G, Dou W. An optical and rapid sandwich immunoassay method for detection of Salmonella pullorum and Salmonella gallinarum based on immune blue silica nanoparticles and magnetic nanoparticles. *Sensors and Actuators B: Chemical.* 2016;226:69–75.
69. Cui X, Das A, Dhawane AN, Sweeney J, Zhang X, Chivukula V, et al. Highly specific and rapid glycan based amperometric detection of influenza viruses. *Chemical Science.* 2017;8(5):3628–34.
70. Huang Y, Zhu F, Guan J, Wei W, Zou L. Label-free amperometric immunosensor based on versatile carbon nanofibers network coupled with Au nanoparticles for Aflatoxin B1 detection. *Biosensors.* 2021;11(1):5.
71. Liv L. Electrochemical immunosensor platform based on gold-clusters, cysteamine and glutaraldehyde modified electrode for diagnosing COVID-19. *Microchemistry Journal.* 2021;168:106445.
72. Shamsipur M, Samandari L, Taherpour AA, Pashabadi A. Sub-femtomolar detection of HIV-1 gene using DNA immobilized on composite platform reinforced by a conductive polymer sandwiched between two nanostructured layers: A solid signal-amplification strategy. *Anal Chim Acta.* 2019;1055:7–16.
73. Pandit C, Alajangi HK, Singh J, Khajuria A, Sharma A, Hassan MS, et al. Development of magnetic nanoparticle assisted aptamer-quantum dot based biosensor for the detection of Escherichia coli in water samples. *Science of The Total Environment.* 2022;831:154857.
74. Hu J, Fu K, Bohn PW. Whole-cell pseudomonas aeruginosa localized surface plasmon resonance aptasensor. *Analytical Chemistry.* 2018;90(3):2326–32.
75. Dabhade AH, Verma RP, Paramasivan B, Kumawat A, Saha B. Development of silver nanoparticles and aptamer conjugated biosensor for rapid detection of E. coli in a water sample. *3 Biotech.* 2023;13(7):244.
76. Shahdost-Fard F, Faridfar S, Keihan AH, Aghaei M, Petrenko I, Ahmadi F, et al. Applicability of a green nanocomposite consisted of spongin decorated Cu2WO4(OH)2 and AgNPs as a high-performance aptasensing platform in Staphylococcus aureus detection. *Biosensors.* 2023;13(2):271.
77. Yang X, Huang R, Xiong L, Chen F, Sun W, Yu L. A colorimetric aptasensor for ochratoxin A detection based on tetramethylrhodamine charge effect-assisted silver enhancement. *Biosensors.* 2023; 13(4).
78. Ma X, Xu X, Xia Y, Wang Z. SERS aptasensor for Salmonella typhimurium detection based on spiny gold nanoparticles. *Food Control.* 2018;84:232–7.
79. Marnissi B, Khalfaoui K, Ebai T, Marques Souza de Oliveira F, Ghram A, Kamali-Moghaddam M, et al. Accurate detection of Newcastle disease virus using proximity-dependent DNA aptamer ligation assays. *FEBS Open Biol.* 2021;11(4):1122–31.
80. Kwon J, Lee Y, Lee T, Ahn JH. Aptamer-based field-effect transistor for detection of avian influenza virus in chicken serum. *Analytical Chemistry.* 2020;92(7):5524–31.
81. Mohsin D, Mashkour M, Fatemi F. Design of aptamer-based sensing platform using gold nanoparticles functionalized reduced graphene oxide for ultrasensitive detection of Hepatitis B virus. *Chemical Papers.* 2020;75:279–95.
82. Zelada-Guillén GA, Sebastián-Avila JL, Blondeau P, Riu J, Rius FX. Label-free detection of Staphylococcus aureus in skin using real-time potentiometric biosensors based on carbon nanotubes and aptamers. *Biosensors and Bioelectronics.* 2012;31(1):226–32.

5 Virus-like Particles

Applications in Vaccine Development and Drug Delivery for Infectious Agents

Saghi Nooraei, Howra Bahrulolum, Camellia Katalani, Mohsen Bidar Ajorloo, Gholamreza Ahmadian, and Abbas Hajizade

5.1 A BRIEF HISTORY OF THE DISCOVERY AND DEVELOPMENT OF VLPS

The term virus-like particles (VLPs) encompasses various biological entities, including unidentified structures displaying viral morphology discovered in biological samples, non-nucleic acid-containing empty structures of viral origin, infectious viruses that have undergone structural alterations through chemical or genetic means, and non-infectious gene products that are self-assembled by the cloning and expression of viral structural genes in various host systems (1).

VLPs have emerged as a pivotal area of study in vaccinology and biotechnology. These protein structures possess the ability to self-assemble into multi-subunit complexes, closely resembling or identical to native viruses. They play a crucial role in stimulating both B-cell receptors and T-cell responses (2, 3). The concept of VLPs originated from observations during investigations into natural virus assembly, where certain viral proteins were found to self-assemble into virus-like structures devoid of viral genetic material. This discovery laid the groundwork for the concept of VLPs.

The exploration of *in vitro* virus self-assembly traces back to the late 1950s with studies on the tobacco mosaic virus (TMV) (4). Initial investigations focused on TMV and filamentous bacteriophage f1 as candidates for developing genetic VLP architectures, providing a foundational understanding for VLP engineering (5, 6). A pivotal achievement in the field of VLP research was the creation of the first VLP-based vaccines targeting the hepatitis B virus (HBV) during the 1980s. It was discovered that the hepatitis B surface antigen (HBsAg) had the ability to self-assemble into VLPs. When the HBV envelope protein was produced within host cells, it generated VLPs that mimicked the exterior of the actual hepatitis B virus. This significant advance paved the way for the FDA's approval in 1986 of Recombivax HB, the inaugural yeast-derived HBsAg vaccine (7).

Following the licensure of Recombivax, additional HBs VLP vaccines produced in various expression hosts obtained approval, including those manufactured in yeast (such as DTP-Hep B, Engerix-B, Euvax B) and CHO cells (such as GenHevac B, Bio-Hep-B) (8–11). HBs VLPs have been utilized as adaptable platforms for displaying foreign epitopes to the immune system, facilitating the creation of vaccines aimed at various infectious diseases (12). Furthermore, advancements in vaccine technology have propelled VLP-based vaccines into their third generation by incorporating multiple epitopes. For instance, Sci-B-VacTM, a third-generation hepatitis B vaccine, includes epitopes from major S antigens, as well as minor pre-S1 and pre-S2 envelope proteins (13).

Expanding on the success of hepatitis B vaccines, researchers have extended their focus to the development of VLPs for other viruses. In the late 1990s and early 2000s, VLP-based vaccines for

DOI: 10.1201/9781003410164-5

human papillomavirus (HPV) were introduced, which have been highly effective in preventing HPV infections and associated cancers (14–16).

In 2006, the FDA approved the second VLP-based vaccine for HPV. Significant vaccines in this field include Cervarix and Gardasil. Cervarix consists of VLPs from HPV16 and HPV18 L1 proteins, produced in insect cells and combined with the $Al_2(SO_4)_3$ adjuvant. Gardasil, a quadrivalent HPV vaccine, includes VLPs from HPV types 6, 11, 16, and 18, produced in *S. cerevisiae* and adsorbed on an amorphous aluminum hydroxyphosphate sulfate adjuvant, and is marketed by Merck (17, 18).

From 2000 to 2021, numerous VLP-based vaccines targeting various viruses, such as norovirus, respiratory syncytial virus (RSV), Norwalk virus, Ebola virus, HIV, rotavirus, SARS-CoV-2 virus, and influenza virus, have undergone diverse clinical trial phases (19). In recent years, the COVID-19 pandemic has reignited interest in VLP technology, with researchers exploring VLPs as potential platforms for COVID-19 vaccine development. Consequently, several VLP-based COVID-19 vaccine candidates are in different stages of clinical trials (20). Among these, Covifenz, a plant-based VLP developed by Medicago Inc., has received approval, as reported by covid19.trackvaccines.org data as of February 2, 2024 (21).

5.2 STRUCTURE AND CLASSIFICATION OF VIRUS-LIKE PARTICLES

Virus-like particles (VLPs) are categorized based on their structural features into two primary types: enveloped VLPs and non-enveloped VLPs (19, 22). Enveloped VLPs have a lipid membrane, whereas non-enveloped VLPs do not possess this lipid layer. Further classification of VLPs is based on their capsid protein arrangement, leading to distinctions between single-capsid protein VLPs or multi-capsid protein VLPs and single-layer VLPs, double-layer VLPs, and triple-layer VLPs (23). The presence of a lipid envelope can significantly influence the stability and immunogenicity of VLPs. Enveloped VLPs may exhibit greater fragility and sensitivity to environmental conditions, yet they closely resemble native viruses, potentially triggering a stronger immune response. Conversely, non-enveloped VLPs are generally more stable and possess a high safety profile due to the absence of a viral genome, making them a promising avenue for vaccine development (3, 23, 24) (Figure 5.1).

Additionally, the structural classification of VLPs is predicated on their morphology, which may exhibit icosahedral, rod-like, or spherical conformations (24). Cryo-electron microscopy (cryoEM) stands as a widely employed technique for elucidating the structure of VLPs (25). VLPs bear a striking resemblance to viruses but are non-infectious owing to their absence of viral genetic components (26). These nanostructures can form naturally or be created by the specific expression of viral structural proteins, which then self-organize into a virus-like configuration (23).

5.2.1 Non-enveloped Virus-like Particles

Non-enveloped VLPs often consist of single or multiple protein structures derived from viruses or pathogens, lacking a host cell membrane (lipid bilayer) (22). These VLPs are primarily formed by the major viral nucleocapsid proteins. They can be categorized as simple (monolayer), bi-layer, or triple-layer structures. Simple VLPs can be expressed using both prokaryotic and eukaryotic expression systems (19, 22). Moreover, monolayer VLP capsids can be folded within cell-free systems, facilitating the production of homogeneous and soluble VLPs. In this method, capsid proteins are initially synthesized in a suitable expression host and then folded within a cell-free environment (27). Typically, the primary structural component of simple VLPs is a monolayer protein encoded by the virus, derived from pathogens such as caliciviruses (3, 22). In contrast, the production of multi-layer VLPs is more intricate and demanding. The complexity of multi-protein non-enveloped VLPs arises from the interactions among multiple capsid proteins (3, 24). These multi-layer VLPs

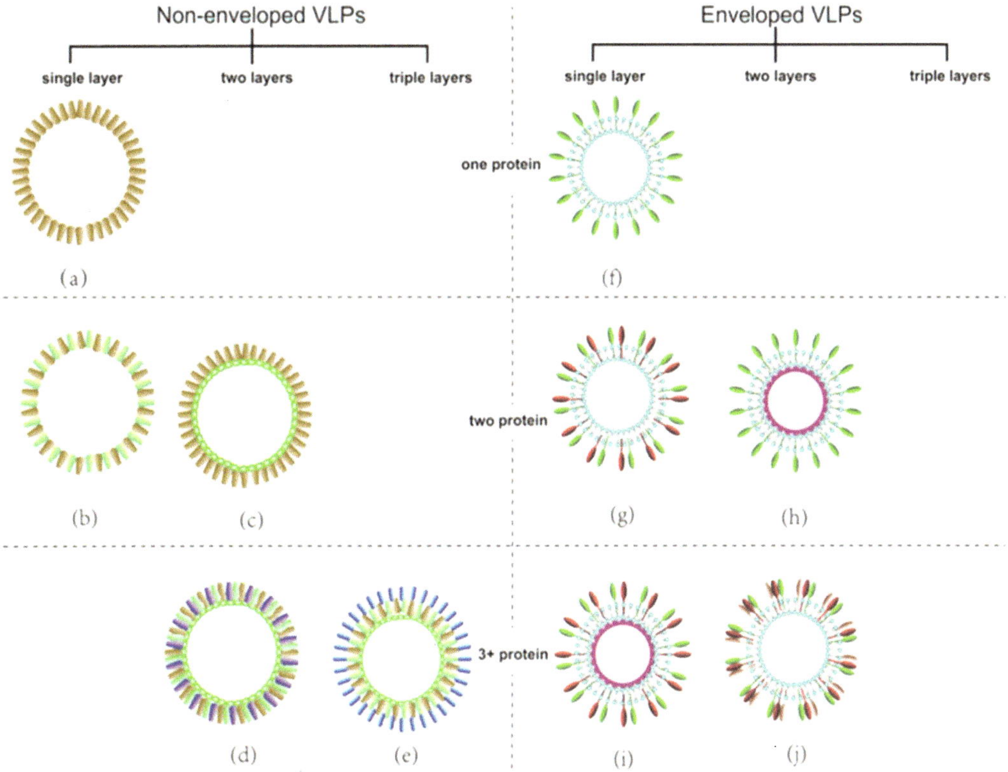

FIGURE 5.1 Schematic diagram of structural classification of virus-like particles based on the presence of lipid envelope. The figure illustrates the structural classification of virus-like particles (VLPs) according to the presence or absence of a lipid envelope surrounding the nanostructure. For non-enveloped VLPs: (a) single-layered non-enveloped VLPs composed of one protein, (b) tsingle-layered non-enveloped VLPs composed of two proteins, (c) two-layered non-enveloped VLPs composed of two proteins, (d) two-layered non-enveloped VLPs composed of multiple proteins, and (e) triple-layered VLPs composed of multiple proteins. For enveloped VLPs: (f) single-layered VLPs consisting of one protein, (g) single-layered VLPs consisting of two proteins; (h) two-layered VLPs consisting of two proteins, and (i) two-layered VLPs consisting of multiple proteins (23).

are produced and folded within eukaryotic expression hosts (22). They exhibit distinctive structural characteristics, comprising multiple complex concentric layers composed of various capsid proteins. Notable examples of multi-capsid protein VLPs include those derived from bluetongue virus, enterovirus, infectious bursal disease virus, poliovirus, and rotavirus (19). Since the production and purification of non-enveloped VLPs are relatively simple, they have emerged as favored candidates for developing subunit vaccines against numerous pathogens (22, 24).

5.2.2 Enveloped Virus-like Particles (eVLPs)

Compared to non-enveloped VLPs, enveloped VLPs (eVLPs) possess a more intricate composition, featuring a lipid membrane acquired during budding from the expressing cell. This category is further classified into single-layer, double-layer, and multi-layer structures (22). Enveloped VLPs are characterized by the presence of a lipid membrane, known as the envelope, with viral proteins embedded on its outer surface. Within this lipid layer, one or more glycoproteins, serving as target antigens, can be embedded to elicit an immune response upon recognition by the host's immune system. eVLPs demonstrate greater versatility as they can target antigenic epitopes from similar or heterologous viruses. For instance, eVLPs containing the Gag protein from simian immunodeficiency virus (SIV) and the Env protein from human immunodeficiency virus (HIV) have been engineered (22, 28).

Different enveloped VLPs may have different origin, composition, and nature, depending on the expression host and cell line utilized during production, assembly, and budding processes (19). eVLPs have been effectively employed in vaccine development against various pathogens, including Hantan virus, retroviruses, influenza A virus, and hepatitis C virus (HCV) (29). However, it is noteworthy that the size of eVLPs typically exceeds 100 nm. Consequently, there is a possibility that they may accumulate at the injection site, potentially disrupting the delivery process of the vaccine to the desired site and antigen presentation (30).

5.2.3 The Distinctions between Enveloped and Non-enveloped VLP Structures

The primary disparities between enveloped and non-enveloped virus-like particles (VLPs) lie in their structural composition and complexity. Here are the key contrasts:

1. Presence of lipid envelope: enveloped VLPs (eVLPs) feature a lipid membrane acquired from the host cell, whereas non-enveloped VLPs lack a lipid envelope altogether.
2. Complexity: eVLPs exhibit greater complexity in composition compared to non-enveloped VLPs, as they incorporate additional host-cell-derived components.
3. Sensitivity to environmental factors: eVLPs tend to be more sensitive to external environmental conditions, including temperature variations, shear force, and chemical treatments, due to the presence of host-cell-derived membranes.
4. Classification: eVLPs can be classified based on their internal structures into single-layer, double-layer, and multi-layer types. In contrast, non-enveloped VLPs can be divided into single-capsid protein VLPs or multi-capsid protein VLPs, as well as single-double-layer or triple-layer VLPs.
5. Applications: both enveloped and non-enveloped VLPs serve as platforms for vaccines and drug delivery vehicles. However, owing to their simpler structure and smaller size, non-enveloped VLPs are often favored for the development of subunit vaccines.

5.3 PRODUCTION OF VIRUS-LIKE PARTICLES

VLPs are non-infectious, self-assembling structures that mimic the shape and function of viruses, yet they lack any viral genetic material.
The production of VLPs involves several key steps, outlined as follows:

1. Selection and cloning of viral structural genes: the first step in the production of VLPs entails the selection and cloning of viral structural genes.
2. Expression of the cloned genes: subsequently, viral structural proteins with self-assembly capability are expressed using suitable expression systems. These systems may include prokaryotic (e.g., bacteria) or eukaryotic (e.g., yeast, plants, mammalian cells, and baculovirus/insect cells) hosts. Selecting the appropriate expression system is crucial for appropriate folding of the expressed proteins, as well as proper post-translational modifications (PTMs).
3. Harvesting and cell lysis: following the expression of viral proteins, cells are harvested and lysed to remove cellular debris and obtain the desired protein products.
4. Further purification: to obtain intact and pure VLPs, additional purification steps, like ultracentrifugation and chromatography techniques, are employed.

5. Polishing: the subsequent purification step, known as polishing, focuses on removing any residual host cell proteins and nucleic acids.
6. Formulation: in the final stage, VLPs undergo filtration and sterile formulation to ensure the production of a safe, efficient, and effective product (19).

Structural proteins from various viruses exhibit the ability to self-assemble, forming virus-like particles (VLPs) with spherical, icosahedral, or rod-like structures. These VLPs, ranging in size from 20 to 200 nm, closely resemble their corresponding viruses in size and morphology. This similarity enables VLPs to efficiently penetrate the lymphatic system and be captured by antigen-presenting cells (31–34).

The development of VLP-based vaccines relies on stable and highly efficient production platforms. Currently, a variety of expression systems and corresponding vectors are available, including prokaryotic and eukaryotic systems. *Escherichia coli* (*E. coli*) is the most commonly utilized prokaryotic expression system, offering cost-effective protein production.

Among eukaryotic expression systems, yeast and insect cell lines are widely employed. Mammalian cells, such as CHO and T293HEK, are also utilized for VLP production. Additionally, plant expression systems have emerged as promising platforms for VLP production. While bacterial systems offer cost advantages, they may not be suitable for proteins requiring post-translational modifications like glycosylation, necessitating the use of eukaryotic systems (35).

The choice of expression system must consider both economic feasibility and product quality. Industrial-scale production requires cost-effective raw materials and favorable economic returns. Moreover, stringent safety regulations govern the production of human medicines, necessitating high product quality. Thus, the selection of the appropriate host system depends on the specific requirements of the target application.

Utilizing recombinant systems to standardize and express desired viral genes in suitable expression hosts addresses challenges associated with large-scale vaccine production.

5.3.1 Expression Systems for VLP Production

As mentioned, expression systems are divided into two categories: prokaryotic and eukaryotic, and we will discuss each of them below.

5.3.1.1 Prokaryotic expression systems

Bacteria, particularly *Escherichia coli* (*E. coli*), are extensively utilized for the production of recombinant proteins, marking significant milestones in drug development. The first recombinant drug approved for diabetes treatment, recombinant human insulin, was expressed in *E. coli*. Notably, numerous proteins, including cytokines, hormones, growth factors, and serine proteases, have been successfully expressed in *E. coli* and licensed for human use. Regulatory organizations such as the FDA and the European Medicines Agency have sanctioned 151 recombinant protein-based drugs, with 29.8% originating from *E. coli* (36).

E. coli remains a prevalent host for recombinant drug production, with 28% of the 174 different virus-like particle (VLP) types successfully produced in bacterial expression platforms. Engineered *E. coli* strains have been developed to enhance protein production performance and quality.

Ensuring proper protein folding is crucial for the clinical efficacy of VLPs. Various *in vitro* analytical methods, complemented by *in vivo* evaluation, have been devised to verify the presence of native epitopes in *E. coli*-derived VLPs, ensuring the effectiveness and efficiency of VLP-based vaccines (36).

While the *E. coli* expression system offers advantages such as cost-effective cultivation, high expression levels, and rapid propagation, it also presents limitations. These include intracellular accumulation of heterologous proteins, improper protein folding, and the absence of post-translational modifications, leading to potential product degradation due to proteases and endotoxin production.

Correct protein folding could position *E. coli* as an ideal expression system for protein production. However, the high cost associated with product development restricts its widespread use, particularly in developing nations. With existing, emerging, and re-emerging infectious diseases posing threats globally, there is a pressing need for novel vaccine production systems capable of rapid and cost-effective antigen production (37).

The enhanced performance of capsid proteins or structural proteins, the fundamental components of VLPs, can significantly streamline vaccine production in terms of time and cost. Thus, microbial platforms hold promise for producing affordable, safe, and effective vaccines, particularly in developing countries (38, 39).

5.3.1.1.1 Advancements in E. coli-derived Vaccines: Prominent Cases

The utilization of *E. coli* as an expression system has led to the development of several significant vaccines. Here are some noteworthy examples:

1. **Hecolin (hepatitis E vaccine)**:

- Hecolin, the first commercial vaccine derived from *E. coli*, targets the prevention of hepatitis E virus (HEV) infection.
- Developed in 1993 by Xiamen Inovio Biotech, this vaccine received human licensure in 2011 and was launched in China in 2012.

2. **Human papillomavirus (HPV) vaccines**:

- Two vaccines against human papillomavirus (HPV), HPV16/18 and HPV6/11, have been produced based on virus-like particles (VLPs) expressed in *E. coli*.
- These vaccines contain the main capsid protein (L1) of HPV16/18 and HPV6/11, respectively.
- The HPV16/18 vaccine is in phase 3 clinical trials, while the HPV6/11 vaccine is in phase 2 clinical trials.

3. **Chimeric Malaria Vax vaccine (ICC1132)**:

- Another notable vaccine produced using the *E. coli* expression system is the chimeric Malaria Vax vaccine (ICC1132).
- This vaccine is a VLP composed of the capsid protein of the hepatitis B virus, expressed and produced in *E. coli*.
- The vaccine displays *Plasmodium falciparum* circumsporozoite protein (CSP) epitopes on its surface, targeting malaria prevention.
- While phase 1 clinical trial results have demonstrated protection against malaria, further clinical data have not been published yet.

These examples highlight the versatility and potential of *E. coli*-based expression systems in vaccine development, offering solutions for various infectious diseases.

5.3.2 Yeast Expression System: Bridging Prokaryotic and Eukaryotic Advantages

Eukaryotic systems, particularly yeast, offer a unique combination of advantages from both prokaryotic and eukaryotic expression systems. Yeast, such as *Saccharomyces cerevisiae* (*S. cerevisiae*) demonstrates high expression levels (10 to 100 times higher) and faster growth, alongside easy maintenance, scalability, and cost-effective culture mediums characteristic of prokaryotic systems (40). Additionally, yeast allows for essential post-translational modifications like protein glycosylation and proper protein folding, typical of eukaryotic systems (19).

S. cerevisiae, in particular, holds significant commercial value and serves as a widely utilized model organism in various research endeavors (40). Its well-characterized biochemical, genetic, and biological features make it an ideal platform for expressing a diverse range of proteins, including vaccine candidates, medicinal proteins, and diagnostic proteins (19). Remarkably, *S. cerevisiae* has been instrumental in enhancing the efficacy of potential antigens for the development of subunit vaccines targeting a variety of bacterial and viral infections. In vaccine development, *S. cerevisiae* has been instrumental in producing 11 approved vaccines against HBV and one vaccine against HPV (19, 40). In both scenarios, the recombinant protein generates highly immunogenic VLPs, fulfilling the criteria for cost-effectiveness and biosafety suitable for human use.The production and secretion of intact proteins by yeast enhance the efficiency of protein production and purification.The efficient inner membrane system of yeast facilitates the secretion of proteins produced within the cell into the extracellular space (40). As a unicellular eukaryote, yeast ensures proper solubility, folding, and essential post-translational modifications necessary for protein functionality. Moreover, the safety and security of the yeast system are ensured by the absence of endotoxin and oncogenes (19).

Despite these benefits, yeast differs from mammalian cells in the way it forms N- and O-linked oligosaccharide structures on target proteins. Specifically, *S. cerevisiae* may improperly glycosylate animal proteins, potentially resulting in excessive glycosylation. However, yeast cells are generally easier to genetically manipulate compared to mammalian cells and can be grown at high cell densities (40).

5.3.3 Insect Cell Expression Systems for VLP Vaccine Development

Insect cell expression systems are widely used for numerous applications, such as producing recombinant proteins, viral vaccines, pesticides, and conducting basic biological research. These systems often achieve high levels of heterologous gene expression, especially for intracellular proteins, compared to other eukaryotic expression systems.

Notably, recombinant proteins are frequently soluble and easily recovered from infected cells late in infection, enhancing their utility in downstream processes (41).

Baculovirus-based insect cell expression systems have significantly advanced the production of recombinant proteins. Baculoviruses, which naturally control insect populations, provide several benefits. They do not infect vertebrates, ensuring safe handling, and many insect cell lines are free from pathogenic viruses, allowing for maintenance under less restrictive conditions. Additionally, the baculovirus genome carries all the genetic information necessary for reproduction across various cell lines or insect larvae, obviating the need for cell lines or helper viruses (41).

Commonly used insect cell lines, such as Sf-9 and Sf-21 from *Spodoptera frugiperda* and BTI-TN-5B-4 (High Five) from *Trichoplusia ni*, alongside insect larvae, facilitate high-level protein production. The unique slow gene expression property of baculoviruses renders them ideal expression vectors for heterologous genes. Moreover, baculovirus/insect expression systems are highly regarded for their ability to process proteins, making them successful tools for producing recombinant glycoproteins (41).

However, this expression system has its limitations. The eventual death of host cells infected with nuclear polyhedrosis virus precludes continuous expression of heterologous genes, necessitating the infection of new insect cells for each round of protein synthesis. Furthermore, differences in glycosylation patterns, oligosaccharide length, and mannose content between insect and mammalian cells can impact the biological activity and immunogenicity of products expressed in insect cells (41).

Despite these limitations, genetic modification of lepidopteran cell lines has led to the creation of transgenic cell lines capable of normal growth properties and efficient baculovirus infection, enabling N-glycan processing and the production of recombinant human glycoproteins (41).

In the realm of vaccine development, insect cell culture systems have been instrumental in producing VLP-based vaccines against HIV and HPV viruses. The Sf-9 cell line infected with baculovirus can produce and process gag gene products, leading to the spontaneous assembly of HIV-VLPs capable of inducing immune responses. Similarly, insect cell expression systems have been utilized for the production of VLPs against HPV (42).

5.3.4 Advancements in Mammalian Cell-based Production of Virus-like Particles

The utilization of mammalian cells for protein production is steadily increasing. Mammalian expression systems offer advantages such as proper protein folding, post-translational modifications, and assembly, crucial for achieving the full biological activity of the target protein. They also facilitate signaling pathways and enable the secretion and glycosylation of proteins, particularly eukaryotic proteins.

Several mammalian cell lines have been employed for protein expression, with 293HEK (human embryonic kidney) and CHO (Chinese hamster ovary) being the most prevalent. These cell lines are typically transfected using polyethyleneimine (PEI) or calcium phosphate. Additionally, other cell lines like baby hamster kidney (BHK) cells, Vero cells, mouse L cells, and myeloma cell lines such as J558L and Sp2/0 are utilized to generate stable transfectants. Despite the high cost, intricate technology, and potential risk of contamination with animal viruses limiting their use in large-scale industrial production, these systems are frequently employed to express various heterologous proteins, including viral structural proteins and bioactive peptides, for specific performance analyses.

Mammalian cells are commonly harnessed to produce complex enveloped virus-like particles (VLPs) comprising multiple structural proteins. CHO cells are extensively utilized in VLP production due to their non-human origin, reducing the risk of human virus contamination. CHO cells have previously been employed to generate hantavirus VLPs, eliciting a specific immune response in mice. Another prevalent mammalian expression platform, the 293HEK cell line, has been widely employed in the production of diverse VLPs such as rabies, HIV, and influenza (43).

5.3.5 Plant Bioreactors: A Sustainable Approach to Protein Production

Plants offer a unique advantage as solar bioreactors for protein production. They possess the ability to perform post-translational modifications necessary for proper protein function, making them suitable for expressing complex proteins. Compared to microbial or animal cell systems, plant-based production is cost-effective and simple, requiring minimal resources. However, challenges such as fluctuations in crop yield, contamination risks, and environmental factors must be addressed. Despite these challenges, the use of genetically engineered plants for protein expression shows promising economic potential and scalability. While commercial production using plant cell culture remains under exploration, several companies are actively investigating its feasibility (44).

5.4 PURIFICATION STRATEGIES FOR VIRUS-LIKE PARTICLES (VLPS)

The purification of virus-like particles (VLPs) follows established steps:

Cell lysis: Synthesized VLPs are released into solution through cell lysis, although it may not always be necessary for systems where VLPs are secreted into the culture medium. Non-secretory systems require cell disruption, achieved through methods such as mechanical operations or enzymatic treatments.

Clarification: The clarification process removes cell debris and large cell masses, ensuring a cleaner solution for further purification steps.

Concentration: VLPs are concentrated to increase their density per unit volume, typically achieved through centrifugation or low-speed ultrafiltration.

Polishing: In the final step, remaining impurities are removed, and the VLPs are purified from any contaminants.

Various techniques are employed depending on the stability and characteristics of the VLPs. These include precipitation with ammonium sulfate or polyethylene glycol, which reduces impurities and volume, facilitating subsequent purification steps such as chromatography.

To protect VLPs during purification, extraction buffers are formulated with reducing and chelating agents, protease inhibitors, and nucleases to minimize host nucleic acids. These measures ensure the preservation of particle integrity and enhance the quality of the final VLP product.

On a laboratory scale, VLP purification often involves cesium chloride or sucrose density gradient ultracentrifugation, which can yield VLPs suitable for subsequent applications. This method is effective, particularly with successive rounds of centrifugation at varying speeds. However, industrial-scale purification using ultracentrifugation has limitations due to VLP accumulation, scalability issues, and increased complexity. Therefore, specific chromatographic methods are preferred to achieve the required purity of VLPs. Depending on VLP characteristics, various chromatography techniques such as size exclusion, ion exchange, and affinity chromatography are employed.

Additional steps may be necessary before column chromatography to achieve desired vaccine purity. For instance, treatment of VLP solutions derived from HBsAg in *Pichia pastoris* yeast with diatomic earth can enhance purity levels. Furthermore, after purification, a dissociation/reassembly step may be needed to improve the functional properties of VLPs. For example, reassembling human papilloma VLPs has been shown to enhance their immunogenicity and stability.

5.5 ANALYTICAL TOOLS FOR VIRUS-LIKE PARTICLE (VLP) CHARACTERIZATION

Characterizing VLPs is crucial for vaccine production, involving a range of modern tools categorized into biochemical, biophysical, and biological methods. Biochemical tools encompass mass spectrometry (17), RP-HPLC, and sodium dodecyl sulfate-polyacrylamide gel electrophoresis (SDS-PAGE) (45). Biophysical methods include transmission electron microscopy, negative agarose gel electrophoresis, cryo-electron microscopy, and analytical ultracentrifugation (45). Biological tools such as ELISA and surface plasmon resonance are also utilized (45).

Thorough characterization of purified VLPs is essential to ensure vaccine efficacy and safety. Understanding the epitope-based function and physical properties, including size and size dispersion, contributes to the immunogenicity of VLP vaccines. To fully understand the structural and functional diversity of engineered VLPs, a comprehensive approach that integrates biochemical, biophysical, and biological techniques is essential. The characterization process employs standard techniques from the biopharmaceutical industry, such as analyzing amino acid composition, molecular weight, and VLP purity (45). These analyses are crucial for guiding the design of VLPs and optimizing vaccine development processes.

Mass spectrometry, a stand-alone tool for protein sequence analysis, determines both the molecular mass and amino acid composition of the protein analyte (17). Protein purity analysis, typically conducted on denatured VLPs, employs SDS-PAGE and RP-HPLC. SDS-PAGE, combined with Western blot analysis, is effective for examining VLPs with multiple capsid proteins. Agarose gel electrophoresis is used to evaluate nucleic acid encapsulation within VLPs, monitor VLP assembly in the presence of RNA, and assess changes in the external surface of VLPs compared to their unmodified counterparts.

Beyond compositional analysis, initial insights into VLP morphology and status are gathered using standard laboratory techniques such as analytical ultracentrifugation, density gradient ultracentrifugation, and transmission electron microscopy.

Analytical molecular weight ultracentrifugation reveals the composition and heterogeneity of a sample, while density gradient ultracentrifugation depicts the concentration and density of a VLP sample. Both techniques, used in conjunction with transmission electron microscopy, confirm the VLP structure. Sample preparation for transmission electron microscopy may alter VLP shape.

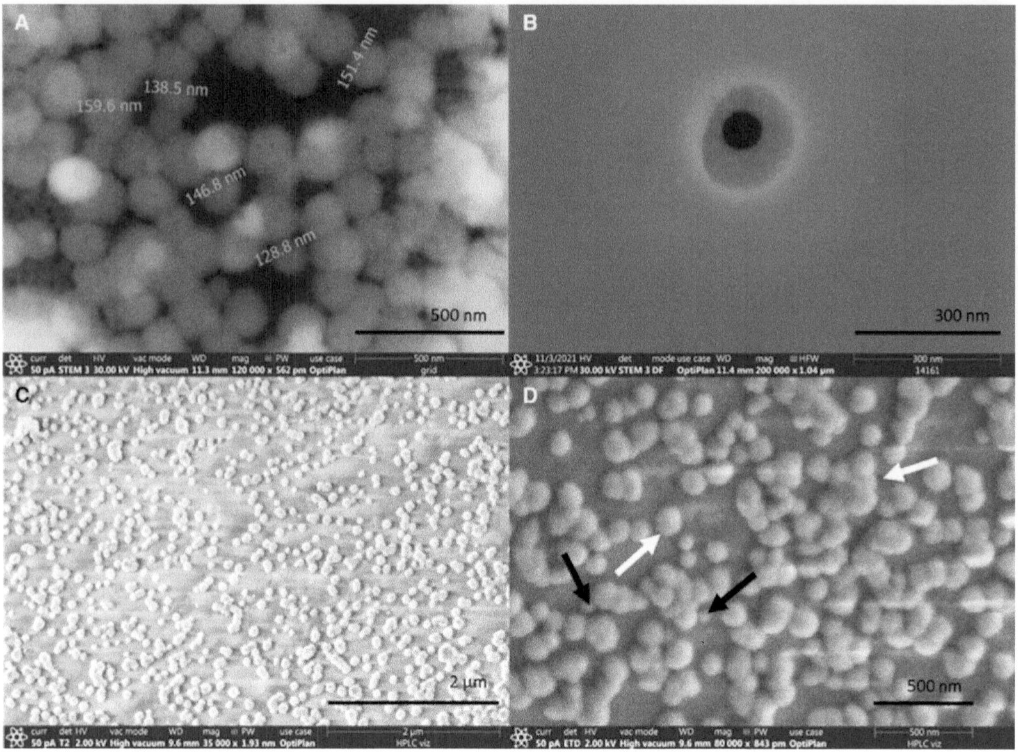

FIGURE 5.2 Scanning electron microscope (SEM) images of the virions. Panel (A): virus suspension dried onto a microscope slide-grid. Panel (B): single virion; the black patch visible in the center of the virus particle was caused by the electron beam. Panels (C) and (D): captured virus particles on the surface of the beads. Single and aggregated virions are indicated by black and white arrows, respectively. Aggregation of virus particles occurred due to the drying step of the sample preparation at room temperature (Adopted from reference (48)).

Cryo-electron microscopy and atomic force microscopy (AFM) offer alternative visual techniques, causing less deformation of VLP structure due to rapid freezing during sample processing and in-solution analysis, respectively. These techniques have been widely employed to characterize HPV and HBsAg VLPs. Biological characterization of VLPs often involves analyzing VLP binding to a known antibody. Methods such as ELISA, dot blotting, immunoprecipitation, or immunodiffusion are commonly used to assess antibody binding to VLPs (45, 46, 47). Figure 5.2 represents the electron microscopy images of VLPs.

5.6 ADVANCED TECHNIQUES FOR ACCURATE QUANTIFICATION OF VIRUS-LIKE PARTICLES

VLP biophysical characterization often relies on techniques such as dynamic light scattering (DLS), centrifugation, electrospray-differential mobility analysis (ES-DMA), and asymmetric field flow coupled with multi-angle light scattering detectors (AF4-MALS) (49). DLS, while straightforward, excels in accurately determining the particle size of monolithic samples but struggles with highly dispersed particles (49). In contrast, ES-DMA and AF4-MALS offer superior size discrimination before measurement, enhancing resolution and accuracy in particle size determination. Studies employing ES-DMA and AF4-MALS, corroborated with transmission electron microscopy (TEM), have showcased the efficacy of these advanced methods (49).

Traditionally, VLP quantification has relied on labor-intensive techniques like electrophoresis, Western blot, ELISA, and flow cytometry (49). These methods, while widely used, often demand large sample volumes, suffer from low specificity and sensitivity, and may yield incomplete or erroneous results (49). To address these limitations, recent advancements have introduced sophisticated technologies such as gel permeation chromatography, capillary electrophoresis, laser ion absorption mass spectrometry coupled with time-of-flight mass spectrometry (MALDI-TOF), and reverse transcription polymerase chain reaction (RT-PCR) (49). These cutting-edge approaches offer enhanced accuracy and efficiency in VLP quantification and characterization.

5.7 APPLICATIONS OF VLP IN MEDICINE

Due to their great potential, VLPs are increasingly used in various medical fields, such as vaccine development, drug delivery, gene therapy, diagnostics, etc. In this chapter, the applications of these nanomaterials in the development of vaccines and their applications in drug delivery are discussed.

5.7.1 VLPs as a Promising Platform for the Development of Vaccines

Human health is a global concern, particularly in countries with weak healthcare systems where most infectious diseases occur. The rise in international travel and trade has increased the transmission of infectious diseases between countries, posing a significant global threat. Diseases such as Ebola, dengue fever, yellow fever, Zika fever, MERS, influenza, tuberculosis, malaria, and COVID-19 persistently threaten public health. Animal diseases like avian influenza, ASFV (African swine fever virus), and COVID-19 have crossed species barriers to humans, underscoring the need for effective prevention and treatment strategies. International organizations like the WHO continuously support studies on vaccines and drugs against pathogens (50, 51). Virus-like particles (VLPs), engineered vaccine candidates produced by expressing viral proteins, offer promising immunogenicity without causing disease. VLP vaccines can stimulate both humoral and cellular immune responses, providing effective immunity against viruses (52).

VLPs possess advantageous features contributing to their high immunogenicity:

1. Structural proteins forming the capsid are presented to the immune system in a multivalent manner, increasing the crosslinking of B-cell receptors to immature B cells. This activation results in the secretion of potent antibodies and the generation of long-lived memory B cells.
2. VLPs possess specific viral epitopes for Th cells, activating T cells. Presentation of these epitopes by antigen-presenting cells (APCs) via MHC class II leads to Th-cell activation, which in turn secretes cytokines, activating B cells, T cells, and macrophages.
3. Most VLPs have sizes ranging from 25 to 100 nanometers, facilitating their entry into the lymphatic system through direct cellular uptake and engagement with immune cells. Additionally, their small size allows efficient trapping by APCs and swift transportation to lymphoid organs for presentation to Th cells.
4. Some VLPs, particularly those derived from bacteriophages, contain single-stranded RNA (ssRNA) acting as an endogenous adjuvant. These ssRNAs, encapsulated during VLP assembly, directly activate immune-stimulatory molecules such as Toll-like receptors (TLRs), thus initiating innate immune responses (Figure 5.3) (53, 54).

5.7.1.1 VLP-based Bacterial Vaccines: Current Status and Future Perspective

VLPs offer a versatile and immunogenic platform for vaccine development, enhancing antigen immunogenicity. This has led to the creation of VLP-based vaccines targeting bacterial infections, showing potential for both preventive and therapeutic use.

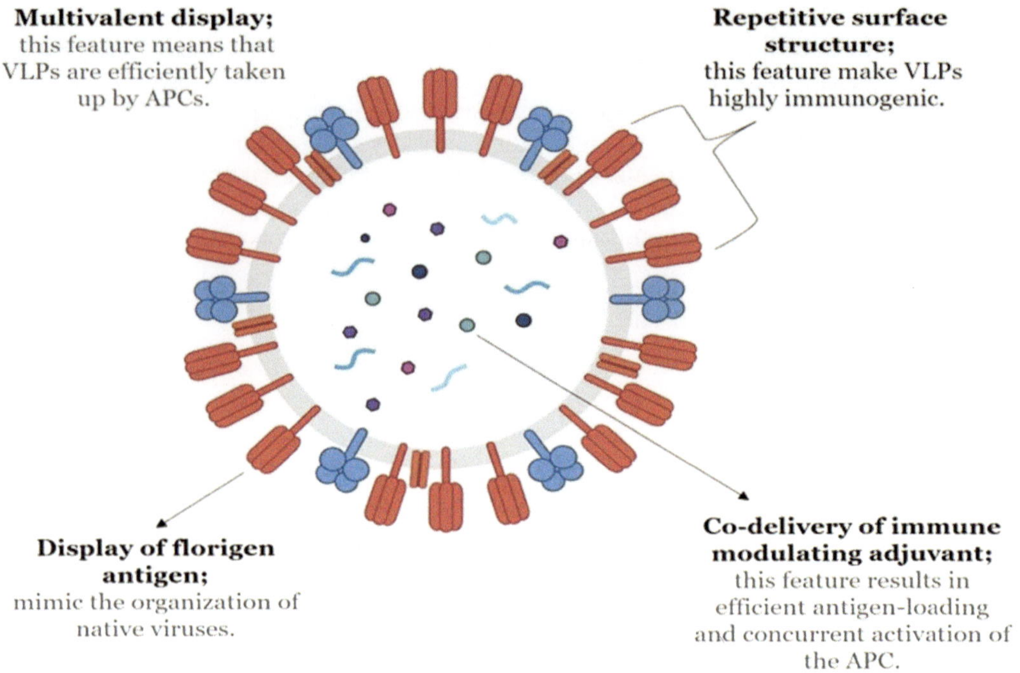

FIGURE 5.3 Immunogenic features of VLPs displaying foreign antigens. The figure demonstrates that the ILP may display multiple antigens (in several copies) along with adjuvant proteins. All of them make them highly immunogenic.

Staphylococcus aureus (*S. aureus*) is a major cause of skin and soft tissue infections (SSTIs) and a driver of antibiotic resistance. *S. aureus* uses its agr quorum sensing system and self-inducing peptides (AIPs) to regulate virulence factor expression for invasive infection. Among the four agr types, type I is more associated with human disease. For AIP1 to function, it must form a thiolactone bond, but this small size and bond have hindered vaccine development targeting AIP1. To solve this problem, a study used the PP7 VLP-based vaccine to display a modified sequence of AIP1 (AIP1S). Vaccination with PP7-AIP1S induced antibodies against AIP1 and limited agr activity *in vivo*. Noticeably, in a mouse model of SSTI using agr type I *S. aureus*, the administration of PP7-AIP1S vaccine reduced the bacterial virulence and increased the bacterial clearance compared to controls. Considering the role of agr type I in human disease, a vaccine targeting AIP1 regulation could successfully fight against antibiotic-resistant *S. aureus* infections like MRSA (55).

Group A *Streptococcus* (GAS), responsible for diseases like pharyngitis and rheumatic fever, poses another challenge. VLPs offer a multivalent approach to vaccine design, incorporating antigens from multiple GAS strains for broader coverage. Safe, non-toxic, and highly conserved antigens, such as the J8 protected α-helical peptide, have shown efficacy in inducing opsonizing antibodies against multiple GAS strains when presented on VLPs (56, 57). The freeze-drying (FD) method emerges as an effective strategy for producing VLP-based GAS vaccines, particularly for pharyngeal infections. It offers advantages such as improved patient compliance, non-invasiveness, and enhanced protein stability, eliminating the need for a cold chain during transport and storage. In a novel approach, sublingual administration of a powder formulation containing the modular carrier peptide J8 (VLP-J8) has been evaluated as a safe and cost-effective GAS vaccine, inducing systemic and mucosal protective immunity. FDJ8-VLPs, combined with the mucosal adjuvant cholera toxin (CT), elicited significant systemic and mucosal antibody responses, particularly beneficial for communities where GAS-related diseases are prevalent (58).

Continued research in VLP-based vaccines, leveraging the success of FDA-approved vaccines against viral and non-viral pathogens, remains crucial for preventing infections caused by important human bacterial pathogens.

5.7.1.2 VLP-based Viral Vaccines: Current Status and Future Perspective

The rise in international travel has facilitated pathogen transmission, particularly viral diseases, resulting in widespread outbreaks. Diseases such as Middle East respiratory syndrome coronavirus (MERS-CoV) and Ebola virus disease have caused significant mortality and economic damage (59). Vaccination is highly effective in preventing viral infections; however, developing vaccines against emerging viruses faces many challenges. Some emerging viruses cause high infection and mortality rates, requiring biosafety level 4 laboratories for study; therefore, live attenuated or inactivated vaccines may not be sufficiently safe. Both of these issues are of particular importance in emerging viral outbreaks (60). Additionally, some vaccines like influenza vaccines do not provide protection against all strains of the causative agent (due to antigenic drift and shift in influenza viruses). This strongly underscores the need for engineered viral vaccine platforms. Engineered viral vaccines are essential, given the inability to culture some viruses outside their natural hosts (61).

VLPs offer several advantages as vaccines against emerging viruses. They effectively display main antigens, stimulating robust humoral and cellular immune responses. VLPs, nanostructures self-assembled from viral structural proteins, possess high immunogenicity and antigenicity due to their virus-like appearance and repetitive surface structure, closely resembling parent viruses. VLPs derived from viral structural proteins are highly effective vaccines capable of eliciting both humoral and cellular responses, uncommon in traditional and subunit vaccines. VLPs are efficiently taken up by antigen-presenting cells (APCs), particularly dendritic cells (DCs), and presented to the adaptive immune system through antigen processing by MHC II molecules, activating CD4+ helper T cells. They are also processed as endogenous antigens in the cytosol of DCs and presented to CD8+ cytotoxic T cells by MHC I molecules, ensuring comprehensive and robust immune responses (62). For instance, the structural protein of HPV self-assembles into a VLP vaccine, generating antibodies that protect against HPV infection, thus preventing cervical cancer. Similarly, surface antigens of the hepatitis B virus can self-assemble as VLP vaccines, effectively preventing HBV infection (63).

Currently, it has been demonstrated that 110 viral proteins from 35 viral families have the intrinsic ability to self-assemble into VLPs. Several VLP-based vaccines for human use, including Recombivax HB and Engerix-B for hepatitis B virus (HBV); Gardasil, Cervarix and Gardasil 9 for human papillomavirus (HPV); and Hecolin for hepatitis E virus (HEV) have been introduced to the market (64). These VLP-based vaccines have advantages over traditional vaccines such as stronger immunogenicity, better safety profile, ease of large-scale production, and thermal stability. Other VLP-based vaccine candidates have also been developed and their efficacy demonstrated in animal models, but they are still undergoing preclinical testing. Promising VLP vaccine candidates include vaccines against human immunodeficiency virus type 1 (HIV-1), severe acute respiratory syndrome coronavirus (SARS-CoV), hepatitis C virus (HCV), respiratory syncytial virus (RSV), rotavirus (RV), dengue virus type 3 (DENV), hantavirus, and so on. The efficacy of VLPs as vaccine candidates against animal diseases has also been investigated. Therefore, VLP technology holds great promise for development of vaccines against other emerging viruses (62). Table 5.1 summarizes the list of VLP-based vaccines available in the market.

5.7.1.3 VLP-based Vaccines against Other Microbes: Current Status and Future Perspective

In recent years, the global population of companion animals, such as cats and dogs, has witnessed significant growth. However, infectious diseases transmitted by these animals have often been overlooked (1). Consequently, new strategies are emerging to develop vaccines targeting these common human-animal or parasitic diseases.

TABLE 5.1
Some Examples of VLP-based Vaccines

Trade Name	Infectious Agent	Target Disease	Clinical Trials Phase	Company Name
Gardasil	Human papillomavirus (HPV)	Human papillomavirus types 6, 11, 16, 18, 31, 33, 45, 52, 58	Approved	Merck
Gardasil9	Human papillomavirus (HPV)	Human papillomavirus types 6, 11, 16, 18, 31, 33, 45, 52, 58	Approved	Merck
Cervarix	Human papillomavirus (HPV)	Human papillomavirus types 16, 18	Approved	Glaxo-SmithKline Inc.
Sci-B-Vac	Hepatitis B virus (HBV)	Hepatitis B	Approved	VBI Vaccines
Still no trade name	Influenza virus	Seasonal flu	Phase 3	Medicago
Still no trade name	Influenza virus	Pandemic flu	Phase 2	Medicago
Still no trade name	Rotavirus	Viral gastroenteritis	Phase 1	Medicago
Still no trade name	SARS-CoV-2 virus	COVID-19	Phase 1	Medicago
NanoFlu	Influenza virus	Seasonal flu	Phase 3	Novavax
ResVax	Respiratory syncytial virus (RSV)	RSV	Phase 3	Novavax
NVX-CoV2373	SARS-CoV-2 virus	COVID-19	Phase 1	Novavax
Ebola GP vaccine	Ebola virus	Ebola	Phase 1	Novavax

Two noteworthy examples of VLP-based vaccines are designed to combat the *Plasmodium falciparum* parasite, the most lethal malaria parasite that affects humans:

1. Mosquirix (RTS, S/AS01): RTS, S/AS01, is currently the most developed malaria vaccine. This advanced VLP vaccine targets the pre-erythrocytic stage of the *Plasmodium falciparum* parasite. It combines a non-viral antigen, the CSP protein, with HBsAg. Clinical trials have demonstrated approximately 50% efficacy in preventing severe malaria in children aged 5 to 17 months. It underwent phase 3 clinical trials in Africa where it was tested in infants aged 5–17 months. Following additional pilot implementation studies, the World Health Organization has now recommended the vaccine for use in parts of Africa. They suggest it be given to young children who are at highest risk of getting infected with *P. falciparum* malaria, to help provide some degree of protection (63).
2. R21: R21 is currently undergoing phase 2/1 clinical trials. Unlike the RTS,S/AS01 vaccine, which presents the CSP antigen in only about 20% of its HBsAg subunits, R21 directly fuses the CSP antigen with HBsAg, resulting in a higher antigenic ratio. In mouse models, low doses of the R21 vaccine have exhibited high immunogenicity. Moreover, the incorporation of potent adjuvants such as Matrix-M and Abisto-100 has augmented protection in mice and elicited robust B- and T-cell responses upon challenge with sporozoite-infected mosquitoes. Furthermore, combining R21 with viral vector-based TRAP has significantly enhanced the vaccine's potency. Presently, R21 is in phase 1 clinical trials with the AS01 adjuvant, while R21 combined with viral vector vaccines ChAd63 and MVA encoding ME-TRAP is in phase 2 clinical trials (63).

Toxoplasma gondii, the causative agent of toxoplasmosis, is a parasite found in warm-blooded animals, including birds and marine creatures, and is prevalent in both humans and animals. It is transmitted through the consumption of raw or undercooked meat and contact with animal feces. This parasite leads to severe complications in pregnant women, immunocompromised patients, and those with AIDS. Urgent research is imperative to develop innovative vaccine candidates against this parasite. *Toxoplasma gondii* subcellular organelles, particularly Surface Antigen 1-4 (SAG1), are pivotal targets for combating toxoplasmosis. A VLP-based vaccine utilizing SAG1 and the influenza virus matrix protein was developed to induce immunity against this parasite. Evaluation in a mouse model demonstrated increased antibody levels and enhanced survival rates post-challenge. Immunization with SAG1-VLP also reduced inflammatory cytokine production, underscoring its efficacy and potential as a vaccine candidate for controlling *Toxoplasma gondii* infection (50). These endeavors contribute significantly to advancing our understanding of vaccine development and protection against infectious diseases.

5.7.2 Drug Delivery Using VLPs: VLPs in the Treatment of Microbial Disease

Although VLPs are most recognized for their immunogenic characteristics, they may be used for a variety of purposes, such as drug delivery. They can function as carriers for targeted delivery of proteins, nucleic acids, or other small molecules to specific cells, tissues, or organs via endocytosis. They can also contain peptides, proteins, or other active molecules on their outer surface. However, lysosomal degradation limits bioavailability, which makes approximately 40% of freshly manufactured pharmaceuticals unusable and poses a hurdle to efficient drug administration. Drug delivery carriers can be created using chemical and physical macromolecule-based techniques; however, these approaches frequently have drawbacks such as toxicity, insufficient cytosolic transfer, and lack of biological breakdown. On the other hand, like natural viruses, VLPs have the unusual capacity to break out from endosomes prior to lysosomal destruction. Their utilization addresses safety concerns associated with administering complete viruses, as VLPs lack the original viral genome, thereby eliminating the risk of virus replication or the formation of live viruses. Consequently, safety issues related to administering complete viruses are significantly reduced (65). The primary objective of targeted drug delivery is to directly administer a drug to a specific tissue or organ, ensuring prolonged interaction for maximal drug release. This approach is important for mitigating side effects and simplifying treatment regimens. Certain VLPs exhibit a natural affinity for specific tissues due to their origin from particular microbes. For example, VLPs derived from the hepatitis B virus (HBV) or the rotavirus show affinity for the liver or intestine, respectively, suggesting potential targeted delivery applications. However, VLPs often exhibit non-specific natural affinities, necessitating prevention of such interactions. Achieving precise targeting involves attaching receptor-binding domains to VLPs, either chemically or genetically, to identify cellular receptors, typically viral receptors, essential for the development of targeted carriers. These receptors facilitate interactions with viral surface proteins, enabling attachment to and entry into the host cell (65).

5.7.2.1 Delivery of Small Molecule Drugs for Antimicrobial Therapy

The evolving resistance of bacteria to current antibacterial treatments emphasizes the need for improved antimicrobial strategies. Among these approaches, drug delivery strategies have emerged as promising avenues, with virus-like particles (VLPs) being engineered to meet this demand. Notably, bacteriophages can be tailored to target specific microbes, facilitated by phage display techniques for selecting microbial peptide ligands. In a seminal study by Yacobi et al., filamentous bacteriophages were proposed as targeted drug carriers to eradicate pathogenic bacteria. These phages were genetically modified to feature a specific surface region capable of delivering a significant amount of cytotoxic drug to the target bacteria. *Staphylococcus aureus* served as the target bacteria, with chloramphenicol as the antibiotic. Filamentous phages linked to chloramphenicol inhibited bacterial growth compared to chloramphenicol alone, highlighting their potential as versatile drug carriers for targeted bacteria elimination (66).

In another study, various macrolide antibiotics, known for their preferential distribution in tissues with higher phagocytic cell populations, were used to target VLP Qβ to lung-resident macrophages for potential treatment of microorganisms like *Mycobacterium tuberculosis* and *Legionella*

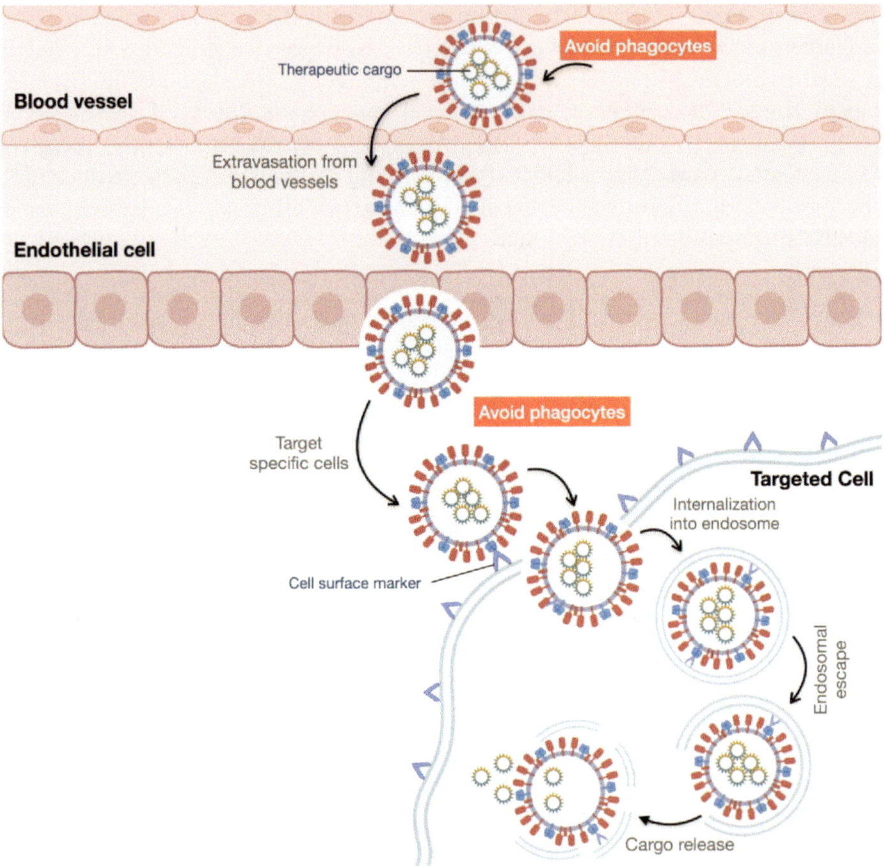

FIGURE 5.4 VLPs first leave the bloodstream and enter the surrounding tissue, targeting specific cell types. The stabilized VLPs can avoid detection by the immune system during this time. Once bound to the target cells, the VLPs are taken inside the cells through endocytosis. At this point, the VLPs must escape the endosome before it is destroyed. Following endosomal escape, the VLPs disassemble and release their cargo into the cell cytoplasm, completing the delivery of the therapeutic agent. The key steps are (1) Extravasation from blood vessels, (2) targeting of specific cell types, (3) internalization through endocytosis, (4) endosomal escape, (5) VLP disassembly, and (6) cargo release.

pneumophila. Results demonstrated enhanced uptake of VLP-macrolide conjugates by macrophage cells in culture (*in vitro*) and lung tissue cells in mice (*in vivo*) (67). The field of VLPs for drug delivery applications has advanced significantly since its inception. Continued growth and progress in this field expand the repertoire of virus-based therapeutic approaches undergoing clinical trials, promising advanced treatments in clinical settings in the near future (67, 68). Figure 5.4 illustrate the mechanism of drug delivery using VLPs.

5.8 CONCLUSION

In conclusion, VLPs represent a versatile and promising platform for various applications, including vaccine development, drug delivery, and nanotechnology. Their ability to mimic the structure and function of native viruses, coupled with their safety profile and ease of production, makes them attractive candidates for biomedical research and therapeutic interventions. The expression systems for VLP production, ranging from microbial to mammalian cell lines, offer unique advantages and challenges, each tailored to specific needs and applications. From prokaryotic systems like bacteria to eukaryotic systems like yeast and mammalian cells, researchers have a diverse array of platforms to choose from based on factors such as protein complexity, post-translational modifications, and scalability. Furthermore, the purification and characterization of VLPs are critical steps in their development and application. Techniques such as chromatography, ultracentrifugation, and mass spectrometry play pivotal roles in isolating and analyzing VLPs, ensuring their purity, integrity, and functionality. Advanced methods like dynamic light scattering and electrospray-differential mobility analysis offer enhanced precision and resolution in characterizing VLP properties, contributing to the optimization of vaccine design and production processes. Overall, the continuous advancements in VLP research and technology underscore their potential as powerful tools in medicine and biotechnology. With ongoing innovations in expression systems, purification methods, and characterization techniques, VLP-based therapies and vaccines hold great promise for addressing a wide range of health challenges and driving advancements in disease prevention and treatment.

REFERENCES

1. Mohsen MO, Zha L, Cabral-Miranda G, Bachmann MF. Major findings and recent advances in virus–like particle (VLP)-based vaccines. *Seminars in Immunology*. 2017;34:123–132.
2. Weedon MN, Schwarz PE, Horikawa Y, Iwasaki N, Illig T, Holle R, Rathmann W, Selisko T, Schulze J, Owen KR, Evans J. Meta-analysis and a large association study confirm a role for calpain-10 variation in type 2 diabetes susceptibility. *The American Journal of Human Genetics*. 2003;73(5):1208–1212.
3. Zeltins A. Construction and characterization of virus-like particles: a review. *Molecular Biotechnology*. 2013;53(1):92–107.
4. Bancroft JB. The self-assembly of spherical plant viruses. *Advances in Virus Research*. 1970;16:99–134.
5. Haynes JR, Cunningham J, Von Seefried A, Lennick M, Garvin RT, Shen SH. Development of a genetically–engineered, candidate polio vaccine employing the self–assembling properties of the tobacco mosaic virus coat protein. *Bio/Technology*. 1986;4(7):637–641.
6. Smith GP. Filamentous fusion phage: novel expression vectors that display cloned antigens on the virion surface. *Science*. 1985;228(4705):1315–1317.
7. Venters C, Graham W, Cassidy W. Recombivax-HB: perspectives past, present and future. *Expert Review of Vaccines*. 2004;3(2):119–129.
8. Delpeyroux F, Chenciner N, Lim A, Malpiece Y, Blondel B, Crainic R, Van der Werf S, Streeck RE. A poliovirus neutralization epitope expressed on hybrid hepatitis B surface antigen particles. *Science*. 1986;233(4762):472–475.
9. Hauser P, Voet P, Simoen E, Thomas HC, Petre J, De Wilde M, Stephenne J. Immunological properties of recombinant HBsAg produced in yeast. *Postgraduate Medical Journal*. 1987;63:83–91.
10. Soulie JC, Devillier P, Santarelli J, Goudeau A, Vermeulen P, Guellier M, Saliou P, Hillion AM, Tron F, Huchet J. Immunogenicity and safety in newborns of a new recombinant hepatitis B vaccine containing the S and pre-S2 antigens. *Vaccine*. 1991;9(8):545–550.
11. Valenzuela P, Coit D, Medina-Selby MA, Kuo CH, Van Nest G, Lyn Burke R, Bull P, Urdea MS, Graves PV. Antigen engineering in yeast: synthesis and assembly of hybrid hepatitis B surface antigen-herpes simplex 1 gD particles. *Bio/Technology*. 1985;3(4):323–326.
12. Vahdat MM, Hemmati F, Ghorbani A, Rutkowska D, Afsharifar A, Eskandari MH, Rezaei N, Niazi A. Hepatitis B core-based virus-like particles: a platform for vaccine development in plants. *Biotechnology Reports*. 2021;29:e00605.
13. Yap I, Guan R, Chan SH. Study on the comparative immunogenicity of a recombinant DNA hepatitis B vaccine containing pre-S components of the HBV coat protein with non pre-S containing vaccines. *Journal of Gastroenterology and Hepatology*. 1995;10(1):51–55.
14. Zhou J, Sun XY, Stenzel DJ, Frazer IH. Expression of vaccinia recombinant HPV 16 L1 and L2 ORF proteins in epithelial cells is sufficient for assembly of HPV virion-like particles. *Virology*. 1991;185(1):251–257.

15. Breitburd F, Kirnbauer R, Hubbert NL, Nonnenmacher B, Trin-Dinh-Desmarquet C, Orth G, Schiller JT, Lowy DR. Immunization with viruslike particles from cottontail rabbit papillomavirus (CRPV) can protect against experimental CRPV infection. *Journal of Virology*. 1995;69(6):3959–3963.
16. Evans TG, Bonnez W, Rose RC, Koenig S, Demeter L, Suzich JA, O'Brien D, Campbell M, White WI, Balsley J, Reichman RC. A Phase 1 study of a recombinant viruslike particle vaccine against human papillomavirus type 11 in healthy adult volunteers. *The Journal of Infectious Diseases*. 2001;183(10):1485–1493.
17. Gellin B, Modlin JF, Barr E, Tamms G. Quadrivalent human papillomavirus vaccine. *Clinical Infectious Diseases*. 2007;45(5):609–617.
18. Monie A, Hung CF, Roden R, Wu TC. Cervarix™: a vaccine for the prevention of HPV 16, 18-associated cervical cancer. *Biologics: Targets and Therapy*. 2008;2(1):107–113.
19. Nooraei S, Bahrulolum H, Hoseini ZS, Katalani C, Hajizade A, Easton AJ, Ahmadian G. Virus-like particles: preparation, immunogenicity and their roles as nanovaccines and drug nanocarriers. *Journal of Nanobiotechnology*. 2021;19:1–27.
20. Yadav T, Kumar S, Mishra G, Saxena SK. Tracking the COVID-19 vaccines: the global landscape. *Human Vaccines & Immunotherapeutics*. 2023;19(1):2191577.
21. Basta NE, Moodie EM. On behalf of the VIPER (Vaccines, Infectious disease Prevention, and Epidemiology Research) Group COVID-19 Vaccine Development and Approvals Tracker Team. COVID-19 vaccine tracker. 2020 Dec 30. Available online: https://covid19.trackvaccines.org/ (accessed on 20 June 2023).
22. Tariq H, Batool S, Asif S, Ali M, Abbasi BH. Virus-like particles: revolutionary platforms for developing vaccines against emerging infectious diseases. *Frontiers in Microbiology*. 2022;12:790121.
23. Yan D, Wei YQ, Guo HC, Sun SQ. The application of virus-like particles as vaccines and biological vehicles. *Applied Microbiology and Biotechnology*. 2015;99:10415–10432.
24. Mejía-Méndez JL, Vazquez-Duhalt R, Hernández LR, Sánchez-Arreola E, Bach H. Virus-like particles: fundamentals and biomedical applications. *International Journal of Molecular Sciences*. 2022;23(15):8579.
25. Snowden JS, Alzahrani J, Sherry L, Stacey M, Rowlands DJ, Ranson NA, Stonehouse NJ. Structural insight into Pichia pastoris fatty acid synthase. *Scientific Reports*. 2021;11(1):9773.
26. Ude CC, Miskon A, Idrus RB, Abu Bakar MB. Application of stem cells in tissue engineering for defense medicine. *Military Medical Research*. 2018;5:1–8.
27. Bundy BC, Franciszkowicz MJ, Swartz JR. Escherichia coli-based cell-free synthesis of virus like particles. *Biotechnology and Bioengineering*. 2008;100(1):28–37.
28. McGonagle D. Novel Coronavirus Information Center. www.elsevier.com/connect/coronavirus-information-center. (Since January 2020 Elsevier has hosted a COVID-19 resource centre with free information in English and Mandarin on the novel coronavirus COVID-19.)
29. Wetzel D, Barbian A, Jenzelewski V, Schembecker G, Merz J, Piontek M. Bioprocess optimization for purification of chimeric VLP displaying BVDV E2 antigens produced in yeast Hansenula polymorpha. *Journal of Biotechnology*. 2019;306:203–212.
30. Keikha R, Daliri K, Jebali A. The use of nanobiotechnology in immunology and vaccination. *Vaccines*. 2021;9(2):74.
31. French TJ, Roy P. Synthesis of bluetongue virus (BTV) corelike particles by a recombinant baculovirus expressing the two major structural core proteins of BTV. *Journal of Virology*. 1990;64(4):1530–1536.
32. Chung YC, Ho MS, Wu JC, Chen WJ, Huang JH, Chou ST, Hu YC. Immunization with virus-like particles of enterovirus 71 elicits potent immune responses and protects mice against lethal challenge. *Vaccine*. 2008;26(15):1855–1862.
33. Kibenge FS, Qian B, Nagy E, Cleghorn JR, Wadowska D. Formation of virus-like particles when the polyprotein gene (segment A) of infectious bursal disease virus is expressed in insect cells. *Canadian Journal of Veterinary Research*. 1999;63(1):49–55.
34. Chaudhary A, Yadav RD. A review on virus protein self-assembly. *Journal of Nanoparticle Research*. 2019;21:1–3.
35. Ong HK, Tan WS, Ho KL. Virus like particles as a platform for cancer vaccine development. *PeerJ*. 2017;5:e4053.
36. Sanchez-Garcia L, Martín L, Mangues R, Ferrer-Miralles N, Vázquez E, Villaverde A. Recombinant pharmaceuticals from microbial cells: a 2015 update. *Microbial Cell Factories*. 2016;15:1–7.

37. Fabich AJ, Jones SA, Chowdhury FZ, Cernosek A, Anderson A, Smalley D, McHargue JW, Hightower GA, Smith JT, Autieri SM, Leatham MP. Comparison of carbon nutrition for pathogenic and commensal Escherichia coli strains in the mouse intestine. *Infection and Immunity*. 2008;76(3):1143–1152.
38. Ghattas M, Dwivedi G, Lavertu M, Alameh MG. Vaccine technologies and platforms for infectious diseases: current progress, challenges, and opportunities. *Vaccines*. 2021;9(12):1490.
39. Liu Y, Wang C, Mueller S, Paul AV, Wimmer E, Jiang P. Direct interaction between two viral proteins, the nonstructural protein 2CATPase and the capsid protein VP3, is required for enterovirus morphogenesis. *PLoS Pathogens*. 2010;6(8):e1001066.
40. Hong JY, Chen TH, Chen YJ, Liu CC, Jan JT, Wu SC. Highly immunogenic influenza virus-like particles containing B-cell-activating factor (BAFF) for multi-subtype vaccine development. *Antiviral Research*. 2019;164:12–22.
41. Yamaji H, Nakamura M, Kuwahara M, Takahashi Y, Katsuda T, Konishi E. Efficient production of Japanese encephalitis virus-like particles by recombinant lepidopteran insect cells. *Applied Microbiology and Biotechnology*. 2013;97:1071–1079.
42. Sarkar B, Islam SS, Zohora US, Ullah MA. Virus like particles-A recent advancement in vaccine development. *The Microbiological Society of Korea*. 2019;55(4):327–343.
43. Bahar MW, Porta C, Fox H, Macadam AJ, Fry EE, Stuart DI. Mammalian expression of virus-like particles as a proof of principle for next generation polio vaccines. *NPJ Vaccines*. 2021;6(1):5.
44. Santi L, Huang Z, Mason H. Virus-like particles production in green plants. *Methods*. 2006;40(1):66–76.
45. Pushko P, Pumpens P, Grens E. Development of virus-like particle technology from small highly symmetric to large complex virus-like particle structures. *Intervirology*. 2013;56(3):141–165.
46. Minkner R, Park EY. Purification of virus-like particles (VLPs) expressed in the silkworm Bombyx mori. *Biotechnology Letters*. 2018;40(4):659–666.
47. Arevalo MT, Wong TM, Ross TM. Expression and purification of virus-like particles for vaccination. *Journal of Visualized Experiments*. 2016;(112):e54041.
48. Jarvas G, Szerenyi D, Jankovics H, Vonderviszt F, Tovari J, Takacs L, Foldes F, Somogyi B, Jakab F, Guttman A. Microbead-based extracorporeal immuno-affinity virus capture: a feasibility study to address the SARS-CoV-2 pandemic. *Microchimica Acta*. 2023;190(3):95.
49. Lothert K, Eilts F, Wolff MW. Quantification methods for viruses and virus-like particles applied in biopharmaceutical production processes. *Expert Review of Vaccines*. 2022;21(8):1029–1044.
50. Choi WH, Park JS. Immunogenicity and protective effect of a virus-like particle containing the SAG1 antigen of toxoplasma gondii as a potential vaccine candidate for toxoplasmosis. *Biomedicines*. 2020;8(4):91.
51. Crisci E, Bárcena J, Montoya M. Virus-like particle-based vaccines for animal viral infections. *Inmunologia*. 2013;32(3):102–116.
52. Roldão A, Mellado MC, Castilho LR, Carrondo MJ, Alves PM. Virus-like particles in vaccine development. *Expert Review of Vaccines*. 2010;9(10):1149–1176.
53. Bárcena J, Blanco E. Design of novel vaccines based on virus-like particles or chimeric virions. *Structure and Physics of Viruses: An Integrated Textbook*. 2013;68:631–665.
54. Yadav R, Zhai L, Tumban E. Virus-like particle-based L2 vaccines against HPVs: where are we today?. *Viruses*. 2019;12(1):18.
55. Daly SM, Joyner JA, Triplett KD, Elmore BO, Pokhrel S, Frietze KM, Peabody DS, Chackerian B, Hall PR. VLP-based vaccine induces immune control of Staphylococcus aureus virulence regulation. *Scientific Reports*. 2017;7(1):637.
56. Brandt ER, Hayman WA, Currie B, Pruksakorn S, Good MF. Human antibodies to the conserved region of the M protein: opsonization of heterologous strains of group A streptococci. *Vaccine*. 1997;15(16):1805–1812.
57. Rivera-Hernandez T, Hartas J, Wu Y, Chuan YP, Lua LH, Good M, Batzloff MR, Middelberg AP. Self-adjuvanting modular virus-like particles for mucosal vaccination against group A streptococcus (GAS). *Vaccine*. 2013;31(15):1950–1955.
58. Seth A, Kong IG, Lee SH, Yang JY, Lee YS, Kim Y, Wibowo N, Middelberg AP, Lua LH, Kweon MN. Modular virus-like particles for sublingual vaccination against group A streptococcus. *Vaccine*. 2016;34(51):6472–6480.
59. De Groot RJ, Baker SC, Baric RS, Brown CS, Drosten C, Enjuanes L, Fouchier RA, Galiano M, Gorbalenya AE, Memish ZA, Perlman S. Commentary: middle east respiratory syndrome coronavirus (mers-cov): announcement of the coronavirus study group. *Journal of Virology*. 2013;87(14):7790–7792.

60. García-Sastre A, Mena I. Novel vaccine strategies against emerging viruses. *Current Opinion in Virology*. 2013;3(2):210–216.
61. Zhao Q, Li S, Yu H, Xia N, Modis Y. Virus-like particle-based human vaccines: quality assessment based on structural and functional properties. *Trends in Biotechnology*. 2013;31(11):654–663.
62. Naskalska A, Pyrć K. Virus like particles as immunogens and universal nanocarriers. *Polish Journal of Microbiology*. 2015;64(1):3–13.
63. Qian C, Liu X, Xu Q, Wang Z, Chen J, Li T, Zheng Q, Yu H, Gu Y, Li S, Xia N. Recent progress on the versatility of virus-like particles. *Vaccines*. 2020;8(1):139.
64. Rodríguez-Limas WA, Sekar K, Tyo KE. Virus-like particles: the future of microbial factories and cell-free systems as platforms for vaccine development. *Current Opinion in Biotechnology*. 2013;24(6):1089–1093.
65. Zdanowicz M, Chroboczek J. Virus-like particles as drug delivery vectors. *Acta Biochimica Polonica*. 2016;63(3):469–473.
66. Yacoby I, Shamis M, Bar H, Shabat D, Benhar I. Targeting antibacterial agents by using drug-carrying filamentous bacteriophages. *Antimicrobial Agents and Chemotherapy*. 2006;50(6):2087–2097.
67. Crooke SN, Schimer J, Raji I, Wu B, Oyelere AK, Finn MG. Lung tissue delivery of virus-like particles mediated by macrolide antibiotics. *Molecular Pharmaceutics*. 2019;16(7):2947–2955.
68. Hajizade A, Ebrahimi F, Salmanian AH, Arpanaei A, Amani J. Nanoparticles in vaccine development. *Journal of Applied Biotechnology Reports*. 2014;1(4):125–134.

6 Nanosensors and Microelectromechanical Systems for Microbial Agents

Javeria Tanvir, Muhammad Ali Syed, Afshan Saleem, and Maimoona Sabir

6.1 INTRODUCTION

Microscopic organisms, often referred to as *microbes* or *microbial agents*, populate the invisible realm that surrounds us, constituting a bustling ecosystem on Earth (1). Microbes were first introduced by Dutch scientist Antonie van Leeuwenhoek, in the 19th century, through a captivating series of letters to the Royal Society of London (2). They encompass bacteria, viruses, fungi, archaea, and protists (3). In spite of their tiny stature and invisibility to an unaided eye, microbes play a fundamental part in shaping our planet, directly affecting numerous domains of our lives, for instance, production of food and consequently our health, industrial activities, and the whole environment (4). Some microbes pose risk and cause various diseases, but most of them are either harmless or beneficial for the overall maintenance of the health of living beings (5).

Microbial agents pose diversity within their kingdom (6). The most abundant type of microbes is bacteria, found in various sizes and shapes, commonly present in our gut which helps in digestion, and many of them cause illness (7). Viruses utilize host cells for their replication and cause numerous diseases from common to high risk, for instance, respiratory illness such as COVID-19 (8). Fungi are larger in size then rest of the microbes, constituting a minuscule filamentous structure, and take part in processes like fermentation, decomposition, production of food, and recycling of nutrients (9). Archaea play an important role in maintaining the biochemical cycles and boom in extreme environmental conditions (10), while protists are mostly single celled, and play a pivotal ecological role as photosynthesizer and also as a predator (11).

At this juncture, we advocate for new technologies for microbial detection which aid in fast and early detection of infectious diseases. Timely diagnosis prevents the escalation of pathogens, and identification and measurement of potential contamination also contribute to the preservation of lives (12). Traditional methods for microbial detection, which include cultivation of microorganisms under controlled settings to boost growth, present lot of constraints such as the necessity of skilled personnel, specialized laboratory equipment, and time-consuming procedures (13, 14). To effectively overcome these limitations and harness the favorable features of microbes, constant development of new technologies is being pursued, which includes notable innovations in nanosensors and MEMS. These technologies are significantly transforming the methodologies of the field of microbial detection (15, 16).

Ability of nanosensors to detect molecules of size one billionth of a meter i.e., at nanometer (nm) range, facilitates precise and accurate microbial detection by enhancing the binding efficiency of the biological elements, generating an electrical signal which indicates the presence of the target microbe (17). MEMS devices are tiny devices, consisting of electrical and mechanical parts, integrated on a single chip, with engraved microscopic channels for transportation of a small

DOI: 10.1201/9781003410164-6

quantity of sample microbes, endorsing probable binding and signal generation by limiting the performance time and possible error (18). In the current chapter, the authors introduce nanosensors and MEMS devices, and their applications in microbial detection and identification.

6.2 NANOSENSORS

Nanosensors offer a next-generation technology in the field of sensor technology. As our global landscape contracts, conventional sensors encounter limitations. Nanosensors open new frontiers of sensing, featuring highly sensitive resolutions, significantly enhanced precision, and miniaturization. This section explores the limitations of conventional sensors and the rise of nanotechnology in sensing, paving the way to nanosensors.

6.2.1 Sensors and the Need for Nanosensors

The history of sensors dates back to ancient times when humans used their senses to perceive the environment (19). The demands of the modern era for improved living standards began with the development of basic sensor devices, for instance, thermometers, barometers, etc., coupled with the progress of modern standards of living driven by emerging technologies and industries such as automotive, aerospace, healthcare, etc. This evolution has triggered dramatic growth in the field of sensor technology, resulting in miniaturization and mass production of sensors and sensor devices (20).

The word "sensor" is derived from a Greek word *sentire* which means *to perceive* (21). Sensors are specialized devices or instruments specially designed to detect and respond to different physical stimuli present in the environment. The fundamental function of a sensor is to first sense the changes or variations occurring in the environment and then convert them all into understandable output (22). This output could be in the form of electrical signal, physical response, or digital data. By providing real-time data, sensors empower decision-making processes and contribute to the optimization of various processes and systems (23). Various possible examples of parameters which necessitate the utilization of sensors for measurement are illustrated in Table 6.1.

In pursuit of enhanced sensor capabilities to address emerging challenges and evolving demands, nanotechnology-based sensors emerged as a solution to overcome the limitations of traditional sensors by leveraging the exclusive properties of nanosensors and all nanoscale devices (24). The limitations of traditional sensors include their application constraints due to large size, low signal to noise ratio, sensitivity, and selectivity (25). Miniaturization of nanoscale sensors allows an integration of nanosensors into a compact device, emerging as a new revolutionary technology exhibiting

TABLE 6.1
Detail of Measurands of Biosensors

No.	Sensor Class	Measurable Properties
1.	Electrical	Voltage, current, resistance, capacitance, inductance, frequency, and impedance
2.	Biological	Concentration, pH level, enzymatic activity, mass changes, and states of biomolecules
3.	Mechanical	Force, pressure, velocity, strain, acceleration, position, mass, weight, density, level, torque, and moment of inertia
4.	Thermal	Thermal conductivity, temperature, thermal response time, heat flux, and thermal resistance
5.	Chemical	Concentration, size, sensitivity, and state of elements
6.	Acoustic	Sensitivity, frequency response, frequency resolution, wave phase, wave velocity, wave amplitude, and wave polarization
7.	Magnetic	Magnetic field, magnetic flux, and magnetic permeability
8.	Optical	Wave phase, wave velocity, wave amplitude, and wave polarization
9.	Radiation	Radiation type, radiation shielding, intensity, and energy

extraordinary sensitivity to various stimuli such as presence of specific molecule (s), changes in the pressure, temperature, or any other physical activity in the surroundings (26). A compact and light-weight sensor facilitates the creation of portable devices which are crucial not only for military and aerospace but also for mobile and palm-top consumer goods (27–29).

Sensors fabricated from nanotechnology base methods are identified as "nanosensors" (30). They own structures which have at least one dimension in nanoscale range i.e., from 1 to 100nm (31). Properties of nanosensors include size of sensor in nanoscale, sensitivity of sensor in nanoscale, or spatial interactions between object and sensor in nanoscale (32). Nanotechnology provides avenues for tailoring the properties of sensors, for instance, their surface engineering, morphology, and electronic structure to enhance the response speed and selectivity of the sensor, which is the top demand in the realm of emerging technologies. Moreover, with ongoing advancements in nanotechnology, the incorporation of nanomaterials into the sensor platform is poised to spur further innovations in sensor technology and meet evolving challenges in sensing and detection (33). Furthermore, starting from an individual molecule and building up to a complex biological system, and utilizing nanoscale phenomena such as quantum confinement effects, molecular recognition, and surface plasmon resonance, nanosensors can achieve remarkable performance across a broad range of targets as their applications span from medical care and environmental monitoring to industrial process control and consumer electronics (34).

6.2.2 Commonly Used Materials for Nanosensor Fabrication

Nanomaterials encompass a broad range of nanostructures and nanocomposites, which have transformed the fabrication of sensors across multiple fields (35). These materials exhibit remarkable properties and functionalities, facilitating the creation of sensors of high strength, fast chemical sensitivity response, precise selectivity response, and high conductivity (36). Here, we explore different nanomaterials commonly used in sensor fabrication along with examples showing their diversified applications.

6.2.2.1 Graphene

Graphene is a two-dimensional honeycomb lattice, made up of a single atom planner layer of carbon atoms bonded together by a sp^2 hybrid bond resulting in intermeshed hexagonal rings of carbon atom. High surface to volume ratio (s/v), excellent conductivity and chemical stability make graphene an attractive material for sensor fabrication (37). Graphene-based biosensors offer rapid and more precise detection of biomolecules such as DNA, proteins, and viruses. Functionalized surface of graphene facilitates specific molecular recognition, enabling applications in medical diagnostics, environmental monitoring, and food safety (38). (Figure 6.1)

6.2.2.2 Nanostructures

The particle size of nanostructures is in nanometer scale, or at least their one dimension is in nanometer range (39). Performance of any material or device can be controlled by altering its dimension(s). The significance of these nanostructures lies in their extremely small size, which considerably increases the surface to volume ratio of particles, thereby intensifying the effectiveness of traditional sensors (40). Nanostructures include nanoparticles (NPs), nanowires, nanorods, nanostars, and many other nanoscale structures, each holding significant potential in the realm of nanosensors (41). NPs consists of three layers; the outermost layer is capable of being modified with different kinds of chemical compounds; the middle layer differs chemically from the central core, which typically represents the nanoparticle itself. Based on attributes like material, shape, size, and properties, NPs are of six types: metallic NPs, polymeric NPs, carbon-based NPs, lipid based NPs, magnetic NPs and dendrites (42). Figure 6.2 illustrates different types of NPs based on their shapes and compositions. The size-dependent properties of these NPs enable versatile sensor designs.

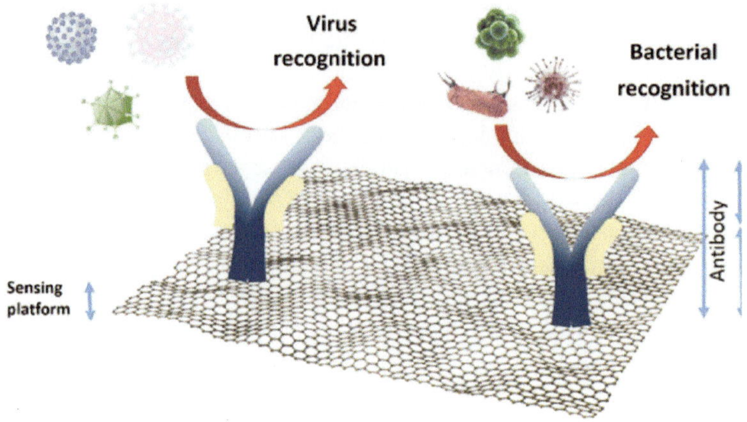

FIGURE 6.1 Illustration of graphene-based immunosensor for pathogen detection (124).

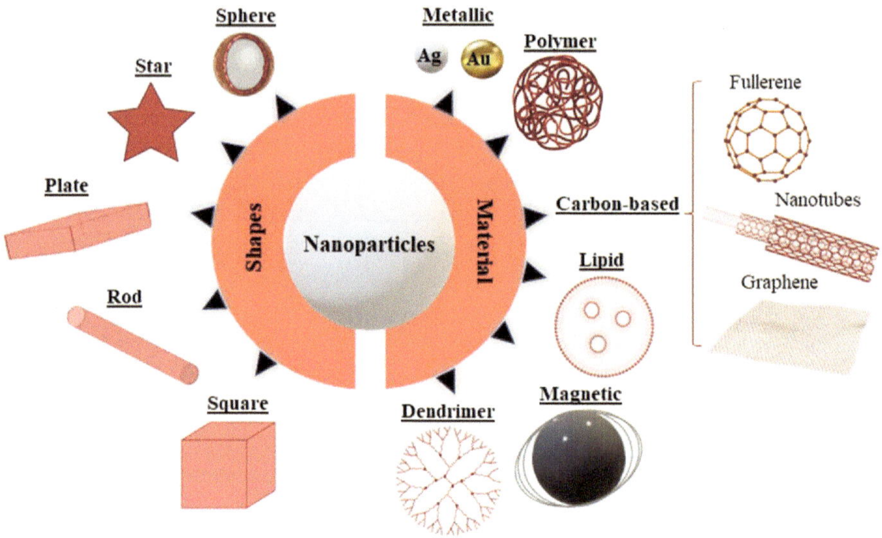

FIGURE 6.2 Illustration of different possible shapes and types of nanomaterial. A variety of nanoparticles catagorized on the basis of their shapes (like nanosphere, nanostar, nanoplate, nanorod, nanosquare) and the materials (like metallic NPs, polymeric NPs, carbon based NPs, lipid based NPs, magnetic NPs and dendrites).

6.2.2.3 Carbon Nanotubes (CNTs)

CNT is a rolled-up sheet of graphene, which is a single-atom carbon layer in hexagonal lattice (43). Their unique graphene-based structure (delocalized electrons across the structure), quantum mechanical effects (ballistic transport), and chirality (44) provide less electrical resistivity and long-range transfer of electrons (45). Functionalizing CNTs with specific receptors enhances their selectivity toward target analytes, enabling detection at low concentrations (46). They exhibit high length-to-diameter ratios and large surface areas which make them an ideal candidate for sensor applications. CNT-based gas sensors demonstrate remarkable sensitivity to gases such as ammonia (NH_3), carbon monoxide(CO), and nitrogen dioxide (NO_2) (47).

6.2.2.4 Quantum Dots (QDs)

QDs are zero-dimensional semiconductor NPs exhibiting quantum mechanically confined charge carriers in all three dimensions (48). The ability to adjust band gap over a wide range by tuning the dot size, and extremely small semiconductor QDs, offer benefits over a conventional device which makes them a promising material for sensing applications. By modifying the surface with biomolecular receptors, it becomes possible to precisely target cellular structures, facilitating the detection of multiple targets with high sensitivity (49).

6.2.3 Integration of Nanosensors with Other Technologies

Integration of nanosensors with other technologies further facilitates them to address complex challenges, and remarkably increases their number of applications. This section explores how integration of nanosensors with other technologies involves gathering a large amount of data for speedy worldwide communication.

6.2.3.1 Artificial Intelligence (AI)

Due to the high sensitivity of nanosensors, they generate a great deal of data and hold immense potential in various fields, for instance, health and medical care, industrial processes, and environmental monitoring. Integration of nanosensors with AI further amplifies the efficacy and ramification of nanosensors. AI techniques and algorithms analyze and process the data from nanosensors, which discern anomalies and minute changes, to create predictive models and make real-time decisions, helping with early decision-making and quality standards (50).

6.2.3.2 Internet of Things (IoT)

IoT technology, its networks and platforms, are recognized as a key technology that interconnects devices, facilitating them in device-to-device communication in everyday businesses. The amalgamation of IoT with nanosensors transforms the surrounding objects into interconnected elements within an extensive and potent network, functioning through the unrestricted exchange of data, consequently greatly streamlining the communication and sharing of worldwide information (51).

6.2.3.3 Internet of Nano-Things (IoNT)

Similar to IoT, there is a new technology called Internet of Nano-things (IoNT) composed of nano-sized interconnected devices like nanosensors, nanorobots, and nano machines, fitting within very small volumes in the network (52). IoNT has revolutionized IoT into a more powerful network with digitized procedures and connectivity with even larger systems (53). IoT and IoNT are collectively named as Internet of Everything (IoE) (54).

6.2.3.4 Edge Computing and Blockchain

Integration with blockchain promises improved security of data, veridicality, and integrity, through its ability to provide impervious recording and traceability of sensor data. Edge computing includes data processing nearer to its origin within the network, significantly enhancing the response time, conserving bandwidth, and magnifying the real-time analysis of data and decision-making in nanosensor implementations (51, 55).

6.2.3.5 Quantum Computing

Quantum computing helps in boosting computational capabilities, which holds immense potential for enabling the more complex simulations and research analysis necessary for handling the large amount of data generated by nanosensors (56).

6.3 NANOSENSORS FOR MICROBIAL DETECTION

Nanosensors can be tweaked and constructed for microbial sensing and chemical recognition. Because of their diverse optical properties, nanosensors have been compared to nanoprobes and nanozymes in microbial detection (57). The main focus of nanosensors is their ability to transform chemical data from biomolecules in a small environment into signals that can be detected. For point-of-care applications, pathogenic agent detection is crucial. Despite their limitations in terms of stability and utilization, several techniques, such as enzyme-linked immunosorbent assay, polymerase chain reaction (PCR), reverse-transcription polymerase chain reaction (RT-PCR), and nanosensor technologies can detect viruses. As more reliable, affordable, and selective nanosensor technologies become more urgently needed for quick diagnosis, several recognition molecules have been studied to create sensing devices. Transducers and recognition components make up nanosensors. The main idea is to capture the target and convert signals into reactions. In the preceding decades, nanosensors have become more important for the identification of analytes in environmental monitoring, food processing, clinical diagnostics, and the battle against bioterrorism. These analytes include explosives, proteins, nucleic acids, carcinogens, bacteria, viruses, and toxins (58).

The three primary components of a nanosensor are a transducer, a receptor, and a detector with a digital output. The biological detecting component uses a process to identify a biological molecule once the target molecule and receptor make contact. The transducer then transforms the changes into a signal that the detector can measure (59). Numerous capabilities are provided by nanosensors, such as remarkable performance, high sensitivity and specificity, quick reaction, compact size, portability, and real-time analysis.

6.3.1 Applications of Nanosensors in Viral Detection

There are a number of applications of nanosensors for virus detection, a few of which are given below.

6.3.1.1 Human Immunodeficiency Virus

One type of lentiviruses that belongs to the retrovirus family is the human immunodeficiency virus (HIV). Since lentiviruses are slow-moving viruses, there is a delay between the onset of the disease and the appearance of symptoms. After entering the bloodstream, HIV infects $CD4^+$ T cells and starts to reproduce (60). One of the major public health issues that results from this is AIDS. An electrochemiluminescence nanosensor for HIV-1 gene detection was studied by Babamiri et al. (2018) (61). They saw a notable increase in signal after the hybridization experiments. They were able to achieve a sensitive detection in the 3.0 fM to 0.3 nM range. When compared to noncomplementary sequences, they demonstrated the high specificity of their nanosensor (61). Similarly, a nanosensor was developed by Lu et al. (2012) (62) to measure HIV-1-related glycoprotein 41 (Gp41). They altered the piezoelectric nanosensor's surface using a synthetic peptide containing Gp41 residues 579–613. They showed that a nanosensor surface can bind Gp41 selectively and has a great affinity for the target peptide. The detection limit was determined to be 2 ng/mL (62). A novel optic nanosensor for HIV-1 detection was demonstrated by Shafiee et al. They reported that, with a resolution of 10 pM, the target was adsorbed and caused a shift. They also investigated a range of 104 to 108 copies/ml (63).

6.3.1.2 Hepatitis B Virus

One of the primary viruses, thought to be responsible for over a million deaths annually from cirrhosis and malignant liver development, is the hepatitis B virus (HBV). Furthermore, liver cirrhosis, hepatocellular cancer, or liver failure will affect 15–40% of infected patients, and tragically, 15–25% of them will pass away.

For identification of HBV using electrochemical impedance spectroscopy, a study on the hybridization of DNA was reported by Hassen et al. (2008) (64). Firstly, they used magnetic nanoparticles to modify DNA probes and then immobilized the nanoparticles within a gold electrode. They demonstrated good DNA immobilization and hybridization with varying amounts of complementary DNA after the characterizations. Additionally, they demonstrated that their nanosensor was able to detect 50 pmol of HBV DNA and attained saturation at 12.65 nM (64). Using an optic nanosensor, Uzun et al. (2009) also found an HBV surface antibody in human serum. They used human serum samples that were positive for HBV surface antibodies to conduct kinetic experiments. They demonstrated that this nanosensor followed the Langmuir adsorption isotherm model and computed the detection limit value (65).

6.3.1.3 Human Papillomavirus

Cervical cancer, the second most common type of cancer in women, has been linked to the human papillomavirus (HPV). Patients can receive treatment at an early stage when they are diagnosed early. Therefore, even with binary positive and negative indications, diagnostics have a high degree of clinical utility.

A microfluidic nanosensor was created by Inan et al. (2017) to identify HPV-16 E7 antibodies. They observed 2.87 ng/mL of detection. When compared to control samples, they produced higher responses after validating the nanosensor in serum samples. They demonstrated how this nanosensor can be used as a pretesting instrument for widespread cancer surveillance linked to HPV (66). A colorimetric nanosensor to screen for synthetic HPV was disclosed by Teengam et al. (2017). They created a multiplex paper-based nanosensor that relies on DNA. By measuring the color shift of silver nanoparticles, which produced a detection limit value of 1.03 nM, the targets were identified. When compared to single-base mismatch, two-base mismatch, and noncomplementary DNA targets, the nanosensor showed excellent selectivity for complementary oligonucleotides (67).

6.3.1.4 Influenza Virus

Influenza is a contagious illness caused by the influenza virus, that is thought to be the root cause of several clinical problems and significant economic disruptions. An article regarding the immobilization of DNA using carbon nanotubes for influenza detection was published by Tam et al. They altered a DNA probe on the surface of the nanosensor and described the basic interaction. They discovered that the limit of detection was 0.5 nM after detecting both the target DNA and the DNA probe hybridization (68). Vollmer et al. investigated the use of optics in influenza virus detection. They conducted single-viron binding studies using discrete shifts in the whispering-gallery mode's resonance frequency. They also used virus-sized NPs to support the sensing method (69).

6.3.1.5 Other Viruses

A paper by Jin et al. (2018) describes a virus diagnostic tool that uses a microfluidic system, an optic nanosensor, and other components to identify human adenovirus. In 30 minutes, they found ten copies of the human adenovirus. Additionally, they confirmed the tool's clinical applicability and stated that it offered a simple, quick, inexpensive, and sensitive detection with a short test time (70). A genetically encoded switch-on fluorescent nanosensor with an adenoviral protease cleavable site made of cyclized green fluorescent protein (cVisensor) was demonstrated by Guerreiro et al. (2019). They started by optimizing the nanosensor, established virus detection through stable expression in mammalian cells, and used it for adenovirus infection live-cell monitoring. 48 hours after infection, they used cVisensor cells to create a flow cytometry-based assay, and the results indicated that the limit of detection was 105 infectious particles/mL (71).

6.3.2 Optical Nanosensors for Biofilm Detection in the Food Industry

Bacteria living in extracellular polymeric substances (EPS) on the contact surfaces of food-processing facilities are known as biofilms. Foodborne bacterial infections encapsulated in EPS are the primary cause of food contamination and a significant health risk to humans. Spectral and conventional microscopic imaging techniques must be developed in order to study the intricate, dynamic, and microcrystalline biofilms. Recently, high-sensitivity and high-spatial-resolution nanosensors have been developed for biofilm optical detection at nanoscale range. These nanosensors can convert biochemical information into optical signals (72).

The capacity of foodborne bacteria to build biofilms determines how long they will remain on surfaces in contact with food. Common foodborne pathogenic species, such as *Escherichia coli* (*E. coli*), *Listeria monocytogenes* (*L. monocytogenes*), *Salmonella* spp., *Pseudomonas aeruginosa* (*P. aeruginosa*), *Pseudomonas fluorescens*, *Staphylococcus aureus* (*S. aureus*), *Vibrio parahemolyticus* (*V. parahemolyticus*), *Bacillus cereus*, and *Cronobacter sakazakii*, can form environmental-resistant biofilms. Other food-processing-related bacterial species that can produce biofilms include *Lactobacillus* spp., *Shewanella* spp. and *Pectobacterium carotovorum*.

6.3.3 Nanosensors for Detection of Human Fungal Pathogens

Infections brought on by human fungal diseases are becoming more and more commonplace worldwide, and they claim around 1.6 million lives each year. The increased use of immunosuppressive medications, antifungals, and climate change, which promotes fungal proliferation, are blamed for the rise in illnesses caused by fungal pathogens. Efficient identification of fungal infections is crucial in reducing their potential harm to the general public health. The current standard methods for detecting fungal diseases involve identifying the causal organism by laboratory techniques after symptom diagnosis. Traditional culture-based techniques take a long time, while techniques like PCR, enzyme-linked immunosorbent assay (ELISA), high-performance liquid chromatography (HPLC), and others call for expensive equipment and knowledgeable workers in lab settings, which delays diagnosis. By making equipment more portable and reducing the size of detecting platforms, nanotechnology can improve diagnostics. Including nanoscale aspects in the physical and material sciences can greatly aid in the creation of nano-tools that enhance handling, signal detection, and sample processing. Portable tools, including lab-on-a-chip, lateral flow assays, nanopore instruments, nanoarrays, nanobarcodes, and nano-detection kits, will improve the quick on-site identification of human fungal infections and the management of those infections in the healthcare industries (73).

For the purpose of detecting *Aspergillus fumigatus*, a magneto-nanosensor biochip was created using arrays of massive magnetoresistive spin-valve sensors (74). The device used secondary antibodies labelled by magnetic nanoparticles as the recognition element, and an antibody coating to capture the protein biomarkers. The chip's 100 pg/mL detection limit allowed it to pick up on minute quantities in solutions. In order to disturb the fluctuating external magnetic field that is captured during the experiment, the magnetic nanoparticles produce a stray magnetic field in real time. When the magnetic nanoparticles attach to the sensor surface, the giant magnetoresistive sensor (GMR) effect modifies spin-valve resistance, resulting in the measurement of the particular biological signal. These magneto-nanosensor biochips show promise as accurate tools for diagnosing fungal infections.

6.3.4 Nanosensors for Microbial Detection in Soil

Since the soil microbiome has an impact on human health in addition to plant growth, researchers have been paying more and more attention to its detection (75).

Because most conventional methods for detecting soil microbiomes depend on culture and the soil contains a wide variety of microbial species that are not culturable in a lab setting, these methods have limited real-time applications. This problem has been resolved, and sensor-based technology has also made use of the special physicochemical characteristics of nanomaterials (76). A significant portion of these tactics relies on the optical measurement of the chemical interaction and its subsequent translation into electrical impulses. Surface plasmon resonance (SPR), fluorescence-based nanosensors, and colorimetric sensors are among the optical methods used to identify the soil microbiome. By measuring or comparing compounds in colored solutions, colorimetric nanosensors make effective use of nanoscale materials for the qualitative or quantitative determination of an analyte. The outcomes are also readily visible to the unaided eye. Even though it is less advanced, this method for microbiological detection is quick. However, SPR-based nanosensors make use of the SPR phenomenon, which is triggered by a corresponding resonant wavelength of light and causes the conduction electrons of metal nanoparticles to oscillate collectively. As a result, a discernible shift in the SPR signal occurs, which can be effectively utilized to identify biomolecular binding interactions. In contemporary biosensing technologies, it is a commonly employed sophisticated tool for the identification of diverse microbial communities.

The field of clinical diagnostic systems research is vast, with numerous unresolved issues that must be resolved in order to produce and market nanosensors. Particularly in clinical applications, they should be simple, stable, robust, inexpensive, and have a high detection capability. A number of analytes have already been found to show the versatility of the nanosensors, and nanosensor feasibility also appears to be moving past the proof-of-concept stage. In some specific instances, the usefulness of nanosensors is strengthened by validation using real samples in clinical circumstances. Although various portable nanosensors have lately been investigated at the research level, there are still some questions to be answered. The deployment of mobile health diagnostics in developing nations has been aided by a notable increase in smartphone technology, albeit at a rather sluggish pace.

6.4 NANOENGINEERED SYSTEMS FOR BIOMOLECULAR ANALYSIS

There are several possible uses for nanotechnology in the biomedical sciences and medicine, including therapy and molecular-level diagnostics of diseases thanks to the special properties of nanomaterials. When it comes to creating the next wave of medications, nanoengineered materials have enormous promise (77).

NPs can be classified into many types according to the type of materials (as discussed in section 6.2.2). The most widely used and researched NPs in diagnostic applications are gold, magnetic, and fluorescent ones. The ability to design polymer particles with desired properties and our growing understanding of how they interact at the bio–nano interface have sparked interest in their potential uses in biomedicine. These applications include, for instance: (i) localized bioreactions; (ii) vaccination and immunostimulation; (iii) medication and gene delivery; and (iv) detection of bacteria and viruses. Templated polymer particles offer a distinct advantage in drug and gene delivery due to the precise control achievable over particle attributes. These carriers can administer and release drugs in response to variations in redox potential, pH levels, or enzyme concentrations (78).

Nanomaterials have become revolutionary components in many fields of study, with significant potential as advanced nano electrical biosensors to identify biomolecules. Nanostructured materials are very promising because of their remarkable sensitivity and simple detection techniques. CNTs, NPs, nanopores, nanowires, nanoclusters, and graphene have all been effectively utilized in sensor development (79). Wet chemistry and flame pyrolysis are used to manufacture huge amounts of metallic NPs, such as Au, Ag, Pt, and Zn and their oxides, each year (80). Because of their antioxidant, anti-inflammatory, and antibacterial qualities, zinc oxides are widely synthesized for use in biomedicine. The production of magnetic nano antioxidants, which are controlled by an external magnetic field, requires engineered iron nanoparticles just as much (81, 82). Enhanced biocompatibility can be achieved by using tailored gold nanoparticles as coating materials and supports due to their self-assembling characteristics. Antioxidant activity is also evident in carbon-based nanostructures like fullerenes and nanotubes. In conclusion, nanosilica is a common inert matrix that is added to biomimetic nano antioxidant structures as a coating or support matrix to improve low toxicity, stability, and biocompatibility (83).

Recent studies have highlighted nanoparticles' potential as labels for detecting biomolecules. Compared to current methods using fluorescent labels for DNA detection and ELISA for protein detection, NPs offer improved sensitivity, selectivity, and the ability to detect multiple targets simultaneously. Among these, gold NPs are widely studied and versatile. QDs can be detected similarly to gold NPs and are good alternatives to organic fluorophores. Although silica NPs show promise, further exploration is needed to fully utilize them as labels. The choice of material will depend on how easily they can be made and handled. Among detection techniques, colorimetric and direct electrical approaches are the simplest, but sensitivity can be improved. Optical methods require bulky and expensive equipment, while electrochemical methods are becoming more popular due to their simplicity, low cost, and portability. However, current electrochemical amplification methods often involve multiple steps, balancing simplicity and sensitivity (84).

6.4.1 Nanoengineered Metallic Dendrite for Clinical Biomolecular Analysis

In recent years, metallic dendrites, a type of three-dimensional nanostructured material, have garnered considerable attention owing to their intricate hierarchical arrangements and unique attributes. They manifest as hierarchical arrays comprising primary, secondary, and terminal branches adorned with numerous edges, ridges, and pointed ends, endowing them with significantly expanded active surface areas. This vast surface area results in heightened catalytic activity and conductivity compared to alternative nanomaterials, making them increasingly favored in sensor fabrication.

Macromolecules are large molecular mass polymeric complexes composed of thousands of monomers. But not all macromolecules have a polymeric structure. DNA, RNA, and proteins represent prevalent examples of polymeric macromolecules, whereas lipid components and macrocycles are characterized as non-polymeric entities. Within biological systems, macromolecules play vital roles as biomarkers, hormones, and biocatalysts. This section outlines various biosensors utilizing nano dendrites for the detection of macromolecules.

Cardiac troponin I (cTnI) stands out as the leading macromolecular biomarker for identifying myocardial damage and necrosis. A sophisticated immunosensor, demonstrating high sensitivity and selectivity, has been created utilizing cutting-edge nanomaterials known as porous trimetallic gold/platinum/palladium fluffy nanodendrites (AuPtPd FNDs).The enhanced catalytic activity of AuPtPd FNDs is attributed primarily to two factors: firstly, the synergistic activity of the tri-metals expedites electron transport, and secondly, the dendritic nanostructures of these metals provide a large surface area with numerous active sites, thereby boosting catalytic activity. In the foreseeable future, this biosensor holds promise for detecting various cardiac biomarkers in real samples (85).

The global proliferation of diabetes presents a significant public health concern. There is a critical need to create a durable, dependable, swift, precise, and user-friendly device tailored for diabetic individuals. A biosensing platform for glucose detection, devoid of templates, has been created. The process involves polishing a graphite rod (GR) electrode with fine emery paper, followed by the electrochemical deposition of gold nanodendrites using a solution comprising HAuCl4 in potassium nitrate (KNO3). This method yields a linear detection range spanning from 0.1 to 9.97 mmol/L for glucose (86).

The presence of toxic chemicals in food poses significant health risks, with melamine and thiram being particularly hazardous. Melamine, a triazine compound, can lead to severe foodborne illnesses including renal failure and kidney stones, while thiram can cause vomiting, nausea, and diarrhea. Biosensors utilizing silver nanoparticles decorated on gold nanodendrites can accurately detect melamine and thiram levels in food products. Thiram and melamine detection in apple juice and milk has been conducted using real samples. Looking ahead, the gold Au nanodendrite (Ag-AuNDs) substrate shows promise as a potential platform for enhancing food safety measures (87).

Heavy metal ions like mercury (Hg (II)), lead (Pb (II)), and copper (Cu (II)) pose significant health risks, leading to various health issues, including neurological disorders like brain function loss and Wilson's disease. Therefore, assessing the levels of these metal ions in water is vital for both human health and environmental evaluation. A sensor utilizing carbon fiber cloth (CFC) embedded with gold Au nanodendrite (AuND@CFC) sensor has been developed, capable of concurrently detecting concentrations of Cu (II), Hg (II), and Pb (II) in water samples (88).

Carcinoembryonic antigen (CEA) is a vital biomarker associated with several cancers, including colon cancer, lung cancer, and pancreatic cancer. Its timely and sensitive detection is essential for prompt clinical diagnosis and treatment. A conducting polymer, polydopamine, and bimetallic nanodendrites were employed to develop an electrochemical aptasensor for measuring CEA levels. Initially, polydopamine@graphene oxide (PDA@Gr) was prepared and applied onto a glassy carbon electrode (GCE), followed by the introduction of aptamer 1 (Apt1) to create Apt1/PDA@Gr (89). Carbohydrate antigen 15-3 (CA15-3) is a prominent marker for diagnosing breast cancer. An immunosensor without the need for labels was created using directing growth agents. This involved synthesizing platinum and cobalt alloyed nanodendrites (PtCo NDs) through a solvothermal process, with L-carnosine acting as a codirecting agent. This dendritic structure provides a large surface area and enhances catalytic active sites for oxygen reduction reactions (85).

Nucleic acid, a crucial macromolecule essential for storing and expressing genetic information, acts as life's genetic blueprint. An electrochemiluminescence (ECL) emission sensor has been developed, employing a DNA tetrahedral scaffold to adjust the distance between cadmium-tellurium nanocrystals (CdTe NCs) and gold nanodendrites (AuNDs), enabling efficient nucleic acid detection. The ECL biosensor utilizing nanodendrites exhibited outstanding sensitivity and selectivity when tested in serum samples. This success suggests promising prospects for detecting additional biomarkers with similarly high sensitivity, opening new avenues for diagnostic applications (90).

Prolactin (PRL), also known as lactotrophin, plays a crucial role in various health conditions. Polycystic ovarian syndrome (PCOS), prolactinoma, and hyperprolactinemia are linked to prolactin, but low PRL levels are linked to a higher risk of metabolic diseases such as insulin resistance and glucose intolerance. Using a metal-organic framework (MOF), a very sensitive sandwich electrochemical biosensor was created to detect prolactin. This immunosensor exhibits superior specificity, sensitivity, and improved stability, indicating significant potential for clinical diagnostics (89).

6.4.2 Nanoengineered Nano Antioxidants

Inspired by natural antioxidants, artificial nano antioxidants mimic their radical scavenging mechanisms using surface functional moieties. Nanoengineering offers diverse fabrication pathways for advanced structures, which hold potential in medicine and biotechnology for preventing or treating cancer, aging-related issues, and neurodegenerative diseases (91). Nanoceria, a rare earth oxide, exhibits the ability to capture, store, and release oxygen from its surface, facilitated by the presence of stable Ce^{3+} and Ce^{4+} oxidation states. This self-regenerating capability, termed oxygen storage capacity (OCS), enables redox-regulated conversion and enzyme-mimicking activity, making nanoceria promising for various applications including antioxidant, antiradical, antibacterial, and anticancer activities (92).

Fe_2O_3 NPs play a crucial role in biomedical technology. A straightforward hydrothermal precipitation method was utilized to create an Fe_2O_3/C hybrid nano antioxidant, incorporating iron oxide NPs and graphitic carbon encapsulation, which endows it with intrinsic superparamagnetic properties (93). Zinc oxides are widely used in biomedicine because of their antibacterial, anti-inflammatory, and antioxidant characteristics. Similarly, engineered iron NPs are significant in the production of magnetic nano antioxidants (81)

6.5 MICROELECTROMECHANICAL SYSTEMS (MEMS)

At the present time modern electronic devices, for instance smartphones, are designed employing a number of MEMS (94). In industrial manufacturing, MEMS began to appear in various commercial products, for instance, a MEMS oscillator is utilized for frequency control in many electronic devices (95). MEMS microphones find out the location of sound source, force sensors are utilized to control the brake system of vehicles, blood pressure sensors are utilized for blood pressure measurements, thermal sensors are utilized to measure variation in the surrounding temperature, accelerometers are employed to detect changes in motion and acceleration of vehicles, and inertial sensors are utilized for motion-sensing in many digital devices (96, 97).

MEMS are a huge dome under which there is a mobile tool box of transduction mechanisms merged with advanced techniques for miniaturized systems at microscale and a physical product built by the integration of many micromachined components into a single system, which controls the physical parameters using computation (98). It is a tiny device which has a complex structure, possessing numerous electrical and mechanical microstructures, microsensors, and microactuators which first detect and then convert the real-life signal or energy into other readable forms (99).

The changes in scaling from macro scale to micro scale make a huge impact on the physical and chemical properties and function of any device: the smaller the scale, the more complex functions the manufactured device can perform, allowing it to be utilized in a wide array of applications (100). Due to microlevel scaling, MEMS have more complex designs with dynamic functions which possess higher sensitivity, low power consumption, and mass production of items as compared with the macroscale devices (101, 102).

6.5.1 Miniaturization Techniques

Out of multivarious techniques used for MEMS fabrication, the core idea is to use light and to draw a tiny blueprint on silicon wafer. To build MEMS, we need a clean silicon wafer onto which a layer of special mask, i.e., a photosensitive material, is deposited. This process hardens specific areas and leaves a shadow on the silicon surface. Then excessive material is removed, engendering a unique pattern on the wafer. Multiple processing steps can be employed to get a tiny pattern which acts as a tiny machine used to augment the MEMS device layer by layer. As a part of the final testing phase, the whole assembly undergoes final packaging (103, 104). The basic manufacturing steps of MEMS are illustrated in Figure 6.3.

6.5.2 Integration of MEMS with Other Technologies

Integration of MEMS with other technologies further maximizes their potential impact by increase the number of applications. This section explores the integration of MEMS with other technologies, which helps them gather a large amount of data for speedy worldwide communication.

6.5.2.1 Integrated Circuits (ICs)

MEMS holds huge potential to sense and control various physical, chemical, and optical properties. The raw output of MEMS is electrical signal; its proper utilization requires processing, which involves filtering, amplification, and conversion of output data. Meanwhile, ICs acts as a main processor which first analyzes the output sensor data and then derives MEMS actuators. Hence, this integration gives a complete microsystem, boosting overall MEMS functionality (105).

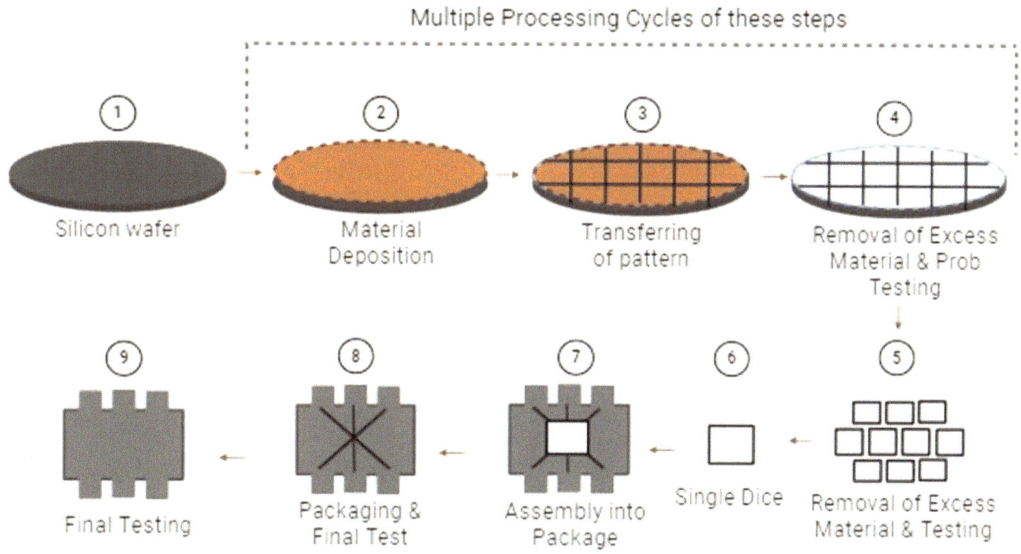

FIGURE 6.3 Illustration of microelectromechanical micromachining. This fabrication technique involves a precise, layer-by-layer creation of a MEMS device through the controlled removal of excess silicon wafer material using repeated etching steps.

6.5.2.2 Internet of Things (IoT)

MEMS, which are compact miniaturized machines composed of sensors and actuators, are perfect for capturing the data from their surroundings. The basic function of IoT is to enable and empower everyday systems to collect and share received data. Hence, this integration creates powerful devices capable of recognizing, gathering, and worldwide sharing of data (106).

6.6 MEMS FOR MICROBIAL DETECTION

MEMS are miniature systems that incorporate both mechanical and electronic components to develop analytical devices. Biomedical microelectromechanical systems (BioMEMS) have been found fascinating by microbiologists due to their unique features such as higher sensitivity, lower detection limit, portability, and reduced assay time.

6.6.1 BioMEMS for Microbial Agents

Since diagnostics is an indispensable component of healthcare systems, enormous attempts have continuously been made to develop robust and improved devices at reduced cost. Diagnosis of infectious diseases is hampered by time taken by diagnostic tests. In many cases microbial agents needs to be cultured in the lab, requiring specific media and growth conditions and sophisticated equipment. Alternative techniques such as nucleic acid amplification-based assays, are also time-consuming, requiring trained personnel and expensive equipment, which may not be available in many diagnostic laboratories, especially in developing countries. Furthermore, many types of microbes are either slow-growing (e.g., *Mycobacterium tuberculosis*, *Bordetella pertussis*, *Brucella* spp. etc.) or unculturable in the lab (e.g., *Treponema pallidum*). Similarly, antimicrobial sensitivity testing is frequently performed to determine the choice the appropriate drugs to cure infections. Therefore, there has always been a need of reliable, rapid, and cost-effective techniques and devices for microbial detection and identification (107, 108).

BioMEMS or lab-on-a-chip (LOC) systems have gained great popularity in the last three decades and have been found to be promising tools for accurate rapid, and point-of-care microbial detection systems as well as for determining antimicrobial sensitivity testing that usually requires long protocols and expertise (107–109). These assays may possess a wide range of applications in microbial detection and identification. In addition to diagnostics, BioMEMS may be promising tools for detection and identification of infectious agents and their toxins in food, the environment, and livestock, as well as drinking water samples. An important application of these devices may be detection of biological warfare agents.

6.6.2 BioMEMS-based Multiplexed Assays

One of the key features of BioMEMS-based assays is detection of different types of microorganisms and their components (e.g., toxins) in the same device in real time. Several groups have reported multiplexed assays using MEMS devices for bacteria and viruses. In addition, such systems offer tremendous advantage over other systems by requiring a very small quantity of sample (Usually nanoliter or microliter volume) as well as possibility of signal enhancement (109, 110).

Recent research has shown that these BioMEMS-based state of the art devices are not only suitable for use in routine diagnostic facilities, but also convenient for remote areas of developing countries or epidemic regions (108, 110).

Developing such devices requires highly a multidisciplinary approach involving a number of scientists from electronics, chemistry, material science, microbiology, material engineering, etc. Dedicated collaborative research carried out by different groups has resulted in breakthrough discoveries in the recent past which include accurate microbial detection from clinical samples with a detection limit of as low as a single cell (111). Integration of photonics, artificial intelligence, and quantum are expected to make further breakthroughs in the near future which may further revolutionize the available techniques of microbial identification (112, 113). BioMEMS-based microfluidics chips have also been employed for testing permissible cell lines to culture viruses in the lab. This will reduce the lengthy, laborious, and expensive protocols associated with this task (113).

6.6.3 BioMEMS-based Systems for the Study of Biofilms

Microbial biofilms are a significant problem in medicine and the food industry. Microbes in biofilms tend to be more drug-resistant and difficult to detect and target. Study of cell adhesion on different types of surfaces, and the kinetics of biofilm formation as well as its inhibition, may provide useful information for control and effective cure of biofilm associated diseases. BioMEMS-based assays have also been used to investigate the presence or formation of bacterial biofilms on the surfaces. The study of multilayered biofilm formation will further facilitate our understanding of bacterial behavior in the biofilms and cell-to-cell interaction (114, 115).

6.6.4 BioMEMS and Microfluidic Devices for Genotyping of Microbes

Simultaneous detection and genotyping of microbes in a single miniaturized assay format has been a dream of microbiologists. Attempts have been made to develop an integrated miniaturized portable device capable of simultaneous detection and genotyping of various clinical bacterial species on the basis of nucleic acid probe hybridization in microfluidic devices to a detection limit of a single cell. Similar attempts are also being made to develop lab-on-a-chip (LOC) devices for simultaneous detection, identification, antibiotic resistance and, genotyping of important pathogenic infectious agents from clinical, water, and food sources in a single assay format (115, 116).

In future focus may be directed to develop integrated, multiplex systems for simultaneous detection of biological and chemical pollutants using a single device. Furthermore, such devices may also be used for cell sorting or identifying antigenic types of pathogenic microbes as well molecular epidemiological studies.

6.7 FUTURE PERSPECTIVES

The technologies we have investigated pioneer the novel realms transforming the study of the invisible realm of the world such as microbial agents, with vast potential to accelerate detection and real-time investigation of microbial cells with only minimal volume samples (117, 118). Future progress of nanosensors and their manufacturing methodologies is expected to uphold a great potential to produce Nanosensors with unparalleled performance and productivity (119). In many diverse environments, nanosensors being an integral part of wearable technologies will enable real-time monitoring of a wide array of numerous microbial samples, revolutionizing the safety of food items, and giving control over infectious diseases and the environment to minimize the spread of harmful microbial agents (120). In future, nanoscale study will focus on highly streamlined single-cell study of biomolecules, promoting the instantaneous detection of multiplexed interactions of biological systems (121). In parallel, MEMS technology continues to grow with further reduction in dimensions and accessibility, and its integration with nanosensors and other technologies will derive more compact and cost-effective sensors and actuators along with large-scale production, increasing the scope of applications in various fields including the environment, food, healthcare, electronics, and auto systems (122, 123).

6.8 CONCLUSION

In conclusion, both nanosensors and MEMS technology offer a flood of revolutionary approaches to uncover microbial agents, showing tremendous potential in pursuit of a significant upheaval in identification, precision, and fast detection. These tools ease the early detection of numerous infectious diseases, assure the evaluation of environmental assessments, and improve the defense against biological threats using sophisticated sensing techniques and meticulous interactions of molecules. However, to improve the above-discussed potential characteristics of nanosensors and MEMS for microbial detection, there are challenges to address, for instance, improvements in scalability and durability, simultaneous multiple detection of targets, and advancement in data analysis techniques and their integration to achieve advanced analytics. Enabling these advancements will serve as a key component in boosting the overall health and medical protection of living beings, agricultural productivity, and environmental defenses.

REFERENCES

1. West SA, Diggle SP, Buckling A, Gardner A, Griffin AS. The social lives of microbes. *Annual Review of Ecology, Evolution, and Systematics*. 2007;38:53–77.
2. Kussell E. Evolution in microbes. *Annual Review of Biophysics*. 2013;42:493–514.
3. Gottschalk G. *Discover the World of Microbes: Bacteria, Archaea, Viruses*. John Wiley & Sons; 2012.
4. Pathak K, Saikia R, Gogoi U, Das A. Potential application of microbes for human health and welfare. *World Journal of Pharmacy and Pharmaceutical Sciences*. 2020;10:514–24.
5. Ojija F, Mayowela F. The critical role and application of microbes towards sustainable development and human wellbeing. *East African Journal of Environment and Natural Resources*. 2022;5(1):231–56.
6. Øvreås L. Population and community level approaches for analysing microbial diversity in natural environments. *Ecology Letters*. 2000;3(3):236–51.
7. Váradi L, Luo JL, Hibbs DE, Perry JD, Anderson RJ, Orenga S, et al. Methods for the detection and identification of pathogenic bacteria: past, present, and future. *Chemical Society Reviews*. 2017;46(16):4818–32.
8. Rossmann MG. Structure of viruses: a short history. *Quarterly Reviews of Biophysics*. 2013;46(2):133–80.
9. Newbound M, Mccarthy MA, Lebel T. Fungi and the urban environment: a review. *Landscape and Urban Planning*. 2010;96(3):138–45.

10. Baker BJ, De Anda V, Seitz KW, Dombrowski N, Santoro AE, Lloyd KG. Diversity, ecology and evolution of Archaea. *Nature Microbiology*. 2020;5(7):887–900.
11. Leander BS. Predatory protists. *Current Biology*. 2020;30(10):R510–R6.
12. Maggi F, Pistello M, Antonelli G. Future management of viral diseases: role of new technologies and new approaches in microbial interactions. *Clinical Microbiology and Infection*. 2019;25(2):136–41.
13. Ferone M, Gowen A, Fanning S, Scannell AG. Microbial detection and identification methods: bench top assays to omics approaches. *Comprehensive Reviews in Food Science and Food Safety*. 2020;19(6):3106–29.
14. Maurer JJ. Rapid detection and limitations of molecular techniques. *Annual Review of Food Science and Technology*. 2011;2:259–79.
15. Ivanov D. BioMEMS sensor systems for bacterial infection detection: progress and potential. *BioDrugs*. 2006;20:351–6.
16. Rolfe P. Micro- and nanosensors for medical and biological measurement. *Sensors and Materials*. 2012;24(6):275–302.
17. Bogue R. Nanosensors: a review of recent research. *Sensor Review*. 2009;29(4):310–5.
18. Abbasian F, Ghafar-Zadeh E, Magierowski S. Microbiological sensing technologies: a review. *Bioengineering*. 2018;5(1):20.
19. Mahbub T, Hoque ME. Introduction to nanomaterials and nanomanufacturing for nanosensors. In: *Nanofabrication for Smart Nanosensor Applications*, ed. K. Pal and F. Gomes. Elsevier; 2020. p. 1–20.
20. Coccia M, Roshani S, Mosleh M. Evolution of sensor research for clarifying the dynamics and properties of future directions. *Sensors*. 2022;22(23):9419.
21. Javaid M, Haleem A, Rab S, Singh RP, Suman R. Sensors for daily life: a review. *Sensors International*. 2021;2:100121.
22. Frank R. *Understanding Smart Sensors*. Artech House; 2013.
23. Sinclair I. *Sensors and Transducers*. Elsevier; 2000.
24. Prajapati S, Padhan B, Amulyasai B, Sarkar A. Nanotechnology-based sensors. In: *Biopolymer-Based Formulations*, ed. K. Pal et al. Elsevier; 2020. p. 237–62.
25. Liu A, Huang Z, Li M, Wan Y, Li W, Han TX, et al. A survey on fundamental limits of integrated sensing and communication. *IEEE Communications Surveys & Tutorials*. 2022;24(2):994–1034.
26. Javaid M, Haleem A, Singh RP, Rab S, Suman R. Exploring the potential of nanosensors: a brief overview. *Sensors International*. 2021;2:100130.
27. Abed MS, Jawad ZA. Nanotechnology for defence applications. In: *Nanotechnology for Electronic Applications*, ed. N.M. Mubarak, S. Gopi, and P. Balakrishnan. Springer; 2022. p. 187–205.
28. Edwards E, Brantley C, Ruffin PB. Overview of nanotechnology in military and aerospace applications. *Nanotechnology Commercialization: Manufacturing Processes and Products*, ed. T.O. Mensah et al. Wiley; 2017. p. 133–76.
29. Gramling HM, Kiziroglou ME, Yeatman EM. Nanotechnology for consumer electronics. *Nanoelectronics: Materials, Devices, Applications*, ed. R. Puers, L. Baldi, M. Van de Voorde, and S.E. van Nooten. Wiley; 2017. p. 501–26.
30. Khanna V. *Nanosensors*. CRC Press; 2016.
31. Genther-Yoshida P, Casassa M, Shull R, Pomrenke G, Thomas I, Price R, et al. *National Science and Technology Council Committee on Technology The Interagency Working Group on Nanoscience, Engineering and Technology September 1999*. About the National Science and Technology Council; 1999.
32. Abdel-Karim R, Reda Y, Abdel-Fattah A. Nanostructured materials-based nanosensors. *Journal of the Electrochemical Society*. 2020;167(3):037554.
33. Nasrollahzadeh M, Sajadi SM, Sajjadi M, Issaabadi Z. An introduction to nanotechnology. *Interface Science and Technology*. 2019;28:1–27.
34. Nasrollahzadeh M, Sajadi SM, Sajjadi M, Issaabadi Z. Applications of nanotechnology in daily life. *Interface Science and Technology*. 2019;28:113–43.
35. Vaseashta A, Dimova-Malinovska D. Nanostructured and nanoscale devices, sensors and detectors. *Science and Technology of Advanced Materials*. 2005;6(3–4):312–8.
36. McNeil SE. Nanotechnology for the biologist. *Journal of Leukocyte Biology*. 2005;78(3):585–94.
37. Bhushan B. Introduction to nanotechnology. In: *Springer Handbook of Nanotechnology*, ed. B. Bhushan. Springer; 2017. p. 1–19.

38. Shahdeo D, Roberts A, Abbineni N, Gandhi S. Graphene based sensors. *Comprehensive Analytical Chemistry*. 2020;91:175–99.
39. Mansoori GA, Soelaiman TF. *Nanotechnology–An Introduction for the Standards Community*. ASTM International; 2005.
40. Muhammed M, Tsakalakos T. Nanostructured materials and nanotechnology: overview. *Journal of the Korean Ceramic Society*. 2003;40(11):1027.
41. Ossai CI, Raghavan N. Nanostructure and nanomaterial characterization, growth mechanisms, and applications. *Nanotechnology Reviews*. 2018;7(2):209–31.
42. Christian P, Von der Kammer F, Baalousha M, Hofmann T. Nanoparticles: structure, properties, preparation and behaviour in environmental media. *Ecotoxicology*. 2008;17:326–43.
43. Kane CL, Mele E. Size, shape, and low energy electronic structure of carbon nanotubes. *Physical Review Letters*. 1997;78(10):1932.
44. Laird EA, Kuemmeth F, Steele GA, Grove-Rasmussen K, Nygård J, Flensberg K, et al. Quantum transport in carbon nanotubes. *Reviews of Modern Physics*. 2015;87(3):703.
45. Aqel A, Abou El-Nour KM, Ammar RA, Al-Warthan A. Carbon nanotubes, science and technology part (I) structure, synthesis and characterisation. *Arabian Journal of Chemistry*. 2012;5(1):1–23.
46. Tasis D, Tagmatarchis N, Bianco A, Prato M. Chemistry of carbon nanotubes. *Chemical Reviews*. 2006;106(3):1105–36.
47. Ibrahim KS. Carbon nanotubes-properties and applications: a review. *Carbon Letters*. 2013;14(3):131–44.
48. Jacak L, Hawrylak P, Wojs A. *Quantum Dots*. Springer Science & Business Media; 2013.
49. Mansuriya BD, Altintas Z. Applications of graphene quantum dots in biomedical sensors. *Sensors*. 2020;20(4):1072.
50. Adir O, Poley M, Chen G, Froim S, Krinsky N, Shklover J, et al. Integrating artificial intelligence and nanotechnology for precision cancer medicine. *Advanced Materials*. 2020;32(13):1901989.
51. Singh KR, Nayak V, Singh J, Singh RP. Nano-enabled wearable sensors for the Internet of Things (IoT). *Materials Letters*. 2021;304:130614.
52. Topel SD, Al-Turjman F. Nanosensors for the internet of nano-things (IoNT): an overview. *Internet of Nano-Things and Wireless Body Area Networks (WBAN)*, ed. F. Al-Turjman. Auerbach; 2019. p. 21–44.
53. Miraz MH, Ali M, Excell PS, Picking R. A review on Internet of Things (IoT), Internet of everything (IoE) and Internet of nano things (IoNT). In: *2015 Internet Technologies and Applications (ITA)*. IEEE; 2015. p. 219–24.
54. Shilpa A, Muneeswaran V, Rathinam DDK, Santhiya GA, Sherin J, editors. Exploring the benefits of sensors in internet of everything (IoE). In: *2019 5th International Conference on Advanced Computing & Communication Systems (ICACCS)*; 2019: IEEE.
55. Rehman KU, Andleeb S, Ashfaq M, Akram N, Akram MW. Blockchain-enabled smart agriculture: enhancing data-driven decision making and ensuring food security. *Journal of Cleaner Production*. 2023;427:138900.
56. Laucht A, Hohls F, Ubbelohde N, Gonzalez -Zalba MF, Reilly DJ, Stobbe S, et al. Roadmap on quantum nanotechnologies. *Nanotechnology*. 2021;32(16):162003.
57. Wang X, Hu Y, Wei H. Nanozymes in bionanotechnology: from sensing to therapeutics and beyond. *Inorganic Chemistry Frontiers*. 2016;3(1):41–60.
58. Saylan Y, Denizli A. Virus detection using nanosensors. In: *Nanosensors for Smart Cities*, ed. B. Han et al. Elsevier; 2020. p. 501–11.
59. Verma N, Bhardwaj A. Biosensor technology for pesticides—a review. *Applied Biochemistry and Biotechnology*. 2015;175:3093–119.
60. Inci F, Tokel O, Wang S, Gurkan UA, Tasoglu S, Kuritzkes DR, et al. Nanoplasmonic quantitative detection of intact viruses from unprocessed whole blood. *ACS Nano*. 2013;7(6):4733–45.
61. Babamiri B, Salimi A, Hallaj R. A molecularly imprinted electrochemiluminescence sensor for ultrasensitive HIV-1 gene detection using EuS nanocrystals as luminophore. *Biosensors and Bioelectronics*. 2018;117:332–9.
62. Lu C-H, Zhang Y, Tang S-F, Fang Z-B, Yang H-H, Chen X, et al. Sensing HIV related protein using epitope imprinted hydrophilic polymer coated quartz crystal microbalance. *Biosensors and Bioelectronics*. 2012;31(1):439–44.
63. Shafiee H, Lidstone EA, Jahangir M, Inci F, Hanhauser E, Henrich TJ, et al. Nanostructured optical photonic crystal biosensor for HIV viral load measurement. *Scientific Reports*. 2014;4(1):4116.

64. Hassen WM, Chaix C, Abdelghani A, Bessueille F, Leonard D, Jaffrezic-Renault N. An impedimetric DNA sensor based on functionalized magnetic nanoparticles for HIV and HBV detection. *Sensors and Actuators B: Chemical*. 2008;134(2):755–60.
65. Uzun L, Say R, Ünal S, Denizli A. Production of surface plasmon resonance based assay kit for hepatitis diagnosis. *Biosensors and Bioelectronics*. 2009;24(9):2878–84.
66. Inan H, Wang S, Inci F, Baday M, Zangar R, Kesiraju S, et al. Isolation, detection, and quantification of cancer biomarkers in HPV-associated malignancies. *Scientific Reports*. 2017;7(1):3322.
67. Teengam P, Siangproh W, Tuantranont A, Vilaivan T, Chailapakul O, Henry CS. Multiplex paper-based colorimetric DNA sensor using pyrrolidinyl peptide nucleic acid-induced AgNPs aggregation for detecting MERS-CoV, MTB, and HPV oligonucleotides. *Analytical Chemistry*. 2017;89(10):5428–35.
68. Tam PD, Van Hieu N, Chien ND, Le A-T, Tuan MA. DNA sensor development based on multi-wall carbon nanotubes for label-free influenza virus (type A) detection. *Journal of Immunological Methods*. 2009;350(1–2):118–24.
69. Vollmer F, Arnold S, Keng D. Single virus detection from the reactive shift of a whispering-gallery mode. *Proceedings of the National Academy of Sciences*. 2008;105(52):20701–4.
70. Jin CE, Lee TY, Koo B, Sung H, Kim S-H, Shin Y. Rapid virus diagnostic system using bio-optical sensor and microfluidic sample processing. *Sensors and Actuators B: Chemical*. 2018;255:2399–406.
71. Guerreiro MR, Freitas DF, Alves PM, Coroadinha AS. Detection and quantification of label-free infectious adenovirus using a switch-on cell-based fluorescent biosensor. *ACS Sensors*. 2019;4(6):1654–61.
72. Pu H, Xu Y, Sun D-W, Wei Q, Li X. Optical nanosensors for biofilm detection in the food industry: principles, applications and challenges. *Critical Reviews in Food Science and Nutrition*. 2021;61(13):2107–24.
73. Ghormade V. Nanosensors for detection of human fungal pathogens. In: *Nanotechnology for Infectious Diseases*, ed. S. Hameed and S. Rehman. Springer; 2022. p. 497–519.
74. Kim D, Wang SX. A magneto-nanosensor immunoassay for sensitive detection of Aspergillus fumigatus allergen Asp f 1. *IEEE Transactions on Magnetics*. 2012;48(11):3266–8.
75. Zia R, Iftikhar M, Rafiq A, Hakim S, Nawaz MS, Imran A, et al. Nanosensors for microbial detection in soil. *Nanosensors for Smart Agriculture*, ed. A. Denizli et al. Elsevier; 2022. p. 367–400.
76. Sashidhar P, Dubey MK, Kochar M. Sensing soil microbes and interactions: how can nanomaterials help? In: *Microbial Nanobionics: Volume 2, Basic Research and Applications*. ed. R. Prasad. Springer; 2019. p. 213–36.
77. Syed MA. Advances in nanodiagnostic techniques for microbial agents. *Biosensors and Bioelectronics*. 2014;51:391–400.
78. Cui J, Richardson JJ, Björnmalm M, Faria M, Caruso F. Nanoengineered templated polymer particles: navigating the biological realm. *Accounts of Chemical Research*. 2016;49(6):1139–48.
79. Kadadou D, Tizani L, Wadi VS, Banat F, Alsafar H, Yousef AF, et al. Recent advances in the biosensors application for the detection of bacteria and viruses in wastewater. *Journal of Environmental Chemical Engineering*. 2022;10(1):107070.
80. He X, Aker WG, Fu PP, Hwang H-M. Toxicity of engineered metal oxide nanomaterials mediated by nano–bio–eco–interactions: a review and perspective. *Environmental Science: Nano*. 2015;2(6):564–82.
81. Scurti S, Caretti D, Mollica F, Di Antonio E, Amorati R. Chain-breaking antioxidant and peroxyl radical trapping activity of phenol-coated magnetic Iron oxide nanoparticles. *Antioxidants*. 2022;11(6):1163.
82. Viglianisi C, Scarlini A, Tofani L, Menichetti S, Baschieri A, Amorati R. Magnetic nanoantioxidants with improved radical-trapping stoichiometry as stabilizers for inhibition of peroxide formation in ethereal solvents. *Scientific Reports*. 2019;9(1):17219.
83. Fragou F, Stathi P, Deligiannakis Y, Louloudi M. Safe-by-design flame spray pyrolysis of SiO2 nanostructures for minimizing acute toxicity. *ACS Applied Nano Materials*. 2022;5(6):8184–95.
84. Tansil NC, Gao Z. Nanoparticles in biomolecular detection. *Nano Today*. 2006;1(1):28–37.
85. Ge X-Y, Feng Y-G, Cen S-Y, Wang A-J, Mei L-P, Luo X, et al. A label-free electrochemical immnunosensor based on signal magnification of oxygen reduction reaction catalyzed by uniform PtCo nanodendrites for highly sensitive detection of carbohydrate antigen 15-3. *Analytica Chimica Acta*. 2021;1176:338750.
86. Ramanaviciene A, German N, Kausaite-Minkstimiene A, Ramanavicius A. Glucose biosensor based on dendritic gold nanostructures electrodeposited on graphite electrode by different electrochemical methods. *Chemosensors*. 2021;9(8):188.

87. Sun Y, Zhai X, Xu Y, Liu C, Zou X, Li Z, et al. Facile fabrication of three-dimensional gold nanodendrites decorated by silver nanoparticles as hybrid SERS-active substrate for the detection of food contaminants. *Food Control*. 2021;122:107772.
88. Dang VH, Yen PTH, Giao NQ, Phong PH, Ha VTT, Duy PK, et al. A versatile carbon fiber cloth-supported Au nanodendrite sensor for simultaneous determination of Cu (II), Pb (II) and Hg (II). *Electroanalysis*. 2018;30(10):2222–7.
89. Zhang F, Huang F, Gong W, Tian F, Wu H, Ding S, et al. Multi-branched PdPt nanodendrites decorated amino-rich Fe-based metal-organic framework as signal amplifier for ultrasensitive electrochemical detection of prolactin. *Journal of Electroanalytical Chemistry*. 2021;882:115032.
90. Li M-X, Feng Q-M, Zhou Z, Zhao W, Xu J-J, Chen H-Y. Plasmon-enhanced electrochemiluminescence for nucleic acid detection based on gold nanodendrites. *Analytical Chemistry*. 2018;90(2):1340–7.
91. Shah ST, Chowdhury ZZ, Simarani K, Basirun WJ, Badruddin IA, Hussien M, et al. Nanoantioxidants: the fourth generation of antioxidants—recent research roadmap and future rerspectives. *Coatings*. 2022;12(10):1568.
92. Seal S, Jeyaranjan A, Neal CJ, Kumar U, Sakthivel TS, Sayle DC. Engineered defects in cerium oxides: tuning chemical reactivity for biomedical, environmental, & energy applications. *Nanoscale*. 2020;12(13):6879–99.
93. Bhattacharya K, Gogoi B, Buragohain A, Deb P. Fe2O3/C nanocomposites having distinctive antioxidant activity and hemolysis prevention efficiency. *Materials Science and Engineering: C*. 2014;42:595–600.
94. Bell DJ, Lu T, Fleck NA, Spearing SM. MEMS actuators and sensors: observations on their performance and selection for purpose. *Journal of Micromechanics and Microengineering*. 2005;15(7):S153.
95. Bogue R. Recent developments in MEMS sensors: a review of applications, markets and technologies. *Sensor Review*. 2013;33(4):300–4.
96. Faudzi AAM, Sabzehmeidani Y, Suzumori K. Application of micro-electro-mechanical systems (MEMS) as sensors: a review. *Journal of Robotics and Mechatronics*. 2020;32(2):281–8.
97. Tanaka M. An industrial and applied review of new MEMS devices features. *Microelectronic Engineering*. 2007;84(5-8):1341–4.
98. Gad-el-Hak M. *MEMS: Introduction and Fundamentals*. CRC Press; 2005.
99. Bogue R. MEMS sensors: past, present and future. *Sensor Review*. 2007;27(1):7–13.
100. Bryzek J, Roundy S, Bircumshaw B, Chung C, Castellino K, Stetter JR, et al. Marvelous mems. *IEEE Circuits and Devices Magazine*. 2006;22(2):8–28.
101. Nagel DJ, Zaghloul ME. MEMS: micro technology, mega impact. *IEEE Circuits and Devices Magazine*. 2001;17(2):14–25.
102. Nguyen CT-C. MEMS technology for timing and frequency control. *IEEE Transactions on Ultrasonics, Ferroelectrics, and Frequency Control*. 2007;54(2):251–70.
103. Judy JW. Microelectromechanical systems (MEMS): fabrication, design and applications. *Smart Materials and Structures*. 2001;10(6):1115.
104. Fedder GK, editor. MEMS fabrication. In: *International Test Conference, 2003 Proceedings ITC 2003*; 2003: IEEE Computer Society.
105. Fischer AC, Forsberg F, Lapisa M, Bleiker SJ, Stemme G, Roxhed N, et al. Integrating mems and ICS. *Microsystems & Nanoengineering*. 2015;1(1):1–16.
106. Huff MA. MEMS: an enabling technology for the internet of things (IoT). In *Internet of Things and Data Analytics Handbook*, ed. H. Geng. Wiley; 2017. p. 147–66.
107. Mandal N, Mitra R, Pramanick B. C-MEMS-derived glassy carbon electrochemical biosensors for rapid detection of SARS-CoV-2 spike protein. *Microsystems & Nanoengineering*. 2023;9(1):137.
108. Pujol-Vila F, Villa R, Alvarez M. Nanomechanical sensors as a tool for bacteria detection and antibiotic susceptibility testing. *Frontiers in Mechanical Engineering*. 2020;6:44.
109. Lillehoj PB, Kaplan CW, He J, Shi W, Ho C-M. Rapid, electrical impedance detection of bacterial pathogens using immobilized antimicrobial peptides. *Journal of Laboratory Automation*. 2014;19(1):42–9.
110. Fande S, Amreen K, Sriram D, Mateev V, Goel S. Electromicrofluidic device for interference-free rapid antibiotic susceptibility testing of Escherichia coli from real samples. *Sensors*. 2023;23(23):9314.
111. Gao J, Chakraborthy A, He S, Yang S, Afsarimanesh N, Nag A, et al. Graphene-based sensors for the detection of microorganisms in food: a review. *Biosensors*. 2023;13(6):579.
112. Katole MK, Balwir MS, Manekar MP. Design of MEMS biosensor for pathogenic bacterial disease detection. *International Journal of Electrical Electronics and Computers*. 2021;6:10–12.

113. Su W, Qiu J, Mei Y, Zhang X-E, He Y, Li F. A microfluidic cell chip for virus isolation via rapid screening for permissive cells. *Virologica Sinica*. 2022;37(4):547–57.
114. Funari R, Shen AQ. Detection and characterization of bacterial biofilms and biofilm-based sensors. *ACS Sensors*. 2022;7(2):347–57.
115. Gorgannezhad L, Stratton H, Nguyen N-T. Microfluidic-based nucleic acid amplification systems in microbiology. *Micromachines*. 2019;10(6):408.
116. Liu RH, Munro SB, Nguyen T, Siuda T, Suciu D, Bizak M, et al. Integrated microfluidic CustomArray device for bacterial genotyping and identification. *JALA: Journal of the Association for Laboratory Automation*. 2006;11(6):360–7.
117. Bata-Vidács I, Adányi N, Beczner J, Farkas J, Székács A. Nanotechnology and microbial food safety. In *Microbial Pathogens and Strategies for Combating Them: Science, Technology and Education*, ed. A. Mendez-Vilas. Formatex; 2013. p. 155–69.
118. Rodrigues SM, Demokritou P, Dokoozlian N, Hendren CO, Karn B, Mauter MS, et al. Nanotechnology for sustainable food production: promising opportunities and scientific challenges. *Environmental Science: Nano*. 2017;4(4):767–81.
119. Adam T, Gopinath SC. Nanosensors: recent perspectives on attainments and future promise of downstream applications. *Process Biochemistry*. 2022;117:153–73.
120. Ileš D, Martinović G, Kozak D. Review of potential use, benefits and risks of nanosensors and nanotechnologies in food. *Strojarstvo: časopis za teoriju i praksu u strojarstvu*. 2011;53(2):127–36.
121. Datta SPA. Future healthcare: bioinformatics, nano-sensors, and emerging innovations. In: *Nanosensors: Theory and Applications in Industry, Healthcare and Defense*, ed. T. Lim. Taylor & Francis; 2016. p. 247.
122. Pryputniewicz RJ. Current trends and future directions in MEMS. *Experimental Mechanics*. 2012;52:289–303.
123. Vigna B, editor. Future of MEMS: An industry point of view. *EuroSime 2006-7th International Conference on Thermal, Mechanical and Multiphysics Simulation and Experiments in Micro-Electronics and Micro-Systems*; 2006: IEEE.
124. Bahamonde J, Nguyen HN, Fanourakis SK, Rodrigues DF. Recent advances in graphene-based biosensor technology with applications in life sciences. *Journal of Nanobiotechnology*. 2018;16:75.

7 Recent Advances in Microbiology Techniques for Imaging Microbes and Their Components

Laiba Hassan, Maria Rukan, Ejaz Mohammad, and Muhammad Ali Syed

7.1 INTRODUCTION

The human visual system, capable of perceiving the optical segment of the electromagnetic spectrum, holds profound implications for our understanding of the world. This unique ability to recognize optical images has not only enriched our daily lives but has also been instrumental in transformative scientific endeavors (1). Between 1665 and 1683, the Royal Society Fellows Robert Hooke and Antonie van Leeuwenhoek laid the foundation for the understanding of microscopic organisms. Hooke's *Micrographia* (1665) featured the initial depiction of a microorganism, Mucor, while Leeuwenhoek later observed and described protozoa and bacteria in rainwater, seawater, and well water. Through his groundbreaking "Leeuwenhoek microscope," he unlocked a hidden realm—individual microbial cells—that had profound implications for our comprehension of life (2, 3). This technological breakthrough not only expanded the boundaries of human knowledge but also paved the way for advancements in medicine, microbiology, and our overall appreciation of the intricacies of the microscopic world, demonstrating the enduring impact of our visual capabilities on the course of human history (1).

Julius Richard Petri in collaboration with Robert Koch in 1877 introduced an innovative method—Petri dishes for examination of microorganisms. These are cylindrical plates which promote microbial growth in which bacterial colonies are visible on surfaces of gel-like culture media. Analysis of optical images of microbes is enhanced by photography to reveal microbial mysteries. However, quantitative methods can be time-consuming which make it difficult for complete extraction of diverse information from optical images, such as spectral distribution, spatial properties, shape, size, quantity, and position of components (1).

Imaging techniques are essential tools across different domains of science for examining complex processes and structures. In medicine and biology, imaging techniques have transformed into a crucial non-destructive method for analysis of objects and samples in three dimensions at different spatial scales. Imaging plays an important role in increasing our understanding of diverse phenomena, ranging from nanometers to micrometers—scrutinizing macromolecules, capturing microbial aggregates at cellular levels, and biofilms—and extending up to centimeters, encompassing microbial mats, organs, tissues, and the human body (4).

The journey toward the microscopic world started with the discovery of magnifying lenses and structured illumination (SIM) optical microscopes. The first step in using the inaugural light microscope was taken by Robert Hooke in 1665. Soon after, Antonie van Leeuwenhoek in 1668 searched microbiological samples, making innovative observations such as the first visual observation of oral

DOI: 10.1201/9781003410164-7

biofilms (5). Conventional light microscopy and its variations have been entrusted companions over the centuries, providing an opportunity for visualizing complex structures of biological specimens and shaping our comprehension of the minute wonders that cannot be visualized by the naked eye (4).

7.2 TRADITIONAL LMAGING TECHNIQUES APPLIED IN MICROBIOLOGY

Over 2,000 years ago, in the first century BC, the magnifying potential of spherical and transparent lenses was recognized. A single convex lens could provide more than tenfold magnification, but it was not sufficient to provide clear details of various tiny objects (6). Robert Hooke (in 1665) took the first step in microscopy by utilizing the first optical microscope, so he is considered the pioneer. Soon thereafter, Antonie van Leeuwenhoek observed oral biofilms for the first time (5). These initial observations, assisted by magnifying glasses and early optical microscopes, led to the beginning of a new era in scientific observation by opening novel avenues for comprehension of complex details of the microscopic domain (4).

Dutch scientist Zacharias Janssen and his son, while dealing with eyeglasses, conducted experiments by inserting multiple lenses into a cylinder. It resulted in enlarged visualization of an object when observed through this cylinder, marking the discovery of the modern microscope and telescope. Thereafter, Janssen designed and constructed the first compound microscope to refine this accidental discovery by using two convex lenses. This novel microscope shared the same principle of enlarging objects, but was different from designs which used only a single convex lens. Leeuwenhoek's exceptional manufacturing skills enabled a uniquely formed single convex lens to attain an unprecedented power of magnification surpassing 300 times (6).

In 1983, scientists Heinrich Rohrer and Gerd Binning developed the scanning tunneling microscope (STM) while conducting research at the laboratory of IBM (7). This innovative microscope departed from the conventional principles, working by "tunnel effect," a unique concept, and avoiding traditional lenses while favoring probes. The tunnel effect takes place when the gap between the observed object and the probe is sufficiently small, mainly in nanometers, which is achieved by providing voltage between them. As electrons traverse this minute gap, a subtle electric current emerges. Fluctuations in the distance between the probe and the object correspondingly alter the current strength, enabling the discernment of the object's three-dimensional shape. The STM achieves an extraordinary spatial resolution, down to the level of a single atom (6). Atomic force microscopes (AFM) are modifications of STM that may be used in imaging biological samples. Both electron microscopes and scanning probe microscopes have played a revolutionary role in life science research.

7.3 THE BASIC CONCEPT OF MICROSCOPIC IMAGING

Various types of microscopes have been developed, yet the core imaging concept and structures are concisely presented in Figure 7.1. The primary components of an optical microscope are an eyepiece and objective lens. The objective lens is responsible for magnifying the object, enabling clear observation. During this process, the sample is positioned near the objective lens focal plane in the object space, forming an initial real image magnified on the intermediate plane. This intermediate plane aligns with the eyepiece focal plane, converting the eyepiece into a magnifier that further amplifies the projected image onto the intermediate plane. Ultimately, the observer perceives a magnified, virtual, and inverted image (6).

7.4 THE MICROSCOPES USED IN THE BIOMEDICAL FIELD

Despite the development and application of different microscope types in various fields of microbiology, they can be effectively classified into three groups: (1) optical microscopes, (2) electron

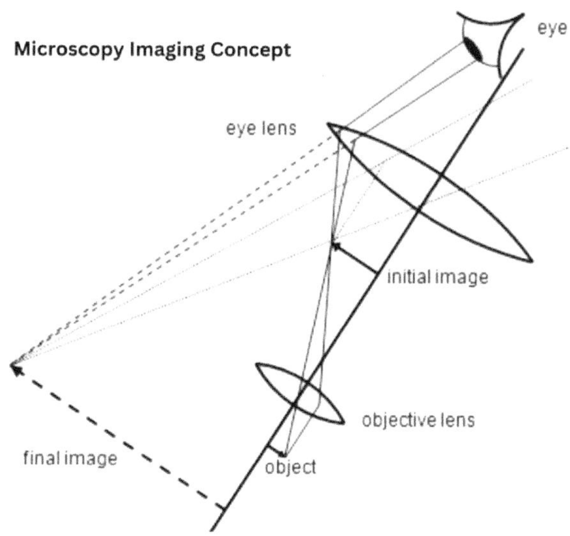

FIGURE 7.1 Imaging concept of conventional light microscopy. It involves the use of an objective lens to collect and focus light from the sample and an eyepiece lens for further magnification of the image (Adopted from ref. 6).

microscopes, and (3) scanning probe microscopes that laid the foundation of advancement in imaging techniques (8).

7.4.1 Light Microscopy

The optical microscope, commonly known as the light microscope, has been a foundational tool for centuries in the observation and study of biological specimens. Traditional light microscopy and its variations served as the primary methods throughout much of the previous century, with the prevailing belief that the evolution of this technique had reached its pinnacle. However, a paradigm shift occurred in the 1980s with the introduction of laser scanning microscopy (LSM), marking a revolutionary phase in the field. This breakthrough technology challenged previous limitations, opening new avenues for high-resolution imaging. Lately, the development of stimulated emission depletion (STED) laser microscopy in 2000, pioneered by Klar et al., represented a substantial leap forward. This innovation has significantly expanded the capabilities of laser microscopy, pushing the boundaries of spatial resolution and imaging precision to unprecedented levels (4).

7.4.2 Electron Microscopy

The electron microscope (EM) is a specialized dynamic tool which uses a beam of electrons to study the morphological surface of cells and subcellular structures of non-biological and biological samples. EM works in a vacuum and a beam of electrons ($\lambda = 0.005$ nm) provides a light source for the magnification of processed specimens by using electromagnetic lenses as compared to visible light wavelengths such as $\lambda = 400$ nm to 700 nm. There is an inverse relationship between revolving power and wavelength in EM. For example, wavelength shortens with the increase in electron velocity, thus resulting in high resolution (9). Investigations for electron microscope development started in the 1920s. Ernst Ruska with Max Knoll's guidance in 1928 designed electron lenses in Germany and built the first functional electron microscope in 1931 for which he received the Nobel

Prize in 1986. Due to its high power of magnification and resolution, the electron microscope began to study a variety of biological specimens (10).

Currently, EM is a significant tool for the diagnosis of ultrastructural pathology. Tobacco mosaic virus was the first virus to be visualized by EM. Electron microscopy is still on the frontline to be used for the diagnosis of infection, specifically when the causative agent is unsuspected or unknown. It is an important technique for the investigation of emerging viruses such as monkey pox viruses and SARS viruses and surveillance of potential agents of bioterrorism. EM studies are ongoing for the development of treatment methods and vaccination strategies (11).

In the bacteriology field, EM has little diagnostic use. However, processed biopsy material is examined by EM for identification of a few bacterial pathogens. EM has been used to study many bacterial structures and functions such as spores, fimbriae, and flagella, and in the identification of bacteriophage (12).

The electron microscope has two basic types: transmission electron microscope (TEM) and scanning electron microscope (SEM) (9).

7.4.2.1 Transmission Electron Microscopy

Transmission electron microscopes (TEM) work by emitting electrons from a tiny source located at the topmost part of the electron microscope. There is a vacuum column in TEM in which electrons travel. An electromagnetic lens is used in this microscope to focus electrons into a beam, instead of the light-focusing lens used in the light microscope. The beam of electrons passes through a specimen and a few electrons disperse from the electron beam based on texture and density of the object. A shadow-like image of the specimen that is being studied is produced based on its density when unscattered electrons collide on the screen. The image can be photographed by the operator and analyzed for further research (13). Transmission electron microscopy (TEM) has proven instrumental in observing temperature-dependent bacteria, such as *Listeria monocytogens*, revealing their characteristic flagellar arrangement (14). In instances of calcium bilirubinate stones with bacterial infection, TEM analysis has demonstrated the presence of mixed microcolonies comprising both gram-positive and gram-negative bacteria embedded in an amorphous anionic matrix (15). Traditional preparation methods for TEM have evolved to Cryo-TEM, employing a high-pressure freezing technique for the measurement and analysis of thin cytoplasmic sections of specific bacteria (16).

7.4.2.2 Scanning Electron Microscopy

Engineered to generate high-spatial-resolution scan images under low acceleration voltage conditions, the scanning electron microscope (SEM) typically employs a raster scan pattern, with the detected signal integrated with the beam's position to produce detailed images. SEM excels in achieving resolutions finer than 1 nanometer, although specimens must be completely dry due to the high vacuum conditions within the specimen chamber. The pathogenicity of *Vibrio cholerae*, a gram-negative bacterium causing cholera, remains elusive. Electron microscopy has unraveled the ultrastructure of bacteria, revealing a distinct nuclear zone and cell wall excretion (17). SEM has been useful in investigations of *Campylobacter pyloridis* histopathology, assisting in diagnosis of the presence of the pathogen in abnormal mucosa which results in gastritis ulceration (18). Moreover, electron microscopy has shown pili and flagella in 21 *Legionella pneumophila* strains (19).

7.4.3 Fluorescence Microscopy

Fluorescence microscopy works on the fluorescence principle. This method utilizes the unique ability of certain atoms and molecules for absorption of light at a specific wavelength and emission of light at a longer wavelength. It is mainly beneficial for detection of biological samples and mapping of specific molecules by the spatial distribution within a structure. Epifluorescence microscopy is mainly useful to count bacterial cells accurately, particularly direct viable bacterial cells (DVC). This method uses a SYBR Green fluorescent stain for counting bacterial cells, allowing us to precisely quantify particles (20).

Fluorescence optical microscopy emerged as a powerful tool in the field of biology, elucidating functional and spatial details about both internal autofluorescence elements and extrinsically labeled structures and molecules. The strategic utilization of fluorescent molecules has enabled researchers to get valuable insights into functional and spatial attributes. Primary light sources encompass mercury vapor or xenon arc lamps equipped with an excitation filter, along with lasers. Within the fluorescence microscope, high-energy light illuminates and stimulates fluorophores within the specimen. Consequently, the stimulated fluorophore emits fluorescent light with lower energy. Filters are strategically employed to isolate the emitted light of lower energy, which is then observable by the naked eye or other detectors. It is noteworthy that fluorescence microscopy distinguishes itself by the specimen emitting its light, as noted by Davidson and Abramowitz in 1999 (21).

7.4.4 Atomic Force Microscopy (AFM)

Atomic force microscopy is a powerful imaging technique that has been used to study lipopolysaccharides found in the Gram-negative outer membrane and lipids found in the inner membrane of both Gram-positive and Gram-negative bacteria (22). In the preceding decade, there has been a significant surge in interest regarding the organization of these biological membranes, which are comprised of lipids, a phenomenon extensively explored through the lens of atomic force microscopy. Biological membranes serve as critical guardians, effectively compartmentalizing biochemical and cellular activities while providing a protective barrier for the cell's inner area against the external surroundings. These membranes orchestrate the selective transport of nutrients, waste, and ions (23, 24).

Between external agents and living cells, the initial interaction unfolds at the external biomembrane before accessing subcellular targets. The delicate nuances of this interaction, especially at the nanometer scale, have been meticulously unraveled through the lens of AFM (25, 26). Beyond this, AFM proves invaluable in unraveling the intricate interplay between lipids, lipid phases, and the nanomechanical properties that define bio membranes. This blend of precision and insight sheds light on the dynamic elegance of biological membranes and their pivotal role in cellular functionality (27, 28).

7.5 RECENT ADVANCES IN IMAGING TECHNIQUES

Recent decades have witnessed significant developments in the field of imaging techniques, which are briefly discussed in the following sections.

7.5.1 Confocal Microscopy

Confocal microscopy is an important imaging technique that totally transforms the discipline of microbiology by providing three-dimensional, high-quality images of microbial structures (29). This process is frequently used for visualizing bacteria due to its capability to preclude out-of-focus light, resulting in crunchy and focused pictures of microbial models (30). Confocal microscopy has been applied to detect various architectures of microbes and the microstructural variations in mats of microbes across different settings (31). This microscopic method has been valuable for examining microbial growth and colonization on diverse substrates, including wetland plants (32). This technique's ability to provide optical sectioning in impenetrable specimens has emerged as a vital asset for inspecting living bacterial cells and biofilms (33). Confocal microscopy has been applied to estimate semen motility, showcasing its adaptability in several biological frameworks (34) (Figure 7.3).

Re-scan Confocal Microscopy (RCM) is a groundbreaking technology that has enhanced the sensitivity and quality of confocal imaging, improving its applicability in biomedical investigation (35, 36). Moreover, integrating other imaging approaches such as fluorescence microscopy with confocal microscopy enhanced its volume for analyzing biological materials (37). Confocal laser scanning microscopy (CLSM) is a recognized technology that is widely used to differentiate between dead and alive bacterial species in dentine infections. CLSM was fruitful in distinguishing between lifeless (red) and alive (green) bacteria in dentin tubules, specifically after staining with dyes like PI/FDA. Acridine orange established metabolic movement in *Enterococcus faecalis* within dentinal tubules by releasing red fluorescence, providing useful insights. CLSM, with acridine orange and PI/FDA, is a valuable technique for measuring the feasibility of bacteria in *E. Faecalis*-infected dentin (38).

CLSM was joined with several methods by Hartman's group, specifically for classifying bacterial species related to plant roots. Oligonucleotide enquiries targeting rRNA and antibodies were used to find rhizosphere germs in wheat and rice rhizospheres, consequential in improved tracking competences. This technique was fruitful in classifying *Pseudomonas* on sugar beet roots. CLSM has been recognized in mycology, utilizing several fluorochromes like FUN-1 for staining fungi. CLSM is specifically used to evaluate the cellular state and analyze the hyphal structural progress of fungi in the arena of mycology. Reid and colleagues in Edinburgh offered an overview of the applicability of CLSM for fungal research (39). Research has explored the connection between fungi and other creatures, namely bacteria, as seen in bacterial–fungal connections with sugar beet roots (40). CLSM imaging reveals straight experimental connections between bacteria and fungi, displaying how bacteria exploit fungal systems to reach soil substrate. CLSM is used to examine the symbiotic connection between fungi and phototrophic cyanobacteria in lichens (41).

7.5.2 Multiphoton Microscopy

Multiphoton microscopy is a crucial imaging method for bacteria and their mechanisms because it can optically divide them into segments and deliver high-quality pictures (42). This technique relies on nonlinear connections with light and matter to attain subcellular imaging in natural tissue with inherent fluorescence, while minimizing photodamage and visual weakening (43). The multiphoton microscopy method is being widely used in a supplementary way in numerous medical research domains to achieve useful insights into cell disease and physiology (44). Most investigation in biological systems uses two-photon excited fluorescence (TPEF) for making images in multiphoton microscopy (45). Multiphoton microscopy permits longer imaging times, making it ideal for detecting dynamic procedures in microorganisms (46). Development in multiphoton microscopy has meaningfully improved the conception of bacteria and their mechanisms. Multiphoton microscopy was used to complete in vivo imaging of bacterial groups of *Streptomyces* (47). Intravital microscopy using ,ultiphoton excitation has been helpful for monitoring of viral contagions in living creatures. MP-IVM investigation has provided valuable information on the growth of the HIV and pseudorabies virus in animal models. MP-IVM research has revealed information regarding the dynamics and development of virological synapses, revealing the murine leukemia virus. Viral particles originating from lymph are transmitted by macrophages to target cells in a process known as real-time trans-infection (48). A multimodal multiphoton microscope design allowed for label-free 3D imaging, allowing non-invasive, chemical-free, and autofluorescence imaging of fungi, bacteria, and their interactions with minerals and plant roots (49). A multiphoton microscopic technique has been devised to study the interactions between the pathogenic fungi *Fusarium oxysporum* and the *Arabidopsis thaliana* plant. This method enables the observation of their interaction in three dimensions and across time using multiphoton and confocal microscopy. It allows for the investigation of the genetic processes linked to *Fusarium* wilt (50). Multiphoton microscopy is a crucial imaging technique for studying microbial structures because it offers high-resolution imaging while minimizing photobleaching and photodamage, making it an excellent tool for studying biological samples' critical functions.

7.5.3 Structured Illumination Microscopy (SIM)

An advanced imaging technique that has transformed super-resolution optical imaging is known as structured illumination microscopy (SIM). SIM works by using various overlapping rays of light to differentiate objects smaller than the limit of diffraction. It was developed for the creation of fluorescence imaging but with the passage of time and techniques extended to non-fluorescent imaging areas, which results in creation of super-resolution images without using tagging of fluorescent molecules (51). As it can increase resolution of imaging by twofold, it is an important technique for live-cell imaging at suitable speed, so serving as an important technique for super-resolution optical imaging (52). In wide-field imaging, SIM has been used to overcome scattering problems, as in HiLo microscopy and structured illumination microscopy optical sectioning microscopy (SIM-OS) (53).

Adaptive optics incorporation with SIM has enabled in-depth imaging with high resolution, presenting its ability for thick samples observation (54). Quasi-periodic patterns have been integrated into SIM to improve optical sectioning and super-resolution imaging (55). SIM helps in providing basic information to achieve optical sectioning and super-resolution imaging and overcome deep tissue image scattering (56). This method utilizes a lightening pattern to overcome the effect of out-of-focus light and blurred imaging and enhance fluorescent imaging quality of the sample (57). SIM has improved spatial resolution twofold in two or three dimensions along with optical sectioning and imaging contrast compared to conventional wide-field fluorescent microscopy with uniform illumination. Therefore, this technique is suitable for live-cell imaging (58). SIM is widely used in biomedical research due to its cost and practicability, making it an effective technique for studying microbial populations and interaction (59).

Stable isotope probing (SIP), fluorescence in situ hybridization for microbial identification, and NanoSIMS imaging have been used in SIM to build a connection between function and identity in microbial communities at the cellular level (60). Secondary-ion mass spectrometry (SIMS), including ToF-SIMS and NanoSIMS, has been used for complete analysis of surface and imaging at cellular level individually for microbiological samples (61).

This innovative SIM technology has improved visualization and imaging for investigation of microorganisms. This advanced technology also enhanced the visualization and imaging of microbial structures by providing complete dynamics, interactions, and microbial morphology. The movement and structure of the cytokinetic Z Ring in *Staphylococcus aureus* and *Bacillus subtilis* cells can be observed by using Fast 3D-SIM (f3D-SIM). A tubulin-like protein FtsZ, which is important for cell division of bacteria, has varying organization in a bead-like structure for Z-ring surrounding. So the exact shape of Z-ring whether it is continuous or fragmented band can be investigated through this study. Utilizing 3D-SIM significantly improves spatial analysis, particularly for small, spherical cells such as *S. aureus*, uncovering the evolving architecture of the Z ring over time (62).

The ExSIM technique merges specimen expansion with SIM to improve spatial resolution to around 30 nm. When examining *Giardia lamblia*'s cytoskeleton, which includes the flagellar axonemes and sticky disc, the study effectively determines the position of disc-associated proteins such as DAP86676 and the functionally unidentified DAP16263. The hybridization method of ExSIM is an effective technique of super-resolution for analysis of fixed samples, representing its applicability for many biological systems (63). The observation of live cells of *Bacillus cereus* under SIM confirmed the suitability of probes for super-resolution microscopy. Using NanoSIMS for analysis of the platinum (II) complex absorption in bacteria demonstrates the ability of its chemical structure for multimodal imaging. The versatility of a variety of bacterial probes is achieved by successful incorporation of NanoSIMS, widening the range of probes of organometallic molecules synthesized to enhance various imaging techniques (64).

A new approach to SIM combines it with algorithms of machine learning to obtain imaging with high resolution and classify biopharmaceutical viruses in large groups. This method gives detailed information about the structure of viruses, providing valuable insights into their functional abilities. This approach allows quick evaluation of the quality of viral production used for oncolytic virotherapy (Newcastle disease virus) and development of vaccines (influenza), removing the time-consuming and complex requirements of traditional methods of testing (65). The utilization of SIM in imaging of microbes has provided useful insights into interactions of microbial communities and dynamics, demonstrating its ability to enhance our understanding of microbial systems.

7.5.4 Revolutionizing Electron Microscopy Techniques

The application of 3D reconstruction approaches in electron microscopy has enhanced it greatly in recent years. Novel electron microscopy techniques continue to evolve, such as cryo-EM and and correlative light and electron microscopy, which are of special interest for microbiologists (66). These techniques are discussed briefly in the following.

7.5.4.1 Cryo-electron Microscopy

Cryo-electron microscopy (cryo-EM) is a strong imaging technology for studying microbes and has greatly progressed the science of structural biology. This technique allows high-resolution imaging of the 3D biomacromolecular structures of microorganisms (67). Cryo-electron microscopy in the form of single particles of biological molecules allows the investigation of molecular assemblies that are too large and flexible for crystallography using X-ray (68). Cryo-electron microscopy has advanced in bacterial imaging by using devices that work at the temperature of liquid nitrogen (69). Advanced cryo-EM is broadly used in structural biology due to its capability for analysis of complex systems which are difficult to investigate with other methods (70). Advancements in imaging processes and microscopic instruments have resulted in an increasing number of maps produced by cryo-EM, enhancing its utilization in imaging microorganisms (71). Cryo-EM has been combined with various imaging techniques to improve its capabilities, beyond its use in structural biology. Correlative imaging combines light microscopy with high-resolution electron microscopy to efficiently target and identify tagged viral and bacterial components (72). Additionally, the advancement of correlated cryogenic photoactivated localization microscopy and cryo-electron tomography has enabled the precise localization of objects within cryotomograms beyond the diffraction limit of the light microscope. This has offered fresh perspectives on the dynamic structures of microbes (73). Cryo-EM has been used to reveal methods of bacterial transcriptional regulation, showcasing its adaptability in investigating microbial activities (74). Recently, developments in cryo-EM have significantly increased the number of structures contributed to the Electron Microscopy Data Bank (EMDB), with a focus on assembled bacterial secretion systems obtained using cryo-EM. Rapisarda et al. (2018) emphasize significant progress in comprehending bacterial secretion systems, focusing on types II, III, IV, VI, and VII, utilizing cryo-EM extensively (75). The approach has been valuable for comprehending the ribosome structure and function in bacteria (76).

Cryo-electron microscopy was used to determine the structure of the AdeB multidrug efflux pump from *Acinetobacter baumannii* embedded in lipidic nanodiscs, attaining a resolution of 2.98 Å. Microscopic investigations showed that when substrates are not present, each AdeB molecule in the trimer typically takes on the resting conformational state (77).

Cryo-scanning electron microscopy (CryoSEM) images showed important morphological characteristics, specifically arbuscules, in fine root endophytes (FRE), the unique arbuscule-producing fungi found beyond Glomeromycotina. Individual hyphae of the fungal root endosphere were seen as tightly interconnected among bundles of hyphae. This work aims to assess nutrient contents in fungal and plant structures of FRE and explore the use of cryoSEM for studying these fungi (78).

Cryo-electron microscopy (cryo-EM) and cryo-electron tomography (cryo-ET) have been crucial in developing structural biology by uncovering intricate structures of viral assemblies and intermediates. The vast cryo-EM data produced has significantly impacted our comprehension and contributed to the development and improvement of virus-based platforms for biotechnological and biological applications (79). Cryo-EM has transformed the visualization of microbes. However, it is crucial to recognize that conventional electron microscopy methods like transmission electron microscopy (TEM), scanning electron microscopy (SEM), or focused ion beam scanning electron microscopy (FIB-SEM) have limitations in preparing biological samples, resulting in significant errors. Cryo-EM provides a viable alternative for high-resolution imaging of microbiological specimens by preserving their original condition and avoiding artifacts from chemical fixation and dehydration. The advancements in cryo-EM have greatly improved the visualization of microorganisms, providing in-depth information about the structure of elements of microbes and their functions (53).

7.5.4.2 Correlative Light and Electron Microscopy

Correlative light and electron microscopy (CLEM) is a powerful and effective system that merges the data obtained from identical material by imaging sensory systems (80). It serves as a link between light microscopy (LM) and electron microscopy (EM) (81). CLEM enables the simultaneous imagining of a precise subcellular structure by means of both fluorescence light microscopy (FLM) and electron microscopy (82). According to Verkade (2015), CLEM consolidates the benefits of light microscopy and electron microscopy in a solo experiment, offering a greater amount of data and interpretation than each technique independently (83). Combining light microscopy and electron microscopy permits quick fluorescence imaging with the thorough resolution of electron microscopy (84). Besides, CLEM is a frontline method that allows for comparing two different forms of information by merging light and electron microscopy in the particular histological area of interest (85). This innovative technique incorporates the interelated strengths of electron microscopy (EM) and fluorescence (light) microscopy (FM) (86).

Correlative imaging is the incorporation of light microscopy and high-resolution electron microscopy to precisely locate and recognize labeled bacterial and viral elements. Advancements in biorthogonal and genetic tagging, enhanced workflows, and thorough data assembling are enhancing the effectiveness of correlative imaging methods (87). A well-organized method for correlative light and scanning electron microscopy (CLSEM) was created to image infection, using gold mesh grids as a substrate for perusing epithelial cell monolayers. This approach enables rapid imaging of live-cell bacterial infections in polarized cells, followed by SEM studies. The invasion mechanisms of *Salmonella enterica*, *Listeria monocytogenes*, and enteropathogenic *Escherichia coli* were studied by using this approach. This study illustrates the diversified use of gold mesh grids for examining the relation of pathogenic bacteria in CLSEM imaging with the polarized apical side of epithelial cells (88).

CLEM has been used to study the antiviral effects of silver nanoparticles (Ag NPs) on BUNV, a well-known virus in the Bunyavirales order. The study contrasted the antiviral properties of Ag NPs with the conventional antiviral ribavirin (RBV). These findings emphasized the importance of imaging techniques in investigating the mechanisms of action for both new and known antiviral medicines. Novel elements of BUNV infection and the antiviral actions of Ag NPs and RBV have been uncovered using both light and electron microscopy. Ag NPs and RBV both disrupted BUNV-induced deconstructed cell nucleoli and hindered the intercellular propagation of large virion groups, a transmission mechanism discovered for the first time in cultivated cells (89). Applying this technique broadened the usage of CLEM for imaging adhered cells. Lectins were used to immobilize yeast cells, providing a powerful method for studying cellular structures at the molecular level by combining CLEM and yeast genetics. Although CLEM is potent, its use with non-adherent cells, particularly in yeast, encounters obstacles. Yeast cells were immobilized on a gridded plate to successfully correlate fluorescence microscopy (FM) and electron microscopy (EM) pictures, which proved to be a beneficial technique for identifying the detected cells. The integration of FM and EM, together with live CLEM, was crucial in linking the localization of target molecules with detailed subcellular architecture, aiding in a comprehensive understanding of cellular structure at subcellular and molecular levels (90). Correlative light and electron microscopy (CLEM) is a technique that enables the concurrent examination of a specific subcellular structure using both fluorescence light microscopy and electron microscopy. It offers a more thorough insight into cellular processes by leveraging the advantages of both imaging techniques.

7.5.6 Super-resolution Microscopy

Super-resolution microscopy is an imaging process in which the resolution limit is exceeded as compared to traditional microscopy by two factors or more. This modern technique has broken down the diffraction limit by modifying the excitation or activation of light temporally or spatially. This technique comprises many types, like stimulated emission depletion microscopy (STED), photoactivated localized microscopy (PALM), and structured illumination microscopy (SIM) (91). These types are discussed as follows.

7.5.6.1 Stimulated Emission Depletion Microscopy for Imaging Microbes

STED is a powerful method for imaging microorganisms with a resolution that exceeds the limits of diffraction. This technique utilizes stimulated emission to allow sub diffraction resolution by decreasing the fluorescence in the surroundings of the spread function of excitation point (92). The cellular activities can be observed fast and efficiently by unlimited resolution of diffraction, which is not possible by the use of an ordinary or traditional microscope (93). It diminishes the effective focus point of cellular structures, which gives highly precise and accurate results (94). Scientists have been working on this technique to further increase its resolution limits for more effective results. Usually, this technique uses solid immersion lenses to produce an image with a resolution limit of about 2.4 ±0.3 nm with a localized precision of about 0.09 nm (95). By using this precise technique on a nanoscale level, the localization of protein can be facilitated. We can clearly examine highly complex structures such as slit diaphragm with the help of this latest method (96). This persuasive technique has overcome the limits of diffraction in fluorescence microscopy to achieve the spatial resolution of a few nm (97). Many biological applications have been shown by this latest technique, for example, living tissues and microorganisms have been well researched and examined by the higher optical resolution standards of this technology (98). Due to the latest technical methods, it has been made available commercially, which shows its vast potential for accessible imaging applications (99).

Recent advances in STED microscopy have allowed for the viewing of intricate structural features of different bacterial species by tagging them with KK114 fluorescent dye. KK114 labeling was useful for visualizing intricate nanoscale cell features, like a helical pattern that resembled the organization of the cytoskeleton of *Bacillus subtilis* (100). STED microscopy, utilizing super-resolution imaging techniques, was used to observe the interactions of LL-37 (an antibiotic peptide that inhibits the growth of *Mycobacterium tuberculosis* at a molecular level). LL-37 was demonstrated to be taken up by both Mtb-infected and uninfected human macrophages, accumulating in lysosomes and early endosomes where mycobacteria are present. LL-37 destroys the cell wall of both extra- and intracellular Mtb, resulting in the elimination of the pathogen. The study emphasizes the use of STED microscopy as a valuable and advanced method for studying interactions between hosts, pathogens, and peptides, beyond the capabilities of traditional confocal microscopy (101). Cell division proteins in bacterial cells, particularly *E. coli* and *B. subtilis*, are observed using stimulated emission depletion (STED) microscopy. Nanobodies that target fluorescent proteins (GFP, mCherry2, and

PAmCherry) coupled to the proteins of interest are used for STED imaging. A study evaluated how several organic fluorescent dyes affect the performance of STED microscopy in bacterial cells, with the goal of offering direction to scientists who are seeing macromolecules in bacterial systems using STED (102).

STED super-resolution microscopy has been used to study the endoplasmic reticulum (ER) structures generated by ZIKV, focusing on convoluted membranes (CM) and vesicular replication factories (RF). STED imaging showed the development of structured, crescent-shaped, dense central ER zones with unique smooth ER tubular matrices. ZIKV RFs are observed to concentrate in these compact areas in a crescent-shaped arrangement. 3D reconstruction using computational methods, confirmed by electron imaging, of brain organoids infected with ZIKV, offered valuable insights into these formations. A viral protein which plays an important role in the replication process has been identified by this microscopic technique along with the discovery of tubular matrices with replication factories induced by Zika virus (103). This has made possible the anatomical examination of microorganisms at a resolution much higher than the diffraction limits of an ordinary optical microscope. It has become a demanding tool for doing research on microorganisms due to its vast resolution capabilities.

7.5.6.2 Photoactivated localization microscopy (PALM) and fluorescence photoactivation localization microscopy (FPALM)

These microscopic techniques facilitate the examination of microorganisms and their components at the level of nanoscale (104). To cross the diffraction limits of a normal optical microscope, these microscopes make use of organic fluorophores and photoactivatable fluorescent proteins which empower the high-resolution imaging of microbes (105). The method of PALM consists of localization of a single molecule with photoactivation fluorescence, which overcomes the limits of diffraction of an ordinary microscope. The biological sample can be examined with high-resolution imaging by using this method (106). This has also been utilized in imaging of living cells in a more advanced resolution process by examining in a focal adhesion (107). PALM microscopy primarily concentrates on the fluorescence molecules of the microorganism to attain a high-resolution imaging along with mutual fluorescent material. When compared to additional techniques like single molecule localization microscopic techniques, the PALM contains diverse pros, such as imaging in sub-diffraction resolution fluorescence (108, 109). PALM, a super-resolution imaging method utilizing single molecules, has been used to investigate the *in vivo* structure of bacterial proteins. The FtsZ protein, functioning as a vital tubulin-like GTPase, is necessary for prokaryotic cell division. *In vivo*, it creates a ring-shaped structure called the Z-ring, which is positioned at the mid cell and examined using PALM in *E. coli* (110).

Past PALM studies have accurately measured the sizes of various structures such as ParA bundles responsible for separating *Caulobacter crescentus* chromosomes, microtubule and actin filaments, clathrin pits, bacterial membrane receptors, histone-like protein clusters, and the budding process of HIV-1 virions. The studies facilitated the measurement of molecular density to investigate focal adhesion, mechanical load, maturation, T cell activation leading to microcluster assembly, cluster formation from bacterial chemotaxis, and membrane protein cluster reorganization due to cholesterol alterations. A study emphasizes the potential of quantitative information in super-resolution microscopy, focusing on fluorophore blinking, a crucial aspect that impacts quantitative density measurements in PALM imaging (111). PALM and FPALM have transformed super-resolution microscopy by providing a potent method for visualizing microorganisms and their components with unparalleled detail and accuracy.

7.5.7 Advancements in Light-Sheet Microscopy

Advancements in light-sheet microscopy have shifted our capabilities for exploring dynamic biological activities in intricate detail. Advancements in these techniques of imaging reduce traditional microscopy limitations such as imaging speed, resolution, and phototoxicity by using advanced illumination techniques, innovative optical systems, and intricate computational algorithms (112). Advanced light-sheet microscopic methods such as selective plane illumination microscopy (SPIM) and oblique plane microscopy (OPM) have enhanced our ability to observe biological specimens with unprecedented clarity and detail (113), as discussed below.

7.5.7.1 Selective Plane Illumination Microscopy (SPIM)

Selective plane illumination microscopy (SPIM), also called light-sheet fluorescence microscopy, has emerged as a powerful imaging tool for observation of microbes due to its capability to provide high-resolution, rapid, and non-invasive imaging of biological specimens in three dimensions (3D) (114). SPIM has developed to address the limitations of conventional wide-field microscopic techniques by restricting excitation to the focal plane near area, resulting in fast operation and effective optical sectioning while decreasing premature photobleaching and out-of-focus background (115). This approach has been used widely in developmental research to visualize biofilms and cellular dynamics at different levels (116, 117). The distinctive approach of SPIM involves selective illumination of just focal plane, enabling high-contrast and rapid image capture with less exposure to light as compared to conventional tools like confocal laser scanning microscopy (118). SPIM allows quick live fluorescence imaging of biological specimens in three dimensions with effective optical sectioning, excellent spatial resolution, and reduced effects of photobleaching and phototoxicity (119).

Multidirectional selective plane illumination microscopy (mSPIM) is an advanced SPIM technique which has improved SPIM by enabling consistent excitation of fluorescence and contrast-improved imaging (120). Plane illumination microscopy using bessel beam has enhanced live fluorescence imaging of cellular dynamics in 3D, enhancing SPIM applications in research on microbial interactions and behavior (121). The utilization of extended depth-of-field light-sheet microscopy improved the imaging of high volumes with huge quantitative aperture, making the SPIM technique appropriate for imaging biofilms and communities of microorganisms (120).

Expansion and light-sheet microscopy were studied for their effectiveness in imaging infected materials of viruses, particularly using human A549 cells which were infected with live attenuated influenza vaccine (LAIV). The imaging technique produced three-dimensional images of whole cells in detail at a very high resolution, which were infected with LAIV, representing the presence of viral nucleoprotein in vesicles of cytosol. Light-sheet microscopy in combination with expansion microscopy is a perfect method for in-depth analysis of the relation between organelles of cells and viral proteins within intact cells (122). Imaging of ribonucleoproteins of influenza virus in real time was obtained by using a dual-view inverted selective plane illumination microscope (diSPIM) in cell lines related to biological samples (123), as demonstrated by Bhagwat et al. in 2018. These advancements in imaging techniques enabled frequent and prolonged imaging sessions of live cells covering more than 24 hours for in-depth examination of the whole infectious cycle of HIV-1 at the individual virus level (124).

A SPIM study focused on examination of the digestive tract of zebrafish larvae, unveiling anatomical structures, commensal activity of microbes, movements of immune cells, and other factors inside the varied and complex environment of the digestive tract (125). The imaging capabilities of SPIM, such as high-resolution, non-invasive 3D and rapid imaging, make it an efficient tool for studying microbial behavior, cellular dynamics, and biofilm formation.

7.5.7.2 Oblique Plane Microscopy (OPM)

Another advanced technique of imaging is oblique plane microscopy (OPM), which has enabled fast volumetric imaging of microorganisms and microbial components at high spatial and temporal resolution (126). OPM utilizes only one objective lens for both detection and illumination without using any extra elements for reflection at the sample space (127). Dunsby developed this technique in 2008, using a single objective of high numerical aperture (NA) to illuminate an oblique plane in the specimen and capture its picture (128). OPM in light-sheet microscopy gives steric access to the sample by its single objective lens (129). This technique produces a three-dimensional image by light-sheet axial scanning, resulting in high spatiotemporal resolution (130). OPM has been used for producing cellular resolution images in optically scattering mouse brain samples at depths of over 330 μm (131). The OPM technique has been improved to resolve particular problems and applications. For example, its 3D imaging has been modified using a stage-scanning approach known as stage-scanning OPM (ssOPM) (132). In addition, structured illumination in multidirection has been used in OPM to achieve a twofold enhanced resolution in light-sheet microscopy (133). OPM has been used for high-speed imaging of live cells, showing its promise and adaptability for dynamic research in biology (134).

Oblique plane microscopy (OPM) offers quick three-dimensional, high-quality imaging for microbial systems like closed habitats, gut microbiota, and bacterial biofilms. The potential of light-sheet techniques in uncovering the dynamics of microbial communities is gaining recognition and showing great promise (135). OPM has been utilized to visualize bacterial structures and their motility. An experiment was conducted employing OPM to directly quantify the propulsion matrix of a microscale helical propeller, specifically focused on the *E. coli* flagellum. The technique was used to evaluate the Brownian fluctuations of individual flagella. OPM's imaging characteristics make it suited for various biological, microbiological, and materials science applications (136).

7.6 IMAGING LIVE MICROORGANISMS

Live-cell imaging of microbes by advanced microscopic approaches has provided unprecedented insight into the dynamic environment of microbial behavior and physiology (137). Imaging of live cells has made possible the real-time observation of microbial cells so researchers can observe the activities of microbes with more detail and clarity. Fundamental microbial processes such as their growth, motility, division, and responses to stimuli within live populations can be visualized by advanced imaging techniques (138), which are discussed in the following.

7.6.1 Microbial Live-Cell Imaging

Recent years have witnessed notable progress in live-cell imaging of microbes, especially in the realm of single-cell microbiology. Consider the non-invasive RMCS-SIP method, which enables real-time cell imaging and signifies significant progress in this area. Musat et al. (2012) emphasize the methodological progress in single-cell imaging that has offered fresh perspectives on microbial abundance, activity, and identification, highlighting the significance of live-cell imaging techniques in comprehending microbial dynamics at the single-cell level (139).

A flexible and quickly deployable tool has been introduced for observing stationary microorganisms and environmental surfaces over time and space, specifically designed for real-time imaging of bacteria attached to surfaces. This tool could be crucial for observing living cells in microbial biofilms and surface-attached microorganisms (140).

High-definition differential interference contrast (DIC) imaging unveils temporary infection cycle phases of pathogens on the stem cell surface of human adults derived from intestinal epithelium. This technique provides information about the pathogens' movement, which is essential for real-time examination of interaction of pathogens with host cells. Van Rijn et al. (2022) emphasize the significance of imaging techniques for live cells in studying microbial dynamics, such as physiological responses, metabolic networks, host-pathogen interactions and biofilm development (141).

7.6.2 Microfluidic Systems for Imaging Live Microbes

Microfluidic technologies have greatly improved the observation of live microbes by allowing for the real-time viewing and control of microorganisms. Microfluidics was introduced for real-time observation of interactions of root bacteria (142). Microfluidic observation of nanoliter-scale human neutrophils at cellular resolution shows the ability of microfluidic system to study host–microbe ineractions (143). Moreover, the utilization of a shallow microfluidic tool for long-term living cell culture and high-resolution imaging without disrupting their dynamics, highlights the importance of microfluidics in producing high-resolution live microbe imaging (144).

Microfluidic systems have played a crucial role in microbial engineering, improving the sensitivity, accuracy, and throughput of microbial phenotyping (145). Additionally, a microfluidic system, RootChip, for live-cell imaging and rapid solution changes, demonstrates the versatility of microfluidics in studying cellular responses in plants (146). The increasing use of microfluidics for studying microorganisms due to its capability for real-time and live-cell imaging, further underlines the significance of microfluidic systems in microbial research (147).

Moreover, a microfluidic platform was presented for studying cellular structure–function relationships through live-cell imaging and cryo-electron tomography, showcasing the potential of microfluidics in preserving and imaging unstained specimens at the cellular level (148). An open-channel microfluidic system for chemical imaging of bacterial biofilms, demonstrating the capability of microfluidics in studying bacterial activity and dynamics with high spatial resolution, has been developed (149). These studies collectively highlight the diverse applications of microfluidic systems in imaging live microbes and studying their behavior.

7.7 IMAGING MICROBIAL COMPONENTS

There are different microbial components such as nucleic acids, lipids, etc. which can be examined by various imaging techniques.

7.7.1 Nucleic Acid Imaging

Nucleic acid imaging of microbes offers essential insights into composition, dynamics within microbial communities, and patterns of gene expression (150). Two effective approaches for observing nucleic acids of microbes are FISH and DNA Microarray (151) which are briefly discussed here.

7.7.1.1 Fluorescence In Situ Hybridization (FISH)

Fluorescence in situ hybridization (FISH) emerges as a widely utilized cytochemical method for genetic detection, identification, and precise localization, facilitated by a nucleic acid probe labeled with fluorescence. This technique specifically finds application in hybridization processes involving nucleic acid sequences of interest (152). The enduring significance of FISH is consistently underscored by advancements in probe design, signal amplification, and multiplexing, thereby broadening the scope of experimental research and clinical diagnoses in this domain (153). Notably, FISH and its variations demonstrate extensive utility in diverse areas such as the investigation of nucleic acids, cellular metabolic studies, diagnostics in oncology, and research in microbiology (154).

FISH characteristics have undergone significant evolution through precise advancements in probe design. These developments encompass heightened signal intensity, increased stability, expanded target range, and enhanced sensitivity. Prominent variants that are frequently employed include identification of individual genes FISH (RING-FISH), catalyzed reporter deposition FISH (CARD-FISH), gene-FISH, combinatorial labeling and spectral imaging FISH (CLASI-FISH), nucleic acid mimics FISH (NAM-FISH), as well as double labeling of oligonucleotide probes for FISH (DOPE-FISH) (155).

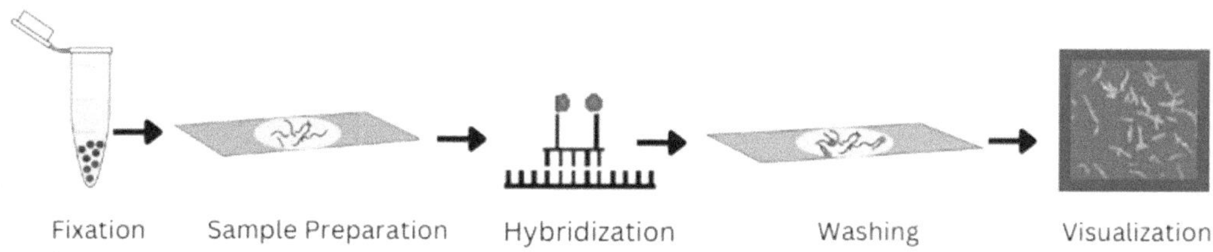

FIGURE 7.2 Illustration of different steps involved in fluorescence in situ hybridization (FISH) microscopy.

The nucleic acid mimics probe has emerged as a highly efficient and promising option, characterized by elevated affinity, specificity, and enhanced stability, effectively resisting enzymatic degradation (156). This particular probe addresses challenges associated with weak signals stemming from the low affinity and susceptibility of traditional DNA or RNA probes. Currently, the employed nucleic acid mimics (NAM) probes encompass peptide nucleic acid (PNA), 2'-O-Methyl-RNAs, and locked nucleic acid (LNA) (157).

The established procedures for fluorescence in situ hybridization (FISH) encompass a series of systematic steps (Figure 7.2): (a) specimen treatment, (b) probe denaturation, (c) hybridization, (d) elution, (e) signal amplification for hybridization signals (especially relevant for biotin-labeled probes), (f) re-staining, (g) encapsulation, and (h) observation of FISH results under a fluorescence microscope (158–160). Depending on specific detection requirements, diverse methods are employed for probe labeling and specimen processing, ensuring optimal clarity in detection outcomes during the subsequent hybridization process (161, 162).

7.7.1.2 DNA Microarray (DNA Chip)

DNA microarrays emerge as a potent device for comprehensive exploration of different aspects of biology in bacteria, facilitating simultaneous monitoring of gene expression across diverse bacterial species. This methodology presents a holistic for the reseach of cellular physiology, complementing conventional "gene-by-gene" methods as outlined by Wildsmith and Elcock in 2001 (163). Coined in publications by DeRisi et al. (1996) and Schena et al. (1995), the term DNA microarray denotes a technique that has evolved from a highly particular technique accessible to only a few individuals into a widely used tool with diverse applications that have become crucial in the field of microbiology (164–169).

The foundational principle of microarray technique involves the concurrent hybridization of a labeled nucleic acid mixture, known as the target, with a multitude of nucleic acid individually referred to as probes. This enables the simultaneous identification of numerous probes, each associated with a different spatial position in a single experiment. The designated position of a specific probe on the array is commonly referred to as a spot or feature. Fixed on a solid support, the probes facilitate the introduction of labeled targets as a solution onto the array for hybridization, a step following fluorescent labeling, as elucidated by Botstein and Brown in 1999 (170).

Numerous studies in the literature have provided comprehensive insights into a variety of fluorescent dyes and alternative labeling agents (171, 172). However, among the options, cyanine dyes Cy-3 and Cy-5 stand out as predominant choices due to their robust fluorescence, same chemical properties, well-differentiated spectra of fluorescence, and minimal affinity to the chip surface. It is noteworthy that despite common perceptions, these dyes do not inherently possess green and red colors; rather, these hues are achieved through computer-generated false coloring during the scanning process. Two primary methodologies for incorporating these dyes into cDNA via reverse transcription (RT) of RNA have been delineated, as outlined by Wildsmith et al. in 2001 (173).

7.7.2 Protein Imaging

Imaging of microbial proteins is important to understanding cellular activities and functions within communities of microbes (174). Protein tagging for fluorescence and mass spectrometry are prominent methods for quantitative analysis and visualization of microbial proteins (175).

7.7.2.1 Fluorescent Protein Tags

The regular practice of monitoring cellular proteins in living organisms using fluorescent tags has become possible with the advancement of GFP and its related derivatives. This is achieved by genetically fusing a fluorescent protein (FP) in-frame with a protein of interest, enabling precise localization in both space and time within particular sub-cellular compartments, cells, and tissues. Over the past decade, there has been notable development in enhancing and diversifying these tags for various biological applications. This advancement provides researchers with a range of options, including different emission and excitation spectra and levels of brightnesss, sensitivity of pH, and photo-physical characteritics for FPs (176). Consequently, a versatile palette of FP proteins is now accessible for genetically "painting" a protein in vivo, especially for light microscopy. Importantly, these tags are generally minimally disruptive when attached to the C- or N- terminal of most proteins, exhibit effective folding within biological temperature ranges, mature rapidly, and maintain fluorescence across numerous sub-cellular compartments (177).

When selecting a fluorescent protein (FP) for use in an imaging experiment, the initial consideration typically centers around the choice of color variant. The widely preferred form of this tag is the improved version (EGFP), which includes additional folding mutations in conjunction with the monomeric A206K substitution, commonly called mGFP (178, 179). Notably, the A206K mutation has been shown to diminish the inherent dimerization of FPs, especially at elevated protein concentrations (40). Recently introduced high-performance GFP mutants, namely superfolder GFP (sfGFP) and Emerald, feature extra mutations aimed at improving folding, enhancing brightness, and reducing sensitivity to environmental conditions (180, 181).

7.7.2.2 Imaging Mass Spectrometry

Since its initial application to microorganisms in 2008, MALDI Imaging Mass Spectrometry (MALDI IMS) has been instrumental in identifying various natural products from bacteria. It has proven its efficacy in probing the biosynthetic origins and processes of natural products across diverse environments, including investigations into marine microorganisms, microorganism–invertebrate assemblages, and complex organism consortia. In a notable instance, the integration of MALDI IMS with epifluorescence microscopy revealed the spatial distribution of a hexachlorinated peptide, 13-demethylisodysidenin, in sponge–cyanobacterium tissue. This application enables two-dimensional representations of natural product distributions in microbial systems, contributing to the creation of comprehensive three-dimensional (3D) IMS models. These models offer insights into molecule distribution, enhancing the understanding of the link between molecular composition and spatial arrangement. Notably, 3D MALDI IMS, unlike TOF-SIMS depth profiling, involves sequential image collection during serial sectioning (182).

7.7.3 Lipid Imaging

Imaging of microbial lipids has a significant role in enhancing our knowledge about microbial activities and their interactions within dynamic environments.

TOF-SIMS and AFM are two prevailing microbiological techniques which promise innovative visions of spatial dispersion of microbial lipids and their configuration at the subcellular level.

7.7.3.1 Time-of-Flight Secondary Ion Mass Spectrometry (TOF-SIMS)

Time-of-flight secondary ion mass spectrometry (ToF-SIMS) is like a microscopic illustrator, carefully revealing the detail of complex biological surfaces. Envision an engrossed high-powered beam slightly engaging with the surface of a sample, prompting it to expose its molecular secrets. Intense chemical descriptions can be created by this advanced technique by capturing the lipid essence directly from a living material's surface. The ion beam, just lika a painter's brush, sweeps across the surface and creates chemical images with high resolution. This makes possible the visualization of tiny lipid molecules by revealing mysteries at attomolar concentrations. The distinct detection mechanism of ToF-SIMS offers a versatile palette of simultaneous exploration, making it an esteemed tool for interpreting the complications of biological landscapes (183).

In the mesmerizing world of imaging mass spectrometry (IMS) techniques for discovering biological materials, we come across not only the notable ToF-SIMS but also the striking companions, matrix-assisted laser desorption ionization (MALDI) and desorption electrospray ionization (DESI). These technologies, much like ToF-SIMS, play a vital role in investigating the mysteries of lipids in the complicated landscape of living matter (184, 185).

Yet, MALDI and DESI go further, offering a broader perspective by detecting a diverse array of biomolecules – from proteins and peptides to nucleotides. It's like expanding the canvas to include an entire spectrum of life's building blocks. This versatility, particularly in directly identifying proteins and peptides, makes them invaluable in the fields of bioanalytical chemistry and biomedicine (183).

Among these techniques, MALDI takes center stage, becoming the go-to method in the realm of medical and bioanalytical research. Its applications are diverse, ranging from clinical diagnostics to pharmaceutical research and the exciting journey of biomarker discovery. In the symphony of IMS techniques, each contributes its unique notes, creating a harmonious melody that resonates through the exploration of life's intricate chemistry (183, 186).

7.8 APPLICATIONS OF ADVANCED IMAGING TECHNIQUES IN MICROBIOLOGY

By applying advanced microbiology imaging techniques, researchers can analyze critical issues, investigate innovative solutions, and promote improvement in various fields such as human health, the environment, and biotechnology.

7.8.1 Microbial Pathogenesis Studies

Microbial pathogenesis studies have greatly improved due to the implementation of advanced techniques of microbial imaging. These techniques have transformed this field by providing high-resolution capabilities with spatiotemporal imaging, enabling the observation of small, delicate, and abruptly occurring processes of biology (187). Advanced techniques of microscopy, encompassing fluorescence microscopy, two-photon fluorescence microscopy, confocal fluorescence microscopy, and super-resolution microscopy, have played a significant role in deep understanding of microbial pathogenesis. They have enabled the observation of both molecular and cellular dynamics (188, 189). Moreover, the implementation of super-resolution microscopy, including PALM, SIM, and STED microscopy, has provided the ability to get resolution at the molecular level, thereby increasing understanding of microbial processes at both molecular and cellular levels (190). Additionally, advanced microscopic techniques and fluorescent probe development offer new insights into biochemistry at the nanometer scale, showing interactions between pathogens and eukaryotic cells. New tools have been provided to researchers for the analysis of dynamic actions of molecules of cells by the integration of new fluorescence labeling and staining techniques with advanced microscopy of fluorescence, thereby further improving microbial pathogenesis studies (191).

Emerging techniques of intravital imaging are bridging the gap for analysis of the interaction between host and pathogen in cultured cells and inside living organisms due to high-resolution capabilities. Microscopic approaches like scanned sheet microscopy, single-plane illumination microscopy, and multiphoton microscopy present novel tools for visualizing the interaction between host and pathogen in tissue (191). Lattice light-sheet microscopy (LLSM) is an advanced technique which has facilitated rapid 4D imaging, enabling studies of microbial pathogenesis with high resolution while reducing photo-damage. LLSM has been used for investigating *Legionella pneumophila* infection which revealed that morphological features such as polarity, migration pattern, and smoothness are correlated with *L. pneumophila* infection both dynamically and statistically (192).

Combination of (cryo) electron microscopy and fluorescence SRM, known as CLEM, in correlative approaches, has significant potential for providing high-resolution data on bacterial molecular processes and their host cell interactions (193). The CLEM application in the biological context of bacterial infection has just emerged. For example, techniques like the combination of EM with SIM, STORM, or super-resolution confocal microscopy, have provided data on various aspects, such as vacuole extensions containing *Brucella* spp. for colocalizing with the endoplasmic reticulum, the architecture of the cell envelope of *Deinococcus radiodurans*, and interactions of *Salmonella typhimurium* with its host (194–196). dSTORM and PALM were also used to investigate the subcellular protein localization on the surface of *Mycoplasma* spp. Super-resolution microscopy has significantly improved our understanding of the biology of these smallest pathogens (197). The pathogenicity of *Salmonella* due to island-2 signaling proteins (SPI-2) was shown by PALM (198). SR-SIM was used to study microbial biofilms which play a significant role in human infection (199).

Translocated or secreted molecules often act as virulence factors in bacterial pathogenesis (200). Advancements in microscopy have enhanced our understanding of the secretory system (201). SIM was used to visualize hemolysin A (HlyA, a substrate of the type I secretion system) on the *E. Coli* surface (202). YopB and YopD are the pore complex proteins of T3SS that were also visualized in *Yersinia enterocolitica* in infected cells (203). Overall, the application of advanced microbiology techniques has advanced the study of microbial pathogenesis, providing tools to researchers for visualization and analysis of complex processes biology at high spatial and temporal resolution, ultimately increasing our understanding of microbial pathogenesis (Figure 7.3).

7.8.2 Drug Discovery and Development

Novel imaging approaches in microbiology have significantly aided drug discovery and development. Imaging techniques *in vivo* have been an effective complement to different platforms for antibiotic resistance and the development of drugs (205). Microscopic techniques have been helpful for imaging antimicrobial agents' activity inside biofilms, which revealed temporal and spatial antimicrobial activity patterns (206). High-content imaging (HCI) has made it possible to consistently differentiate bacteria with different modes of action by evaluating and detecting minute morphological traits in individual bacteria that are exposed to several classes of antimicrobials (207).

With the help of advanced microbiology imaging and SRM techniques, like single-molecule localization microscopy (SMLM), as shown by random optical reconstruction microscopy (STORM) and photoactivated localization microscopy (PALM), structured illumination microscopy (SIM) and stimulated emission depletion microscopy (STED) (208), researchers are not only enabled to identify and analyze drug actions at subcellular level but they also monitor changes in drug behavior within various organelles and their impacts on interactions with organelles (209). It involves monitoring the convenience and entry route of drugs into the transmembrane for evaluation of drug accumulation and metabolism in different organelles and identification of the main proteins involved in regulation and specific targeting of organelles (210). With the help of SRM technology, the route of entry of drugs into the cell and the content of drugs in subcellular and intracellular targets can be visualized. The kinetic methods of drug absorption, transmission, distribution, metabolism, and removal can

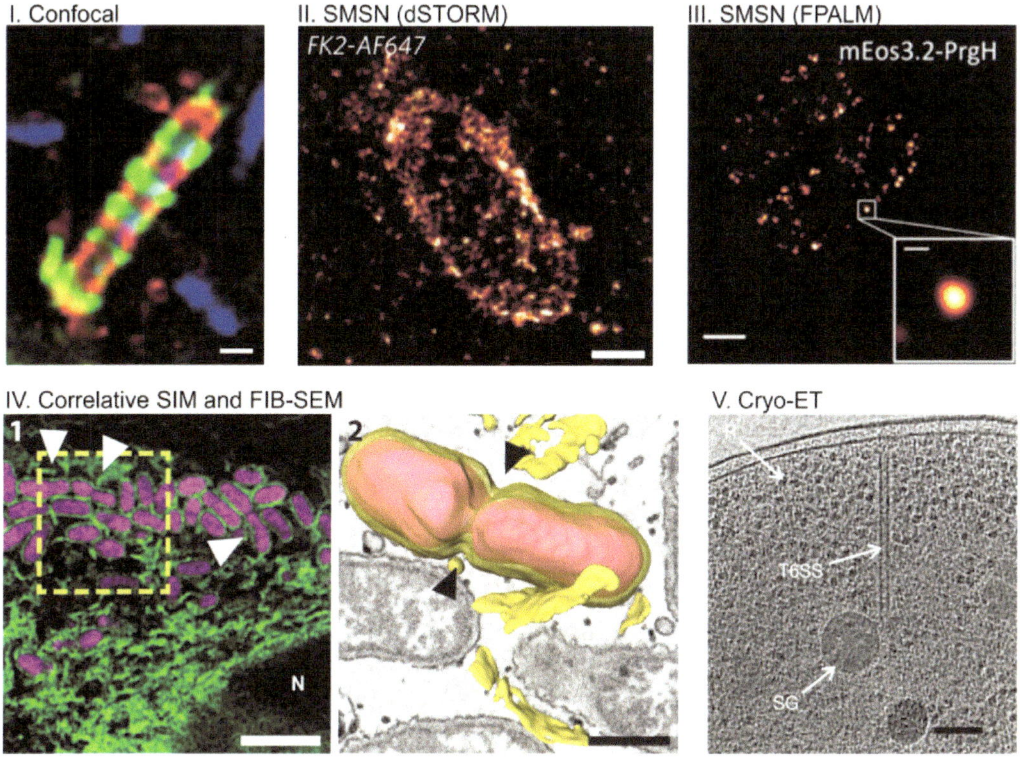

FIGURE 7.3 Recent advances in imaging techniques to study the infection process. I. Confocal microscopy image of *Shigella flexneri* bacteria entrapped in septin cage. II. dSTORM image of cytocolic *Salmonella typhimurium*. III. FPALM visualization of *S. typhimurium* T3SS. IV. Correlative SIM showing infection of 1. *Brucella abortus* in HeLA cells and 2. rBCV (replicative *Brucella* containing vacule). V. Cryo-ET image of *Vibrio cholerae* T6SS (204).

also be examined by SRM techniques. These approaches help researchers to comprehend their technical benefits and applications in development of drugs and pharmacokinetics (211).

SIM technique has enabled the visualization of recently developed teixobactin antibiotic, its structural organization and interactions with the Gram-positive cell wall, by labeling it fluorescently. It enabled understanding of the mechanism of teixobactin action and the discovery of similar antibiotics (212). Cryo-EM is contributing significantly to the development of drugs in the upcoming years. This technique defines the mechanism of binding and mode of action of commonly used drugs as well as helping in designing and discovering new compounds (213). A large number of cryo-EM studies have revealed the binding of ribosomes to anti-protozoals, antibiotics, and different compounds from various organisms. The natural products include avilamycin, erythromycin, mefloquine, emetine, and paromomycin, while the synthetic products are cadazolid and linezolid. Understanding the ribosome's binding mechanism of action with drugs as shown by cryo-EM studies is essential for the development of drugs (214). Recent cryo-EM studies revealed the integrase strand-transfer inhibitors (INSTIs) such as dolutegravir and bictegravir binding to integrase, unveiling the two Mg2+ ions' coordination with these molecules at the active site (215). Understanding the binding mode of these antiretroviral drugs to the HIV intasome is helpful in the development of more potent INSTIs and in designing more antiretroviral drugs (216).

Confocal Raman microscopy has recently been utilized to investigate Erlotinib, a therapeutic drug for lung cancer, by using an intracellular marker such as the alkyl bond. The mechanism of drug delivery has been analyzed by spontaneous Raman spectroscopy. The emergence of Raman spectroscopy is an alternative for understanding the topical delivery of drugs (217). SRS was utilized by Xie and Guy to optimize the topical delivery of drugs for skin diseases (218). SRS microscopy helps in understanding the dermato-pharmacokinetics, which provides new insights into various aspects such as intracellular diffusion and different rates of penetration through hair follicles. Studies on toxicity for topically used drugs may be improved by monitoring the morphology of dermal cells through SRS imaging after administration of the drug (219). The application of advanced microbiology imaging techniques has significantly advanced drug development. These techniques have enabled the visualization of antimicrobial action, high-throughput phenotyping of antimicrobial effects on individual bacteria, and the development of cost-effective imaging systems, all of which are crucial for the discovery and development of antimicrobial drugs.

7.8.3 Applications in Environmental Microbiology

Advancements in imaging techniques in microbiology have highly contributed to environmental microbiology by enhancing our comprehension of the communities of microorganisms and their roles in various ecological systems. These techniques have made possible the examination of biofilm composition, structure, and function (220). One of the commonly used technique in environmenta; microbiology is fluorescence in situ hybridization (FISH), enabling detection and identification of microorganisms in their natural habitat (221). In addition, the manufacturing of new classes of fluorescent probe has improved the fluorescence toolkit for imaging and observing microbial communities in cell biology (222). The ribosomal-RNA (rRNA) technique has become significant in evolution and ecology of microorganisms, providing a powerful tool for single-cell detection in communities of microorganisms (223).

Moreover, imaging applications have been revolutionized by the incorporation of nanotechnology, allowing real-time subwavelength imaging of various processes taking place within cells and nanobiosensor synthesis of next-generation sensing applications (224). Raman spectroscopy has emerged as a significant tool for monitoring the presence of pigments in communities of microbes, giving important information about the chemical composition in samples of the environment (225). Furthermore, flow cytometry has been employed frequently in environmental microbiology research, making it possible to analyze microbial populations in diverse ecosystems (226).

Advanced techniques of imaging such as neural networks of convolution and analysis of images have been used to detect variations and analyze microbes, which further improves understanding of microbial communities (227). The implementation and miniaturization of fluorescence microscopy have enhanced the potential uses of light microscopy, mainly efficient screening, portability, and diagnosis based on imaging (228). Confocal X-ray fluorescence tomography and XRF imaging have been used for three-dimensional microanalysis of trace elements, demonstrating the multiple applications of advanced microbiological techniques of imaging in environmental microbiology (229).

Overall, the applications of advanced microbiological imaging techniques have significantly contributed to the comprehensive understanding of microbial communities in various environmental settings, offering valuable insights into their composition, structure, and functions.

7.8.4 Biotechnological Spplications

The implementation of advanced microbiological imaging techniques has greatly contributed to the development of biotechnology applications. This allows a deeper comprehension of biological systems and offers innovative solutions for different industries. Microbiological approaches have been extensively applied in various processes of biotechnology, including plant crops and animal optimization, supplements and food production, therapy development for various diseases using microorganisms, and the bioremediation of polluted areas (230). Advancements in microscopic techniques for imaging, like confocal laser scanning microscopy, have become important biotechnology tools, facilitating the visualization and analysis of biological samples in detail (231). These microscopic techniques have helped to study photosynthesis in living cells and have applications in diverse fields such as solid-state physics, automation engineering, and biotechnology (232).

Moreover, the utilization of photoactivatable complexes of metal has made it possible for physiological imaging and analysis of tissues and cells, enabling an important understanding of biological functions at the molecular level (233). Furthermore, microbiology and metabolomics studies have shown the potential benefits of imaging using mass spectrometry by providing novel opportunities to understand the spatial organization of molecules inside biological samples (234). Advancements in microbiological imaging techniques have taken biotechnological research to new heights.

7.9 FUTURE PERSPECTIVES

Recent advancements in microbiology imaging techniques have created interesting opportunities for future research in different areas. Microbiology is revolutionizing due to the integration of data handling and automation, as well as the enhanced resolution power of microscopic techniques (235).

Despite the recent emergence of super-resolution microscopy, it has had a great impact on biological research, with the expectation of further advancements in fluorescent markers, analytical tools, labeling techniques, and equipment. Advanced techniques need to be commercialized for greater accessibility. Designers need to integrate user-friendly algorithms for super-resolution microscopic imaging, focusing on accessibility and adaptability. The implementation of deep knowledge and machine learning is recognized as a powerful approach for extracting new biological information from different images of super-resolution microscopy (236). More precise structural details of microorganisms will be obtained without using specialized instruments of super-resolution by further advancing expansion microscopy which will enlarge the biological specimens physically (237).

Implementing the combination of various imaging techniques such as super-resolution microscopy with live-cell imaging or electron microscopy provides a complete understanding of microbial life in dynamic settings at high resolution (238). The development of miniaturized microscopes will provide new opportunities in neuroscience and other related fields, having numerous versions which are already available (239). Ongoing advancements in ultra-high-resolution methods of imaging, such as MINFLUX and other similar techniques, have the potential to extend the resolution boundaries to the atomic level, providing comprehensive insights into molecular machinery found inside microbes (240). Development of *in vivo* non-invasive imaging techniques will provide the ability to observe communities of microorganisms in their natural habitats inside living hosts, which will provide insights into microbial pathogenesis, ecology, and interactions between microbe and host (241).

The application of machine learning and AI for the analysis of images has the potential to revolutionize the interpretation of complex sets of data by enabling high throughput and automated analysis of imaging data. This would greatly enhance the accuracy and efficiency of understanding and identification of microbial interactions and behaviors (242). Prospects of imaging technology include "smart microscopes" which are completely automatic and will adapt to biological specimens (243). Combining analysis of neural networks with low-resolution imaging reflectance shifts toward intelligent and adaptive microscopy to target high-resolution microscopic imaging (244). The integration of AI with microscopic imaging techniques enables automated analysis of images, minimizes human error, and reduces image artifacts (245). The screening of large populations of microorganisms will be made possible at high resolution by advancements in high-throughput and automated imaging equipment, which will facilitate the investigation of expression patterns of genes, microbial diversity, and phenotypic responses to environmental stimuli (246). The combination of microbial biosensors and micro or nanotechnologies is expected to result in significant improvement and development in microbial biosensors (247). Advancements in these sensing technologies are expected to have a large number of applications in microbiology in the future (248).

These future perspectives emphasize the ongoing evolution of microbiological imaging techniques toward automated, more integrated, and high-resolution approaches, highlighting the potential for innovative discoveries in microbiology and its applications.

7.10 CONCLUSION

Advancements in microbiology imaging techniques have enhanced our knowledge about microorganisms at the molecular and cellular levels. These innovative imaging technologies, ranging from revolutionizing electron microscopy to super-resolution microscopy and various other imaging techniques, have provided novel insights into structures, functions, and interactions of microbial life. As these techniques are continually evolving, they provide tremendous opportunities for further research in microbiology, resulting in novel discoveries as well as their applications in various fields that range from medicine to environmental microbiology. The ongoing implementation of emerging imaging technologies is expected to increase our understanding of the intricate details and complexities of microbes, offering opportunities for targeted interventions, discoveries, and advancements in a variety of scientific fields.

REFERENCES

1. Puchkov E. Image analysis in microbiology: a review. *Journal of Computer and Communications*. 2016;4(15):8.
2. Gest H. The discovery of microorganisms by Robert Hooke and Antoni Van Leeuwenhoek, fellows of the Royal Society. *Notes and Records of the Royal Society of London*. 2004 May 22;58(2):187–201.
3. Wainwright M, Lederberg J, Lederberg J. History of microbiology. *Encyclopedia of Microbiology*. 1992;2:419–37.
4. Neu TR, Manz B, Volke F, Dynes JJ, Hitchcock AP, Lawrence JR. Advanced imaging techniques for assessment of structure, composition and function in biofilm systems. *FEMS Microbiology Ecology*. 2010 Apr 1;72(1):1–21.
5. Wainwright M. An alternative view of the early history of microbiology. *Advances in Applied Microbiology*. 2003 Jan 1;52:333–56.
6. Chen X, Zheng B, Liu H. Optical and digital microscopic imaging techniques and applications in pathology. *Analytical Cellular Pathology*. 2011 Jan 1;34(1–2):5–18.
7. Binnig G, Rohrer H. Scanning tunneling microscopy. *Surface Science*. 1983 Mar 2;126(1–3):236–44.
8. Drummen GP. Fluorescent probes and fluorescence (microscopy) techniques—illuminating biological and biomedical research. *Molecules*. 2012 Nov 28;17(12):14067–90.
9. Lakshman M. A review on application of conventional electron microscopy in disease diagnosis and research. *Microscopy Research*. 2022 Jul 28;46 :1–12.
10. Gelderblom HR, Krüger DH. Helmut Ruska (1908–1973): his role in the evolution of electron microscopy in the life sciences, and especially virology. *Advances in Imaging and Electron Physics*. 2014 Jan 1;182 :1–94).
11. Goldsmith CS, Miller SE. Modern uses of electron microscopy for detection of viruses. *Clinical Microbiology Reviews*. 2009 Oct;22(4):552–63.
12. Curry A, Appleton H, Dowsett B. Application of transmission electron microscopy to the clinical study of viral and bacterial infections: present and future. *Micron*. 2006 Feb 1;37(2):91–106.
13. Khodavirdipour A, Mehregan M, Rajabi A, Shiri Y. Microscopy and its application in microbiology and medicine from light to quantum microscopy: a mini review. *Avicenna Journal of Clinical Microbiology and Infection*. 2019 Dec 31;6(4):133–7.
14. Bartholomew JW, Levin R. The structure of Saccharomyces carlsbergensis and S. cerevisiae as determined by ultra-thin sectioning methods and electron microscopy. *Microbiology*. 1955 Jun;12(3):473–7.
15. Leung JW, Sung JY, Costerton JW. Bacteriological and electron microscopy examination of brown pigment stones. *Journal of Clinical Microbiology*. 1989 May;27(5):915–21.

16. Matias VR, Al-Amoudi A, Dubochet J, Beveridge TJ. Cryo-transmission electron microscopy of frozen-hydrated sections of Escherichia coli and Pseudomonas aeruginosa. *Journal of Bacteriology*. 2003 Oct 15;185(20):6112–8.
17. Chatterjee SN, Das J. Electron microscopic observations on the excretion of cell-wall material by Vibrio cholerae. *Microbiology*. 1967 Oct;49(1):1–1.
18. Price AB, Levi J, Dolby JM, Dunscombe PL, Smith A, Clark J, Stephenson ML. Campylobacter pyloridis in peptic ulcer disease: microbiology, pathology, and scanning electron microscopy. *Gut*. 1985 Nov;26(11):1183.
19. Rodgers FG, Greaves PW, Macrae AD, Lewis MJ. Electron microscopic evidence of flagella and pili on Legionella pneumophila. *Journal of Clinical Pathology*. 1980 Dec;33(12):1184.
20. Mishra M, Chauhan P. Applications of microscopy in bacteriology. *Microscopy Research*. 2016 Jan 5;4(1):1–9.
21. Jensen EC. Types of imaging, Part 2: an overview of fluorescence microscopy. *The Anatomical Record: Advances in Integrative Anatomy and Evolutionary Biology*. 2012 Oct;295(10):1621–7.
22. Swana KW, Nagarajan R, Camesano TA. Atomic force microscopy to characterize antimicrobial peptide-induced defects in model supported lipid bilayers. *Microorganisms*. 2021 Sep 17;9(9):1975.
23. Morandat S, Azouzi S, Beauvais E, Mastouri A, El Kirat K. Atomic force microscopy of model lipid membranes. *Analytical and Bioanalytical Chemistry*. 2013 Feb;405:1445–61.
24. Bagatolli LA, Ipsen JH, Simonsen AC, Mouritsen OG. An outlook on organization of lipids in membranes: searching for a realistic connection with the organization of biological membranes. *Progress in Lipid Research*. 2010 Oct 1;49(4):378–89.
25. Johnston LJ. Nanoscale imaging of domains in supported lipid membranes. *Langmuir*. 2007 May 22;23(11):5886–95.
26. El Kirat K, Morandat S, Dufrêne YF. Nanoscale analysis of supported lipid bilayers using atomic force microscopy. *Biochimica et Biophysica Acta (BBA) – Biomembranes*. 2010 Apr 1;1798(4):750–65.
27. Butt HJ, Franz V. Rupture of molecular thin films observed in atomic force microscopy. I. Theory. *Physical Review E*. 2002 Sep 5;66(3):031601.
28. Garcia-Manyes S, Sanz F. Nanomechanics of lipid bilayers by force spectroscopy with AFM: a perspective. *Biochimica et Biophysica Acta (BBA) – Biomembranes*. 2010 Apr 1;1798(4):741–9.
29. Schermelleh L, Heintzmann R, Leonhardt H. A guide to super-resolution fluorescence microscopy. *Journal of Cell Biology*. 2010 Jul 26;190(2):165–75.
30. Lue N, Choi W, Badizadegan K, Dasari RR, Feld MS, Popescu G. Confocal diffraction phase microscopy of live cells. *Optics Letters*. 2008 Sep 15;33(18):2074–6.
31. De Los Ríos A, Ascaso C, Wierzchos J, Fernández-Valiente E, Quesada A. Microstructural characterization of cyanobacterial mats from the McMurdo Ice Shelf, Antarctica. *Applied and Environmental Microbiology*. 2004 Jan;70(1):569–80.
32. Verma B, Robarts RD, Headley JV. Effect of tetracycline on the colonization and growth of microbes on Scirpus lacustris litter in oligotrophic and eutrophic waters. *Aquatic Microbial Ecology*. 2007 Apr 3;47(1):91–8.
33. Kim SW, Ki MS, Kim CL, Hwang IS, Jeon IS. A simple confocal microscopy-based method for assessing sperm movement. *Development & Reproduction*. 2017 Sep;21(3):229.
34. De Luca GM, Desclos E, Breedijk RM, Dolz-Edo L, Smits GJ, Nahidiazar L, Bielefeld P, Picavet L, Fitzsimons CP, Hoebe R, Manders EM. Configurations of the re-scan confocal microscope (RCM) for biomedical applications. *Journal of Microscopy*. 2017 May;266(2):166–77.
35. De Luca GM, Breedijk RM, Brandt RA, Zeelenberg CH, de Jong BE, Timmermans W, Azar LN, Hoebe RA, Stallinga S, Manders EM. Re-scan confocal microscopy: scanning twice for better resolution. *Biomedical Optics Express*. 2013 Nov 1;4(11):2644–56.
36. Wang CY, Tsai JC, Chuang CC, Hsieh YS, Sun CW. Aorta fluorescence imaging by using confocal microscopy. *International Scholarly Research Notices*. 2011;2011:215627.
37. Zapata RO, Bramante CM, de Moraes IG, Bernardineli N, Gasparoto TH, Graeff MS, Campanelli AP, Garcia RB. Confocal laser scanning microscopy is appropriate to detect viability of Enterococcus faecalis in infected dentin. *Journal of Endodontics*. 2008 Oct 1;34(10):1198–201.
38. Neu TR, Lawrence JR. Investigation of microbial biofilm structure by laser scanning microscopy. In: *Productive Biofilms*. 2014;146:1–51.

39. Hickey PC, Swift SR, Roca MG, Read ND. Live-cell imaging of filamentous fungi using vital fluorescent dyes and confocal microscopy. *Methods in Microbiology*. 2004 Jan 1;34:63–87.
40. Zachow C, Fatehi J, Cardinale M, Tilcher R, Berg G. Strain-specific colonization pattern of Rhizoctonia antagonists in the root system of sugar beet. *FEMS Microbiology Ecology*. 2010 Oct 1;74(1):124–35.
41. Aschenbrenner IA, Cernava T, Berg G, Grube M. Understanding microbial multi-species symbioses. *Frontiers in Microbiology*. 2016 Feb 18;7:180.
42. Zoumi A, Lu X, Kassab GS, Tromberg BJ. Imaging coronary artery microstructure using second-harmonic and two-photon fluorescence microscopy. *Biophysical Journal*. 2004 Oct 1;87(4):2778–86.
43. Pavlova I , Hume KR, Yazinski SA, Flanders J, Southard TL, Weiss RS, Webb WW. Multiphoton microscopy and microspectroscopy for diagnostics of inflammatory and neoplastic lung. *Journal of Biomedical Optics*. 2012;17(3):036014.
44. Dunn KW, Sutton TA. Functional studies in living animals using multiphoton microscopy. *ILAR Journal*. 2008 Jan 1;49(1):66–77.
45. Zoumi A, Yeh A, Tromberg BJ. Imaging cells and extracellular matrix in vivo by using second-harmonic generation and two-photon excited fluorescence. *Proceedings of the National Academy of Sciences*. 2002 Aug 20;99(17):11014–9.
46. Tauer U. Advantages and risks of multiphoton microscopy in physiology. *Experimental Physiology*. 2002 Nov;87(6):709–14.
47. Fernández A, Classen A, Josyula N, Florence JT, Sokolov AV, Scully MO, Straight P, Verhoef AJ. Simultaneous two-and three-photon deep imaging of autofluorescence in bacterial communities. *Sensors*. 2024 Jan 20;24(2):667.
48. Sewald X. Visualizing viral infection in vivo by multi-photon intravital microscopy. *Viruses*. 2018 Jun 20;10(6):337.
49. Lee J, Hestrin R, Nuccio EE, Morrison KD, Ramon CE, Samo TJ, Pett-Ridge J, Ly SS, Laurence TA, Weber PK. Label-free multiphoton imaging of microbes in root, mineral, and soil matrices with time-gated coherent Raman and fluorescence lifetime imaging. *Environmental Science & Technology*. 2022 Jan 14;56(3):1994–2008.
50. Kim HS, Park SY, Kang S, Czymmek KJ. Time-lapse imaging of root pathogenesis and fungal proliferation without physically disrupting roots. In: *Fusarium Wilt: Methods and Protocols*, ed. Jeffrey Coleman. Springer; 2022. p. 153–70.
51. Bai J, Zhang L, Ma H, Shi Z, Liu L, Ma K, Ren H, Chai L. Super-resolution imaging of submicron natural defects on optics component by structured illumination microscopy. *Journal of Physics: Conference Series*. 2023 Mar 1;2464:012011.
52. Christensen CN, Ward EN, Lu M, Lio P, Kaminski CF. ML-SIM: universal reconstruction of structured illumination microscopy images using transfer learning. *Biomedical Optics Express*. 2021 May 1;12(5):2720–33.
53. Shi R, Zhang Y, Zhou T, Kong L. Hilo based line scanning temporal focusing microscopy for high-speed, deep tissue imaging. *Membranes*. 2021 Aug 17;11(8):634.
54. Lin R, Kipreos ET, Zhu J, Khang CH, Kner P. Subcellular three-dimensional imaging deep through multicellular thick samples by structured illumination microscopy and adaptive optics. *Nature Communications*. 2021 May 25;12(1):3148.
55. Xu D, Ding J, Peng L. Structured illumination imaging with quasi-periodic patterns. *Journal of Biophotonics*. 2020 Jun;13(6):e201960209.
56. Xu D, Zhou W, Peng L. Three-dimensional live multi-label light-sheet imaging with synchronous excitation-multiplexed structured illumination. *Optics Express*. 2017 Dec 11;25(25):31159–73.
57. Park JK, Joo KN. High-resolution 3D imaging by structured illumination microscopy. *Proceedings of the 9th International Conference of Asian Society for Precision Engineering and Nanotechnology*, ed. NML Sharon and AS Kumar. Research Publishing; 2022. p. 507–509.
58. Smith CS, Slotman JA, Schermelleh L, Chakrova N, Hari S, Vos Y, Hagen CW, Müller M, van Cappellen W, Houtsmuller AB, Hoogenboom JP. Structured illumination microscopy with noise-controlled image reconstructions. *Nature Methods*. 2021 Jul;18(7):821–8.
59. Gao L. QSIM: quantitative structured illumination microscopy image processing in ImageJ. *Biomedical Engineering Online*. 2015 Dec;14(1):1–9.

60. Musat N, Musat F, Weber PK, Pett-Ridge J. Tracking microbial interactions with NanoSIMS. *Current Opinion in Biotechnology*. 2016 Oct 1;41:114–21.
61. Jin GQ, Sun DE, Xia X, Jiang ZF, Cheng B, Ning Y, Wang F, Zhao Y, Chen X, Zhang JL. Bioorthogonal lanthanide molecular probes for near-infrared fluorescence and mass spectrometry imaging. *Angewandte Chemie International Edition*. 2022 Oct 24;61(43):e202208707.
62. Turnbull L, Strauss MP, Liew AT, Monahan LG, Whitchurch CB, Harry EJ. Super-resolution imaging of the cytokinetic Z ring in live bacteria using fast 3D-structured illumination microscopy (f3D-SIM). *JoVE (Journal of Visualized Experiments)*. 2014 Sep 29;91:e51469.
63. Halpern AR, Alas GC, Chozinski TJ, Paredez AR, Vaughan JC. Hybrid structured illumination expansion microscopy reveals microbial cytoskeleton organization. *ACS Nano*. 2017 Dec 26;11(12):12677–86.
64. Ranieri AM, Vezzelli M, Leslie KG, Huang S, Stagni S, Jacquemin D, Jiang H, Hubbard A, Rigamonti L, Watkin EL, Ogden MI. Structure illumination microscopy imaging of lipid vesicles in live bacteria with naphthalimide-appended organometallic complexes. *Analyst*. 2021;146(12):3818–22.
65. Laine RF, Goodfellow G, Young LJ, Travers J, Carroll D, Dibben O, Bright H, Kaminski CF. Structured illumination microscopy combined with machine learning enables the high throughput analysis and classification of virus structure. *eLife*. 2018 Dec 13;7:e40183.
66. Miranda K, Girard-Dias W, Attias M, de Souza W, Ramos I. Three-dimensional reconstruction by electron microscopy in the life sciences: an introduction for cell and tissue biologists. *Molecular Reproduction and Development*. 2015 Jul;82(7–8):530–47.
67. Benjin X, Ling L. Developments, applications, and prospects of cryo-electron microscopy. *Protein Science*. 2020 Apr;29(4):872–82.
68. Frank J. Single-particle imaging of macromolecules by cryo-electron microscopy. *Annual Review of Biophysics and Biomolecular Structure*. 2002 Jun;31(1):303–19.
69. Milne JL, Subramaniam S. Cryo-electron tomography of bacteria: progress, challenges, and future prospects. *Nature Reviews Microbiology*. 2009 Sep;7(9):666–75.
70. Nogales E. The development of cryo-EM into a mainstream structural biology technique. *Nature Methods*. 2016 Jan;13(1):24–7.
71. He J, Huang SY. EMNUSS: a deep learning framework for secondary structure annotation in cryo-EM maps. *Briefings in Bioinformatics*. 2021 Nov;22(6):bbab156.
72. Howes SC, Koning RI, Koster AJ. Correlative microscopy for structural microbiology. *Current Opinion in Microbiology*. 2018 Jun 1;43:132–8.
73. Chang YW, Chen S, Tocheva EI, Treuner-Lange A, Löbach S, Søgaard-Andersen L, Jensen GJ. Correlated cryogenic photoactivated localization microscopy and cryo-electron tomography. *Nature Methods*. 2014 Jul;11(7):737–9.
74. Wood DM, Dobson RC, Horne CR. Using cryo-EM to uncover mechanisms of bacterial transcriptional regulation. *Biochemical Society Transactions*. 2021 Dec 17;49(6):2711–26.
75. Rapisarda C, Tassinari M, Gubellini F, Fronzes R. Using cryo-EM to investigate bacterial secretion systems. *Annual Review of Microbiology*. 2018 Sep 8;72:231–54.
76. Razi A, Britton RA, Ortega J. The impact of recent improvements in cryo-electron microscopy technology on the understanding of bacterial ribosome assembly. *Nucleic Acids Research*. 2017 Feb 17;45(3):1027–40.
77. Su CC, Morgan CE, Kambakam S, Rajavel M, Scott H, Huang W, Emerson CC, Taylor DJ, Stewart PL, Bonomo RA, Yu EW. Cryo-electron microscopy structure of an Acinetobacter baumannii multidrug efflux pump. *MBio*. 2019 Aug 27;10(4):10–128.
78. Albornoz FE, Hayes PE, Orchard S, Clode PL, Nazeri NK, Standish RJ, Bending GD, Hilton S, Ryan MH. First cryo-scanning electron microscopy images and X-ray microanalyses of mucoromycotinian fine root endophytes in vascular plants. *Frontiers in Microbiology*. 2020;11:2018.
79. Luque D, Castón JR. Cryo-electron microscopy for the study of virus assembly. *Nature Chemical Biology*. 2020 Mar;16(3):231–9.
80. Spiegelhalter C, Laporte JF, Schwab Y. Correlative light and electron microscopy: from live cell dynamics to 3D ultrastructure. *Methods in Molecular Biology*. 2014;1117:485–501.
81. Joosten B, Willemse M, Fransen J, Cambi A, Van den Dries K. Super-resolution correlative light and electron microscopy (SR-CLEM) reveals novel ultrastructural insights into dendritic cell podosomes. *Frontiers in Immunology*. 2018 Aug 22;9:1908.

82. Vicidomini G, Gagliani MC, Canfora M, Cortese K, Frosi F, Santangelo C, Di Fiore PP, Boccacci P, Diaspro A, Tacchetti C. High data output and automated 3D correlative light–electron microscopy method. *Traffic*. 2008 Nov;9(11):1828–38.
83. Verkade P. Important steps in a correlative light electron microscopy experiment. *Microscopy and Microanalysis*. 2015 Aug;21(S3):387–8.
84. Reddick LE, Alto NM. Correlative light and electron microscopy (CLEM) as a tool to visualize microinjected molecules and their eukaryotic sub-cellular targets. *JoVE (Journal of Visualized Experiments)*. 2012 May 4;63:e3650.
85. Sato K, Kurita T, Sato F, Ohta K, Sato K, Chitose SI, Ono T, Umeno H. Three-dimensional fine structures of the maculae flavae of the human vocal fold using correlative light and electron microscopy. *Laryngoscope Investigative Otolaryngology*. 2022 Dec;7(6):1943–50.
86. Schwertner M, Stacey D. Cryo-correlative light and electron microscopy (Cryo-CLEM): specimen workflow paths and recent instrument developments. *Microscopy and Microanalysis*. 2015 Aug 1;21:1565.
87. Howes SC, Koning RI, Koster AJ. Correlative microscopy for structural microbiology. *Current Opinion in Microbiology*. 2018 Jun 1;43:132–8.
88. Kommnick C, Lepper A, Hensel M. Correlative light and scanning electron microscopy (CLSEM) for analysis of bacterial infection of polarized epithelial cells. *Scientific Reports*. 2019 Nov 19;9(1):17079.
89. Garcia-Serradilla M, Risco C. Light and electron microscopy imaging unveils new aspects of the antiviral capacity of silver nanoparticles in bunyavirus-infected cells. *Virus Research*. 2021 Sep 1;302:198444.
90. Asakawa H, Hiraoka Y, Haraguchi T. A method of correlative light and electron microscopy for yeast cells. *Micron*. 2014 Jun 1;61:53–61.
91. Galbraith CG, Galbraith JA. Super-resolution microscopy at a glance. *Journal of Cell Science*. 2011 May 15;124(10):1607–11.
92. Blom H, Widengren J. Stimulated emission depletion microscopy. *Chemical Reviews*. 2017 Jun 14;117(11):7377–427.
93. Pellett PA, Sun X, Gould TJ, Rothman JE, Xu MQ, Corrêa IR, Bewersdorf J. Two-color STED microscopy in living cells. *Biomedical Optics Express*. 2011 Aug 1;2(8):2364–71.
94. Kim D, Kim T, Lee J, Shim SH. Amplified expansion stimulated emission depletion microscopy. *ChemBioChem*. 2019 May 15;20(10):1260–5.
95. Wildanger D, Patton BR, Schill H, Marseglia L, Hadden JP, Knauer S, Schönle A, Rarity JG, O'Brien JL, Hell SW, Smith JM. Solid immersion facilitates fluorescence microscopy with nanometer resolution and sub-Ångström Emitter localization. *Advanced Materials*. 2012 Nov 20;24(44):OP309–13.
96. Unnersjö-Jess D, Scott L, Blom H, Brismar H. Super-resolution stimulated emission depletion imaging of slit diaphragm proteins in optically cleared kidney tissue. *Kidney International*. 2016 Jan 1;89(1):243–7.
97. Dyba M, Hell SW. Photostability of a fluorescent marker under pulsed excited-state depletion through stimulated emission. *Applied Optics*. 2003 Sep 1;42(25):5123–9.
98. Bethge P, Chéreau R, Avignone E, Marsicano G, Nägerl UV. Two-photon excitation STED microscopy in two colors in acute brain slices. *Biophysical Journal*. 2013 Feb 19;104(4):778–85.
99. Schröder J, Benink H, Dyba M, Los GV. In vivo labeling method using a genetic construct for nanoscale resolution microscopy. *Biophysical Journal*. 2009 Jan 7;96(1):L1–3.
100. Lucidi M, Hristu R, Nichele L, Stanciu GA, Tranca DE, Holban AM, Visca P, Stanciu SG, Cincotti G. STED nanoscopy of KK114-stained pathogenic bacteria. *Journal of Biophotonics*. 2020 Sep;13(9):e202000097.
101. Deshpande D, Grieshober M, Wondany F, Gerbl F, Noschka R, Michaelis J, Stenger S. Super-resolution microscopy reveals a direct interaction of intracellular Mycobacterium tuberculosis with the antimicrobial peptide LL-37. *International Journal of Molecular Sciences*. 2020;21(18):6741.
102. Cramer K, Bolender AL, Stockmar I, Jungmann R, Kasper R, Shin JY. Visualization of bacterial protein complexes labeled with fluorescent proteins and nanobody binders for STED microscopy. *International Journal of Molecular Sciences*. 2019 Jul 10;20(14):3376.
103. Long RK. 3D Stimulated Emission Depletion super resolution microscopy reveals novel association between replication factories and organized smooth tubular matrices during Zika virus infection (Doctoral dissertation, University of British Columbia). 2019.
104. Puchkov EO. Computerized fluorescence microscopy of microbial cells. *World Journal of Microbiology and Biotechnology*. 2021 Nov;37(11):189.

105. Ishitsuka Y, Nienhaus K, Nienhaus GU. Photoactivatable fluorescent proteins for super-resolution microscopy. In: *Photoswitching Proteins: Methods and Protocols*. 2014:239–60.
106. Wijesooriya CS, Nyamekye CK, Smith EA. Optical imaging of the nanoscale structure and dynamics of biological membranes. *Analytical Chemistry*. 2018 Oct 19;91(1):425–40.
107. Deschout H, Lukes T, Sharipov A, Szlag D, Feletti L, Vandenberg W, Dedecker P, Hofkens J, Leutenegger M, Lasser T, Radenovic A. Complementarity of PALM and SOFI for super-resolution live-cell imaging of focal adhesions. *Nature Communications*. 2016 Dec 19;7(1):13693.
108. Wolter S, Sauer M. Follow-up to paper by Wolter S, Schüttpelz M, Tscherepanow M, van de Linde S, Heilemann M, Sauer M. Entitled real-time computation of subdiffraction-resolution fluorescence images. *Journal of Microscopy*. 2011 Oct 25;245(1):109.
109. Klementieva NV, Bozhanova NG, Mishina NM, Zagaynova EV, Lukyanov KA, Mishin AS. Common fluorescent proteins for single-molecule localization microscopy. *Progress in Biomedical Optics and Imaging—Proceedings of SPIE*. 2015 Jun 21;9536:953609.
110. Fu G, Huang T, Buss J, Coltharp C, Hensel Z, Xiao J. In vivo structure of the E. coli FtsZ-ring revealed by photoactivated localization microscopy (PALM). *PLoS One*. 2010 Sep 13;5(9):e12680.
111. Coltharp C, Kessler RP, Xiao J. Accurate construction of photoactivated localization microscopy (PALM) images for quantitative measurements. *PLoS One*. 2012 Dec 12;7(12):e51725.
112. Chen BC, Legant WR, Wang K, Shao L, Milkie DE, Davidson MW, Janetopoulos C, Wu XS, Hammer III JA, Liu Z, English BP. Lattice light-sheet microscopy: imaging molecules to embryos at high spatiotemporal resolution. *Science*. 2014 Oct 24;346(6208):1257998.
113. Stelzer EH, Strobl F, Chang BJ, Preusser F, Preibisch S, McDole K, Fiolka R. Light sheet fluorescence microscopy. *Nature Reviews Methods Primers*. 2021 Nov 3;1(1):73.
114. Jemielita M, Taormina MJ, Burns AR, Hampton JS, Rolig AS, Guillemin K, Parthasarathy R. Spatial and temporal features of the growth of a bacterial species colonizing the zebrafish gut. *MBio*. 2014 Dec 31;5(6):10–128.
115. Planchon TA, Gao L, Milkie DE, Davidson MW, Galbraith JA, Galbraith CG, Betzig E. Rapid three-dimensional isotropic imaging of living cells using Bessel beam plane illumination. *Nature Methods*. 2011 May;8(5):417–23.
116. Huisken J, Stainier DY. Selective plane illumination microscopy techniques in developmental biology. *Development*. 2009 Jun 15;136(12):1963–75.
117. Neu TR, Lawrence JR. Innovative techniques, sensors, and approaches for imaging biofilms at different scales. *Trends in Microbiology*. 2015 Apr 1;23(4):233–42.
118. Keomanee-Dizon K, Jones M, Luu P, Fraser SE, Truong TV. Extended depth-of-field light-sheet microscopy improves imaging of large volumes at high numerical aperture. *Applied Physics Letters*. 2022 Oct 17;121(16).
119. Gao L. Optimization of the excitation light sheet in selective plane illumination microscopy. *Biomedical Optics Express*. 2015 Mar 1;6(3):881–90.
120. Glaser AK, Chen Y, Yin C, Wei L, Barner LA, Reder NP, Liu JT. Multidirectional digital scanned light-sheet microscopy enables uniform fluorescence excitation and contrast-enhanced imaging. *Scientific Reports*. 2018 Sep 17;8(1):13878.
121. Gao L, Shao L, Chen BC, Betzig E. 3D live fluorescence imaging of cellular dynamics using Bessel beam plane illumination microscopy. *Nature Protocols*. 2014 May;9(5):1083–101.
122. Mascheroni L, Scherer KM, Manton JD, Ward E, Dibben O, Kaminski CF. Combining sample expansion and light sheet microscopy for the volumetric imaging of virus-infected cells with super-resolution. *Biomedical Optics Express*. 2020;11(9):5032–44.
123. Bhagwat AR, Le Sage V, Lakdawala SS. Live imaging of influenza viral ribonucleoproteins using light-sheet microscopy. *Methods in Molecular Biology*. 2018;1836:303–27.
124. Francis AC, Melikyan GB. Live-cell imaging of early steps of single HIV-1 infection. *Viruses*. 2018 May 19;10(5):275.
125. Baker RP, Taormina MJ, Jemielita M, Parthasarathy R. A combined light sheet fluorescence and differential interference contrast microscope for live imaging of multicellular specimens. *Journal of Microscopy*. 2015;258(2):105–12.
126. Lamb JR, Ward EN, Kaminski CF. Open-source software package for on-the-fly deskewing and live viewing of volumetric lightsheet microscopy data. *Biomedical Optics Express*. 2023 Feb 1;14(2):834–45.

127. Yang B, Lange M, Millett-Sikking A, Zhao X, Bragantini J, VijayKumar S, Kamb M, Gómez-Sjöberg R, Solak AC, Wang W, Kobayashi H. DaXi—high-resolution, large imaging volume and multi-view single-objective light-sheet microscopy. *Nature Methods*. 2022 Apr;19(4):461–9.
128. Kumar M, Kishore S, Nasenbeny J, McLean DL, Kozorovitskiy Y. Integrated one-and two-photon scanned oblique plane illumination (SOPi) microscopy for rapid volumetric imaging. *Optics Express*. 2018 May 14;26(10):13027–41.
129. Kumar M, Kishore S, McLean DL, Kozorovitskiy Y. Crossbill: an open access single objective light-sheet microscopy platform. *BioRxiv*. 2021 May 1:2021-04.
130. Chen B, Chang BJ, Roudot P, Zhou F, Sapoznik E, Marlar-Pavey M, Hayes JB, Brown PT, Zeng CW, Lambert T, Friedman JR. Resolution doubling in light-sheet microscopy via oblique plane structured illumination. *Nature Methods*. 2022 Nov;19(11):1419–26.
131. Kumar M, Kozorovitskiy Y. Tilt-invariant scanned oblique plane illumination microscopy for large-scale volumetric imaging. *Optics Letters*. 2019 Apr 1;44(7):1706--9.
132. Maioli V, Chennell G, Sparks H, Lana T, Kumar S, Carling D, Sardini A, Dunsby C. Time-lapse 3-D measurements of a glucose biosensor in multicellular spheroids by light sheet fluorescence microscopy in commercial 96-well plates. *Scientific Reports*. 2016 Nov 25;6(1):37777.
133. Chen B, Chang BJ, Zhou FY, Daetwyler S, Sapoznik E, Nanes BA, Terrazas I, Gihana GM, Castro LP, Chan IS, Conacci-Sorrell M. Increasing the field-of-view in oblique plane microscopy via optical tiling. *Biomedical Optics Express*. 2022 Nov 1;13(11):5616–27.
134. Kumar S, Wilding D, Sikkel MB, Lyon AR, MacLeod KT, Dunsby C. Application of oblique plane microscopy to high-speed live cell imaging. *Proceeding of SPIE—The International Society for Optical Engineering*. 2011;8086:80860V.
135. Parthasarathy R. Monitoring microbial communities using light sheet fluorescence microscopy. *Current Opinion in Microbiology*. 2018 Jun;43:31–7.
136. Djutanta F, Brown PT, Nainggolan B, Coullomb A, Radhakrishnan S, Sentosa J, Yurke B, Hariadi RF, Shepherd DP. Decoding the hydrodynamic properties of microscale helical propellers from Brownian fluctuations. *Proceedings of the National Academy of Sciences of the United States of America*. 2023 May 30;120(22):e2220033120.
137. Liu YN, Lv ZT, Lv WL, Liu DF, Liu XW. Label-free optical imaging of the electron transfer in single live microbial cells. *Nano Letters*. 2023 Jan 3;23(2):558–66.
138. Ducret A, Maisonneuve E, Notareschi P, Grossi A, Mignot T, Dukan S. A microscope automated fluidic system to study bacterial processes in real time. *PLoS One*. 2009 Sep 30;4(9):e7282.
139. Musat N, Foster R, Vagner T, Adam B, Kuypers MM. Detecting metabolic activities in single cells, with emphasis on nanoSIMS. *FEMS Microbiology Reviews*. 2012 Mar;36(2):486–511.
140. Kiyokawa T, Usuba R, Obana N, Yokokawa M, Toyofuku M, Suzuki H, Nomura N. A versatile and rapidly deployable device to enable spatiotemporal observations of the sessile microbes and environmental surfaces. *Microbes and Environments*. 2017;32(1):88–91.
141. Van Rijn JM, Eriksson J, Grüttner J, Sundbom M, Webb DL, Hellström PM, Svärd SG, Sellin ME. High-definition DIC imaging uncovers transient stages of pathogen infection cycles on the surface of human adult stem cell-derived intestinal epithelium. *mBio*. 2022 Feb 22;13(1):e0002222.
142. Massalha H, Korenblum E, Malitsky S, Shapiro OH, Aharoni A. Live imaging of root–bacteria interactions in a microfluidics setup. *Proceedings of the National Academy of Sciences of the United States of America*. 2017 Apr 25;114(17):4549–54.
143. Ellett F, Jalali F, Marand AL, Jorgensen J, Mutlu BR, Lee J, Raff AB, Irimia D. Microfluidic arenas for war games between neutrophils and microbes. *Lab on a Chip*. 2019;19(7):1205–16.
144. Kozgunova E, Goshima G. A versatile microfluidic device for highly inclined thin illumination microscopy in the moss Physcomitrella patens. *Scientific Reports*. 2019 Oct 23;9(1):15182.
145. Kou S, Cheng D, Sun F, Hsing IM. Microfluidics and microbial engineering. *Lab on a Chip*. 2016;16(3):432–46.
146. Hani S, Cuyas L, David P, Secco D, Whelan J, Thibaud MC, Merret R, Mueller F, Pochon N, Javot H, Faklaris O. Live single-cell transcriptional dynamics via RNA labelling during the phosphate response in plants. *Nature Plants*. 2021 Aug;7(8):1050–64.
147. Takahashi K, Li X, Kunoh T, Nagasawa R, Takeshita N, Utada AS. Novel insights into microbial behavior gleaned using microfluidics. *Microbes and Environments*. 2023;38(5).

148. Fuest M, Schaffer M, Nocera GM, Galilea-Kleinsteuber RI, Messling JE, Heymann M, Plitzko JM, Burg TP. In situ microfluidic cryofixation for cryo focused ion beam milling and cryo electron tomography. *Scientific Reports*. 2019 Dec 13;9(1):19133.
149. Holman HY, Miles R, Hao Z, Wozei E, Anderson LM, Yang H. Real-time chemical imaging of bacterial activity in biofilms using open-channel microfluidics and synchrotron FTIR spectromicroscopy. *Analytical Chemistry*. 2009 Oct 15;81(20):8564–70.
150. Rastogi G, Sani RK. Molecular techniques to assess microbial community structure, function, and dynamics in the environment. In: *Microbes and Microbial Technology: Agricultural and Environmental Applications*, ed. I Ahmad, F Ahmad, and J Pichtel. Springer; 2011. p. 29–57.
151. Newby DT, Marlowe EM, Maier RM. Nucleic acid-based methods of analysis. In: *Environmental Microbiology*, ed. RM Maier, IL Pepper, and CP Gerba. Academic Press; 2009. p. 243–84.
152. Guimarães NM, Azevedo NF, Almeida C. FISH variants. In: *Fluorescence In-Situ Hybridization (FISH) for Microbial Cells: Methods and Concepts*, ed. NF Azevedo and C Almeida. Springer; 2021. p. 17–33.
153. Levsky JM, Singer RH. Fluorescence in situ hybridization: past, present and future. *Journal of Cell Science*. 2003 Jul 15;116(14):2833–8.
154. Veselinyová D, Mašlanková J, Kalinová K, Mičková H, Mareková M, Rabajdová M. Selected in situ hybridization methods: principles and application. *Molecules*. 2021 Jun 24;26(13):3874.
155. Gu J, Wang H, Zhang M, Xiong Y, Yang L, Ren B, Huang R. Application of fluorescence in situ hybridization (FISH) in oral microbial detection. *Pathogens*. 2022 Dec 1;11(12):1450.
156. Frickmann H, Zautner AE, Moter A, Kikhney J, Hagen RM, Stender H, Poppert S. Fluorescence in situ hybridization (FISH) in the microbiological diagnostic routine laboratory: a review. *Critical Reviews in Microbiology*. 2017 May 4;43(3):263–93.
157. Fontenete S, Carvalho D, Guimarães N, Madureira P, Figueiredo C, Wengel J, Azevedo NF. Application of locked nucleic acid-based probes in fluorescence in situ hybridization. *Applied Microbiology and Biotechnology*. 2016 Jul;100(13):5897–906.
158. Behnam F, Vilcinskas A, Wagner M, Stoecker K. A straightforward DOPE (double labeling of oligonucleotide probes)-FISH (fluorescence in situ hybridization) method for simultaneous multicolor detection of six microbial populations. *Applied and Environmental Microbiology*. 2012 Aug 1;78(15):5138–42.
159. Al-Ahmad A, Roth D, Wolkewitz M, Wiedmann-Al-Ahmad M, Follo M, Ratka-Krüger P, Deimling D, Hellwig E, Hannig C. Change in diet and oral hygiene over an 8-week period: effects on oral health and oral biofilm. *Clinical Oral Investigations*. 2010 Aug;14:391–6.
160. Alraies A, Canetta E, Waddington RJ, Moseley R, Sloan AJ. Discrimination of dental pulp stem cell regenerative heterogeneity by single-cell Raman spectroscopy. *Tissue Engineering Part C: Methods*. 2019 Aug 1;25(8):489–99.
161. Brinig MM, Lepp PW, Ouverney CC, Armitage GC, Relman DA. Prevalence of bacteria of division TM7 in human subgingival plaque and their association with disease. *Applied and Environmental Microbiology*. 2003 Mar;69(3):1687–94.
162. Chen Y, Wong WK, Seneviratne JC, Huang S, McGrath C, Hagg U. Associations between salivary cytokines and periodontal and microbiological parameters in orthodontic patients. *Medicine*. 2021 Mar 3;100(10):e24924.
163. Wildsmith SE, Elcock FJ. Microarrays under the microscope. *Molecular Pathology*. 2001 Feb;54(1):8.
164. DeRisi J, Penland L, Bittner ML, Meltzer PS, Ray M, Chen Y, Su YA, Trent JM. Use of a cDNA microarray to analyse gene expression. *Nature Genetics*. 1996 Dec;14:457–60.
165. Schena M, Shalon D, Davis RW, Brown PO. Quantitative monitoring of gene expression patterns with a complementary DNA microarray. *Science*. 1995 Oct 20;270(5235):467–70.
166. Bowtell DD. Options available—from start to finish—for obtaining expression data by microarray. *Nature Genetics*. 1999 Jan;21(1):25–32.
167. Cheung VG, Morley M, Aguilar F, Massimi A, Kucherlapati R, Childs G. Making and reading microarrays. *Nature Genetics*. 1999 Jan;21(1):15–9.
168. DeRisi JL, Iyer VR, Brown PO. Exploring the metabolic and genetic control of gene expression on a genomic scale. *Science*. 1997 Oct 24;278(5338):680–6.
169. Dharmadi Y, Gonzalez R. DNA microarrays: experimental issues, data analysis, and application to bacterial systems. *Biotechnology Progress*. 2004;20(5):1309–24.
170. Brown PO, Botstein D. Exploring the new world of the genome with DNA microarrays. *Nature Genetics*. 1999 Jan;21(1):33–7.

171. Badiee A, Eiken HG, Steen VM, Løvlie R. Evaluation of five different cDNA labeling methods for microarrays using spike controls. *BMC Biotechnology*. 2003 Dec;3(1):1–5.
172. Schena M, Davis RW. Technology standards for microarray research. *Microarray Biochip Technology*, ed. M Schena. Eaton Publishing; 2000. p. 1–8.
173. Wildsmith SE, Archer GE, Winkley AJ, Lane PW, Bugelski PJ. Maximization of signal derived from cDNA microarrays. *Biotechniques*. 2001 Jan;30(1):202–8.
174. Whidbey C, Wright AT. Activity-based protein profiling—enabling multimodal functional studies of microbial communities. In: *Activity-Based Protein Profiling*, ed. BF Cravatt, K Hsu, and E Weerapana. Springer; 2019. p. 1–21.
175. Smits AH, Vermeulen M. Characterizing protein–protein interactions using mass spectrometry: challenges and opportunities. *Trends in Biotechnology*. 2016 Oct;34(10):825–34.
176. Day RN, Davidson MW. The fluorescent protein palette: tools for cellular imaging. *Chemical Society Reviews*. 2009;38(10):2887–921.
177. Crivat G, Taraska JW. Imaging proteins inside cells with fluorescent tags. *Trends in Biotechnology*. 2012 Jan;30(1):8–16.
178. Zacharias DA, Violin JD, Newton AC, Tsien RY. Partitioning of lipid-modified monomeric GFPs into membrane microdomains of live cells. *Science*. 2002 May 3;296(5569):913–6.
179. Cubitt AB, Woollenweber LA, Heim R. Understanding structure-function relationships in the Aequorea victoria green fluorescent protein. *Methods in Cell Biology*. 1998;58:19–30.
180. Tsien RY. Building and breeding molecules to spy on cells and tumors. *FEBS Letters*. 2005 Feb 7;579(4):927–32.
181. Pédelacq JD, Cabantous S, Tran T, Terwilliger TC, Waldo GS. Engineering and characterization of a superfolder green fluorescent protein. *Nature Biotechnology*. 2006 Jan;24(1):79–88.
182. Shih CJ, Chen PY, Liaw CC, Lai YM, Yang YL. Bringing microbial interactions to light using imaging mass spectrometry. *Natural Product Reports*. 2014;31(6):739–55.
183. Passarelli MK, Winograd N. Lipid imaging with time-of-flight secondary ion mass spectrometry (ToF-SIMS). *Biochimica et Biophysica Acta (BBA) – Molecular and Cell Biology of Lipids*. 2011 Nov;1811(11):976–90.
184. Fuchs B, Süß R, Schiller J. An update of MALDI-TOF mass spectrometry in lipid research. *Progress in Lipid Research*. 2010 Oct;49(4):450–75.
185. Wu C, Ifa DR, Manicke NE, Cooks RG. Molecular imaging of adrenal gland by desorption electrospray ionization mass spectrometry. *Analyst*. 2010;135(1):28–32.
186. Wisztorski M, Lemaire R, Stauber J, Ait-Menguellet S, Jardin-Mathé O, Salzet M, Fournier I.MALDI imaging: a new technology to discover and validate new biomarkers. *Médecine/Sciences*. 2007;23(1):31–6.
187. Rella CE, Ruel N, Eugenin EA. Development of imaging techniques to study the pathogenesis of biosafety level 2/3 infectious agents. *Pathogens and Disease*. 2014 Dec 1;72(3):167–73.
188. Chen Y, Liang CP, Liu Y, Fischer AH, Parwani AV, Pantanowitz L. Review of advanced imaging techniques. *Journal of Pathology Informatics*. 2012 Jan 1;3(1):22.
189. Haagensen JA, Regenberg B, Sternberg C. Advanced microscopy of microbial cells. *High Resolution Microbial Single Cell Analytics*. 2011;124:21–54.
190. Bond C, Santiago-Ruiz AN, Tang Q, Lakadamyali M. Technological advances in super-resolution microscopy to study cellular processes. *Molecular Cell*. 2022 Jan 20;82(2):315–32.
191. Hoppe AD, Seveau S, Swanson JA. Live cell fluorescence microscopy to study microbial pathogenesis. *Cell Microbiology*. 2009 Apr;11(4):540–50.
192. Yi X, Miao H, Lo JK, Elsheikh MM, Lee TH, Jiang C, Zhang Y, Segelke BW, Overton KW, Bremer PT, Laurence TA. Tailored approach to study Legionella infection using a lattice light sheet microscope (LLSM). *Biomedical Optics Express*. 2022 Aug 1;13(8):4134–59.
193. Kooger R, Szwedziak P, Böck D, Pilhofer M. CryoEM of bacterial secretion systems. *Current Opinion in Structural Biology*. 2018 Oct 1;52:64–70.
194. Sedzicki J, Tschon T, Low SH, Willemart K, Goldie KN, Letesson JJ, Stahlberg H, Dehio C. 3D correlative electron microscopy reveals continuity of Brucella-containing vacuoles with the endoplasmic reticulum. *Journal of Cell Science*. 2018 Feb 15;131(4):jcs210799.

195. Sexton DL, Burgold S, Schertel A, Tocheva EI. Super-resolution confocal cryo-CLEM with cryo-FIB milling for in situ imaging of Deinococcus radiodurans. *Current Research Structural Biology*. 2022 Jan 1;4:1–9.
196. Van Elsland DM, Pujals S, Bakkum T, Bos E, Oikonomeas-Koppasis N, Berlin I, Neefjes J, Meijer AH, Koster AJ, Albertazzi L, van Kasteren SI. Ultrastructural imaging of Salmonella–host interactions using super-resolution correlative light–electron microscopy of bioorthogonal pathogens. *ChemBioChem*. 2018 Aug 16;19(16):1766–70.
197. Rideau F, Villa A, Belzanne P, Verdier E, Hosy E, Arfi Y. Imaging minimal bacteria at the nanoscale: a reliable and versatile process to perform single-molecule localization microscopy in mycoplasmas. *Microbiology Spectrum Journal*. 2022 Jun 29;10(3):e00645–22.
198. Liew AT, Foo YH, Gao Y, Zangoui P, Singh MK, Gulvady R, Kenney LJ. Single cell, super-resolution imaging reveals an acid pH-dependent conformational switch in SsrB regulates SPI-2. *Elife*. 2019 Apr 29;8:e45311.
199. Wang Y, Fu M, Wu B, Huang M, Ma T, Zang H, Jiang H, Zhang Y, Li C. Insight into biofilm-forming patterns: biofilm-forming conditions and dynamic changes in extracellular polymer substances. *Environmental Science and Pollution Research*. 2022 Dec;29(59):89542–56.
200. Galán JE. Common themes in the design and function of bacterial effectors. *Cell Host & Microbe*. 2009 Jun 18;5(6):571–9.
201. Lunelli M, Kamprad A, Bürger J, Mielke T, Spahn CM, Kolbe M. Cryo-EM structure of the Shigella type III needle complex. *PLoS Pathogens*. 2020 Feb 24;16(2):e1008263.
202. Beer T, Hänsch S, Pfeffer K, Smits SH, Weidtkamp-Peters S, Schmitt L. Quantification and surface localization of the Hemolysin A type I secretion system at the endogenous level and under conditions of overexpression. *Applied Environmental Microbiology*. 2022 Feb 8;88(3):e01896–21.
203. Carsten A, Wolters M, Aepfelbacher M. Super-resolution fluorescence microscopy for investigating bacterial cell biology. *Molecular Microbiology*. 2024;121:646–58.
204. López-Jiménez AT, Mostowy S. Emerging technologies and infection models in cellular microbiology. *Nature Communication*. 2021 Nov 19;12(1):6764.
205. Khan SN, Khan AU. Breaking the spell: combating multidrug resistant 'superbugs. *Frontiers in Microbiology*. 2016 Feb 18;7:174.
206. Takenaka S, Trivedi HM, Corbin A, Pitts B, Stewart PS. Direct visualization of spatial and temporal patterns of antimicrobial action within model oral biofilms. *Applied and Environmental Microbiology*. 2008 Mar 15;74(6):1869–75.
207. Sridhar S, Forrest S, Warne B, Maes M, Baker S, Dougan G, Bartholdson Scott J. High-content imaging to phenotype antimicrobial effects on individual bacteria at scale. *Msystems*. 2021 Jun 29;6(3):10–128.
208. Sun N, Jia Y, Bai S, Li Q, Dai L, Li J. The power of super-resolution microscopy in modern biomedical science. *Advances in Colloid and Interface Science*. 2023 Mar 13;314:102880.
209. Lefman J, Cheuk T, Metaferia B, Scott K, Khan J, Stranick SJ. Super-resolution imaging of drug delivery to live cells using real-time structured illumination microscopy. *Microscopy and Microanalysis*. 2010 Jul;16(S2):666–7.
210. Cardellini J, Balestri A, Montis C, Berti D. Advanced static and dynamic fluorescence microscopy techniques to investigate drug delivery systems. *Pharmaceutics*. 2021 Jun 11;13(6):861.
211. Zhang C, Tian Z, Chen R, Rowan F, Qiu K, Sun Y, Guan JL, Diao J. Advanced imaging techniques for tracking drug dynamics at the subcellular level. *Advanced Drug Delivery Reviews*. 2023 Jun 28;199:114978.
212. Carsten A, Wolters M, Aepfelbacher M. Super-resolution fluorescence microscopy for investigating bacterial cell biology. *Molecular Microbiology*. 2024;121:646–58.
213. Van Drie JH, Tong L. Cryo-EM as a powerful tool for drug discovery. *Bioorganic & Medicinal Chemistry Letters*. 2020 Nov 15;30(22):127524.
214. Wong W, Bai XC, Brown A, Fernandez IS, Hanssen E, Condron M, Tan YH, Baum J, Scheres SH. Cryo-EM structure of the Plasmodium falciparum 80S ribosome bound to the anti-protozoan drug emetine. *Elife*. 2014 Jun 9;3:e03080.
215. Cook NJ, Li W, Berta D, Badaoui M, Ballandras-Colas A, Nans A, Kotecha A, Rosta E, Engelman AN, Cherepanov P. Structural basis of second-generation HIV integrase inhibitor action and viral resistance. *Science*. 2020 Feb 14;367(6479):806–10.

216. Passos DO, Li M, Jóźwik IK, Zhao XZ, Santos-Martins D, Yang R, Smith SJ, Jeon Y, Forli S, Hughes SH, Burke Jr TR. Structural basis for strand-transfer inhibitor binding to HIV intasomes. *Science*. 2020 Feb 14;367(6479):810–4.
217. Tipping WJ, Lee M, Serrels A, Brunton VG, Hulme AN. Stimulated Raman scattering microscopy: an emerging tool for drug discovery. *Chemical Society Reviews*. 2016;45(8):2075–89.
218. Saar BG, Contreras-Rojas LR, Xie XS, Guy RH. Imaging drug delivery to skin with stimulated Raman scattering microscopy. *Molecular Pharmaceutics Journal*. 2011 Jun 6;8(3):969–75.
219. Belsey NA, Garrett NL, Contreras-Rojas LR, Pickup-Gerlaugh AJ, Price GJ, Moger J, Guy RH. Evaluation of drug delivery to intact and porated skin by coherent Raman scattering and fluorescence microscopies. *Journal of Controlled Release*. 2014 Jan 28;174:37–42.
220. Neu TR, Manz B, Volke F, Dynes JJ, Hitchcock AP, Lawrence JR. Advanced imaging techniques for assessment of structure, composition and function in biofilm systems. *FEMS Microbiology Ecology*. 2010 Apr 1;72(1):1–21.
221. Kubota K. CARD-FISH for environmental microorganisms: technical advancement and future applications. *Microbes and Environments*. 2013;28(1):3–12.
222. Giepmans BN, Adams SR, Ellisman MH, Tsien RY. The fluorescent toolbox for assessing protein location and function. *Science*. 2006 Apr 14;312(5771):217–24.
223. Amann R, Fuchs BM. Single-cell identification in microbial communities by improved fluorescence in situ hybridization techniques. *Nature Reviews Microbiology*. 2008 May;6(5):339–48.
224. Fixler D, Leary JF, Wei X. Nanoscale imaging and sensing for biomedical applications. *Cytometry A*. 2017 Aug;91(8):758–9.
225. Jehlička J, Edwards HG, Oren A. Raman spectroscopy of microbial pigments. *Applied and Environmental Microbiology*. 2014 Jun 1;80(11):3286–95.
226. Porter J, Deere D, Hardman M, Edwards C, Pickup R. Go with the flow–use of flow cytometry in environmental microbiology. *FEMS Microbiology Ecology*. 1997 Oct 1;24(2):93–101.
227. Sarvamangala DR, Kulkarni RV. Convolutional neural networks in medical image understanding: a survey. *Evolutionary Intelligence*. 2022 Mar;15(1):1–22.
228. Ghosh KK, Burns LD, Cocker ED, Nimmerjahn A, Ziv Y, Gamal AE, Schnitzer MJ. Miniaturized integration of a fluorescence microscope. *Nature Methods*. 2011 Oct;8(10):871–8.
229. Vincze L. Confocal X-ray fluorescence imaging and XRF tomography for three-dimensional trace element microanalysis. *Microscopy and Microanalysis*. 2005 Aug;11(S02):682–3.
230. Estevinho LM, Combarros-Fuertes P, Paula VB. Recent advances in applied microbiology. *Microorganisms*. 2020 Sep 7;8(9):1364.
231. Chang GW, Twu MJ, Lin YH, Liao CC, Kuo HZ. Optomechanical analysis for confocal laser scanning microscopes. *Optomechanics*. 2005; 5877 :347–55.
232. Grigoryeva N, Chistyakova L. Confocal laser scanning microscopy for spectroscopic studies of living photosynthetic cells. In: *Color Detection*, ed. L Zeng and S Cao. IntechOpen; 2020. p. 39–64.
233. Smith NA, Sadler PJ. Photoactivatable metal complexes: from theory to applications in biotechnology and medicine. *Philosophical Transactions: Mathematical, Physical and Engineering Sciences*. 2013 Jul 28;371(1995):20120519.
234. Clench MR. Advances in mass spectrometry imaging. *Proteomics*. 2016 Jun;16(11–12):1605–6.
235. De Boer P, Giepmans BN. State-of-the-art microscopy to understand islets of Langerhans: what to expect next? *Immunology & Cell Biology*. 2021 May;99(5):509–20.
236. Bond C, Santiago-Ruiz AN, Tang Q, Lakadamyali M. Technological advances in super-resolution microscopy to study cellular processes. *Molecular Cell*. 2022 Jan 20;82(2):315–32.
237. Chen F, Tillberg PW, Boyden ES. Expansion microscopy. *Science*. 2015 Jan 30;347(6221):543–8.
238. Jeong D, Kim D. Recent developments in correlative super-resolution fluorescence microscopy and electron microscopy. *Molecules Cells*. 2022 Jan 1;45(1):41–50.
239. Glas A, Hübener M, Bonhoeffer T, Goltstein PM. Benchmarking miniaturized microscopy against two-photon calcium imaging using single-cell orientation tuning in mouse visual cortex. *PLoS One*. 2019 Apr 4;14(4):e0214954.
240. Balzarotti F, Eilers Y, Gwosch KC, Gynnå AH, Westphal V, Stefani FD, Elf J, Hell SW. Nanometer resolution imaging and tracking of fluorescent molecules with minimal photon fluxes. *Science*. 2017 Feb 10;355(6325):606–12.

241. Welling MM, Hensbergen AW, Bunschoten A, Velders AH, Scheper H, Smits WK, Roestenberg M, van Leeuwen FW. Fluorescent imaging of bacterial infections and recent advances made with multimodal radiopharmaceuticals. *Clinical and Translational Imaging*. 2019 Apr 1;7:125–38.
242. Moen E, Bannon D, Kudo T, Graf W, Covert M, Van Valen D. Deep learning for cellular image analysis. *Nature Methods*. 2019 Dec;16(12):1233–46.
243. Shin YH, Gutierrez-Wing MT, Choi JW. Recent progress in portable fluorescence sensors. *Journal of The Electrochemical Society*. 2021 Jan 7;168(1):017502.
244. Jeckel H, Drescher K. Advances and opportunities in image analysis of bacterial cells and communities. *FEMS Microbiology Reviews*. 2021 Jul;45(4):fuaa062.
245. Vásárhelyi L, Kónya Z, Kukovecz Á, Vajtai R. Microcomputed tomography–based characterization of advanced materials: a review. *Materials Today Advances*. 2020 Dec 1;8:100084.
246. Siu DM, Lee KC, Chung BM, Wong JS, Zheng G, Tsia KK. Optofluidic imaging meets deep learning: from merging to emerging. *Lab on a Chip*. 2023;23(5):1011–33.
247. Lim JW, Ha D, Lee J, Lee SK, Kim T. Review of micro/nanotechnologies for microbial biosensors. *Frontiers in Bioengineering and Biotechnology*. 2015 May 11;3:61.
248. Abbasian F, Ghafar-Zadeh E, Magierowski S. Microbiological sensing technologies: a review. *Bioengineering*. 2018 Mar 2;5(1):20.

8 Quantum Microbiology

Javeria Tanvir, Syeda Fatima Zehra, and Muhammad Ali Syed

8.1 INTRODUCTION

Various pivotal transformations that have profoundly expanded the comprehension of the universe have influenced the development of scientific understanding (1). The emergence of quantum mechanics, a field of physics, challenged the intuitive and deterministic theories of classical mechanics and exposed a world composed of probabilities, wave-particle duality, and uncertainty (2). It was one of the most crucial developments in the early 20th century. Initially, the principles of quantum mechanics were confined to the study of subatomic particles, but its principles are now applied across numerous industrial and scientific domains. Biology, especially at the microscopic level, is one of the most exciting and rapidly developing intersections influenced by quantum mechanics, giving rise to the emergence of quantum microbiology (3).

The advanced specialty of quantum microbiology investigates how the laws of quantum mechanics affect and control the behavior of microorganisms (4). This chapter aims to deliver a detailed introduction to quantum microbiology, beginning with foundational concepts of quantum mechanics and extending to explore the prospective implications for microbiology.

The opening section of this chapter introduces our readers to the core concept of quantum microbiology, starting with a basic overview of the foundational concept of quantum mechanics. This overview of quantum features a timeline tracing the progression of the quantum world, from Max Planck's seminal quantum hypothesis to the establishment of quantum theory. Understanding these essential concepts is vital for recognizing how quantum effects might influence biological systems. This groundwork prepares readers for the subsequent discussion on quantum microbiology.

The chapter then explores the difference between classical and quantum mechanics, emphasizing that quantum mechanics dominates at the atomic and subatomic levels while classical mechanics is effective for macroscopic phenomena. It highlights the peculiar behavior of quantum mechanics, such as superposition, tunneling, and entanglement. The chapter then moves to the real-world implications of quantum mechanics, showcasing the advancements in atomic and molecular physics, quantum computing, and quantum cryptography. Building on these concepts, the chapter transitions to quantum biology where recent research highlights the significance of quantum effects in different biological processes, which led to the development of quantum microbiology, a field which mainly focuses on how quantum events influence the behavior of microbes. The chapter concludes with a discussion on the anticipated impact of quantum microbiology on medical advancements and biotechnology, encouraging our readers to reflect on the radical potential of the innovative field of quantum microbiology.

8.2 QUANTUM PHYSICS: THE BASICS

The domain of physics that investigates the behavior of the tiniest particles of the universe, for instance, electrons, protons, and photons, is identified as quantum physics, commonly referred

DOI: 10.1201/9781003410164-8

to as quantum mechanics or quantum theory. These tiny particles demonstrate behavior that commonly contradicts traditional physics principles, unlike the deterministic and predictable trajectories of larger entities in our everyday environment (5). At the quantum level, particles can simultaneously occupy multiple states, a phenomenon termed superposition, or they become entangled, which allows the one particle state to immediately influence other particles regardless of the distance separating them. This fundamentally paradoxical nature of the field of quantum mechanics or quantum particles has altered our perception of the whole physical universe (6).

The notion of wave-particle duality is crucial in the field of quantum mechanics, claiming that particles, for instance electrons and photons, can manifest the characteristics of both the wave and the particle depending on the nature of their observation (7). This duality results in phenomena such as interference patterns, where particles that are expected to behave independently demonstrate wave-like collective behavior (8). Another tenet of quantum mechanics is Heisenberg's uncertainty principle proposed by Werner Heisenberg, which asserts that certain combinations of physical attributes, for instance, position and momentum cannot be measured simultaneously with complete precision. This inherent uncertainty is not a consequence of limitations in our measuring technology but it is a fundamental property of the quantum universe. This illustrates the essence of all quantum phenomena (9).

Beyond theoretical understanding, quantum mechanics holds significance that affects a diverse array of real-world applications (10). Quantum mechanics is the cornerstone of numerous modern technologies like magnetic resonance imaging, lasers, and semiconductors, for instance. Furthermore, it acts as a driving force behind the development of innovative technologies like quantum computing which holds the potential to transform the fields, for instance, of material science, cryptography, and complex systems modeling by exploiting the distinctive unique behavior of quantum bits also termed qubits, that can occupy in multiple states concurrently (11). By confronting our classical concepts and revealing the complexities of the quantum realm, quantum not only deepens our scientific understanding of the physical world but also paves the way for revolutionary groundbreaking innovations in technology (12).

8.2.1 Evolutionary Timeline of Quantum Mechanics

Quantum mechanics developed its roots in the late 19th and early 20th centuries, an era when Newtonian mechanics and Maxwell's equation emerged as the central models for the explanation and investigation of the physical universe (13). But when the researchers explored atomic and subatomic particles, they came across experimental findings that opposed the established theories. This led to the creation of the groundbreaking theory of quantum mechanics, that fundamentally modified our basic concepts of the cosmos (14).

A major challenge to classical mechanics was first highlighted by the study of black body radiation (15). According to classical perceptions, the energy of a black body would infinitely rise with increasing radiation frequency, particularly in the ultraviolet range. This anomaly is known as "Ultraviolet catastrophe" (16). Max Planck addressed this problem by proposing that electromagnetic energy could only be emitted or absorbed in discrete packets known as "quanta." This concept introduced the idea that the energy is quantized with each quantum of energy being directly proportional to the frequency of radiation. This proposal was a major shift in the continuous energy framework of classical mechanics (17).

The study of the photoelectric effect offered more evidence supporting the quantum nature of light (18). To further explain this phenomenon, in 1905, Albert Einstein extended Planck's quantum hypothesis and suggested that light could be viewed as particles or photon-packets of energy, each possessing a quantum of energy (19). Furthermore, he suggested that dependency on the frequency of the incoming light might be explained in experiments if these photons have enough energy to expel electrons from the surface of the metal. His contribution, for which Einstein won a Nobel Prize in physics in 1921, represented compelling evidence for the particle nature of light (20).

To explain the structure of an atom and the hydrogen spectral lines, employing the concepts of quantum mechanics, in 1913 Niels Bohr presented a novel model of the atom. According to Bohr's theory, electrons can only gain or lose energy by hopping between distinct, quantized orbits around the nucleus, and absorbing and releasing the quanta of energy during the process (21). Bohr's model was a major step toward a quantum mechanical description of all atomic systems, even if it was only able to explain basic atomic systems and had limits for more complex systems (22).

Quantum mechanics reached the heights of advancement to evolve as a theory in the 1920s. Experiments such as electron diffraction supported Louis de Broglie's theory which explains that a particle may behave like a wave. Schrödinger presented wave mechanics and his wave equation became crucial to the quantum theory. Heisenberg established the uncertainty principle, which highlights the probabilistic nature of quantum mechanics, and sets it apart from classical mechanics by asserting some physical attributes, such as position and momentum, which cannot be measured simultaneously with precision (23).

The deterministic model of classical mechanics was significantly challenged by the advent of quantum mechanics during the early 20th century, leading to an advanced understanding of the microscopic world. The progress in modern physics was greatly driven by the pioneering work of scientists, for instance, Bohr, Planck, Einstein, Schrdöinger, and Heisenberg. Their work affected a wide range of disciplines through technological advancements, ranging from quantum computing to materials science. The advent of quantum mechanics has reshaped our grasp of the physical world, uncovering a world where particles exhibit wave-like properties, energy is discrete, and certainty yields probabilistic outcomes (24).

8.2.2 Dichotomy between Classical and Quantum Mechanics

Classical mechanics, founded on Newton's laws, establishes a foundational basis for understanding the movements of physical objects, for instance, automobiles, airplanes, projectiles, and planets (25). It is predicated on the idea of determinism, which suggests that if the initial conditions, like position and velocity, are known, the future state of the system can be accurately measured. Classical mechanics dictated that the motion of objects is greatly influenced by forces such as gravity, friction, and tension, leading to trajectories that are both continuous and well-defined. Assuming the energy as a continuous quantity enables us a smooth transition between different states of motion. This framework is a cornerstone of classical mechanics and works potentially well for macroscopic systems (26).

In quantum mechanics, other specific phenomena include quantum entanglement, where particles are instantly correlated regardless of their mutual distance and superposition, which permits particles to be in multiple states simultaneously until detected. These phenomena reveal the counterintuitive nature of quantum mechanics and lack analogs in classical mechanics (27).

8.3 APPLIED INNOVATIONS OF QUANTUM MECHANICS

The applied innovations of quantum mechanics have revolutionized several scientific and technological domains by providing groundbreaking solutions. This section explores its influence on atomic and molecular physics, as well as the field of quantum computing.

8.3.1 Atomic and Molecular Physics

In atomic physics, quantum mechanics is a key to understanding the intrinsic structure of an atom. It describes how electrons reside inside the atom in specific energy levels around the nucleus, characterized by quantum numbers, for instance, principal quantum number, angular momentum, magnetic, and spin quantum numbers. These quantized energy levels restrict electrons to certain orbits or states, and their spatial distribution influences the chemical properties of the elements (28).

In molecular physics, these concepts are applied to how atoms connect to form molecules through quantum mechanics. The formation of molecular orbits which lead to bonding and anti-bonding interactions arises from the overlap of the atomic orbitals leading to the formation of molecular orbits. Techniques like molecular orbital theory and valence band theory are used to predict the bond lengths, strength, and angles (29). Quantum mechanics is crucial for drug design, reaction pathway analysis, and catalyst enhancement (30).

8.3.2 Quantum Computing

Utilizing the principles of quantum, quantum computing represents a groundbreaking advancement in computational technology and offers potential advantages over traditional computers in numerous specific applications (31). Quantum computing uses qubits which exist in the superposition of states, which differ from the classical bits. This property allows qubits to perform numerous calculations simultaneously (32). Quantum algorithms propose more effectively advanced approaches for certain problems. This has significant implications for cryptography since many encryption systems depend on the ability to factor in large numbers. For instance, Shor's algorithm processes large integers more efficiently as compared to the most widely used classical methods (33).

8.3.3 Quantum Cryptography

Quantum cryptography employs the foundational principles of quantum mechanics to advance secure communication methods that are potentially immune to eavesdropping (34). Quantum key distribution (QKD) is one of the most important applications, which enables two parties to exchange cryptographic keys securely by using quantum superposition and quantum entanglement. During the key distribution process, qubits encoded in distinct states are transmitted and any attempt to observe or measure, or intercept these qubits results in changes in the key (35). Moreover, quantum cryptography encompasses procedures that utilize quantum entanglement to facilitate secret key exchanges which makes any interception easily detectable. With the advancements in quantum technology, quantum cryptography provides exceptional security for information exchange, addressing weaknesses in previous cryptographic techniques, while paving the way for broader implementation of quantum technologies (36).

8.4 QUANTUM BIOLOGY

It was a difficult task to apply mathematical calculations and physical principles to biological systems before the twentieth century. We know that all living organisms, including microorganisms, have complex cellular organization and biologists have developed their tools and approaches to explore the processes and phenomena occurring in living cells and subcellular components (3). The lecture series and book by Schrödinger *What is Life?* raised the question about the possible role of quantum mechanics in living organisms, since the study of biological processes at atomic and subatomic levels applying theories of physics and mathematical calculations or using computer models at this level was a difficult and challenging task (37). Nevertheless, it is now becoming clear that all chemical reactions involve quantum mechanics. Quantum mechanics (QM) is a revolutionary concept that has been in use for over a century since scientists realized that laws of classical physics may not be sufficient to explain matter at the atomic level. The world of living organisms, including the microbial world, consists of matter, therefore, the theories of quantum mechanics may be applied to a deep understanding of biological systems (38, 39).

Over the last few decades, results of a range of experiments suggest that quantum mechanics may be harnessed for biological advantage (38, 39). A complex biological phenomenon like random mutations in DNA, the orientation of birds while migrating, or how photons of light are captured and energy transferred to the reaction center in photosynthesis, etc. may be addressed by applying quantum mechanics. Similarly, the effects of enzymes and consciousness may also be explained in terms of quantum mechanics. Such developments lead to paradigm shifts in the understanding of biomolecules at atomic and subatomic levels (3, 38, 39).

As stated before, most of the mechanisms occurring in living systems, including microbial cells, may not be explained by trivial laws of physics. For instance, one of the most notable roles of quantum mechanics in biology may be the transport of excitation of energy for photosynthesis by long-lived quantum coherence (3).

Most of the attempts made so far to apply quantum mechanics in biology have been merely theoretical. In quantum biology, we study the quantum realm of atoms and molecules such as coherence, tunneling, entanglement, or spin, which disappear at the macro scale. Due to the unavailability of sophisticated tools, techniques, and approaches, it has not been feasible to experimentally investigate the dynamics of biological processes by applying quantum mechanics (40). Most of the studies on the application of quantum to understanding biological phenomena are theoretical and little experimental work has been done due to the delicate nature of biomolecules (3, 40).

8.5 FUNDAMENTALS OF QUANTUM MICROBIOLOGY

An emerging field of research called "quantum microbiology" investigates how microorganisms behave and function by merging the principles of microbiology with quantum mechanics. Unlike traditional microbiology, which primarily focuses on the biomedical and genetic characteristics of microbes, quantum microbiology utilizes quantum physics to gain a better understanding of bacterial systems. This interdisciplinary approach posits that quantum effects may play a pivotal role in fundamental biological functions, for instance, photosynthesis and cellular respiration, and other essential biological mechanisms, offering new perspectives on these phenomena (41).

Though the idea of quantum effects influencing biological systems is not entirely novel, it has developed recently. Quantum mechanical phenomena like tunneling and superposition may play a crucial role in microorganisms (42). For instance, quantum effects that boost the survival and metabolic efficiency of bacteria and archaea could explain well the behavior of protons and electrons within their enzymes. Exploring these processes through a quantum perspective could revolutionize synthetic biology, as the quantum-based designs might yield more robust microbes for applications in industry, environmental management, and healthcare (43).

8.6 EXPLORING THE REALM OF QUANTUM MICROBIOLOGY

The world of microorganisms is extremely diverse. The biochemical reactions occurring inside microbial cell like bacteria and fungi, as well as viral infections, are fast, highly efficient, and genetically and environmentally controlled. Quantum microbiology may offer a unique opportunity for a deep understanding of chemical reactions at atomic scale and their overall impact, especially on selective advantage and existence (3, 38–40).

Microbiology is comparatively a new field and most of the breakthrough developments occurred after advancements in microscopy and culturing techniques as well as advances in molecular techniques. In the past, scientists applied culture-based techniques to study the properties of microorganisms. In such an approach, the behavior of microorganisms was studied as a community (44). However, scientists later found themselves to be more interested in the behavior of individual cells. Thus, a new branch of microbiology came into being called "single-cell microbiology"

(45). Exploration of biochemical reactions/ interactions/processes at atomic and subatomic scales will open new doors for further investigation. Apart from strengthening our understanding of biomolecular interactions, one may utilize such knowledge for advancements in diagnostics, therapeutics, and biotechnological applications (46).

A number of microbial processes may well be explained in terms of quantum mechanics. For example, the movement of magnetostatic bacteria according to the earth's magnetic field, ligand-receptor interaction and initiation of cellular response, signal transduction in a microbial cell, activation of Tc or TH cells by major histocompatibility complex I or II (MHC I or II)- peptide complex, random mutations in DNA, excitation of photosynthetic proteins, etc. Since quantum microbiology is in its infancy, the sophistication and complexity of the problems require extensive focus by scientists using a multidisciplinary approach (47).

A well-studied example of quantum coherent energy transfer in bacteria is the light Fenna-Matthews-Olson (FMO) harvesting machinery of green and purple bacteria. They have an efficient energy-harvesting system of chlorosome antennae that enables them to thrive in low light conditions (48). The energy captured by the chlorosome is then transferred to the reaction center through a protein complex called Fenna-Matthews-Olson (FMO) complex. The FMO protein complex has attracted great research attention due to its efficiency of energy transfer (nearly 100%) and quantum coherence (49). The body of evidence suggests that quantum coherence is non-negligible at room temperature (39).

In the discussion of quantum biology, viruses offer a case study, since they are at the interface of living and non-living. Scientists have attempted to investigate the possible role of quantum mechanics in microbial infections, including viral infections (50). Quantum theory of infection may be applied to study the mechanism of invasion by microbes such as SARS-CoV-2 into the host. As stated previously, quantum tunneling has been observed in enzyme-substrate interactions. Since SARS-CoV-2 uses its enzymes to invade the host cells, quantum mechanics may enhance our understanding of the mechanism of this infection. A recent study (51) by Adams et al. (2022) illustrated this with a simple model with vibronic mode. They observed that the vibronic mode of spike proteins may enhance charge transfer from spike protein receptors. They proposed that possible intervention for SARS-Cov-2 may be identified through vibronic spectra of spike proteins.

Similarly, quantum mechanics is also playing a significant role in adaptive immune response, hence a new field may emerge known as "quantum immunology" (52). Adaptive immune response is a dynamic, self-organized function of the immune system of vertebrates. Presentation of foreign antigenic peptides by major histocompatibility complex (MHC) I or II to T cell receptors and development of immune response is highly specific and may not be explained with biochemistry alone. To bridge this gap, several groups have applied quantum models to study T-cell activation. One such model suggests that protein activation during T-cell activation is not a continuous process, but it occurs as quanta of phosphates (53).

One of the unique bacterial structures responsible for the locomotion of bacteria is the flagellar motor. It is the largest motor in bacteria. This structure is about 40 nm in size and is embedded in the cell membrane of some groups of bacteria. The maximum bacterial rotation speed of the flagellum is 1,700 resolutions per second. The flagellar motor converts the electrochemical potential difference of cations across the cell membrane into mechanical work required to rotate the flagellum with nearly 100% efficiency. A quantum mechanics model has been developed to describe the energy transfer in the flagellar motor and its functioning (54).

Attempts are being made to investigate the optimized quantum conditions to improve microbial detection by controlled quantum conditions using optical instruments such as spectrophotometers. The superiority of quantum light in early detection, discrimination, and tracking of bacteria using spectrophotometers has also been investigated (55).

8.7 FUTURE PERSPECTIVES

The organization of living organisms is unique and different from the world of mathematics and physics. Living organisms are open systems that are continuously exchanging energy and matter with their environments; nevertheless, they are thought to have originated from inanimate matter during evolution. Quantum mechanics may be the appropriate approach to study the origin of microbial life, genetic code and molecular events such as genetic mutations, DNA repair, genetic recombination and storage of genetic information, etc. (56).

Much of the work carried out so far is theoretical and experimental studies have been left behind. Much more needs to be done to explore biological processes at the level of quantum mechanics. The coarse-grained knowledge of biological processes often fails to answer many questions that require a deeper understanding of the quantum level (57).

Despite huge expectations for future breakthroughs in scientific knowledge and innovations, quantum mechanics applications in biology still suffer from a few challenges (58). As stated previously, quantum microbiology is an interdisciplinary field that requires strong collaborative work of physicists, mathematicians, chemists, and microbiologists. Nevertheless, all of them have their own way of studying living organisms (59). They even have different languages and terminologies. Furthermore, quantum literacy is another challenge faced by scientists. Quantum mechanics is different from biological sciences, and therefore many microbiologists may not be familiar with it. To overcome this issue, we need a dedicated generation of scientists who will be working at the interface of physics, mathematics, and microbiology. The real challenge may be developing systems for tractable quantum mechanics events (60).

On one hand, scientists are trying to study biological processes at the quantum level which will offer a deep understanding of the quantum behaviors of atoms and molecules and their overall effect on the organization and behavior of the microbes. On the other hand, bioinspired systems may be developed that may offer unique opportunities to devise smart devices, systems, machines, etc. that may better help us in fulfilling our energy needs. Nevertheless, quantum mechanics studies require highly sensitive tools for studying quantum processes in living cells, which are expensive and inaccessible for many (61).

8.8 CONCLUSION

There is growing evidence of quantum mechanics' role in microbial life. As the field has been established and expanded globally, there is standardization of the methodology, tools, and communication that will not only facilitate scientists but also attract other stakeholders for collaborative research work.

Nanoscale quantum dynamic events may have a significant impact on overall macroscopic organization and fitness. The mysterious and most diverse world of microbes may be best studied at the quantum level. From cell signaling, enzymatic action, spontaneous mutations in DNA, and ligand-receptor interactions, to energy transfer in photosynthetic pigments in bacteria, quantum mechanics plays a significant role. Future challenges demand interdisciplinary work at the interface of physics, mathematics, chemistry, molecular biology, microbiology, and computation. Moreover, future research may reveal whether these quantum mechanical processes offer any selective advantage to the organisms or not. Furthermore, as the research progresses, bioinspired devices may outperform their classical counterparts.

REFERENCES

1. Davies P. *Cosmic blueprint: new discoveries in natures ability to order universe*: Templeton Foundation Press; 2004.

2. Bhattacharya G. *Quantum mechanics and relativity: a conceptual study. History of science and philosophy of science: a historical perspective of the evolution of ideas in science*, ed. PK Sengupta. Longman; 2010; p. 225.
3. McFadden J, Al-Khalili J. The origins of quantum biology. *Proceedings of the Royal Society A*. 2018;474(2220):20180674.
4. Brunk E, Rothlisberger U. Mixed quantum mechanical/molecular mechanical molecular dynamics simulations of biological systems in ground and electronically excited states. *Chemical Reviews*. 2015;115(12):6217–63.
5. Binney J, Skinner D. *The physics of quantum mechanics*: Oxford University Press; 2013.
6. Levin FS. *An introduction to quantum theory*: Cambridge University Press; 2002.
7. Marburger J. *Constructing reality: quantum theory and particle physics*: Cambridge University Press; 2011.
8. Vaccaro JA. Particle-wave duality: a dichotomy between symmetry and asymmetry. *Proceedings of the Royal Society A: Mathematical, Physical and Engineering Sciences*. 2012;468(2140):1065–84.
9. Mehra J. Niels Bohr's discussions with Albert Einstein, Werner Heisenberg, and Erwin Schrödinger: the origins of the principles of uncertainty and complementarity. *Foundations of Physics*. 1987;17(5):461–506.
10. Laloë F. Do we really understand quantum mechanics? Strange correlations, paradoxes, and theorems. *American Journal of Physics*. 2001;69(6):655–701.
11. Nandhini S, Singh H, Akash U. An extensive review on quantum computers. *Advances in Engineering Software*. 2022;174:103337.
12. Williams CP. *Explorations in quantum computing*: Springer Science & Business Media; 2010.
13. Golub R, Lamoreaux SK, Lamoreaux S. *The historical and physical foundations of quantum mechanics*: Oxford University Press; 2023.
14. MATTER COFD. Evolution of cosmos and life on earth. *Journal "Scientific Israel–Technological Advantages*. 2019;21(2):3.
15. Boyer TH. Derivation of the blackbody radiation spectrum without quantum a;ssumptions. *Physical Review*. 1969;182(5):1374.
16. Boyer TH. Blackbody radiation in classical physics: a historical perspective. *American Journal of Physics*. 2018;86(7):495–509.
17. Wagh SM. *The light quantum. Subtlety in relativity*: Jenny Stanford Publishing; 2017. p. 3–88.
18. Ossiander M, Riemensberger J, Neppl S, Mittermair M, Schäffer M, Duensing A, et al. Absolute timing of the photoelectric effect. *Nature*. 2018;561(7723):374–7.
19. Arvind A, Dorai K, Chaturvedi S, Mukunda N. The development of quantum mechanics: A story of people, places and philosophies. *Resonance*. 2018;23:1077–1100.
20. Kumar M. *Quantum: Einstein, Bohr and the great debate about the nature of reality*: Icon Books Ltd; 2008.
21. Kragh H. *Niels Bohr and the quantum atom: the Bohr model of atomic structure 1913–1925*: Oxford University Press; 2012.
22. Kuehn K. *A student's guide through the great physics texts: Volume IV: Heat, Atoms and Quanta*. Springer; 2016.
23. Coben S. The scientific establishment and the transmission of Quantum Mechanics to the United States, 1919–32. *The American Historical Review*. 1971;76(2):442–66.
24. Whitaker A. *Einstein, Bohr and the quantum dilemma: from quantum theory to quantum information*: Cambridge University Press; 2006.
25. Virk HS. Classical physics versus quantum physics: an overview. *OmniScience: A Multi-Disciplinary Journal*. 2014;4(2):36–42.
26. Lynds P. Time and classical and quantum mechanics: indeterminacy versus discontinuity. *Foundations of Physics Letters*. 2003;16(4):343–55.
27. Rosenblum B, Kuttner F. Consciousness and quantum mechanics: the connection and analogies. *The Journal of Mind and Behavior*. 1999;20:229–56.
28. Tichy MC, Mintert F, Buchleitner A. Essential entanglement for atomic and molecular physics. *Journal of Physics B: Atomic, Molecular and Optical Physics*. 2011;44(19):192001.
29. Bader RF. A quantum theory of molecular structure and its applications. *Chemical Reviews*. 1991;91(5):893–928.
30. Ahn S, Hong M, Sundararajan M, Ess DH, Baik M-H. Design and optimization of catalysts based on mechanistic insights derived from quantum chemical reaction modeling. *Chemical Reviews*. 2019;119(11):6509–60.

31. Shor PW. Quantum computing. *Documenta Mathematica*. 1998;1(1000):467–86.
32. Steane A. Quantum computing. *Reports on Progress in Physics*. 1998;61(2):117.
33. Mavroeidis V, Vishi K, Zych MD, Jøsang A. *The impact of quantum computing on present cryptography*. arXiv preprint arXiv:180400200; 2018.
34. Bennett CH, Brassard G, Ekert AK. Quantum cryptography. *Scientific American*. 1992;267(4):50–7.
35. Alléaume R, Branciard C, Bouda J, Debuisschert T, Dianati M, Gisin N, et al. Using quantum key distribution for cryptographic purposes: a survey. *Theoretical Computer Science*. 2014;560:62–81.
36. Broadbent A, Schaffner C. Quantum cryptography beyond quantum key distribution. *Designs, Codes and Cryptography*. 2016;78:351–82.
37. Symonds N. What is life?: Schrodinger's influence on biology. *The Quarterly Review of Biology*. 1986;61(2):221–6.
38. Cronin O, Barton W, Skuse P, Penney NC, Garcia-Perez I, Murphy EF, et al. A prospective metagenomic and metabolomic analysis of the impact of exercise and/or whey protein supplementation on the gut microbiome of sedentary adults. *MSystems*. 2018;3(3):10–128.
39. Lambert N, Chen Y-N, Cheng Y-C, Li C-M, Chen G-Y. 323 Franco Nori. *Quantum Biology Nature Physics*. 2013;9(1):10–8.
40. Kim Y, Bertagna F, D'souza EM, Heyes DJ, Johannissen LO, Nery ET, et al. Quantum biology: an update and perspective. *Quantum Reports*. 2021;3(1):80–126.
41. Waring S. Quantum biology: a scientific revolution in our understanding of biological systems. *Biology System Open Access*. 2018;7(185):2.
42. Devault D. Quantum mechanical tunnelling in biological systems. *Quarterly Reviews of Biophysics*. 1980;13(4):387–564.
43. Goh BH, Tong ES, Pusparajah P. Quantum biology: does quantum physics hold the key to revolutionizing medicine? *Progress in Drug Discovery & Biomedical Science*. 2020;3(1).
44. Vitorino LC, Bessa LA. Technological microbiology: development and applications. *Frontiers in Microbiology*. 2017;8:827.
45. Brehm-Stecher BF, Johnson EA. Single-cell microbiology: tools, technologies, and applications. *Microbiology and Molecular Biology Reviews*. 2004;68(3):538–59.
46. Phogat N, Kohl M, Uddin I, Jahan A. *Interaction of nanoparticles with biomolecules, protein, enzymes, and its applications. Precision medicine*: Elsevier; 2018. p. 253–76.
47. Razin S, Yogev D, Naot Y. Molecular biology and pathogenicity of mycoplasmas. *Microbiology and Molecular Biology Reviews*. 1998;62(4):1094–156.
48. Richards G. *Quantum coherent effects in the light-harvesting complex Phycocyanin*. PhD thesis, Swinburne University of Technology, 2014.
49. González Soria B. *Quantum effects in the efficiency of Fenna-Matthews-Olson light-harvesting complexes*. Masters thesis, Instituto Tecnológico y de Estudios Superiores de Monterrey, 2020.
50. Cochran GM, Ewald PW, Cochran KD. Infectious causation of disease: an evolutionary perspective. *Perspectives in Biology and Medicine*. 2000;43(3):406–48.
51. Adams B, Sinayskiy I, van Grondelle R, Petruccione F. Quantum tunnelling in the context of SARS-CoV-2 infection. *Scientific Reports*. 2022;12(1):16929.
52. Germenis AE, Manousakis M, Antipas GS. The dawn of quantum immunology. *SRL Immunology and Immunotherapy*. 2016;1:3–6.
53. Manjili MH, Manjili SH. The quantum model of T-cell activation: revisiting immune response theories. *Scandinavian Journal of Immunology*. 2024;100:e13375.
54. Melkikh A, Bondar V. Mechanisms and models of movement of protocells and bacteria in the early stages of evolution. *Progress in Biophysics and Molecular Biology*. 2022;175:3–13.
55. Spedalieri G, Piersimoni L, Laurino O, Braunstein SL, Pirandola S. Detecting and tracking bacteria with quantum light. *Physical Review Research*. 2020;2(4):043260.
56. Koonin EV, Novozhilov AS. Origin and evolution of the universal genetic code. *Annual Review of Genetics*. 2017;51(1):45–62.
57. Noid WG. Perspective: advances, challenges, and insight for predictive coarse-grained models. *The Journal of Physical Chemistry B*. 2023;127(19):4174–207.

58. Fakruddin M, Hossain Z, Afroz H. Prospects and applications of nanobiotechnology: a medical perspective. *Journal of Nanobiotechnology*. 2012;10:1–8.
59. Green EJ. *Quantum evolution: how physics' Weirdest theory explains life's biggest mystery*: JSTOR; 2002.
60. Marais A, Adams B, Ringsmuth AK, Ferretti M, Gruber JM, Hendrikx R, et al. The future of quantum biology. *Journal of the Royal Society Interface*. 2018;15(148):20180640.
61. El Morsalani M. Quantum sensing: basics, algorithms, applications and the German ecosystem. 2024;preprint.

9 Omics Technologies in Microbiology

Faouzia Tanveer, Farah Alam Gill, and Fariha Masood Siddiqui

9.1 INTRODUCTION

Microorganisms play a significant role in maintaining environmental and human health. They generate a huge array of products of biotechnological importance, along with specific metabolites that can impact humans in both beneficial and detrimental ways (1). The term "omics" is derived from the Greek word "ome," meaning "all" or "every" (2). It refers to the investigation and analysis of extensive datasets portraying the anatomy and physiology of the whole composition of a specific biological system at an individual level. This technology has significantly transformed our approaches to investigating biological systems (3). It implies the utilization of high-throughput technologies to scrutinize diverse components of biological molecules as well as systems on a worldwide scale (4).

Omics methodologies encompass genomics, transcriptomics, proteomics, and metabolomics. These methodologies have transformed our comprehension of microorganisms and their therapeutic, industrial, and environmental utilization. Genomics pertains to a comprehensive analysis of the whole DNA, giving insights into the potential of an organism. Transcriptomics explores RNA, while proteomics investigates proteins, providing a close-up of the functioning informational molecules. This technique has identified diverse genes and their respective products implicated in signaling and additional pathways. Metabolomics focuses on the analysis of metabolites, representing the final product as well as giving evidence of the topical metabolic state. It frequently displays the impact of changes in the levels of expression of specific pathways. These techniques empower researchers to investigate the genetic, protein, and metabolic make-up of microorganisms, enabling useful perceptions of their biotechnological potential, responses to environmental components, and physiological interplays. An emerging strategy entails incorporating data from these omics tools to attain a comprehensive knowledge of composite microbiological systems (5, 6).

The omics technologies are high-throughput techniques that substantially enhance the simultaneous detection of a substantial number of genes/proteins, rendering the interconnection of complex combinations with intricate outcomes in the form of gene/protein expression profiles. The principal aim of omics technologies is the untargeted finding of entire gene products including transcripts, proteins, and metabolites found in a specific biological specimen (7). This chapter focuses on the applications of omics-based technologies in interpreting the application of microbes in environment, industry, and therapeutics. There are several key omics disciplines, each focusing on a specific type of biological molecule or system (Figure 9.1). A complete list of omics disciplines is implausible because of the rapidly developing new omics technologies. Some of the major omics fields discussed in this chapter include:

DOI: 10.1201/9781003410164-9

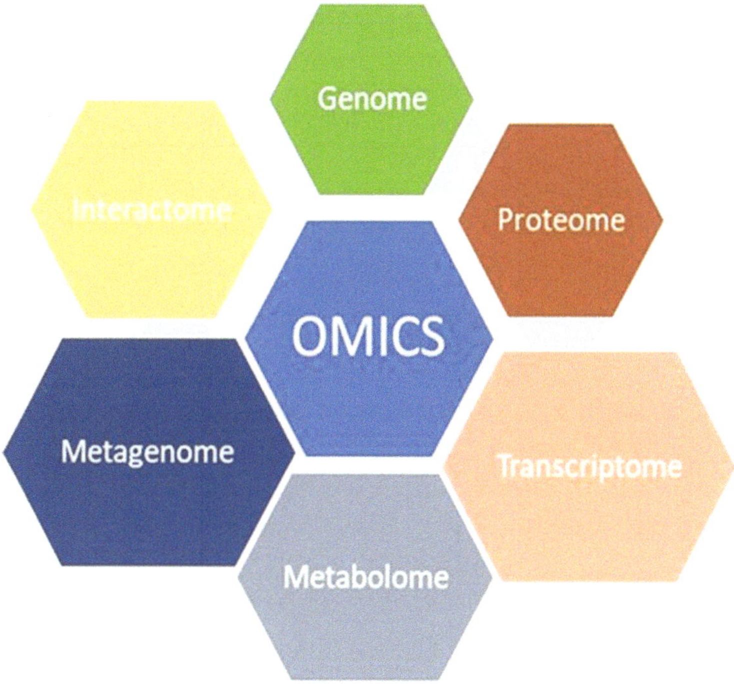

FIGURE 9.1 Diagram showing key disciplines of OMICS based on specific biomolecules/systems.

a. **Genomics:** Genomics entails the investigation of the entire set of genes (the genome) within an organism. It includes the sequencing, mapping, and analysis of DNA to understand the anatomy and physiology of genes, as well as the variations between individuals.
b **Transcriptomics:** Transcriptomics focuses on the examination of the entire array of RNA transcripts produced by the genome (the transcriptome). It provides comprehension of the patterns of gene expression, alternative splicing, and the dynamics of RNA molecules.
c. **Proteomics:** Proteomics involves the comprehensive analysis of the entire array of proteins within a cell, tissue, or organism (the proteome). It comprises the identification, quantification, and functional characterization of proteins to understand cellular processes and functions.
d. **Metabolomics:** Metabolomics explores the entire array of compact molecules (metabolites) involved in cellular processes. It provides information about metabolic pathways, identifying biomarkers, and understanding how metabolites contribute to the overall function of cells. Metabolomics involves quantifying the concentrations and spatial distribution of all metabolites within cells or tissues. Metabolites refer to the small molecules that undergo transformations during metabolic processes, typically serving as substrates and products of enzymes (8).
e. **Epigenomics:** Epigenomics studies the DNA modifications and related proteins that regulate gene expression without altering the fundamental DNA sequence. It involves the study of DNA methylation, histone modifications, and non-coding RNAs (2).
f. **Metagenomics:** Metagenomics concentrates on examining the complete microbial community linked to a specific environment or host organism. This encompasses bacteria, viruses, fungi, and other microorganisms, and is especially pertinent in fields such as human microbiome research (9). It is essentially the analysis of microbial communities without the involvement of cultural techniques.

9.2 OMICS-BASED TECHNIQUES

Methodological and technical advances have permitted the concurrent profiling of genome, epigenome, transcriptome, proteome, and other emerging omics modalities to better understand genotype-to-phenotype relationships and underlying biological mechanisms (10). High-throughput mass spectrometry and sequencing techniques represent fundamental investigational tools that aid in our exploration of the omics of a specific biological system. While approaches based on sequencing are suitable for studying the genome, transcriptome, epitomes, and interactomes involving DNA or RNA, techniques based on mass spectrometry are employed to investigate the proteome, metabolome, and interactomes without the requirement of DNA or RNA (3).

A range of available high-throughput methodologies enable the large-scale analysis of biological molecules, providing comprehensive insights into the structure, function, and interactions within biological systems. Omics technologies encompass the cells, tissues, and organisms, with the integration of information obtained from various omics-based platforms aiding greatly in the data interpretation (4). Following is an overview of some of the key omics technologies:

9.2.1 Genomics Technologies

DNA sequencing technologies determine the order of nucleotides in DNA, allowing for the identification of genes, variations, and genomic structure. Traditional technologies like Sanger sequencing as well as advanced next-generation sequencing (NGS) are widely used (11).

Whole genome sequencing (WGS) involves sequencing the complete genome of an organism and is the keystone genomics technique. It depends on the next-generation sequencing (NGS) methodologies that generally involve the acquisition of short or long DNA reads which are then assembled into longer, continuous sequences (contigs) with the help of bioinformatics tools. NGS platforms, e.g., Illumina MiSeq, single-molecule real-time sequencing (Pacific Bioscience), and Nanopore sequencing (Oxford Nanopore), allow rapid sequencing of millions of DNA fragments with enhanced sequence accuracy as well as reduced time, low cost, and availability of improved data analysis tools. NGS can provide detailed information about the genomes, including structure, genetic variations, and gene activity (12). Moreover, genotyping arrays facilitate the simultaneous analysis of multiple genetic variants across the genome, aiding in the identification of genetic markers associated with traits or diseases (9).

9.2.2 Transcriptomics Technologies

Microarrays enable the simultaneous detection of gene expression levels for thousands of genes. These platforms help identify patterns of gene expression under different conditions (9). cDNA microarrays consist of small chips or slides with short nucleotide probes, assessing the relative abundance of mRNA by hybridization of fluorescently labeled mRNA transcripts to these probes. Relative abundance of transcripts is indicated by the fluorescence intensity at each sample location on the microarray. While traditional microarrays primarily focus on the transcriptomics of a single organism, a single array can evaluate targets from different organisms. These techniques are applied to comprehend mechanisms of virulence, dynamics of polymicrobial infections, optimum delivery of antimicrobial drugs, and the identification of new drug targets (9, 13).

Transcriptomics techniques have been further integrated with microfluidics, resulting in "lab-on-a-chip (LOC)" technologies. These compact testing devices allow extraction, identification, and analysis of samples in small volumes, eliminating the need for large laboratory equipment. These devices can be utilized for the evaluation of treatment biomarkers or differentiation between an active or latent infection (14). These microfluidic technologies have greatly facilitated the detection, manipulation, and categorization of microbial populations at a single-cell level, advancing our understanding of the role microorganisms play in the microbial ecosystem (15).

RNA sequencing (RNA Seq) provides a high-resolution view of the transcriptome by sequencing and quantifying RNA molecules. It allows the identification of alternative splicing, novel transcripts, and non-coding RNAs (16). RNA Seq technology combines high-throughput sequencing methods with computational methods to study the transcripts found in an extracted RNA. This technology allows deep sampling of the transcriptome by generation of several short fragments which are then combined to reconstruct the original RNA transcript by alignment of sequence reads to a reference genome or their de novo assembly. As compared to microarrays, RNA-Seq has a key advantage in terms of lower requirement for RNA input amount (nanograms), allowing finer investigation of cellular structure at the single-cell level (13).

9.2.3 Proteomics Technologies

In microbiology, proteomics is utilized for the analysis of a multitude of proteins expressed in an organism under specific culture conditions. Identification of microorganisms using the method of time-of-flight mass spectrometry has been widely used in various fields of microbiology for more than a decade. Despite its rather narrow area of applicability compared to other areas of proteomic analysis, this technology has developed significantly, making it a feasible competitor to the traditional bacteriological method. Proteomics has predominantly depended on two key approaches for the separation and visualization of proteins: (i) two-dimensional polyacrylamide gel electrophoresis (2D-PAGE) and (ii) gel-free profiling methods, which include the liquid chromatographic fractionation of proteins coupled with automated tandem mass spectrometry (LC-MS/MS) (17, 18).

Mass spectrometry (MS) technique identifies and quantifies proteins by measuring their mass and charge. Liquid chromatography-mass spectrometry (LC-MS) and matrix-assisted laser desorption/ionization (MALDI) are common techniques (19). Two-dimensional gel electrophoresis (2D-PAGE) is also employed in microbiological research to separate proteins based on isoelectric point and molecular weight, allowing for the visualization of protein expression patterns (20).

MALDI-TOF-MS (MALDI time-of-flight mass spectrometry) is an outstanding example of a technique of proteomics being widely used in research and clinical laboratories. Using MALDI-TOF has assisted in the rapid and accurate identification of bacterial and fungal species as well as decreased the cost in comparison to classical morphological and biochemical assays. In the initial stage, a bacterial colony or an extracted protein from bacteria is shifted onto a metal plate and ionized with a laser. The mass spectrum is obtained from bacterial samples depending on the time it takes for the proteins to travel through the equipment, with smaller proteins flying more rapidly than the larger ones. For identification of the organism, the spectrum of an unknown patient isolate is compared with the bacterial spectra database. Through augmentation of this technology, researchers are exploring potential applications for the assessment of antimicrobial susceptibility and identification of toxins and virulence factors (21–23).

9.2.4 Metabolomics Technologies

Metabolomics offers an insight into the physiological condition of a cell by providing a snapshot of all the metabolites within a microorganism. The advancement in mass spectrometry-based technologies and the development of improved informatics tools have boosted microbial metabolome research, revealing important information about an unforeseen diversity of microbial metabolites and metabolic networks (24). Metabolomics has been extensively employed in a number of microbiological procedures, including identification of microorganisms, screening of mutants, gene function research, elucidation of metabolic pathways, and genetic engineering of microbes (25).

Mass spectrometry (MS) is a very well-established analytical method used to identify and quantify metabolites in a high-throughput manner and it offers high selectivity and sensitivity (26, 27). A standard metabolomics technique involves these key steps: firstly, metabolites are extracted from the samples using a chosen extraction method. Subsequently, depending on the study design and aims, targeted or untargeted metabolomics analysis is performed. Routine detection and quantification of metabolites generally utilize techniques such as gas chromatography-MS (GC-MS), liquid chromatography-MS (LC-MS), and capillary electrophoresis-MS (CE-MS). These hyphenated separation techniques have widened the applications of MS techniques (28). In general, nuclear magnetic resonance (NMR) spectroscopy and liquid chromatography-mass spectroscopy (LC-MS) are the most favorable techniques used for metabolome analysis (8).

NMR typically gives information on the structure and quantity of metabolites in a sample (29). It can also provide a general view of metabolites under certain conditions, providing rich structural information (25). It offers the advantage of direct quantification but is characterized by relatively low sensitivity and throughput. In contrast, MS is a more sensitive technique, albeit facing challenges in quantification and its destructive nature. Specifically, an NMR-based metabolomic study typically provides details on 50–200 identified metabolites with concentrations exceeding 1 μM. On the other hand, a typical LC-MS-based metabolomic study can yield details on over 1,000 identified metabolites with concentrations ranging from >10 to 100 nM (30–32).

9.2.5 Epigenomics Technologies

Epigenomics involves a change in phenotype without altering the genotype of an organism. It typically employs techniques such as bisulfite sequencing and chromatin immunoprecipitation (ChIP). Bisulfite sequencing detects DNA methylation patterns by treating DNA with bisulfite, which converts unmethylated cytosines to uracil while leaving methylated cytosines unchanged (9). Chromatin immunoprecipitation (ChIP) identifies DNA regions associated with specific proteins, including those involved in histone modifications (33).

9.2.6 Metagenomic Technologies

Metagenomics techniques entail shotgun sequencing of the genomic DNA directly from an environmental sample. Organisms can be investigated in their natural environment and these techniques are utilized for microbiome, environmental, and drug discovery research. This sequencing approach has enabled the detection of novel pathogens that would not be identifiable by standard culture techniques (34, 11).

Two techniques are commonly employed, specifically marker gene analysis and shotgun sequencing. In the marker gene approach, primers are designed to bind to the highly conserved regions of a gene of interest. 16S rRNA sequencing targets the 16S rRNA gene for profiling bacterial communities and identifying microbial species whereas shotgun sequencing analyzes the collective genomes of entire microbial communities, providing insights into genetic diversity and functional potential. Shotgun sequencing may present the details about the potential functions of a whole microbial community at the gene level based on databases of previously sequenced genomes (35).

9.3 APPLICATIONS OF OMICS TECHNOLOGIES IN MICROBIOLOGICAL RESEARCH

Omics technologies have revolutionized our apprehension of microbes and their role in keeping environmental balance and improving human health. In this section, we will explore various applications of omics technologies in microbiological research, ranging from environmental and agricultural to medical, food, and industrial microbiology (Figure 9.2).

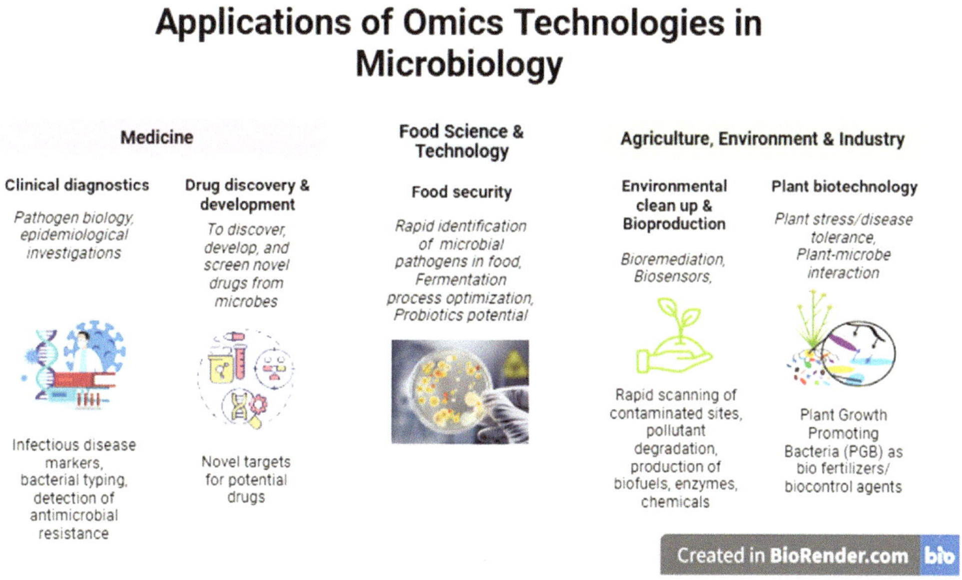

FIGURE 9.2 Diagram depicting various applications of omics technologies in microbiology, including diagnostics, industry, infections, diseases, epidemiology, agriculture, etc.

9.3.1 Medical Microbiology

In microbiology, proteomic analysis is either used for the structural and functional characterization of proteins/peptides or for quantifying the protein content in cells and tissues. Specifically, in medical microbiology, proteomics-based techniques are being employed for the identification of infectious disease markers, searching for protein antigens and vaccine candidates, elucidation of pathogenicity mechanisms, as well as determining the influence of different factors on the expression proteins in microorganisms. In bacteriology and virology, these have been used to study the characteristics of an infectious pathogen as well as disease pathogenesis. Moreover, the study of the protein content is an imperative step in the taxonomic classification of a species, investigation of the pathogenicity of an infectious agent, and assessment of therapeutic effectiveness against an infectious agent (36, 37).

In clinical microbiology, different methodologies and techniques are used for reliable, direct detection of infectious pathogens in human body fluids. MS-based procedures, especially MALDI-TOF-MS, have been commonly employed in the routine clinical depiction of pathogens. An example is the detection of *Klebsiella pneumoniae* proteins in urinary tract infections (38). In microbiological and clinical research, whole genome sequences (WGS) have been significant in establishing the phylogenetic associations among families of clinically related organisms. WGS not only provides accurate identification of microbial isolates but has also been useful for the identification of novel microbes, detection of microorganisms that are slow-growing, and identification of microorganisms in mixed cultures (39–41). WGS is widely used in public health, reference, and hospital-affiliated infection control laboratories for purposes such as bacterial typing for epidemiological investigation, detection of antimicrobial resistance, and searching for genetic markers of virulence (42).

9.3.2 Environmental and Agricultural Microbiology

Environmental omics contribute to a better understanding of the metabolic pathways of microorganisms and complex microbial communities to enhance phenotype-genotype linkages.

Omics technologies can provide new insights into the key molecular players in response to environmental stresses and vital particulars of microbial communities. A wide diversity of applications of omics in environmental microbiology includes the bioremediation of pollutants, design of innovative biosensors, screening for novel catalysts and therapeutic drugs, and production of novel materials and chemicals in microorganisms (43). Omics technologies including metagenomics, metabolomics, transcriptomics, and proteomics are becoming quick and more reliable options for the rapid scanning of contaminated sites to study specific degradative processes and to look for microorganisms playing an important role in pollutant degradation. These omics technologies may aid in the evaluation of bioremediation efficiency and the development of novel bioremediation techniques (44).

In agriculture, seed-coating agents such as fludioxonil (FL) and metalaxyl-Mfludioxonil azoxystrobin (MFA) are used for the control of cotton seedling diseases. In a study aimed at assessing the effects of FL and MFA on endophytic microbes of cotton seed, it was observed that the application of these coatings altered the bacterial and fungal community composition. They also noted that the number of beneficial microorganisms was reduced at the same time, enriching the microbial community of pollutant degraders. Furthermore, soil metabolomic analysis showed a strong link between soil microbes, metabolites, and soil enzymatic activities (45).

Metagenomics, which is based on examining the environmental sample's genetic pool, has been significantly employed to examine the diversity of microbial communities and their functional extent. Using metagenomics, the genomic diversity of these microbial communities and their role in ecological functions including nutrient cycling and contaminants degradation has been revealed (6). In a relevant study, metagenomic and network analyses revealed important players involved in nitrification in upland agricultural soils (46). Similarly, another study examined the key role bacteria play in phosphate solubilization in agricultural soil. The study employed genome-resolved metagenomic techniques to identify the specific genes conferring phosphate solubilization in bacteria (47). Furthermore, information garnered from metagenomics in combination with RNA sequencing (metatranscriptomics) and protein analysis (metaproteomics) has also provided deeper insights into microbial processes governing nutrient cycling in different ecosystems (48).

9.3.3 Industrial Microbiology

Omics techniques have been extensively used in the field of industrial microbiology. For instance, genomics allowed the targeted manipulation of industrially important microbes for improved production of various bioproducts, including biopharmaceuticals, enzymes, and biofuels (6). A human intestinal bacterium *Cytobacillus oceanisediminis* NB2 was isolated and characterized and further subjected to whole genome sequencing. The genomic analysis of this strain identified the presence of gene clusters for diverse bio-catalytic activity, stress response, and antimicrobial activity as well as pointing toward the absence of pathogenic gene islands. The strain native to the human gut also displayed probiotic potential as its genetic repertoire coded for functional characteristics such as anti-amylase, anti-lipase, prolyl endopeptidase, bile salt hydrolase, lactase, glutenase, cholesterol oxidase, and anti-pathogenic activity (49).

Transcriptome of the microbes has been studied to optimize the fermentative production of numerous bulk chemicals. In a relevant study, succinate production was investigated in *Corynebacterium glutamicum*. Higher production of succinate had shown harmful effects on the rate of glucose consumption and succinate production and affected the overall cellular fitness of the strain. Expression profiling was performed using microarrays which revealed the upregulation or downregulation of certain metabolic pathways, transcriptional regulation, and DNA repair at higher succinate concentrations. Upon analysis of downregulated genes, a few were found associated with the restoration of glucose consumption rate. Various experimental investigations led to the conclusion that overexpression of a particular transcriptional regulator NCgl0275 reinstated cellular fitness and resulted in an increased succinate production (by 38%) (50). In another study, 1-butanol production in *E. coli* was improved by adjusting the CoA balance. With the help of metabolomic analysis, it was found that conversion of butanoyl-CoA to butanal in the presence of alcohol dehydrogenase (AdhE2) is a rate-limiting step of the overall reaction. Adjustment of AdhE2 expression resulting in the subsequent release of free CoA and the addition of cysteine improved the 1-butanol production (51).

9.3.4 Food Microbiology

Mass spectrometry techniques are now being widely employed in food microbiology. In 2018, the Food and Drug Administration approved the application of MALDI-TOF platforms for rapid identification of microbial pathogens in food samples (52). Similarly, WGS has been used to decipher the probiotic potential of microbes. A study reported the probiotic action and several associated genes in a probiotic strain, *Limosilactobacillus reuteri* PNW1 (previously named *Lactobacillus reuteri* PNW1), including stress tolerance, lactic acid production, mucosal adhesion, and production of therapeutically active peptides (53).

Metagenomic studies have confirmed the transfer of microbes from food to the human gut environment and their persistence in that environment (54). A study used genomic analysis to establish the relationship of probiotic bacteria with the gut microbiome. Comparative genomic analysis revealed the great diversity of lactic acid bacteria (LAB) and showed the relationship between LAB found in food and the gut microbiome wherein fermented food was found to be the most frequent niche of these bacteria in nature (55).

In food microbiology, the two most commonly employed NGS technologies include amplicon sequencing and shotgun metagenomics. Amplicon sequencing, otherwise called metabarcoding, relies on sequencing of specific marker genes, whereas shotgun metagenomics analyzes the whole genome content of the microbial communities. Metabarcoding gives a snapshot of the overall taxonomic composition of microbial communities and is extensively used in food microbial ecology. On the other hand, the shotgun approach also determines the functional capability of microbial communities in addition to taxonomical classification (e.g., identifying the occurrence and abundance of specific genes involved in spoilage and/or virulence) as well as the reconstruction of the associated metabolic pathways (56).

9.3.5 Integrated Omics: Recent Trends in Microbiology

The integration of data from multiple omics technologies has provided information about the biological mechanisms and translational characteristics of microbes. Precisely, this data integration enables an all-inclusive view of a biological phenomenon, starting from information contained in DNA transcribed into RNA which is then translated into proteins. Proteins (enzymes) in turn are responsible for conversion/production of varied metabolites (30).

Integrated omics methods are rapidly spreading across various microbiological research labs, with promising prospects regarding the detection of previously undetected features of microbial cells and the study of various molecular components in different cellular capacities. Single-omics investigation cannot reveal a complete picture of the complexities of microbial systems (29). Integrated omics-based strategies have been incorporated into different microbiology applications ranging from biofuel production and biopharmaceutical manufacturing to medical microbiology and host–pathogen interactions. However, the integration of such massive and diverse data presents a great challenge to bioinformaticians to manage and present this data in useful form (57).

9.4 FUTURE PERSPECTIVES

In the era of omics, microbiology has transformed from single omics to multiomics, advancing from investigating individual microorganisms to exploring whole communities. Most importantly, this transition entails capturing microbial functions not only in controlled laboratory conditions, but also in natural environments. Studies based on multiomics are swiftly widening our comprehension of microbial function across multiple fields like biotechnology, ecology, environment, and medicine. Omics-based systemic technologies hold immense promise in addressing crucial global challenges such as ecological preservation, food production and security, sustainable fuel, and health crises. Despite their great potential, omics approaches present significant challenges in interpreting complex and large datasets to extract relevant insight into biological phenomena. Consequently, the establishment of user-friendly integration tools and the creation of mathematical models are indispensable to perceiving these composite biochemical processes. This necessitates the expansion and utilization of advanced bioinformatics tools for thorough analysis, integration, and interpretation of the data (58).

9.5 CONCLUSION

Over time, extensive data has been produced by omics technologies, providing useful insights into the microbial world. Omics approaches play a crucial role in areas such as personalized medicine, drug discovery, and the comprehension of intricate biological processes. The ongoing progress of omics technologies continues to augment our capacity to unravel the complexities of living systems and address diverse biological inquiries.

REFERENCES

1. Akinsemolu AA. The role of microorganisms in achieving the sustainable development goals. *Journal of Cleaner Production*. 2018 May;182:139–55.
2. Darshna D, Tiwari SS. Basic of omics and its applications. In *Industrial Microbiology and Biotechnology: Emerging concepts in Microbial Technology*, ed. P Verma. Springer Nature; 2023. p. 1–22.
3. Dai X, Shen L. Advances and trends in omics technology development. *Frontiers in Medicine*. 2022 Jul 1;9:911861.
4. Priyanka Narad, Vasanthan KS. *Introduction to Omics*. Springer eBooks; 2018. p. 1–10.
5. Singh J, Saini DK, Kashyap R, Kumar S, Chopra Y, Sandhu KS, et al. Omics technologies for agricultural microbiology research. In *Trends of Applied Microbiology for Sustainable Economy*, ed. R Soni et al. Academic Press; 2022. p. 343–94.
6. Sood U, Pandey J, Patel SK, Verma H. Omics techniques in deciphering environmental, industrial and therapeutic applications of microbes. *Frontiers in Microbiology*. 2023 Nov 28;14:1327368.
7. Debnath M, Prasad GBKS, Bisen PS. *Molecular Diagnostics: Promises and Possibilities*. Springer; 2009. p. 467–82.
8. Marchesi JR, Ravel J. The vocabulary of microbiome research: a proposal. *Microbiome*. 2015 Dec;3:1–3.
9. Micheel CM, Nass SJ, Omenn GS, Trials C on the R of OBT for PPO in C, Services B on HC, Policy B on HS, et al. *Committee on the Review of Omics-Based Tests for Predicting Patient Outcomes in Clinical Trials*. National Academies Press (US); 2012 (cited 2024 Apr 14). Available from: www.ncbi.nlm.nih.gov/books/NBK202160/
10. Alev Baysoy, Bai Z, Rahul Satija, Fan R. The technological landscape and applications of single-cell multi-omics. *Nature Reviews Molecular Cell Biology*. 2023 Jun 6;24:695–713.
11. Kim M, Lee KH, Yoon SW, Kim BS, Chun J, Yi H. Analytical tools and databases for metagenomics in the next-generation sequencing era. *Genomics & Informatics*. 2013;11(3):102.
12. Satam H, Joshi K, Mangrolia U, Waghoo S, Zaidi G, Rawool S, Thakare RP, Banday S, Mishra AK, Das G, Malonia SK. Next-generation sequencing technology: current trends and advancements. *Biology*. 2023 Jul 13;12(7):997.

13. Lowe R, Shirley N, Bleackley M, Dolan S, Shafee T. Transcriptomics technologies. *PLoS Computational Biology*. 2017 May 18;13(5):e1005457.
14. Yager P, Edwards T, Fu E, Helton K, Nelson K, Tam MR, Weigl BH. Microfluidic diagnostic technologies for global public health. *Nature*. 2006 Jul;442(7101):412–8.
15. Song Y, Yin J, Huang W, Li B, Yin H. Emerging single-cell microfluidic technology for microbiology. *TrAC Trends in Analytical Chemistry*. 2023 Nov 18;170:117444.
16. Wang Z, Gerstein M, Snyder M. RNA-Seq: a revolutionary tool for transcriptomics. *Nature Reviews Genetics*. 2009 Jan;10(1):57–63.
17. Baggerman G, Vierstraete E, De Loof A, Schoofs L. Gel-based versus gel-free proteomics: a review. *Combinatorial Chemistry & High Throughput Screening*. 2005 Dec 1;8(8):669–77.
18. Nie L, Wu G, Zhang W. Statistical application and challenges in global gel-free proteomic analysis by mass spectrometry. *Critical Reviews in Biotechnology*. 2008 Dec 16;28(4):297–307.
19. Wei X, Li L. Mass spectrometry-based proteomics and peptidomics for biomarker discovery in neurodegenerative diseases. *International Journal of Clinical and Experimental Pathology*. 2009;2(2):132.
20. Carrette O, Burkhard PR, Sanchez JC, Hochstrasser DF. State-of-the-art two-dimensional gel electrophoresis: a key tool of proteomics research. *Nature Protocols*. 2006 Jul 20;1(2):812–23.
21. Alby K, Gilligan PH, Miller MB. Comparison of matrix-assisted laser desorption ionizaton–time of flight (MALDI-TOF) mass spectrometry platforms for the identification of gram-negative rods from patients with cystic fibrosis. *Journal of Clinical Microbiology*. 2013 Aug 21;51(11):3852–4.
22. Dingle TC, Butler-Wu SM. MALDI-TOF mass spectrometry for microorganism identification. *Clinics in Laboratory Medicine*. 2013 Sep 1;33(3):589–609.
23. Richter S, Sercia L, Branda JA, Burnham CAD, Bythrow M, Ferraro M, et al. Identification of Enterobacteriaceae by matrix-assisted laser desorption/ionization time-of-flight mass spectrometry using the VITEK MS system. *European Journal of Clinical Microbiology and Infectious Diseases*. 2013 Jul 2;32(12):1571–8.
24. Ali A, Mughal A, Ahmad S, Abbas S, Mumtaz A, Ahmed I. Microbial metabolomics: recent advancements and applications in infectious diseases and drug discovery. In *Recent Advances and Future Perspectives of Microbial Metabolites*, ed. S De Mandal. Academic Press; 2023. p. 269–335.
25. Xu YJ, Wang C, Ho WE, Ong CN. Recent developments and applications of metabolomics in microbiological investigations. *TrAC Trends in Analytical Chemistry*. 2014 Apr;56:37–48.
26. Lei Z, Huhman DV, Sumner LW. Mass spectrometry strategies in metabolomics. *Journal of Biological Chemistry*. 2011 Jun 1;286(29):25435–42.
27. Weckwerth W. Metabolomics in system biology. *Annual Review of Plant Biology*. 2003 Jun;54(1):669–89.
28. Tyagi R, Kumar P, Sharma U. Metabolomics techniques: a brief update. In *Epigenetics and Metabolomics*, ed. PK Agrawala and P Rana. Academic Press; 2021. p. 1–29.
29. Zhang R, He X, Fu W, Chen T, Sun P, Li B, Ding D. Efficient identification of different types of carbons in organic solids by 2D solid-state NMR spectroscopy. *The Journal of Physical Chemistry*. 2011;A115(42):11665–70.
30. Arıkan M, Muth T. Integrated multi-omics analyses of microbial communities: a review of the current state and future directions. *Molecular Omics*. 2023 Jan 1;19(8):607–23.
31. Van Belkum A, Welker M, Erhard M, Chatellier S. Biomedical mass spectrometry in Today's and Tomorrow's clinical microbiology laboratories. *Journal of Clinical Microbiology*. 2012 Feb 22;50(5):1513–7.
32. Emwas AH, Roy R, McKay RT, Tenori L, Saccenti E, Gowda GN, Raftery D, Alahmari F, Jaremko L, Jaremko M, Wishart DS. NMR spectroscopy for metabolomics research. *Metabolites*. 2019 Jun 27;9(7):123.
33. Singh P, Szabó PE. Chromatin immunoprecipitation to characterize the epigenetic profiles of imprinted domains. *Methods in Molecular Biology*. 2012;925:159–72.
34. Bragg LM, Tyson GW. Metagenomics using next-generation sequencing. Methods in Molecular Biology. 2014 Jan 1;1096:183–201.
35. Daliri EBM, Ofosu FK, Chelliah R, Lee BH, Oh DH. Challenges and perspective in integrated multi-omics in gut microbiota studies. *Biomolecules*. 2021 Feb 17;11(2):300.
36. Khodadadi E, Zeinalzadeh E, Taghizadeh S, Mehramouz B, Kamounah FS, Khodadadi E, et al. Proteomic applications in antimicrobial resistance and clinical microbiology studies. *Infection and Drug Resistance*. 2020 Jun;13:1785–806.
37. Sharov TN, Viktorov DV, Toporkov AV. Proteomic analysis in microbiology. *Molecular Genetics, Microbiology and Virology*. 2023 Mar;38(1):1–7.

38. Svetličić E, Dončević L, Ozdanovac L, Janeš A, Tustonić T, Štajduhar A, et al. Direct identification of urinary tract pathogens by MALDI-TOF/TOF analysis and De Novo peptide sequencing. *Molecules*. 2022 Aug 25;27(17):5461.
39. Hasman H, Saputra D, Sicheritz-Ponten T, Lund O, Svendsen CA, Frimodt-Møller N, Aarestrup FM. Rapid whole-genome sequencing for detection and characterization of microorganisms directly from clinical samples. *Journal of Clinical Microbiology*. 2014 Jan;52(1):139–46.
40. Long SW, Williams D, Chandni Valson, Cantu C, Cernoch PL, Musser JM, et al. A genomic day in the life of a clinical microbiology laboratory. *Journal of Clinical Microbiology*. 2013 Apr 1;51(4):1272–7.
41. Reuter S, Ellington MJ, Cartwright EJ, Köser CU, Török ME, Gouliouris T, Harris SR, Brown NM, Holden MT, Quail M, Parkhill J. Rapid bacterial whole-genome sequencing to enhance diagnostic and public health microbiology. *JAMA Internal Medicine*. 2013 Aug 12;173(15):1397–404.
42. Kwong JC, McCallum N, Sintchenko V, Howden BP. Whole genome sequencing in clinical and public health microbiology. *Pathology*. 2015 Apr 1;47(3):199–210.
43. Martínez-Espinosa RM, Armengaud J, Matallana-Surget S, Olaya-Abril A. Environmental omics and their biotechnological applications. *Frontiers in Microbiology*. 2023 Mar 9;14:1165558.
44. Shyam K, Kumar N, Chandel H, Dogra AS, Sharma G, Saxena G. *Omics Technologies in Environmental Microbiology and Microbial Ecology*. Wiley; 2023. p. 433–54.
45. Ma Z, Feng H, Yang C, Ma X, Li P, Feng Z, et al. Integrated microbiology and metabolomics analysis reveal responses of cotton rhizosphere microbiome and metabolite spectrum to conventional seed coating agents. *Environmental Pollution*. 2023 Sep 1;333:122058–8.
46. Zhao H, Li Y, Liu M, Song X, Liu B, Ju X. Metagenomic and network analyses reveal key players in nitrification in upland agricultural soils. *Environmental Microbiology*. 2023 Nov;25(11):2636–40.
47. Wu X, Cui Z, Peng J, Zhang F, Liesack W. Genome-resolved metagenomics identifies the particular genetic traits of phosphate-solubilizing bacteria in agricultural soil. *ISME Communications*. 2022 Feb 16;2(1).
48. Taş N, de Jong AE, Li Y, Trubl G, Xue Y, Dove NC. Metagenomic tools in microbial ecology research. *Current Opinion in Biotechnology*. 2021 Feb;67:184–91.
49. Yadav M, Kumar T, Kanakan A, Maurya R, Pandey R, Chauhan NS. Isolation and characterization of human intestinal bacteria Cytobacillus oceanisediminis NB2 for probiotic potential. *Frontiers in Microbiology*. 2022 Jul 13;13:932795.
50. Ohnishi, S. Mitsuhashi, M. Hayashi, J. A novel methodology employing Corynebacterium glutamicum genome information to generate a new L -lysine-producing mutant. *Applied Microbiology and Biotechnology*. 2002 Jan 1;58(2):217–23.
51. Ohtake T, Pontrelli S, Laviña WA, Liao JC, Putri SP, Fukusaki E. Metabolomics-driven approach to solving a CoA imbalance for improved 1-butanol production in Escherichia coli. *Metabolic Engineering*. 2017 May;41:135–43.
52. Commissioner of the. *FDA Authorizes New Use of Test, First to Identify the Emerging Pathogen Candida auris*. FDA; 2020. Available from: www.fda.gov/news-events/press-announcements/fda-authorizes-new-use-test-first-identify-emerging-pathogen-candida-auris
53. Alayande KA, Aiyegoro OA, Nengwekhulu TM, Katata-Seru L, Ateba CN. Integrated genome-based probiotic relevance and safety evaluation of Lactobacillus reuteri PNW1. *PLOS ONE*. 2020 Jul 20;15(7):e0235873.
54. Ferrocino I, Rantsiou K, Cocolin L. Microbiome and -omics application in food industry. *International Journal of Food Microbiology*. 2022 Sep;377:109781.
55. Pasolli E, De Filippis F, Mauriello IE, Cumbo F, Walsh AM, Leech J, et al. Large-scale genome-wide analysis links lactic acid bacteria from food with the gut microbiome. *Nature Communications*. 2020 May 25;11(1):2610.
56. Sequino G, Valentino V, Villani F, De Filippis F. Omics-based monitoring of microbial dynamics across the food chain for the improvement of food safety and quality. *Food Research International*. 2022 Jul;157:111242.
57. Fondi M, Liò P. Multi -omics and metabolic modelling pipelines: challenges and tools for systems microbiology. *Microbiological Research*. 2015 Feb;171:52–64.
58. Dey D, Mekala LP, Mohammed M. Integrating omics technologies to understand microbial systems. *Omics Technologies for Sustainable Agriculture and Global Food Security*. 2021;1:191–212.

10 Computational Biology and Its Role in Life Science Research

Maria Rukan, Mohammad Ejaz, Muhammad Hamza, and Hafiz Muhammad Waqas Munir

10.1 INTRODUCTION

Researchers have consistently found inspiration in nature, exploring its diverse aspects in various ways. The majority of nature-inspired algorithms are derived from distinctive features of biological systems. Consequently, the predominant category among these algorithms is that of biology-inspired. While highlighting the source of inspiration, there are still diverse approaches to categorize nature-inspired algorithms, depending on the specific details and sub-sources utilized. Computer science and biology have shared a long-standing and mutually beneficial history, engaging in a productive collaboration for several decades. Over the years, computer scientists have developed numerous computational methods and algorithms inspired by biological systems. In order to accomplish tasks such as acquiring and interpreting biological data, biologists have begun to utilize frameworks from the field of computer science. Beyond the pursuit of successful applications and accurate models using computer algorithms, there is a broader aim of fostering a more intricate understanding of biology, paving the way for new insights in the biological sciences (1, 2).

Computational biology employs mathematical and computational methods to tackle theoretical and exploratory inquiries within the realm of biology (1). Molecular data derived from sequencing of DNA, RNA, and proteins serves as a primary source for this field. Over the past few decades, numerous sequencing technologies have been developed for each major macromolecule, contributing to the accessibility of biological data and facilitating more straightforward and comprehensive analyses of molecular data (2). Bioinformatics integrates computer programming, big data, and biological principles to empower scientists in identifying and comprehending crucial patterns within biological data. As a result, massive datasets are more efficiently organized, which makes DNA and genome sequence research much easier. Extensively utilized in the field of biotechnology, it greatly increases the entire domain, promoting a more profound comprehension of certain topics (3). Conversely, computational biology not only sharpens biological concepts, making them more focused and conclusive, but also offers a roadmap that harmonizes diverse perspectives. This aids researchers in sequencing the human genome and constructing precise models of fundamental biological elements, contributing to enhanced understanding and excellence. Evidently, computational biology empowers both audiences and researchers to explore diverse aspects of biology across a broad spectrum (2).

In recent years, significant transformations in the realms of biology and medicine have transpired, driven by the swift expansion of high-throughput data collection in genomics and biomedical imaging. Computational biologists play a pivotal role in navigating the intersection of computation and biology, aiming to seamlessly integrate these components into integral aspects of both biology and medicine. This domain is experiencing rapid growth not only within academic circles but also in the industrial sector (3, 4).

DOI: 10.1201/9781003410164-10

Bioinformatics encompasses the research, development, or application of computational tools and methods to utilize biological, medical, or behavioral data for the in-depth analysis of subject matter in biology. This not only assists researchers in acquiring biological data but also empowers them to interpret the data with success and efficiency. The objective of computational medicine is to advance healthcare by constructing computational models of pertinent subjects for improved understanding and accelerated progress (4). A correlation exists between artificial intelligence and computational biology, facilitating the transformation of extensive data into valuable and more accessible knowledge, and analytical programming skills can decipher large datasets for enhanced comprehension. This chapter aims to provide a comprehensive discussion of the computational methodologies utilized for interpreting biological data and their application in life-science-related data and research.

10.2 COMPUTATIONAL BIOLOGY APPROACHES

Computational biology is recognized as a multidisciplinary domain that integrates computer science, statistics, and mathematics to address biological problems. In essence, it employs computational approaches and mathematical tools to tackle both experimental and theoretical challenges within the field of biology that encompass the development of algorithms, theoretical models, computational simulations, and mathematical models (5). Leveraging these methodologies, researchers can gain a more comprehensive understanding of various aspects of biology. Dependent on computers, computational biology experiences significant impacts from advanced technologies, including machine learning. Presently, one of its widely utilized applications involves computer-based mathematical modeling for the identification and comprehension of disease patterns, contributing to enhanced and rapid treatment. Particularly prevalent in the medical industry, this application aims to elevate healthcare standards and effectively prevent diseases.

Utilizing computational models of diseases involves tailoring these models by integrating patient data and applying them to enhance and expedite disease diagnosis. A computational biologist enhances and refines resources and tools, thereby aiding both biomedical researchers and the general public in understanding and interpreting genome sequence data; besides, artificial intelligence has transformed large amounts of raw data into knowledge that can be accessed easily, which has greatly influenced the field of computational biology. Moreover, it helps scientists to discover new drugs, making certain treatments quicker to act upon (6). When machine learning and AI, for instance, are merged with other technologies, the error chances are reduced greatly and an increased possibility of success is assured. This way, it is possible for scientists to harvest, compile, and interpret data at scales that have never been witnessed before, hence making it easier for them to come up with universally applicable prognostic models featuring different subdomains (7).

Computational biology helps scientists detect patterns arising from complex biological phenomena, which allows them to resolve them quickly. It allows for the formation of chromosome fragment maps that aid in recognizing genes that encode protein products. Peptide sequencing methods entail particular types of proteins coding DNA while clustering algorithms are applied to define gene expression patterns (8). Together, these processes are part of what contributes to a more profound understanding of biological substances and thus they pave the way for more improvements in medicine that are expected to be highly advanced. Over the years, technology has caused a revolution in biological context, thus offering new approaches in addressing biological issues, thus there is a bright future ahead in this field (2).

10.3 DATA ACQUISITION AND PREPROCESSING

Initially, computational science largely dealt with the development and application of mathematical models in the simulation of natural and artificial phenomena. Advances in computer performance

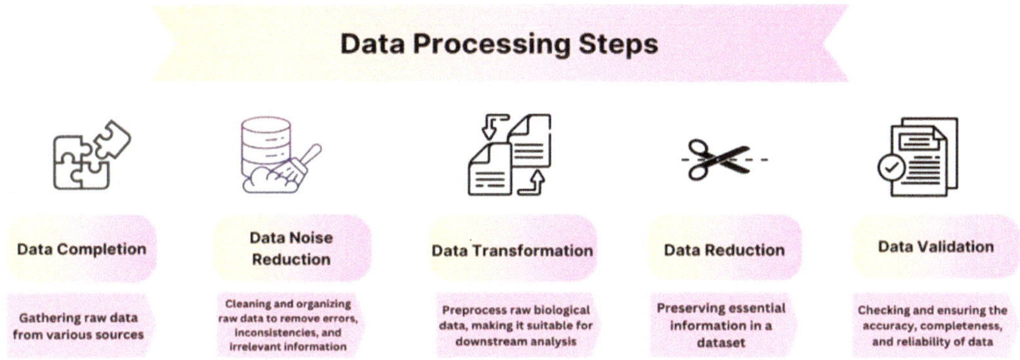

FIGURE 10.1 Illustration of different steps of data preprocessing.

and data storage capacity have altered the way information in every field can be expressed through data. This shift extends the initial goals of computational science beyond computation itself, now including data management and analysis (2). The surge in data volume has recently reached an overwhelming scale, continually accelerating and intensifying, that has given rise to a new concept known as big data (9).

From a scientific standpoint, big data is defined in connection with the gathering, storage, and examination of datasets characterized by a high degree of complexity or size. The big data definition not only involves data volume but also encompasses its variety, velocity, and value (10). Consequently, conventional tools and methods to handle and analyze this data have become a difficult task. Information technology (IT) plays a crucial role in ensuring data availability and facilitating analysis. Data is deemed as big when its reuse contributes to the formulation of novel insights. In light of this, IT has been instrumental in supporting emerging models of science and the establishment of newly formed interdisciplinary fields of study (11).

Considerable volumes of information are present in the real world, and these data can be collected and processed for practical applications. Data acquisition involves the sampling of signals that measure physical conditions in the real world, converting these samples into digital numeric values that computers can manipulate. In a broader sense, data processing refers to the collection and manipulation of data to generate meaningful information (12, 13). In studies with human subjects, the data acquisition process commonly encompasses participant recruitment, screening, consent, and data collection. The data preprocessing phase typically comprises five steps: reviewing data, entering and verifying, cleaning, editing, and documentation (Figure 10.1) (14).

10.4 COMPUTATIONAL BIOLOGY TOOL APPROACHES IN LIFE SCIENCE RESEARCH

Computational biology heavily depends on the examination of biological data, with one of the primary sources being molecular data derived from sequencing three major types of macromolecules: DNA, RNA, and protein. Sequencing involves the determination of nucleotide sequences in the cases of DNA and RNA, as well as the sequence of amino acids for proteins. Over the past few decades, numerous sequencing technologies have been developed for each main macromolecule, including Sanger sequencing, next-generation sequencing, DNA microarray, and RNA sequencing (15). The modern sequencing technologies generate an immense volume of biological data from diverse living species. Consequently, there has been a substantial increase in the range of biological inquiries addressable through mathematical, statistical, and computational methods. Computational biology has emerged as a vast interdisciplinary field that both develops and applies computational techniques to analyze extensive sets of biological data, enabling the formulation of new predictions and the discovery of novel biological insights. This field encompasses a wide array of approaches, including various analytical methods, mathematical modeling, and simulation, spanning several key areas such as the analysis of protein and nucleic acid sequences, structures, and functions, evolutionary genomics and proteomics, systems biology, population genomics, regulatory and metabolic networks, biomedical image analysis and modeling, gene–disease associations, and the study of the development and spread of diseases. The advancements in these technologies have made biological data more accessible, thereby facilitating simpler and more comprehensive analyses of molecular data (2).

In contrast to the abundance of DNA and RNA sequencing technologies, the options for protein sequencing are notably fewer. The inaugural automated technology for protein sequencing, known as the protein sequenator relying on the Edman degradation reaction, was introduced in 1967 by Edman and Begg (16). Furthermore, the introduction of mass spectrometry (MS) techniques has had a profound and revolutionary effect on proteomics studies (17). Continual enhancements in the sensitivity and resolution of mass spectrometry instrumentation, coupled with advancements in technologies for sample preparation, have resulted in the most comprehensive and versatile technology for large-scale proteomics available today. The primary techniques within modern MS technologies include matrix-assisted laser desorption ionization (MALDI) and electrospray ionization (ESI) (18, 19). Exploring the application of nanopore-based technology for proteins holds the potential to usher in novel techniques for protein sequencing (20, 21).

10.5 COMPUTATIONAL MODELING

The field of systems biology has experienced significant advancements in the past decade through the integration of computational models and methodologies, traditionally utilized in computer science for program correctness and safety evaluation. Within this framework, crafting a biological model is akin to programming a computer. Various programming languages, often tailored for the biological domain, offer a means of articulating the sequence of instructions that dictates the control flow of a biological process. The language's syntax establishes the permissible ways in which symbols can be combined to generate well-formed sentences or instructions. While this specification is commonly presented textually (e.g., process calculus, rule-based systems), certain cases involve a graphical representation (e.g., Petri nets, state charts) that aids users in visualizing processes through diagrams illustrating species flow in reactions or changes in internal states (22, 23).

Semantics elucidate the meaning of syntactically valid instructions by detailing the model's behavior and how it should be executed by a computer. It is noteworthy that a model specified in a particular language syntax can be executed with different language semantics (24). For instance, a set of rules governing chemical reactions can be executed with either a continuous semantics or a stochastic semantics, depending on the desired level of approximation or complexity. COPASI is one of the tools for numerically simulating and analyzing biochemical networks, accommodating both continuous and stochastic dynamics (25, 26).

10.5.1 Process Algebras

In recent years, researchers in computer science have delved deeply into the application of process algebras (PAs) for modeling and analyzing biological systems. The expressive capabilities of PAs allow for formal specification without ambiguity regarding interactions, communications, and synchronizations among a set of concurrent processes, also referred to as agents. The interest in employing PAs in systems biology stems from the notion that biological systems can be viewed as concurrent reactive systems, where biological species function as processes interacting with one another (27).

An additional noteworthy attribute of PAs, particularly in modeling complex biological systems that are often multiscale, is their compositional nature. This enables the definition of the entire system based on the specifications of its subcomponents. Moreover, PAs typically facilitate formal reasoning about equivalences between processes. A few of the widely used PAs in computational systems biology include Beta-Binders/BlenX, SPiM, Bio-PEPA, sCCP, and BioShape (28). PA specifications commonly serve as intermediate models, subsequently executed or translated into other computational models using diverse semantics such as continuous differential (ODEs), stochastic (Continuous Time Markov Chains), or abstract (transition systems) (29, 30).

10.5.2 Rule-based Systems

Rule-based systems have garnered significant attention from biologists due to their notation's striking similarity to the chemical reaction representation employed in systems biology for modeling biochemical interactions among molecular species. A significant feature of this modeling approach is that rules, unlike equations, are standalone units, making it simple to modify or adjust them. Moreover, the straightforward syntax of rule-based models can be stored in a file as human-readable text, and it can be edited and visualized using a graph representation. This user-friendly aspect makes rule-based modeling accessible to individuals without specialized mathematical or computer science skills. Rule-based models can then be translated, employing different semantics, to generate other computational models, offering either a quantitative prediction (e.g., the species amount over time) or a qualitative understanding of the system's emergent behavior where time is abstracted away (30, 31).

Due to these advantages, numerous rule-based modeling languages and tools, such as BIOCHAM, Kappa, BioNetGen, have gained popularity among systems biologists in recent years and have been extensively applied in concrete case studies (32). For a more comprehensive review of rule-based modeling, please refer to (33).

10.5.3 Petri Nets

A Petri net is a directed graph featuring vertices divided into two distinct sets (a bipartite graph). These sets comprise nodes known as "transitions" (representing events like reactions, depicted graphically as bars) and nodes referred to as "places" (representing conditions for a reaction to occur, such as the presence of a molecule, depicted graphically as circles). Arrows connect these nodes, indicating the flow direction, adhering to a key rule: a place node can only connect to a transition node and vice versa. Data, typically species, are portrayed as "tokens," denoted by black marks. Tokens are consumed from input places through transitions and subsequently generated in output places. A transition "fires" when enabled by the presence of tokens in a place directly connected to it. Concurrent semantics delineate the temporal evolution of the token distribution (23, 34).

Originally introduced by Carl Adam Petri in 1962 to describe chemical processes, this modeling framework has found extensive use in computer science for specifying and analyzing concurrent and distributed systems. Unsurprisingly, the intuitive and graphical nature of this modeling style is popular among computational systems biologists for describing biochemical reaction systems, where tokens are interpreted as individual molecules of the involved species (23, 35, 36). The Petri net formalism provides a natural framework in which both qualitative (informed by the static structural topology of Petri nets) and quantitative (informed by the time evolution of the token distribution) analyses are tightly integrated (36). Key tools for Petri nets in computational biology include Snoopy, MARCIE, GreatSPN, and Pathway Logic Assistant (28).

10.5.4 Boolean/Qualitative Networks

Boolean networks are commonly employed to approximate the dynamics of genetic regulatory networks, where genes are considered either activated (true state) or deactivated (false state) (37).

A Boolean network is characterized by Boolean variables, each updated by a Boolean function determining the subsequent truth value state based on inputs from a subset of these variables (38). Its model is used only approximately given that it does not account for intermediate internal states but it is widely accepted due to its application in assessing the flexibility and survival ability of genetic regulatory networks. It is feasible to identify single-point attractors (what is referred to as fixed points) hence showing stability of a system by employing this model that involves selection of starting random states followed by simulation. Among the tools that are available for analysis involving Boolean networks when dealing with biology systems are GINsim, BoolNet, and BNS (39).

Recently, qualitative networks have improved the Boolean network by letting its components accept multiple values. Biologists are thus allowed more flexibility beyond binary values and the capability of modeling a wider variety of behaviors using this way. Qualitative networks are best modeled and scrutinized with Bio Model Analyzer (BMA) (40).

10.5.5 State Charts

An alternative and inherent method for modeling the dynamics of a biological system involves detailing the sequence of states that characterize its behavior. For instance, the addition of a phosphate group to certain proteins can alter their functional behavior, transitioning them to a phosphorylated state and facilitating potential protein–protein interactions. A biological system dwells in a particular state until the occurrence of an event, such as the activation or inhibition of a gene, which prompts a shift in its internal behavior from one state to another. This distinctive feature renders a biological system a multiscale reactive system, where event-driven concurrent interactions take place at various levels (molecular, cellular, tissue, organ, or population levels) or between levels. These interactions, occurring with different timing and order, collectively determine the system's emergent behavior (41).

The state charts notation provides a suitable formalism for visually representing the interdependence among the states of a reactive system. Different versions of state diagrams with varied semantics have been proposed. The state charts introduced by Harel have gained prominence among biologists due to their adept handling of the complexity inherent in modeling biological systems (42). Traditional state charts notation, if used without modifications, would necessitate specifying every conceivable combination of parameters as a distinct state, resulting in an unwieldy proliferation of states. Notably, among the tools for state charts in systems biology, IBM Rational Rhapsody holds particular relevance (28).

10.5.6 Hybrid Systems

Hybrid systems expand upon the state-based discrete representation mentioned earlier by incorporating continuous dynamics, typically described by ordinary differential equations (ODEs), within each state or mode (43). Hybrid modeling techniques are gaining increasing attention in systems biology due to their efficacy in capturing the behavior of biological systems featuring distinct switching characteristics. Sigmoidal switches, ubiquitous in biological models at various levels (molecular, cellular, tissue, organ, and population), exemplify the applicability of hybrid modeling (44). This approach is generally well-suited for integrating both qualitative (stemming from the discrete state) and quantitative (stemming from continuous dynamics) information (45).

Over the past decade, several methods for hybrid system identification (hybridization) have been proposed in the literature (46–48). These methods aim to approximate complex nonlinear dynamics using piecewise-linear or piecewise-multi-affine functions rendering such models amenable to formal analysis and enhancing large-scale simulation of multicellular ensembles (49, 50). Notably, widely used mathematical platforms such as Matlab and Simulink empower users to model and simulate hybrid systems (51, 52). Additional noteworthy tools for hybrid systems modeling in biology include Rovergene, BioDivine, Breach, dReach, and S-TaLiRo (28).

10.5.7 Spatio-Temporal Models

Continuous state deterministic spatio-temporal systems are typically expressed as "reaction-diffusion" systems formulated through semi-linear parabolic partial differential equations (PDEs) (53). In discrete state scenarios, models based on compartments (e.g., membrane computing), agent-based models, and lattice-based computational models (e.g., cellular automata, cellular Potts) have been utilized to simulate the collective behavior of cellular structures. All these models exhibit a diverse array of emergent behaviors resulting from both local and nonlocal interactions, such as the manifestation of traveling waves (e.g., in cardiac tissue), Turing patterning, and the formation of spirals (54).

10.6 PHYLOGENETIC AND EVOLUTIONARY ANALYSIS

In the pursuit of understanding the intricate tapestry of life on Earth, scientists turn to two foundational pillars of biological exploration: phylogenetic and evolutionary analysis. These methodologies serve as the guiding compass, unveiling the stories of the past, deciphering the relationships between organisms, and shedding light on the dynamic processes that have sculpted the diversity of life over eons (55). Phylogenetic and evolutionary analysis play a crucial role in elucidating the relationships between biological sequences, understanding their evolutionary history, and extracting meaningful insights from genomic data (56). Phylogenetic analysis is intimately connected to the exploration and reconstruction of evolutionary history. The relationship between phylogenetic analysis and evolutionary history is profound, as phylogenetics serves as a key tool for unraveling the branching patterns and interrelationships among species over time (57).

In its essence, phylogenetics is the method of reconstructing the ancestral relationships among organisms, like creating a grand familial tree for all living beings. This computational technique measures the evolutionary changes and links between protein sequences from various species, and through the lens of molecular data or morphological traits, scientists endeavor to chart the branching paths that connect species to a common ancestry (58, 59). Phylogenetic trees, resembling intricate family trees, emerge as visual representations of these evolutionary connections. Molecular phylogenetics employs DNA, RNA, or protein sequences, while morphological phylogenetics relies on physical characteristics, collectively guiding researchers in the quest to organize life into a cohesive and informative framework (60).

Complementary to phylogenetics, evolutionary analysis delves into the mechanisms and patterns that have orchestrated the dance of life across epochs. It explores the forces of natural selection, genetic drift, mutation, and the splendor of speciation events (61). By scrutinizing the distribution of genetic variation within populations and comparing the genomic landscapes of diverse species, scientists unlock the secrets embedded in the code of life. Molecular clocks become instruments for estimating the temporal tapestry, allowing the calculation of the ages of divergent lineages. Evolutionary analysis, a multidisciplinary endeavor, employs bioinformatics tools, statistical methods, and computational models to decipher the genomic narratives written by the hand of evolution (62).

10.6.1 Applications of Phylogenetic and Evolutionary Analysis

Integrating phylogenetic and evolutionary analyses into bioinformatics not only enhances understanding of biological processes and genetic relationships but also fosters breakthroughs in diverse scientific domains. These tools are helpful in regenerating ancestral relations, showing journeys of genes and species through time and helping break down sickness into genetic terms. In addition to taxonomy, they have helped comparative genomics by pointing at functional parts and how living things adapt. The applications of the technologies have gone far and wide, helping scientists, as they understand how ecosystems are evolving and assisting in the fight against diseases, to improve on genetics, as well their use as a medication (63).

10.6.1.1 Species Classification and Taxonomy

When you integrate phylogenetic analysis with bioinformatics you come up with ways used by scientists worldwide for classifying the animals and plants we see today. This enables them but to understand how different species evolved from one another and what makes them similar/different from each other. This method not only offers a way of correctly classifying different types of species but also gives directions to structuring the vast variety of living organisms. The cooperation involving phylogenetic research and bioinformatics helps us to differentiate nomenclatures and processes of evolution that have resulted in the diversity of life forms on earth. (61).

10.6.2 Drug Discovery and Design

For the wise design of new pharmaceutical approaches, it is highly important that the history of how evolution proceeded among parasites is understood and also behavior in regard to drug resistance. In these efforts phylogeny integrated with evolutionary analysis and bioinformatics plays a key role. By use of advanced bioinformatics equipment, researchers can carefully sift through enormous amounts of genome-wide information in order to locate as well as determine which among these might be most useful drugs targets in pathogenic organisms. Furthermore, the evolutionary outcomes accessible through these tools can help us predict emerging drug resistance among pathogens. Although it assists in producing potent and specific drugs, this comprehensive strategy also allows us to learn how microorganisms change, thus enabling us to take better and more intentional directions during development of medications (55).

10.6.3 Comparative Genomics

Bioinformatics tools delve deep into the intricacies of genomic data, making it possible for researchers to look for parts of genetic material that have the same meaning across all animals using them; harnessing its power in integrating phylogenetic and evolutionary analysis enhances comparative genomics by allowing comprehensive exploration of whole genomes from different species. Going beyond only just comparing genomes, this approach provides more information about how genomic structures change over time and what causes variation between organisms by natural selection. Also, with assistance from computational biology, comparative genomics can help in identifying known genes for characteristic features or illnesses, hence learning more about different ways they operate within cells, and contributing to advancements in personalized medicine and targeted therapeutic interventions (61, 64).

10.6.4 Functional Annotation of Genomes

Phylogenetic analyses in the complex world of genomics are tools that are crucial in the search for the functional significance of genes. Bioinformatics tools assist in deducing the roles of genes based on their evolutionary history, thereby allowing for a more complex comprehension of their functions in the larger context of biological science. This method has the dual advantage of facilitating the functional annotation of genomes as well as helping in the understanding of the molecular pathways and processes shaped by these genes over evolutionary time frames. More important is the intertwining of phylogenetic analysis with bioinformatics tools, especially so when dealing with large volumes of genomic data from high-throughput sequencing technology, thus enabling expeditious and comprehensive exploration of the functional genomics landscape (65).

10.6.5 Phylogenomics

The merger of genomic data with phylogenetic analysis, as seen in the field of phylogenomics, serves as a revolutionary strategy that transcends classical methods of phylogenetics, enabling a more comprehensive understanding of evolution. This combined approach enhances an all-inclusive view of evolutionary relationships by going beyond single genes toward analyzing full genomes. Advanced bioinformatics tools allow phylogenomic approaches to be applied in different areas, which include reconstructing the intricate Tree of Life, understanding the nuances of genome evolution, and identifying genetic innovations shaping life diversity. This enhances researchers' understanding of evolutionary processes that have shaped the living world by using genome information in the context of phylogenetic frameworks. (61).

10.6.6 Evolutionary Ecology

Using phylogenetic data analysis in bioinformatics advances evolutionary ecology. Here, it's possible to examine ecological relationships and adaptations in subtle ways. This integration allows scientists to merge phylogenetic information with other ecological datasets using bioinformatics tools in order to understand various aspects involving species evolution based on environment responses and their interactions with other organisms. From the adaptive radiations of species to the co-evolutionary patterns that characterize ecological communities, this integrative approach provides a comprehensive perspective on the evolutionary dynamics shaping ecosystems. The insights derived from phylogenetic analysis in the context of bioinformatics contribute significantly to elucidating the ecological and evolutionary processes that govern biodiversity and ecosystem functioning (66).

10.6.7 Prediction of Protein Structure and Function

When there is a relationship between two or more things, it will be easier for us to predict what might happen in future because in ancient times when some species were evolving, they had to adapt new environments. This means such organisms used different ways to survive and as result were forced make some changes. When you examine the evolutionarily conserved regions of proteins, these resources become quite sophisticated tools for the identification of crucial functional domains on these proteins and the rendering of three-dimensional protein structures. Understanding the molecular basis for biological processes plays a central part in drug discovery besides helping us understand the intricate relationship between the evolution of proteins and the function of molecules. This contribution enhances our understanding of biological processes at a molecular level. In essence, integrating evolutionary analysis into bioinformatics reveals a more thorough understanding of the intricate association among protein structure, functionality, as well as the evolutionary past (67).

10.6.8 Biogeography and Historical Migration

Within the field of bioinformatics, mixing both phylogenetic alongside evolutionary analyses allows for detailed tracking of historical migration patterns linking biogeography. Reconstruction of species movement history over time relies on bioinformatics tools to determine species dispersion history across different regions. The way this comprehensive approach assists in illuminating the evolutionary dynamics that helped shape biodiversity is the same way it gives us a preview of how various factors cause the distribution of different species. When used in combination with bioinformatics, phylogenetic analysis becomes one of the strongest means through which geographical patterns can be analyzed, drawing scientists back firmly along an evolutionary path for organisms at different times across the ever-changing landscapes of our planet (64).

10.6.9 Conservation Biology

In the realm of conservation biology, it is essential to understand the evolutionary relations between endangered species, so as to come up with effective conservation strategies. The use of bioinformatics tools is very important because they help assess genetic differences, identity evolutionarily unique populations, and guide decisions on how to conserve species. Utilization of phylogenetic and evolutionary analyses enables researchers to have deep insights into the specific evolutionary paths followed by the rare species, hence providing information on their predisposition to environmental changes, and potential adaptation strategies. Bioinformatics then helps in integrating this information which is critical for planning-focused biodiversity conservation actions aimed at prioritizing against loss of evolutionarily important populations, ultimately contributing to the conservation of biodiversity in the face of ongoing environmental challenges (68).

10.6.10 Prediction of Zoonotic Diseases

To predict and understand zoonotic diseases, we must appreciate the vital relationship between parasites and their hosts; this is where bioinformatics combined with phylogeny and evolution becomes essential. Assorted software within bioinformatics helps locate areas where we can get all the information needed for saving different species as well as evaluate how to prevent animals from infecting each other. Researchers could predict the appearance of zoonotic diseases and know what promotes cross-species transmission by analyzing how pathogens and hosts evolved in relation to each other. Thus, public health planning needs this forward-looking outlook which uses bioinformatics to be able to avoid or contain any pandemic caused by a zoonotic disease (69).

10.6.11 Understanding Evolutionary Dynamics in Microbial Communities

It has been reported that various applications of phylogenetic and evolutionary investigations are used in unravelling complex microbial communities, for which bioinformatics tools take a central place. Researchers understand how microorganisms adapt with changes through evolution, for instance microbes living on human skin. Bioinformatics facilitates the analysis of large-scale genomic data, allowing for a detailed examination of microbial diversity, evolution, and ecological interactions. The convergence of data simplifies our comprehension of what influences the growth of a community of microbes, including but not limited to ways gene are transferred across species, and following the ways different kinds of microorganisms living in one ecosystem function. The insights derived from phylogenetic and evolutionary analysis of microbial communities contribute to broader ecological and environmental studies, influencing fields ranging from biogeochemistry to biotechnology (66).

10.7 PHYLOGENETIC TREE CONSTRUCTION METHODS

A crucial factor in a full understanding of the evolutional relations among species or genes is the creation of a phylogenetic tree. There exist different ways and algorithms used to draw phylogenetic trees grounded on molecular data. The building of phylogenetic trees plays a central role in unraveling the complicated lacework of evolutionary connections between species or genes. The diversity of life can be understood by looking at the ancestral relationship, genetic differences, and historical pathways that characterize it in this important stage. To understand the complex patterns of genetic evolution, a number of methods and tools have been developed in the area of bioinformatics in order to draw upon molecular data in constructing phylogenetic trees. These methods encompass a spectrum of approaches, ranging from distance-based methods that compute genetic dissimilarities to model-based methods that incorporate explicit evolutionary models. Distance-based methods such as Neighbor-Joining and UPGMA calculate pairwise genetic distances that are used in the construction of trees, while explicit models for molecular evolution are employed to find the most probable tree by model-based methods such as Maximum Likelihood and Bayesian methods. (70). The development of advanced bioinformatics tools has enabled the integration of large-scale molecular datasets, including DNA and protein sequences, allowing researchers to choose methods that best suit the characteristics of their data and the complexities of the evolutionary processes under investigation. Ongoing advancements make the field rapidly changing, and with time it brings up new methodologies that serve to improve the precision and effectiveness of phylogenetic tree construction, resulting in better comprehension of complex biological landscapes. (71). Here are some commonly used phylogenetic tree construction methods.

10.7.1 Distance-Based Methods

The neighbor-joining (NJ) algorithm is a common distance-related method used to create phylogenetic trees. The algorithm works by combining sequences with the highest similarity to each other step by step and so on until we have the whole tree constructed this way. The major advantage of this technique is its rapid computation, hence it works well on big datasets. The algorithm's foundation lies in the estimation of pairwise genetic distances, allowing it to discern the closeness of sequences or taxa in terms of evolutionary divergence. Through connecting neighbors with the lowest distances first, the NJ algorithm integrates the hierarchical relations among the analyzed entities. Despite being simple, neighbor-joining has shown itself to be strong and elastic enough to serve as an important method in a wide range of areas such as genomics, evolutionary biology, and microbial ecology. Researchers often turn to this method when navigating the intricate landscape of molecular data to gain insights into the evolutionary tapestry of species or genes (72).

10.7.2 Character-based Methods

Describing the search for the most simple inexpensive method, maximum parsimony (MP) is a method of phylogenetic tree construction that is based on character. This method has the main aim of identifying a tree that needs least evolutionary changes such as mutations in order to explain the observed sequence data. Generally speaking MP tries to find out the least wasteful justification for the evolutionary connections between the objects being studied. This involves searching for the tree configuration with the least total branch length, a criterion that aligns with the principle of Occam's razor. The MP technique has been used in a number of biological fields and has presented information on how genes, species, and entire ecosystems change over time, mostly by distribution-packing biological sequences into hypothetical patterns that are then analyzed by statistical algorithms. Although such applications involve the problem of how to generate optimal distribution-packing schemes, recent developments in computational tools have made it possible to implement MP for solving real-world problems related to the internal structure of molecular data. (73).

Maximum likelihood (ML) is a robust method for inferring phylogenetic trees in the area of character-based approaches. Taking a probabilistic attitude, ML develops the likelihood of observing data given certain assumptions concerning the process of evolution and the model of evolution. This approach is all about determining the tree configuration that maximizes this probability—the one tree that fits best statistically to the observed sequence data. ML is versatile and can accommodate complex evolutionary models, making it well-suited for capturing the nuances of molecular evolution. Its application spans a wide array of disciplines, including genomics, population genetics, and evolutionary biology. ML has attained a cornerstone status as computational resources have advanced to levels where it provides a rigorous framework for disentangling the complexities of evolutionary histories, thereby giving a strong underpinning for other forms of analysis or interpretation. It is rooted in methodological rigor and statistical underpinnings, making it one of the most widely used methods in modern phylogenetics due to its requirement for accuracy and precision in reconstructing the actual phylogeny (61).

10.7.3 Bayesian Methods

In phylogenetic analysis, Bayesian inference (BI) is a method that uses probabilities to estimate phylogenetic trees. This method is more advanced than regular statistical methods since it combines past knowledge and what is seen to determine the posterior chances of various tree topologies. The initial step for the BI is setting up prior probabilities which reflect what the investigator thinks about connections between different species or genes (74). These prior probabilities are then combined with the likelihood of the observed data under a given tree topology, reflecting the probability of the data given the proposed evolutionary model. By applying Bayes' theorem, BI can calculate the posterior probabilities so eloquently that it gives a comprehensive probabilistic representation of the uncertainty around several tree topologies. Prior knowledge incorporation is for injecting significant biological information into analysis when one has minimal data or hard evolutionary cases. BI has the capability to create a collection of trees with associated likelihoods, giving rise to a more elaborate view that is not confined to just one "optimal" tree. Thus, BI has transformed into a powerful means of investigating phylogeny. This computational problem has been solved by MCMC algorithms and now allows for analyzing large tree spaces with a BI for analyzing any type of biological data (75, 76).

10.7.4 Clustering Methods

Single linkage clustering is one of the key methods for building phylogenetic trees within clustering techniques. It is a technique of hierarchical clustering whereby sequences are grouped based on their pairwise distances and the tree is constructed by merging the most similar sequences in series. The essence of single linkage clustering is the idea that the most immediate ancestors of objects are clustered. A tree form is generated by this type of clustering, in which merging is done on the basis of minimum pairwise distances. Single linkage clustering is known to be efficient in its ability to handle big data, although it can falter in terms of chaining, whereby externally dispersed data might interfere in the clustering. Despite this consideration, Single linkage clustering has found applications in diverse fields, from genomics to ecological studies, providing a valuable tool for researchers seeking to explore evolutionary relationships among sequences in a hierarchical and visually interpretable manner. For scientists requiring a combination of phylogenetic tree development and cluster generation approaches, comprehending the advantages and shortcomings of the single linkage clustering technique is crucial. Situations where understanding how sequences' hierarchies are significant in understanding the evolutionary dynamics of what is being studied are those in which such kind of method becomes most relevant (77).

The unweighted pair group method with arithmetic mean (UPGMA) stands as a foundational hierarchical clustering method, playing a pivotal role in the construction of phylogenetic trees. This approach works on the basis of the successive union of the most similar groups of sequences. This hierarchical structure denotes evolutionary relationships among entities. Most importantly, UPGMA works with an unweighted approach that treats each sequence equally in the clustering process. The construction of a tree is usually done by joining iteratively the closest sequences or groups. The length of the branches shows average pairwise distances of the entities merged. UPGMA has this advantage of simplicity and computational efficiency, and these factors make it a favourite while handling large datasets. But most importantly, it assumes a constant evolution rate on each of the branches, and this might not be the actual case while dealing with the complexities that occur in biological evolution. For example, the researcher shall use UPGMA in those situations where a fast and interpretable representation of hierarchical relationships is needed, such as in the preliminary exploration of genetic data or while dealing with entities that evolve at rates considered almost uniform. For this reason, understanding the subtleties of UPGMA, its assumptions, and its applicability is thus important for researchers dealing with this diversified phylogenetic tree construction methodology (78).

10.7.5 Evolutionary Models

Phylogenetic tree construction methods: many of these methods rely on evolutionary models to characterize the intricate processes of substitutions occurring over time. These models are mathematically defined so as to describe the changes in sequences of nucleotides or amino acids as they evolve along the branches of a phylogenetic tree. Several models, therefore, offer varying complexities of sequence evolution.

Jukes-Cantor model: an elementary model, Jukes-Cantor considers equal rates of substitution for all nucleotide positions and all types of substitutions to be equally probable. This model is characterized by its simplicity and is often applied to situations where a simple presentation of evolutionary processes will suffice.

Kimura model: the Kimura model extends from the Jukes-Cantor model by accounting for transition and transversion rates and taking into consideration the fact that perhaps transitions, being purine to purine or pyrimidine to pyrimidine, may occur at different rates from transversions, that is, purine to pyrimidine or vice versa. This brings in another dimension of complexity while presenting a further nuanced picture of the patterns of nucleotide substitutions.

General time reversible (GTR) model: moving to a greater level of complication, the GTR model assumes that distinct rates are characteristic for different pairs of nucleotide substitution. This model is specifically powerful when working with heterogeneous sequence datasets, where differences in substitution rates may obviously appear across genomic regions or lineages.

These evolutionary models form a cornerstone of likelihood-based methods, such as maximum likelihood, where the objective is to find the tree that maximizes the likelihood of the observed data given the assumed model. The appropriate evolutionary model is thus dependent on the characteristics of the data and the studied biological processes. This, however, is a sophisticated landscape, and researchers enter into it with a well-informed understanding of the assumptions and complexities of each model and, hence, of capturing all the minute intricacies of sequence evolution in the quest for the discovery of true phylogeny (79).

10.7.6 Hybrid Methods

Maximum parsimony with bootstrap, in short MP-Bootstrap, represents the most advanced technique developed in the framework of phylogenetic tree construction to increase the robustness and reliability of inferred tree topologies. MP-Bootstrap combines principles of maximum parsimony with the statistical resampling technique of the bootstrap. This hybrid methodology tries to solve uncertainties in tree reconstruction by assessing the stability of single branches.

In the MP-Bootstrap approach, one makes use of bootstrap resampling, which comes from generating a number of datasets through randomly sampling the original data with replacement. For each one of those bootstrap datasets, a phylogenetic tree is constructed, again using the method of maximum parsimony. The set of generated trees now offers a distribution of possible tree topologies, reflecting the variability brought on by the sampling.

Thus, in the consensus tree, the bootstrap values are attributed to branches as percentages and denote just how frequently a certain branch is supported in the bootstrap replicates. The higher the bootstrap values, the more robust it is, which refers to the fact that respective branches are repeatedly recovered during resamplings of the data set. For this reason, researchers could interpret the bootstrap values to reveal reliability and stability of certain branches in the phylogenetic tree. (80).

The MP-Bootstrap hybrid method is important in phylogenetic studies that require a proper assessment of the uncertainty in tree topology. It not only embeds the parsimony principle for the search for the most economical tree but also offers a statistical framework for the assessment of confidence in the inferred evolutionary relationships. This hybrid approach is exemplary of the synergy of algorithmic precision and statistical rigor, offering the researcher a nuanced tool for navigating the complexities of phylogenetic tree construction. (81).

10.7.7 Network Methods

Among the phylogenetic network methods, split decomposition arises as one of the more developed and powerful tools that build networks, offering one of the most refined pictures of complex evolutionary histories. Specially designed to construct the networks in the presence of reticulate events, such as recombination or hybridization, split decomposition is a methodological power tool that is able to disentangle the complex patterns of genetic exchange among and within species. Following the core idea of identifying and representing the conflicting signals inherent in the data, split decomposition extends beyond traditional tree-based methods by accounting for the cases in which a single bifurcating tree-like structure might not be sufficient to capture the underlying complexity of evolutionary dynamics. Instead, this method divides the original data into a network of interconnected splits, where each one represents a different signal or relationship in the dataset. Being more flexible and inclusive, this network-like representation offers an encompassing framework that embraces reticulate events and unravels the mosaic-like structure of evolutionary histories. One of the most unique features of split decomposition is its capability to capture the conflicting phylogenetic signals—mostly coming from reticulate events—that lead to incongruencies not captured by traditional tree-like methods. In a time when genetic exchange and hybridization have become recognized as integral parts of evolutionary processes, the use of split decomposition is emblematic of the changing landscape of methodologies, as it offers researchers a fine-grained lens through which to disentangle the interlaced pathways of evolutionary history across diverse biological systems (82).

10.7.8 Coalescent-Based Methods

Among these coalescent-based methods, species tree methodologies are very much the main ways of handling the frequent incongruence between gene trees and species trees, often resulting from incomplete lineage sorting. These methods, based on coalescent theory from population genetics, focus on inferring the underlying species tree by carefully considering the stochastic coalescence events of gene lineages over time. Species tree methods become important tools to reconcile these conflicts in historical processes, such as rapid speciation or fluctuation of effective population sizes, which can lead to discordance among gene trees. These methods provide a much more refined and accurate way to infer the complex interplay of gene and species histories by probabilistically modeling genealogies and using statistical approaches for estimation of parameters that control the coalescent process. Such an improved toolkit in the growing field of phylogenetic studies allows the researcher to untangle the puzzle of evolutionary relationships in the presence of genetic diversity, giving insight into the real evolutionary dynamics at the overarching species level (79).

10.8 DISEASE RESEARCH AND DRUG DISCOVERY

Disease research is defined by systematic and scientific studies undertaken to enrich our understanding of diseases. It encompasses several disciplines: biological, medical, genetic, epidemiological, and bioinformatics, among others. Disease research is the application of computational and statistical methods to biological and biomedical data for an in-depth understanding of the disease process. Computational biology becomes an important component in analyzing, interpreting, and combining vast amounts of biological information generated in disease studies. Following are the key aspects of disease research (83).

10.8.1 Genomic Analysis and Disease Research

Computational resources have made genomic analysis another key area within research into diseases, changing our understanding of the genetic basis for health and disease. With high-throughput DNA sequencing, bioinformatics helps in the identification of genetic variations, including mutations, SNPs, and structural variations that are important in disease susceptibility. This is widely applied in cancer genomics, whereby driver mutations are identified, investigation of rare Mendelian disorders, facilitation of pharmacogenomics for individualized treatments, and genome-wide association studies that identify complex disease genetic underpinnings (85).

10.8.2 Transcriptomic Profiling and Disease Research

Transcriptomic profiling, powered by computational methods such as RNA sequencing, is crucial in disease research. These tools analyze gene expression data and give insight into the regulation of genes in health and disease states. The computational analysis allows the identification of biomarkers that are useful in disease diagnosis and prognosis. Identification of altered pathways gives researchers a better understanding of the molecular mechanisms associated with diseases. This approach favors precision medicine, as it unearths individualized treatment strategies based on the pattern of gene expression. The approach enriches our understanding of the pathogenesis of diseases and has allowed for the discovery of new therapeutic targets. It thus forms part of the frontier in the study of transcriptional signatures and the dynamic landscape of gene regulation. The incorporation of computational techniques with transcriptomic data enriches our capabilities to understand the complexity of diseases at a molecular level. This eventually gives lift-off to transformational advances in personalized healthcare and targeted therapeutic interventions (86, 87, 88).

10.8.3 Proteomic Investigations and Disease Research

Proteomic investigations, enhanced by computational biology, are pivotal in disease research. Many computational techniques allow the analysis of protein expression, its post-translational modification, and interaction with partners; this information is important for grasping the molecular details of diseases. Knowledge of the protein roles in disease processes is essential to the identification of putative therapeutic targets. By analyzing post-translational modifications, computational methods uncover regulatory mechanisms that control protein function. Moreover, the study of protein–protein interactions helps to detail the large-scale networks that govern cellular functions in health and disease. This integrated approach gives a deeper understanding of disease pathophysiology and accelerates the discovery of targeted therapies. Computational biology and proteomic studies have the potential to improve and develop precision medicine and personalized treatment strategies (89).

10.8.4 Metabolomics Studies and Disease Research

Metabolomics studies combined with bioinformatics hold the key to unravel the complex metabolic landscape related to various diseases. The analysis of small molecule metabolites uses computational tools in order to elucidate the complex world of cellular metabolism. Using bioinformatics methodologies, the composition and dynamics of metabolic pathways can be deciphered, and light shed on the complex interconnected biochemical processes governing cellular function. Identification of disease-specific metabolite signatures emerges as a key outcome, providing valuable biomarkers that indicate a wide range of pathological conditions. In metabolomics, bioinformatics helps in the interpretation of huge metabolomic datasets and leads to the discovery of metabolic alterations in diseases. This integrated approach thus not only furthers our understanding of mechanisms related to disease but also opens doors toward finding diagnostic tools and therapeutic interventions that target the metabolic basis of diseases. The interplay between metabolomics and bioinformatics illustrates a potent strategy in the search for precision medicine and personalized healthcare, tailored to the unique metabolic profiles associated with different diseases (90).

10.8.5 Systems Biology Approaches in Disease Research

In disease research, the synergy of bioinformatics with systems biology facilitates the modeling and simulation of intricate biological systems. Bioinformatics tools empower researchers to comprehend the dynamic interactions among genes, proteins, and biomolecules in the context of diseases. This systems biology approach transcends reductionism, allowing for the exploration of emergent properties in disease processes. Through computational modeling, intricate molecular networks are simulated, offering a comprehensive understanding of diseases at a systems level. The integration of bioinformatics enables the incorporation of diverse omics data, enhancing the accuracy and applicability of these models. This interdisciplinary fusion deepens our insights into the interconnected nature of biological components and provides a holistic perspective on disease mechanisms. The collaborative efforts of bioinformatics and systems biology are pivotal in advancing precision medicine and personalized therapeutic strategies (90).

10.8.6 Disease Biomarker Discovery and Disease Research

In the domain of disease research, computational biology is at the core of disease biomarker discovery—molecular indicators crucial to early diagnosis, prognosis, and treatment monitoring. Using state-of-the-art computational tools, bioinformatics trawls through the complexity of biological datasets to identify robust biomarkers that indicate disease presence or progression. Besides allowing for early disease detection, such biomarkers also contribute to the development of personalized treatment strategies. Accurate identification and reliability of biomarkers are achieved through bioinformatics algorithms and statistical analyses, enhancing our ability to predict disease outcomes and to tailor interventions to maximize patient care. This integration of bioinformatics into biomarker discovery represents a transformative leap in empowering precision medicine with nuanced diagnostics and targeted therapeutic approaches (91).

10.8.7 Machine Learning and Predictive Modeling in Disease Research

Within the realm of disease research, integration with machine learning offers a sea change for bioinformatics. In bioinformatics, machine learning algorithms are harnessed for predictive modeling, pattern recognition, and classification to facilitate the identification of intricate patterns within biological data. Driven by artificial intelligence, these sophisticated techniques excel in the discerning of subtle relationships and the prediction of disease outcomes. This approach guides the navigation through vast datasets for a deep understanding of complex disease patterns. Machine learning offers insights into potential trajectories and patient prognoses. The synergy empowers the researcher to uncover nuanced insights for an improved ability to predict and understand the multifaceted nature of diseases for better, more personalized healthcare strategies (91).

10.9 DRUG DISCOVERY

Drug discovery, illuminated by the developments in computational biology, is dynamic and fast-paced, and it integrates computational approaches with biological insights toward streamlining and optimizing the development of new therapeutic interventions. At its very core, computational approaches help significantly in the identification and validation of potential drug targets through the analysis of vast biological datasets; pinpointing those proteins or genes which are associated with diseases, and assessing their importance in disease pathways. Virtual screening and molecular docking simulations, enabled by bioinformatics, help in the prediction of interactions between small molecules and target proteins for the fast screening of promising drug candidates. The field has also played a huge role in chemical information management and compound library design, and contributed to cheminformatics. Furthermore, computational biology embraces systems pharmacology, enabling an understanding of how drugs impact the biological network. In the personalized medicine era, bioinformatics is key to pharmacogenomics, tailoring drug treatments to individual genetic variations. This approach—aided by the integration of big data analytics and omics technologies—offers a comprehensive view of disease mechanisms and drug responses. Additionally, bioinformatics supports drug repurposing efforts, clinical trial design, and biomarker discovery for more efficient and targeted drug development. By predicting adverse drug reactions and fostering data sharing and collaboration, bioinformatics is placed at the forefront of shaping the very landscape of drug discovery and fostering innovation toward the optimization of therapeutic outcomes (91).

10.10 GENOME SEQUENCING AND ASSEMBLY

Genome sequencing and assembly together encompass the elaborate process of determining and rebuilding the complete genetic information encoded in an organism's DNA. Genome sequencing and assembly are fundamental processes that involve the use of computational tools and algorithms to analyze the vast amount of data generated during DNA sequencing (92). When we combine genome sequencing and assembly, it becomes a process through which the entirety of the genetic information encoded in an organism's DNA is determined and then reconstructed. In contrast, genome assembly involves the computational reconstruction of the entire DNA sequence from the short fragments obtained during sequencing and organizing them into longer, continuous sequences known as contigs. This intricate process, carried out with the aid of bioinformatics tools and algorithms, seeks to accurately represent the structure and organization of an organism's genome, including the genes, their regulatory elements, and other features of the genome. Genome sequencing and assembly are indispensable elements that expand our understanding of genetics, allowing the study of genetic variation and the identification of functional elements, and thereby contributing to many applications in medicine, agriculture, and evolutionary biology (93).

10.10.1 Genome Sequencing

Genome sequencing is the process of determining the precise order of nucleotides (adenine, thymine, cytosine, and guanine, represented as A, T, C, and G) in an organism's DNA. This provides the comprehensive map of the genetic material that allows researchers to analyze and understand the structure and function of genes within the genome (31).

10.10.1.1 Methods of Genome Sequencing

Genome sequencing, the process of determining the order of nucleotides in an organism's DNA, employs several methods that are in evolution. One is the traditional Sanger sequencing, known for its accuracy, but limited by its laborious and time-consuming nature. Next-generation sequencing represents a radical shift by providing high-throughput sequencing with short read lengths (30).

10.10.1.2 Sanger Sequencing

Sanger sequencing, otherwise called chain-termination sequencing, works by the differential incorporation of chain-terminating dideoxynucleotides (ddNTPs) during DNA synthesis. During DNA replication, synthesis occurs in the presence of both normal and chain-terminating nucleotides, generating fragments of all possible lengths, each ending with a ddNTP. These fragments are then separated according to size by means of gel electrophoresis, and the DNA sequence is determined by detection of the labeled ddNTP at each fragment's end. Sanger sequencing has very high accuracy and is therefore reliable in DNA analysis. However, it traditionally returns somewhat shorter reads compared to some NGS technologies. Despite such limitations in terms of read length, Sanger sequencing has been extensively applied in smaller-scale sequencing projects, DNA sequencing of specific genes, and in the validation of results obtained from NGS methods. However, it must be understood that Sanger sequencing is rather more laborious, time-consuming, and expensive in relation to the newer and more effective methods of NGS (94).

10.10.1.3 Next-Generation Sequencing

NGS, also referred to as high-throughput sequencing, is a suite of revolutionary technologies that have changed the dimension of the study of genomics by enabling the rapid and inexpensive sequencing of large amounts of DNA or RNA. The NGS methods are quite contrary to the Sanger sequencing and offer speed, scalability, and efficiency. This game-changing technology has a wide range of applications in genomics, including whole-genome sequencing, transcriptome analysis, study of epigenetics, and metagenomics (95).

10.10.1.4 Illumina Sequencing

NGS technologies have been transformative for studies in genomics, and among these, Illumina stands out as one of the most popular and effective methods. The key feature of the technology is its high throughput, allowing for the parallel sequencing of millions of DNA fragments in one run. Illumina sequencing is especially distinguished by its potential to produce short reads—usually between 100 and 300 base pairs—in a remarkably efficient manner.

The DNA in a sequencing process is first fragmented into short pieces, and each fragment gets adapters attached. These adapted fragments are then amplified on a solid surface and sequenced in parallel, such that a large number of DNA fragments are analyzed at the same time. This method of sequencing, which records the fluorescent signals to give information that is later used to determine the sequence of DNA fragments, makes Illumina sequencing a very high-throughput technique that enables it to be widely used for a lot of applications in genomics by making short reads (96).

10.10.1.5 PacBio and Oxford Nanopore sequencing

PacBio and Oxford Nanopore sequencing are the most advanced methods in NGS, recognized for their single, innovative approaches and capabilities. PacBio, provided by Pacific Biosciences, employs single-molecule sequencing and real-time sequencing with circular DNA templates. On the contrary, Oxford Nanopore sequencing sequences DNA by passage of DNA through a nanopore, which measures the changes in the electric current. One of the distinguishing features of both PacBio and Oxford Nanopore technologies is that they both generate longer reads than traditional methods, such as Illumina sequencing.

This advantage of producing long reads is useful for resolving complex genomic regions, passing through repetitive elements, and identifying structural variations within the genome. The extended read lengths enhance the ability to capture comprehensive genomic information in a more contiguous manner. Hence, PacBio and Oxford Nanopore sequencing can be used in de novo genome assembly when no reference genome is available, and in the study of long-range genomic interactions. Their ability to yield long reads renders these technologies applicable tools for the untangling of genomic architecture, contributing to a deeper understanding of the complexity of genetic information (95).

10.10.2 Genome Assembly

Genome assembly is the computational process of reconstructing the complete DNA sequence of an organism from the short DNA fragments generated during sequencing. The goal is to assemble these fragments into longer, continuous sequences called contigs, which represent different regions of the genome (97). In genome assembly, two main approaches can be followed: de novo assembly and reference-based assembly.

10.10.2.1 De Novo Assembly

In this process, the reconstruction of a genome is done without the use of a pre-existing reference. This is particularly useful in the case of non-model organisms, which lack a previously sequenced genome. However, this is not without challenges, including dealing with repetitive regions and managing the variations, including heterozygosity (98).

10.10.2.2 Reference-Based Assembly

In this process, the assembly of a genome takes advantage of using a known genome as the reference. In this assembly, the newly sequenced reads are aligned to the reference template. Reference-based assembly is useful in organisms where genomes have already been well-annotated, or when refining an existing assembly. It is faster and simpler; thus, it will be useful for applications such as variant calling and population studies. Bioinformatic tools applied to each approach are unique. Notable de novo sssembly tools include SPAdes, designed for short-read sequencing data, and Canu, specialized in assembling long-read sequencing data. These tools facilitate the complex task of reconstructing genomes in a huge contribution to genomics research and the understanding of the genetic makeup of diverse organisms. (99).

10.10.2.3 Reference-Based Genome Assembly

Reference-based genome assembly is a computational method employed in genomics to reconstruct a complete DNA sequence through the mapping and alignment of short DNA sequences obtained through sequencing technologies to a pre-existing reference genome. This procedure relies on the presence of a known reference genome for facilitating a more efficient and straightforward assembly process compared to de novo methods. Reference-based assembly is most applicable in the study of organisms with well-characterized genomes, or when refining existing assemblies (100). It is crucial for variant calling, in the identification of genetic differences between the sequenced genome and the reference, and in the case of population studies, where multiple genomes are compared against a common reference. Alignment tools like Bowtie, BWA, and SAMtools are the common applications in reference-based assembly, which contribute to accurate mapping and the identification of genetic variations, increasing the understanding of the structure and genomic diversity (101).

10.11 PROTEIN STRUCTURE PREDICTION

Protein structure prediction is an important aspect of computational biology and refers to the computational inference of the three-dimensional arrangement of atoms in a protein from its amino acid sequence. This field is central to understanding protein function, dynamics, and interactions, as the structure of a protein often determines its biological activity. Protein structure prediction is critical in computational biology for drug discovery, understanding disease mechanisms, and designing proteins with specific functions. This is due to continual advances in computational methods and integration with experimental data that improve the accuracy and reliability of predicted protein structures. Thus, knowledge of biological systems is furthered (102). Predicting protein structures computationally is challenging because proteins can explore a large conformational space. Several methods are employed within bioinformatics for protein structure prediction, such as homology modeling, ab initio methods, and hybrid methods.

10.11.1 Homology Modeling

Homology modeling, also known as comparative modeling, is one of the most widely used approaches in the prediction of protein structure within the scope of bioinformatics. Through this method, the knowledge of experimentally determined structures of homologous proteins is used as templates for predicting the three-dimensional structure of a target protein. The underlying assumption is that similarly structured proteins have similar functions. In homology modeling, the amino acid sequence of the target protein is aligned with the sequence of a known template protein. By extrapolation of the structural information from the template onto the target protein, computational methods are able to generate a model that approximates the native structure of the target. Homology modeling has proved particularly effective when precise templates with high sequence similarity are available. This makes it a very useful tool in the prediction of protein structures and in the understanding of their functions and for guiding experimental studies (103, 104).

10.11.2 Ab Initio Methods

Ab initio, or de novo methods, in protein structure prediction are a computationally challenging yet utterly important aspect of bioinformatics. It purports to predict protein structures in the absence of any pre-existing templates. In distinction to homology modeling, which uses known structures of homologous proteins as a guide, ab initio methods start with the amino acid sequence and build the protein structure de novo. These techniques are of a rather complex computational nature; hence, they involve energy minimization and molecular dynamics simulations, among other techniques like optimization algorithms. The idea is to span the vast conformational space of the protein and find the lowest energy native state that would represent the most thermodynamically stable and probable structure. Ab initio methods are particularly important in the situation of proteins without close homologous templates and add much to our understanding of the physical principles behind protein folding. However, the complexity of protein folding and the immense requirements in computational resources make the accuracy of de novo prediction a big challenge in the field of structural bioinformatics (105).

10.11.3 Hybrid Methods

Hybrid approaches in protein structure prediction are an advanced approach, through which the best features of homology modeling and ab initio methods combine to give the best results. In this regard, hybrid methods help achieve the best predictive outcome by combining the complementary pieces offered by these approaches. One key characteristic involves the use of experimental data, such as NMR and cryo-EM, in computational predictions. Such experimental information aids in refining and validating structural models resulting from the computations. In the domain of computational and experimental methods, hybrid approaches intend to reach more reliable and biologically relevant predictions of protein structures. This is particularly useful in circumstances where the available experimental data would be able to provide insight into the actual conformation of the protein in question, thus being important to understand the function and structure of such a protein (103).

10.11.4 Fold Recognition

Fold recognition is a unique approach applied in protein structure prediction in the wide area of bioinformatics. It works by first giving a putative fold to a protein through the threading of its amino acid sequence onto already known structures. The threading process is actually the alignment of the target protein sequence with the structures of proteins already experimentally characterized. Threading is most helpful in the scenario where one has no close homologous template, which makes the ordinary homology modeling methods difficult to use. Using the algorithms that are based on threading, one can outline the complex protein folds and correlate them with known structural themes, even when there are no direct homologous matches. This has greatly aided the expansion in knowledge on protein structure, as it provides a very important tool for the prediction of protein architectures with sequences that are unique or very divergent and thus resist template-based modeling (106).

10.11.5 Machine Learning and Deep Learning

In computational prediction of protein structure, the use of machine learning and deep learning methodologies represents a cutting-edge approach in the field of bioinformatics. This is because this methodology uses multiple algorithms and higher-order neural networks that extract subtle patterns related to amino acid sequences and their respective protein structures. Machine learning exploits these patterns and features present in vast datasets for generalization and making predictions on learned associations. Deep learning, as the subset of machine learning, introduces complex neural network architectures, such as convolutional neural networks (CNNs) and recurrent neural networks (RNNs). These deep learning models are capable of picking up subtle dependencies within amino acid sequences, thus enabling them to decipher structural subtleties that might otherwise go undetected through traditional methods. Machine learning and deep learning applied to protein structure prediction show great promise for the development of accuracy and offer insight into the complex relationship between sequence and structure across diverse biological contexts (106, 107).

10.11.6 Critical Assessment of Structure Prediction

Critical Assessment of Structure Prediction (CASP) stands as a pivotal community-driven initiative within the realm of bioinformatics, serving as an extensive experiment designed to assess and appraise various protein structure prediction methods. CASP has been adopted as an important component of critical evaluation of the efficacy and accuracy of a wide range of computational approaches used in protein structure predictions across the globe. It acts like a collaborative testing platform for rigorously testing methods and benchmarking them against each other and against experimental data. The systematic analysis of the different prediction methods' outputs makes CASP a gauge of the actual state of affairs and accelerates progress by revealing strengths and weaknesses of methods. This team effort renders protein structure predictions more reliable and accurate, contributing to the continued improvement and advancement of computational methods in this complex area of structural bioinformatics (108).

10.12 METAGENOMICS AND MICROBIOME ANALYSIS

Metagenomics and microbiome analysis are new revolutionary fields of research in the study of microbial communities, shedding light on extensive, intricate ecosystems that have been identified in almost every environment. Metagenomics involves the direct sequencing and analysis of genetic material recovered directly from environmental samples, hence bypassing the step of isolation and cultivation of individual organisms. It is through this approach that the genomic diversity of whole communities is revealed, and novel species, genes, and functions can be discovered (109).

Microbiome analysis, on the other hand, specifically aims at describing the collective genomes of microorganisms in a given niche, like the human gut, soil, or oceans. Advances in high-throughput sequencing technologies and bioinformatics have revolutionized our ability to decipher the composition and functional potential of microbiomes. In computational biology, a cornerstone in metagenomics and microbiome analysis deals with the high-throughput sequencing data. Mainly, this consists of taxonomic profiling, which describes the kinds of microorganisms present, and functional annotation, which predicts the biological functions encoded in their genomes. Moreover, metagenomic assembly techniques would serve to reconstruct genomes from complex mixtures of genetic material (110).

In the context of the human microbiome, for instance, metagenomics studies have uncovered the intermingled relationships between the microbial communities that inhabit and surround the human body and all aspects of health and disease. In particular, the gut microbiome has been implicated in functions that go beyond simple nutrient metabolism to immunomodulatory roles in health and wellness (111). Over a trillion microorganisms live in the human body and are essential for our health and growth. High-throughput sequencing tools have transformed our understanding of these indigenous bacteria (112). Such approaches produce count data, for example, gene read counts in RNA-seq and operational taxonomic unit (OTU) counts in 16S rRNA gene sequencing, which may be used to gain insights into the structure and functions of microbial communities (113). DNA microarrays, such as the Virochip, which has probes for over 1,500 viruses, have proven useful in detecting viruses in pathology samples (114).

Chips such as the Glomics GeoChip 4.0 zero in on RNA expression from bacteria within the human microbiome, encompassing bacteriophages (115). Similarly, PathGen Dx has introduced the PathChip kit, utilizing an Affymetrix microarray inclusive of all identified viruses and a diverse array of bacteria (116). The PathoChip array incorporates probes for publicly available virus sequences and numerous pathogenic microbes like bacteria, fungi, and helminths, offering comprehensive coverage of microbial pathogens in an efficient setup. Employing multiple probes targeting independent sections of the genome enhances the precision and effectiveness of detection. These technologies underscore the pressing need for swift advancement in downstream analysis methods. Various analysis methods like ALDEx2, ANCOMBC (117), and the ZINB omnibus test (118) have emerged for handling sequencing count data. However, unlike sequencing count data, certain technologies like the PathoChip array generate continuous data for the microbiome.

Metagenomic analysis primarily involves two key approaches. The first, known as direct metagenomic classification, compares microbiota reads from a sample to a public genome database to discern taxonomic composition or gene expression. This method stands out for its advantages over meta-taxonomics as it mitigates biases from marker genes and identifies organisms across all life domains. However, when the sample contains species without known genomes in the database, another approach becomes necessary: metagenomic assembly. While assembling a single genome poses challenges, doing so for a mixed sample with numerous species in varying abundances is notably more complex and demands specialized assembly algorithms (119). Uneven sequencing depth across different organisms within a metagenomic sample presents a significant hurdle. Additionally, a metagenomic sample typically lacks deep coverage of more than a couple of species, meaning that the assembly results won't match those of a single organism's assembly. Nonetheless, metagenomic assembly often successfully consolidates numerous reads into contigs, facilitating easier alignment with reference genomes or analysis without relying on reference genomes. All of these techniques for microbiome analysis share some steps that are employed in the majority of available technologies. Marker gene classifiers utilize a concise database of specific marker genes that identify distinct species, allowing for a rapid estimation of metagenomic sample diversity. However, this method

TABLE 10.1
Tools Used for Analysis of Data Originating from Microbiome Research

Paper Title	Authors	Method /Tool	Focus
Cascabel: An Automated Amplicon Analysis Pipeline	(123)	Cascabel	Amplicon analysis, Snakemake workflow for processing massive data
NG-Tax 2.0: ASV Clustering for Amplicon Sequencing	(124).	NG-Tax 2.0	ASV approach, demultiplexing, read cleaning, taxonomic classification
Improving Taxonomic Resolution with Multiple 16S rRNA Regions	(125)	Computer simulations	Combined 16S rRNA regions for enhanced taxonomic classification
ATLAS: Sub-genus Level Taxonomic Annotation	(126)	ATLAS	Sub-genus level taxonomic annotation for 16S rRNA data
MetaErg: User-friendly Metagenomic Data Exploration	(127)	MetaErg	Annotation, visualization, and interpretation of metagenomic datasets
MetaTOR: Metagenome Binning using Chromosomal Contact Data	(128)	MetaTOR	Reconstruction of MAGs from meta3C data for metagenome binning
Metabolic Overlap Metric for Evaluating Functional Redundancy	(129)	Metabolic Overlap (MO)	Evaluating functional redundancy in metagenomes using MAGs
EviMass: Mining Microbial Associations from Literature	(130)	EviMass	Query system linking microbial associations to biomedical literature
Benchmarking Gene-Targeted Assembly Tools	(131)	Gene-targeted assembly tools	Efficiency, sensitivity, specificity, and chimera rate assessment
Computational Methods for Macroecology of Microbiomes	(132)	Macroecological investigation methods	Biodiversity, spatial-temporal scales, predictive modeling
Advances in Metatranscriptome Analysis	(133)	Metatranscriptomics tools	Preprocessing, assembly, annotation, differential expression analysis
Time-Series Analysis Primer for Microbial Ecology	(134)	Time-series analysis modules	R and Matlab tutorials for analyzing time-series data
Bioinformatics Challenges in Industrial Microbiome Applications	(135)	Industrial microbiome applications	Challenges and technological developments in food, cosmetics, and health industries

compares reads to a smaller database, leaving many reads unclassified because it only identifies a few genes per genome.

In contrast, all-reads classifiers analyze the entire sequenced dataset against databases containing complete genomes. These classifiers, utilizing a larger percentage of sequences, offer insights into the abundances of organisms within the sample. Some direct classifiers using reference databases perform an additional step: assembling metagenomic reads into longer contigs before classification. The next chapter will review software tools implementing these diverse approaches in metagenomic research. Table 10.1 presents a breakdown of microbiota research methods alongside corresponding tools for each approach (120, 121).

The findings, comprising 13 papers, have been compiled to provide comprehensive insights into innovative methods for microbiome analysis. These papers cover a wide spectrum of computational tools and approaches, ranging from amplicon analysis to metatranscriptomics, metagenomics, time-series analysis, and applications in industrial microbiology. Each paper contributes unique and high-quality insights, collectively enhancing our understanding of microbiome analysis through computational methodologies (122).

10.12.1 Classification of Computational Methods in Microbiome–Host Interactions

The most researched interaction types in interspecies cross-talks from a mechanistic standpoint are microbial metabolite-mediated networks, protein–protein interactions (PPIs), and RNA-mediated interactions; and integrated pipelines. As a result, many of the computational tools created to analyze microbe–host interactions have focused on the three types of interactions indicated. Integrated pipelines, as a fourth technique, incorporate several microbial and host-omic data types and networks to infer the cumulative functional consequences of inter-species interactions/communication on the host (136).

10.13 FUTURE PERSPECTIVES

The influx of available information has brought forth new challenges in comprehending biological systems. Analyzing observed data and predicting its behavior through appropriate mathematical modeling has become pivotal. This is fortified by statistical tools, ensuring the accuracy of recorded data. The incorporation of soft computing and the support of fuzzy logics allow for the exploration of diverse hypotheses across different models. The complexity of biological systems is a topic of investigation—whether they consist of elaborate architectures of simple system components or are intrinsically complex structures. The amazing plasticity of biological systems often depicts apparently simple architectures as hugely complex. For example, DNA, the most structured and yet least understood molecule, has been referred to in some quarters as "junk DNA" or non-transcribed DNA. DNA sequences have nucleotide patterns that follow ordered distributions or variations owing to insertions and deletions. The concept of the epigenome further expands this interest as this information—which is only partially translated through mRNA to proteins—is important. The ability to investigate biochemical reactions from single cells to multi-celled systems enables the identification of functional units responsible for specific biological phenomena. These units include enzymes, cofactors, signaling molecules, transport systems, and others. Each functional biomolecule has active sites, modulation, or regulatory elements.

This requires deep mining to define functional features and to draw correlations across DNA, RNA, and protein levels with their primary structures. These features may be nucleotide patterns, their arrangements, or structured distributions of amino acids as motifs. Thus, an effort targeted at developing a database of functional features relevant to various cellular biochemical processes is a prerequisite. There is an exciting possibility of arranging functional features in other ways or predicting novel functions for some biomolecules with the use of various databases. Though these novel functions need not work with the efficiency predicted, they would surely elicit desired stress responses in cells, and that is their inherent plasticity.

A wide variety of omics requires diverse data mining and organizational options that computational tools need. The future interpretation of such datasets is visualized as a QR code; an arrangement of identical functional elements in different two- or three-dimensional configurations would point to different cellular events. Integration of data from different datasets has already begun, but stringent rules are essential for combining this information. The sole integration of tools is not enough, and a systematic pipeline concatenating these tools is required. By observing different levels of cellular organization and making inferences using the signatures of specific biochemical events, we can predict the final course of cellular decision-making processes.

Furthermore, systems biology sought to understand complex biological interactions. Once significant computational capacity had developed, research began to orient to amalgamating disparate omics-based data into a new platform that provided perspectives through systems biology. Most research usually studies cell responses or signaling mechanisms in isolation. This, even without consideration of their effects on peripheral and integral pathways. From a whole-system approach, achieved through systematic interlinked networks, much more insight into general physiology may be learned. It is necessary to develop a model plan for such complex interlinked systems. Indeed, work has already started in this regard; for example, pathway tools organize genomic, metabolic, and regulatory data to enable the study of the biological network being examined. These tools have expanded to encompass energy and flux modeling systems, facilitating more accurate predictions of metabolic pathways and potential interactions and reactions.

REFERENCES

1. Ramsden J. *Bioinformatics: an introduction*: Springer Nature; 2023.
2. Chelly Dagdia Z, Avdeyev P, Bayzid MS. Biological computation and computational biology: survey, challenges, and discussion. *Artificial Intelligence Review*. 2021;54:4169–235.
3. Kashyap H, Ahmed HA, Hoque N, Roy S, Bhattacharyya DK. Big data analytics in bioinformatics: a machine learning perspective. arXiv preprint. arXiv:150605101. 2015.
4. Louis DN, Feldman M, Carter AB, Dighe AS, Pfeifer JD, Bry L, et al. Computational pathology: a path ahead. *Archives of Pathology & Laboratory Medicine*. 2016;140(1):41–50.
5. Zeigler BP, Praehofer H, Kim TG. *Theory of modeling and simulation*: Academic Press; 2000.
6. Secinaro S, Calandra D, Secinaro A, Muthurangu V, Biancone P. The role of artificial intelligence in healthcare: a structured literature review. *BMC Medical Informatics and Decision Making*. 2021;21:1–23.
7. Nagarajan N, Yapp EK, Le NQK, Kamaraj B, Al-Subaie AM, Yeh H-Y. Application of computational biology and artificial intelligence technologies in cancer precision drug discovery. *BioMed Research International*. 2019;2019:8427042.
8. Kim T, Chen IR, Lin Y, Wang AY-Y, Yang JYH, Yang P. Impact of similarity metrics on single-cell RNA-seq data clustering. *Briefings in Bioinformatics*. 2019;20(6):2316–26.
9. Haig BD. Big data science: a philosophy of science perspective. In *Big Data in Psychological Research*, ed. SE Woo, L Tay, and RW Proctor. American Psychological Association; 2020. p. 15–33.
10. Chen CP, Zhang C-Y. Data-intensive applications, challenges, techniques and technologies: a survey on Big Data. *Information Sciences*. 2014;275:314–47.
11. Callebaut W. Scientific perspectivism: a philosopher of science's response to the challenge of big data biology. *Studies in History and Philosophy of Science Part C: Studies in History and Philosophy of Biological and Biomedical Sciences*. 2012;43(1):69–80.
12. Rüegg J, Gries C, Bond-Lamberty B, Bowen GJ, Felzer BS, McIntyre NE, et al. Completing the data life cycle: using information management in macrosystems ecology research. *Frontiers in Ecology and the Environment*. 2014;12(1):24–30.
13. Jørn Nielsen H, Hjørland B. Curating research data: the potential roles of libraries and information professionals. *Journal of Documentation*. 2014;70(2):221–40.
14. Dai L, Gao X, Guo Y, Xiao J, Zhang Z. Bioinformatics clouds for big data manipulation. *Biology Direct*. 2012;7(1):1–7.
15. Pareek CS, Smoczynski R, Tretyn A. Sequencing technologies and genome sequencing. *Journal of Applied Genetics*. 2011;52:413–35.
16. Begg G, Edman P. A protein sequenator. *European Journal of Biochemistry*. 1967;1:80–91.
17. Karas M, Bachmann D, Bahr U, Hillenkamp F. Matrix-assisted ultraviolet laser desorption of non-volatile compounds. *International Journal of Mass Spectrometry and Ion Processes*. 1987;78:53–68.
18. Karas M, Hillenkamp F. Laser desorption ionization of proteins with molecular masses exceeding 10,000 daltons. *Analytical Chemistry*. 1988;60(20):2299–301.
19. Fenn JB, Mann M, Meng CK, Wong SF, Whitehouse CM. Electrospray ionization for mass spectrometry of large biomolecules. *Science*. 1989;246(4926):64–71.
20. Timp W, Nice AM, Nelson EM, Kurz V, McKelvey K, Timp G. Think small: nanopores for sensing and synthesis. *IEEE Access*. 2014;2:1396–408.
21. Rosen CB, Rodriguez-Larrea D, Bayley H. Single-molecule site-specific detection of protein phosphorylation with a nanopore. *Nature Biotechnology*. 2014;32(2):179–81.
22. Dematte L, Priami C, Romanel A. The Beta Workbench: a computational tool to study the dynamics of biological systems. *Briefings in Bioinformatics*. 2008;9(5):437–49.
23. Heiner M, Herajy M, Liu F, Rohr C, Schwarick M, editors. Snoopy–a unifying Petri net tool. *Application and Theory of Petri Nets: 33rd International Conference, PETRI NETS 2012, Hamburg, Germany, June 25-29, 2012 Proceedings 33*; 2012: Springer.

24. Harel D, Rumpe B. Meaningful modeling: what's the semantics of" semantics"? *Computer*. 2004;37(10):64–72.
25. Hoops S, Sahle S, Gauges R, Lee C, Pahle J, Simus N, et al. COPASI—a complex pathway simulator. *Bioinformatics*. 2006;22(24):3067–74.
26. Gay S, Soliman S, Fages F. A graphical method for reducing and relating models in systems biology. *Bioinformatics*. 2010;26(18):i575–i81.
27. Bergstra JA, Middelburg CA. Process algebra for hybrid systems. *Theoretical Computer Science*. 2005;335(2–3):215–80.
28. Bartocci E, Lió P. Computational modeling, formal analysis, and tools for systems biology. *PLoS Computational Biology*. 2016;12(1):e1004591.
29. Bartocci E, Corradini F, Di Berardini MR, Merelli E, Tesei L. Shape calculus. A spatial mobile calculus for 3D shapes. *Scientific Annals of Computer Science*. 2010;20:1–31.
30. Bartocci E, Cacciagrano DR, Di Berardini MR, Merelli E, Tesei L. Timed operational semantics and well-formedness of shape calculus. *Scientific Annals of Computer Science*. 2010;20:33.
31. Faeder JR, Blinov ML, Hlavacek WS. Rule-based modeling of biochemical systems with BioNetGen. *Systems Biology*. 2009;500:113–67.
32. Chylek LA, Stites EC, Posner RG, Hlavacek WS. Innovations of the rule-based modeling approach. In *Systems Biology: Integrative Biology and Simulation Tools*, ed. A Prokop and B Csukas. Springer; 2013. p. 273–300.
33. Hlavacek WS, Faeder JR, Blinov ML, Posner RG, Hucka M, Fontana W. Rules for modeling signal-transduction systems. *Science's STKE*. 2006;2006(344):re6-re.
34. Chaouiya C. Petri net modelling of biological networks. *Briefings in Bioinformatics*. 2007;8(4):210–9.
35. Cordero F, Horváth A, Manini D, Napione L, De Pierro M, Pavan S, et al. Simplification of a complex signal transduction model using invariants and flow equivalent servers. *Theoretical Computer Science*. 2011;412(43):6036–57.
36. Blätke MA, Heiner M, Marwan W, editors. Predicting phenotype from genotype through automatically composed Petri nets. *International Conference on Computational Methods in Systems Biology*; 2012: Springer.
37. Kauffman SA. Metabolic stability and epigenesis in randomly constructed genetic nets. *Journal of Theoretical Biology*. 1969;22(3):437–67.
38. Dubrova E, Teslenko M, Tenhunen H. A computational scheme based on random Boolean networks. *Transactions on Computational Systems Biology X*; 2008: Springer. p. 41–58.
39. Kotiang S, Eslami A. Boolean factor graph model for biological systems: the yeast cell-cycle network. *BMC Bioinformatics*. 2021;22(1):1–27.
40. Benque D, Bourton S, Cockerton C, Cook B, Fisher J, Ishtiaq S, et al., editors. BMA: visual tool for modeling and analyzing biological networks. *Computer Aided Verification: 24th International Conference, CAV 2012, Berkeley, CA, USA, July 7-13, 2012 Proceedings 24*; 2012: Springer.
41. Fisher J, Harel D. *On statecharts for biology. Symbolic systems biology*: Theory and Methods Jones & Bartlett Publishers; 2010.
42. Harel D. Statecharts: a visual formalism for complex systems. *Science of Computer Programming*. 1987;8(3):231–74.
43. Alur R, Courcoubetis C, Henzinger TA, Ho P-H, editors. Hybrid automata: an algorithmic approach to the specification and verification of hybrid systems. *International Hybrid Systems Workshop*; 1991: Springer.
44. Bartocci E, Bortolussi L, Smolka SA. Hybrid systems and biology. *Information and Computation*. 2014;236:1–2.
45. Fromentin J, Eveillard D, Roux O. Hybrid modeling of biological networks: mixing temporal and qualitative biological properties. *BMC Systems Biology*. 2010;4:1–11.
46. Asarin E, Dang T, Girard A. Hybridization methods for the analysis of nonlinear systems. *Acta Informatica*. 2007;43:451–76.
47. Dang T, Maler O, Testylier R. Accurate hybridization of nonlinear systems. In *Proceedings of the 13th ACM International Conference on Hybrid Systems: Computation and Control*. ACM; 2010. p. 11–20.
48. Dang T, Testylier R. Hybridization domain construction using curvature estimation. In *Proceedings of the 14th International Conference on Hybrid Systems: Computation and Control*. ACM; 2011. p. 123–32.
49. Nguyen A-T, Sugeno M, Campos V, Dambrine M. LMI-based stability analysis for piecewise multi-affine systems. *IEEE Transactions on Fuzzy Systems*. 2016;25(3):707–14.

50. Pasquini M, Angeli D. Study of piecewise multi-affine models for genetic regulatory networks via a Lyapunov approach: an LMI framework. *IFAC-PapersOnLine*. 2020;53(2):16739–44.
51. Giordano AA, Levesque AH. *Modeling of digital communication systems using SIMULINK*: John Wiley & Sons; 2015.
52. Sobie EA. An introduction to MATLAB. *Science Signaling*. 2011;4(191):tr7-tr.
53. Eberl HJ, Parker DF, Vanloosdrecht MC. A new deterministic spatio-temporal continuum model for biofilm development. *Computational and Mathematical Methods in Medicine*. 2001;3(3):161–75.
54. Walgraef D. *Spatio-temporal pattern formation: with examples from physics, chemistry, and materials science*: Springer Science & Business Media; 2012.
55. Som A. Causes, consequences and solutions of phylogenetic incongruence. *Briefings in Bioinformatics*. 2015;16(3):536–48.
56. Brinkman FS, Leipe DD. Phylogenetic analysis. *Bioinformatics: A Practical Guide to the Analysis of Genes and Proteins*. 2001;2:349.
57. Prosdocimi F, Chisham B, Pontelli E, Thompson JD, Stoltzfus A. Initial implementation of a comparative data analysis ontology. *Evolutionary Bioinformatics*. 2009;5:S2320.
58. Morrison DA. Phylogenetic networks: a review of methods to display evolutionary history. *Annual Research & Review in Biology*. 2014;4:1518–43.
59. Bienstock RJ, Beard WA, Wilson SH. Phylogenetic analysis and evolutionary origins of DNA polymerase X-family members. *DNA Repair*. 2014;22:77–88.
60. Schoch CL, Ciufo S, Domrachev M, Hotton CL, Kannan S, Khovanskaya R, et al. NCBI Taxonomy: a comprehensive update on curation, resources and tools. *Database*. 2020;2020:baaa062.
61. Kapli P, Yang Z, Telford MJ. Phylogenetic tree building in the genomic age. *Nature Reviews Genetics*. 2020;21(7):428–44.
62. Atri B, Lichtarge O. Computational approaches to studying molecular phylogenetics. *Bioinformatics: Sequences, Structures, Phylogeny*, ed. A Shanker. Springer; 2018. p. 173–90.
63. Huson DH, Bryant D. Application of phylogenetic networks in evolutionary studies. *Molecular Biology and Evolution*. 2006;23(2):254–67.
64. Rokas A. Phylogenetic analysis of protein sequence data using the Randomized Axelerated Maximum Likelihood (RAXML) Program. *Current Protocols in Molecular Biology*. 2011;96(1):19.1. 1-.1. 4.
65. Pagel M. Inferring the historical patterns of biological evolution. *Nature*. 1999;401(6756):877–84.
66. Sakamoto T, Ortega JM. TaxOnTree: a tool that generates trees annotated with taxonomic information. *BioRxiv*. 2020:2020.12. 24.424364.
67. Cooper ED. Overly simplistic substitution models obscure green plant phylogeny. *Trends in Plant Science*. 2014;19(9):576–82.
68. Roy SS, Dasgupta R, Bagchi A. A review on phylogenetic analysis: a journey through modern era. *Computational Molecular Bioscience*. 2014;4(03):39.
69. Liu Z, Meng J, Sun X. A novel feature-based method for whole genome phylogenetic analysis without alignment: application to HEV genotyping and subtyping. *Biochemical and Biophysical Research Communications*. 2008;368(2):223–30.
70. Van de Peer Y, Salemi M. Phylogenetic inference based on distance methods. *The Phylogenetic Handbook*. 2009;2:142–59.
71. Torres A, Goloboff PA, Catalano SA. Assessing topological congruence among concatenation-based phylogenomic approaches in empirical datasets. *Molecular Phylogenetics and Evolution*. 2021;161:107086.
72. Löytynoja A, Goldman N. Uniting alignments and trees. *Science*. 2009;324(5934):1528–9.
73. Roch S, Wang K-C. Sufficient condition for root reconstruction by parsimony on binary trees with general weights. *Electronic Communications in Probability*. 2021;26:1–13.
74. Holder M, Lewis PO. Phylogeny estimation: traditional and Bayesian approaches. *Nature Reviews Genetics*. 2003;4(4):275–84.
75. Neureiter N, Ranacher P, Efrat-Kowalsky N , Kaiping GA, Weibel R, Widmer P, et al. Detecting contact in language trees: a Bayesian phylogenetic model with horizontal transfer. *Humanities and Social Sciences Communications*. 2022;9(1):1–14.
76. Nei M, Tajima F, Tateno Y. Accuracy of estimated phylogenetic trees from molecular data: II. Gene frequency data. *Journal of Molecular Evolution*. 1983;19:153–70.

77. Hellmuth M, Schaller D, Stadler PF. Clustering systems of phylogenetic networks. *Theory in Biosciences*. 2023;142(4):301–58.
78. Yoshida R, Paul L, Nesbitt P. Stochastic safety radius on UPGMA. *Algorithms*. 2022;15(12):483.
79. Tateno Y, Nei M, Tajima F. Accuracy of estimated phylogenetic trees from molecular data: I. Distantly related species. *Journal of Molecular Evolution*. 1982;18:387–404.
80. Felsenstein J. Confidence limits on phylogenies: an approach using the bootstrap. *Evolution*. 1985;39(4):783–91.
81. Deen E, van Iersel L, Janssen R, Jones M, Murakami Y, Zeh N. A near-linear Kernel for two-parsimony distance. arXiv preprint arXiv:221100378. 2022.
82. Bandelt H-J, Dress AW. Split decomposition: a new and useful approach to phylogenetic analysis of distance data. *Molecular Phylogenetics and Evolution*. 1992;1(3):242–52.
83. Forman MR, Greene SM, Avis NE, Taplin SH, Courtney P, Schad PA, et al. Bioinformatics: tools to accelerate population science and disease control research. *American Journal of Preventive Medicine*. 2010;38(6):646–51.
84. Zheng M, Allington G, Vilarinho S. Genomic medicine for liver disease. *Hepatology*. 2022;76(3):860–8.
85. Mitchell JA. Medical informatics and bioinformatics. In *Bioinformatics: Concepts, Methodologies, Tools, and Applications*, ed. Information Resources Management Association. Medical Information Science Reference; 2013. p. 195.
86. Hu B, Sajid M, Lv R, Liu L, Sun C. A review of spatial profiling technologies for characterizing the tumor microenvironment in immuno-oncology. *Frontiers in Immunology*. 2022;13:996721.
87. Cockrum C, Kaneshiro KR, Rechtsteiner A, Tabuchi TM, Strome S. A primer for generating and using transcriptome data and gene sets. *Development*. 2020;147(24):dev193854.
88. Veeranki SR, Varshney M. Comparative analysis of thyroid disease and predict them using machine learning techniques. *International Journal of Health Sciences*. 2022;6:11005–14.
89. Daniel C, Tannier X, Kalra D. Clinical research informatics. *Yearbook of Medical Informatics*. 2022;31(01):161–4.
90. Devarapalli D, Anusha C, Srikanth P. Identification of deleterious SNPs in TACR1 gene using genetic algorithm. In *Computational Intelligence Techniques for Comparative Genomics: Dedicated to Prof Allam Appa Rao on the Occasion of His 65th Birthday*, ed. NB Muppalaneni and VK Gunjan. Springer; 2015. p. 87–97.
91. Olson RD, Assaf R, Brettin T, Conrad N, Cucinell C, Davis JJ, et al. Introducing the bacterial and viral bioinformatics resource center (BV-BRC): a resource combining PATRIC, IRD and ViPR. *Nucleic Acids Research*. 2023;51(D1):D678–D89.
92. Koren S, Walenz BP, Berlin K , Miller JR, Bergman NH, Phillippy AM. Canu: scalable and accurate long-read assembly via adaptive k-mer weighting and repeat separation. *Genome Research*. 2017;27(5):722–36.
93. Bradnam KR, Fass JN, Alexandrov A, Baranay P, Bechner M, Birol I, et al. Assemblathon 2: evaluating de novo methods of genome assembly in three vertebrate species. *Gigascience*. 2013;2(1):2047-217X-2-10.
94. DiGuistini S, Liao NY, Platt D, Robertson G, Seidel M, Chan SK, et al. De novo genome sequence assembly of a filamentous fungus using Sanger, 454 and Illumina sequence data. *Genome Biology*. 2009;10(9):1–12.
95. Quail MA, Kozarewa I, Smith F, Scally A, Stephens PJ, Durbin R, et al. A large genome center's improvements to the illumina sequencing system. *Nature Methods*. 2008;5(12):1005–10.
96. Ishige T, Itoga S, Matsushita K, Nomura F. Locked nucleic acid probe enhances Sanger sequencing sensitivity and improves diagnostic accuracy of high-resolution melting-based KRAS mutational analysis. *Clinica Chimica Acta*. 2016;457:75–80.
97. Davey JW, Hohenlohe PA, Etter PD, Boone JQ, Catchen JM, Blaxter ML. Genome-wide genetic marker discovery and genotyping using next-generation sequencing. *Nature Reviews Genetics*. 2011;12(7):499–510.
98. Kolmogorov M, Yuan J, Lin Y, Pevzner PA. Assembly of long, error-prone reads using repeat graphs. *Nature Biotechnology*. 2019;37(5):540–6.
99. Dierckxsens N, Mardulyn P, Smits G. NOVOPlasty: de novo assembly of organelle genomes from whole genome data. *Nucleic Acids Research*. 2017;45(4):e18–e.
100. Li R, Zhu H, Ruan J, Qian W, Fang X, Shi Z, et al. De novo assembly of human genomes with massively parallel short read sequencing. *Genome Research*. 2010;20(2):265–72.

101. Tamazian G, Dobrynin P, Krasheninnikova K, Komissarov A, Koepfli K-P, O'Brien SJ. Chromosomer: a reference-based genome arrangement tool for producing draft chromosome sequences. *Gigascience*. 2016;5(1):s13742-016-0141-6.
102. Jumper J, Evans R, Pritzel A, Green T, Figurnov M, Ronneberger O, et al. Highly accurate protein structure prediction with AlphaFold. *Nature*. 2021;596(7873):583–9.
103. Webb B, Sali A. Comparative protein structure modeling using MODELLER. *Current Protocols in Bioinformatics*. 2016;54(1):5.6. 1–5.6. 37.
104. Senior AW, Evans R, Jumper J, Kirkpatrick J, Sifre L, Green T, et al. Improved protein structure prediction using potentials from deep learning. *Nature*. 2020;577(7792):706–10.
105. Bradley P, Misura KM, Baker D. Toward high-resolution de novo structure prediction for small proteins. *Science*. 2005;309(5742):1868–71.
106. Torrisi M, Pollastri G, Le Q. Deep learning methods in protein structure prediction. *Computational and Structural Biotechnology Journal*. 2020;18:1301–10.
107. Mirabello C, Wallner B. rawMSA: end-to-end deep learning using raw multiple sequence alignments. *PloS One*. 2019;14(8):e0220182.
108. Moult J, Fidelis K, Kryshtafovych A, Schwede T, Tramontano A. Critical assessment of methods of protein structure prediction (CASP)—Round XII. *Proteins: Structure, Function, and Bioinformatics*. 2018;86:7–15.
109. Fishburn CS. Translational research: the changing landscape of drug discovery. *Drug Discovery Today*. 2013;18(9–10):487–94.
110. Pirogova E, Istivan T, Gan E, Cosic I. Advances in methods for therapeutic peptide discovery, design and development. *Current Pharmaceutical Biotechnology*. 2011;12(8):1117–27.
111. Sahu NK, Sahu S, Kohli DV. Novel molecular targets for antimalarial drug development. *Chemical Biology & Drug Design*. 2008;71(4):287–97.
112. Johnson CL, Versalovic J. The human microbiome and its potential importance to pediatrics. *Pediatrics*. 2012;129(5):950–60.
113. Fernandes AD, Reid JN, Macklaim JM, McMurrough TA, Edgell DR, Gloor GB. Unifying the analysis of high-throughput sequencing datasets: characterizing RNA-seq, 16S rRNA gene sequencing and selective growth experiments by compositional data analysis. *Microbiome*. 2014;2:1–13.
114. Chen EC, Miller SA, DeRisi JL, Chiu CY. Using a pan-viral microarray assay (Virochip) to screen clinical samples for viral pathogens. *JoVE (Journal of Visualized Experiments)*. 2011;50:e2536.
115. Baldwin DA, Feldman M, Alwine JC, Robertson ES. Metagenomic assay for identification of microbial pathogens in tumor tissues. *MBio*. 2014;5(5). doi: 10.1128/mbio.01714-14
116. Wong CW, Heng CLW, Wan Yee L, Soh SW, Kartasasmita CB, Simoes EA, et al. Optimization and clinical validation of a pathogen detection microarray. *Genome Biology*. 2007;8:1–13.
117. Lin H, Peddada SD. Analysis of compositions of microbiomes with bias correction. *Nature Communications*. 2020;11(1):3514.
118. Chen J, King E, Deek R, Wei Z, Yu Y, Grill D, et al. An omnibus test for differential distribution analysis of microbiome sequencing data. *Bioinformatics*. 2018;34(4):643–51.
119. Cao MD, Ganesamoorthy D, Elliott AG, Zhang H, Cooper MA, Coin LJ. Streaming algorithms for identification pathogens and antibiotic resistance potential from real-time MinION™ sequencing. *Gigascience*. 2016;5(1):s13742-016-0137-2.
120. Niu S-Y, Yang J, McDermaid A, Zhao J, Kang Y, Ma Q. Bioinformatics tools for quantitative and functional metagenome and metatranscriptome data analysis in microbes. *Briefings in Bioinformatics*. 2018;19(6):1415–29.
121. Marić J, Šikić M, editors. Approaches to metagenomic classification and assembly. *2019 42nd International Convention on Information and Communication Technology, Electronics and Microelectronics (MIPRO)*. IEEE; 2019.
122. Setubal JC, Stoye J, Dutilh BE. *Computational methods for microbiome analysis*: Frontiers Media SA; 2020. p. 623897.
123. Abdala Asbun A, Besseling MA, Balzano S, Van Bleijswijk JD, Witte HJ, Villanueva L, et al. Cascabel: a scalable and versatile amplicon sequence data analysis pipeline delivering reproducible and documented results. *Frontiers in Genetics*. 2020;11:489357.

124. Poncheewin W, Hermes GD, Van Dam JC, Koehorst JJ, Smidt H, Schaap PJ. NG-Tax 2.0: a semantic framework for high-throughput amplicon analysis. *Frontiers in Genetics.* 2020;10:1366.
125. Pinna NK, Dutta A, Monzoorul Haque M, Mande SS. Can targeting non-contiguous V-regions with paired-end sequencing improve 16S rRNA-based taxonomic resolution of microbiomes?: an in silico evaluation. *Frontiers in Genetics.* 2019;10:653.
126. Shah N, Meisel JS, Pop M. Embracing ambiguity in the taxonomic classification of microbiome sequencing data. *Frontiers in Genetics.* 2019;10:1022.
127. Dong X, Strous M. An integrated pipeline for annotation and visualization of metagenomic contigs. *Frontiers in Genetics.* 2019;10:999.
128. Baudry L, Foutel-Rodier T, Thierry A, Koszul R, Marbouty M. MetaTOR: a computational pipeline to recover high-quality metagenomic bins from mammalian gut proximity-ligation (meta3C) libraries. *Frontiers in Genetics.* 2019;10:753.
129. Hester ER, Jetten MS, Welte CU, Lücker S. Metabolic overlap in environmentally diverse microbial communities. *Frontiers in Genetics.* 2019;10:989.
130. Srivastava D, Baksi KD, Kuntal BK, Mande SS. "EviMass": a literature evidence-based miner for human microbial associations. *Frontiers in Genetics.* 2019;10:849.
131. Guo J, Quensen JF, Sun Y, Wang Q, Brown CT, Cole JR, et al. Review, evaluation, and directions for gene-targeted assembly for ecological analyses of metagenomes. *Frontiers in Genetics.* 2019;10:957.
132. Mascarenhas R, Ruziska FM, Moreira EF, Campos AB, Loiola M, Reis K, et al. Integrating computational methods to investigate the macroecology of microbiomes. *Frontiers in Genetics.* 2020;10:1344.
133. Shakya M, Lo C-C, Chain PS. Advances and challenges in metatranscriptomic analysis. *Frontiers in Genetics.* 2019;10:904.
134. Coenen AR, Hu SK, Luo E, Muratore D, Weitz JS. A primer for microbiome time-series analysis. *Frontiers in Genetics.* 2020;11:310.
135. Van den Bogert B, Boekhorst J, Pirovano W, May A. On the role of bioinformatics and data science in industrial microbiome applications. *Frontiers in Genetics.* 2019;10:721.
136. Sudhakar P, Machiels K, Verstockt B, Korcsmaros T, Vermeire S. Computational biology and machine learning approaches to understand mechanistic microbiome-host interactions. *Frontiers in Microbiology.* 2021;12:618856.

11 Computation and Disease Prediction Models

Furqan Awan and Muhammad Muddassir Ali

11.1 INTRODUCTION

Computation plays a crucial role in disease prediction models, enabling researchers and healthcare professionals to forecast the spread and impact of various illnesses (1). Using advanced computational techniques, such as machine learning (ML) algorithms and data analytics, researchers can enhance the accuracy and reliability of disease prediction models (2). To control a disease's spread, researchers need to improve their understanding of the disease and integrate their micro-level and macro-level approaches with computational algorithms. For instance, at the micro-level, genetic sequences of causative agents (F\for example in case of infectious disease), should be coupled with algorithms and bioinformatics. Such couplings and integrations will not only improve disease control but will also improve other aspects of disease management (3).

Computational algorithms have opened new avenues of research and policy-making in the disciplines of public health. By analyzing vast amounts of data, including demographic information, environmental factors, and health records, these models can identify patterns and trends that help predict the likelihood of disease outbreaks (4). Various types of modeling techniques and algorithms are used to develop predictive models (5). These predictive models are instrumental in guiding public health interventions and resource allocation (6). They allow for early detection of potential outbreaks, estimation of casualties, and economic burden of any outbreak such as measles. This timely understanding of disease patterns helps policymakers in designing public policies and health interventions. Non-pharmaceutical interventions designed during COVID-19 are among the recent examples of public health policies (7, 8).

Clinical medicine is another discipline being supported by computational algorithms in day-to-day life. Recently developed convolutional neural networks have greatly helped in improving diagnosis, prescription, and surgical procedures (9, 10). Along with the development of cellular phones and handheld computer technologies, data collection for these algorithms has become easier (9). Furthermore, small android applications based on these algorithms provide individuals with rapid diagnosis and data-driven decision-making. Although accuracy is still a big problem, big data-trained algorithms will eventually overcome these issues. The current chapter highlights some basics and applications of the use of computation in disease prediction with a special focus on infectious diseases.

11.2 OVERVIEW OF DISEASE PREDICTION MODELS

Artificial intelligence (AI) facilitates the automatization of analytical modeling and contributes to prediction, diagnostics, and treatment of diseases. Artificial intelligence, machine learning (ML), and deep learning (DL) can be perceived as a series of nested terminologies. AI encompasses both symbolic and connectionist approaches. ML can be understood as the prevailing branch of connectionism, which amalgamates biological (neural networks) and statistical (data-driven learning

DOI: 10.1201/9781003410164-11

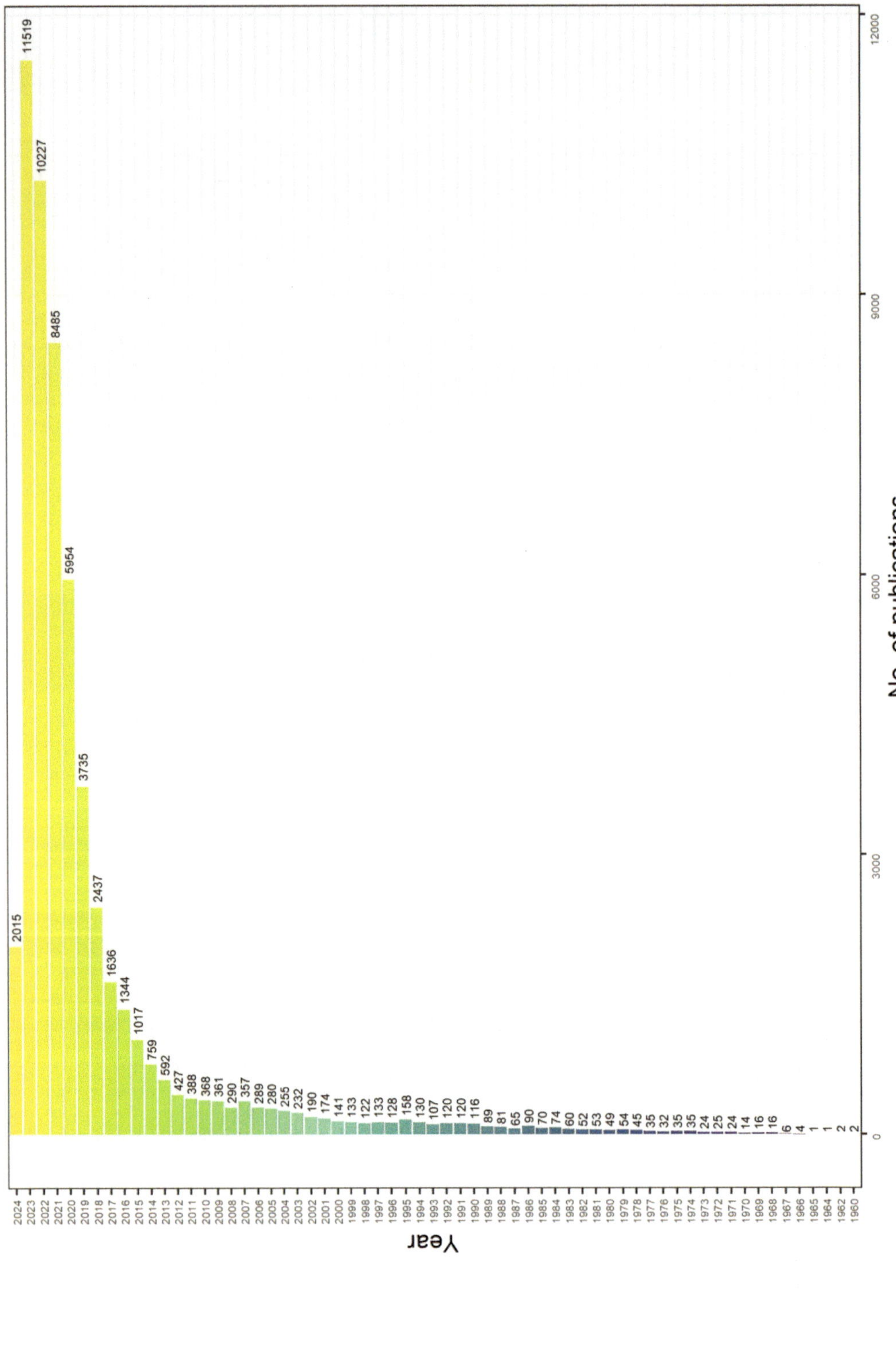

FIGURE 11.1 A diagram showing total number of studies on in-silico models catalogued in PubMed during 1960 and 2024.

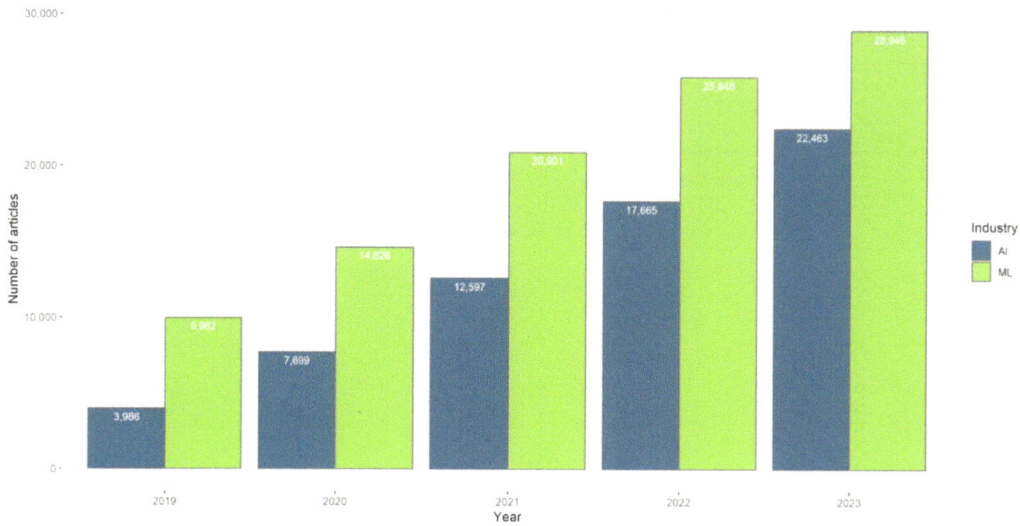

FIGURE 11.2 A diagram showing comparison over time of the keywords AI and ML in studies catalogued in PubMed during 2019–2023.

theory) influences. DL primarily focuses on extensive neural networks, exhibiting specific functionalities in processing images, sounds, videos, and other related data (11). In this era of technology, science disciplines are using AI, ML, and DL as techniques to develop their in-silico or virtual models or predictions. A huge number of publications on in-silico models, around 48,000, have been catalogued in PubMed, an online database, during 1960–2024 (Figure 11.1). Use of these techniques/algorithms has increased exponentially during the last ten years.

A comparison over time of the number of studies catalogued in PubMed with the keywords AI and ML revealed that recently published research articles are using more ML and AI (Figure 11.2). However, there is an urgent need for a comprehensive study that can compare the various methodologies and analyze them in terms of the dataset, number of images, number of classes, algorithms used, and overall performance achieved.

Disease prediction models based on ML algorithms have shown high accuracy compared to traditional diagnostic methods. The use of risk prediction models has been reported for various diseases, such as rheumatoid arthritis (12), lung cancer screening, chronic kidney disease, cardiovascular disease, and colon cancer (13, 14). A huge diversity of ML algorithms is available to understand and predict diseases. Furthermore, in order to meet the full promise of precision medicine, research is attempting to leverage increasing genomic understanding and further develop personalized medical healthcare through ever more accurate disease risk prediction models (15). The use of ML algorithms in disease prediction offers several advantages, including more accurate and timely analysis, saving of time, and providing warnings about health conditions before it is too late (16).

11.3 IMPORTANCE OF COMPUTATION IN DISEASE PREDICTION

The importance of computation in disease prediction and other aspects of disease management is a topic that has gained significant attention in recent years. To control infectious diseases in a timely way, computational models are being used to understand etiology, pathogenesis, spread, drug–host interactions, protein–protein interactions, and cellular interactions with infectious agents (17, 18).

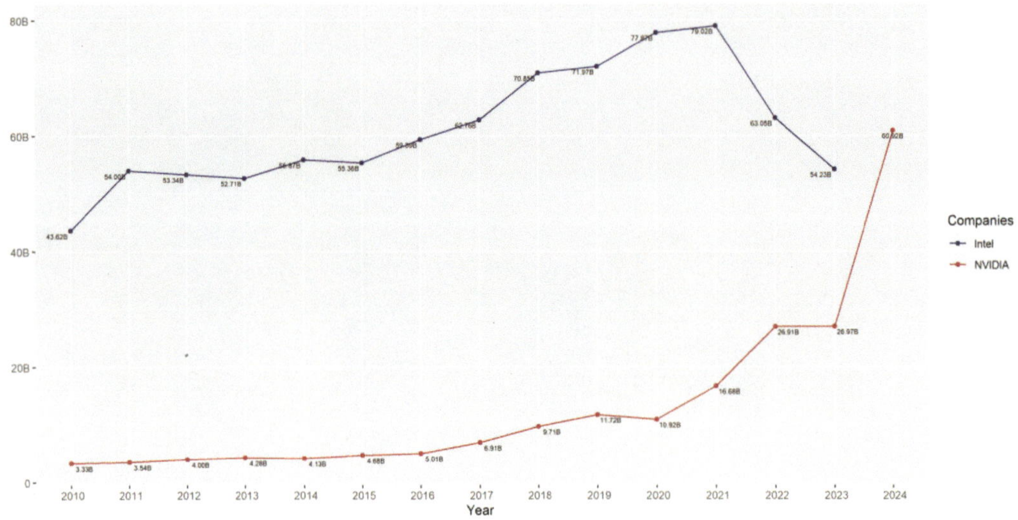

FIGURE 11.3 Comparison of stock shares of Intel and NVIDIA during 2010–2024. Stocks data obtained from https://stockanalysis.com/stocks/nvda/revenue.

Improvement in medical analysis is occurring day by day using computational modeling. A major example is the use of intelligent computational models for the early and accurate detection and diagnosis of cardiovascular disease (19). Computational epigenomics has also been highlighted for predicting potential miRNA-disease associations and risk assessment such as the development of Path-Based MiRNA-Disease Association (PBMDA) (20) and Ensemble of Kernel Ridge Regression based MiRNA-Disease Association prediction (EKRRMDA) (21). Additionally, the development of effective computational models for predicting lncRNA-disease association relationships has been emphasized as extremely important (22). The computational interpretation of genetic variants and their role in predicting phenotype, such as disease risk or differential drug response, has been recognized as an important capability (23).

In the last 10–15 years, with the advent of machine learning algorithms, industries/companies dealing with high-performance computation have progressed financially also (24). Two companies, Intel and Nvidia, are top in the domain of computation. CPU processors are manufactured by Intel while GPUs are manufactured by Nvidia. GPUs have been proved better in performing calculations as compared to CPU processors (25). This has caused GPU sales to rise exponentially. Figure 11.3 shows a comparison of Intel and Nvidia stock prices.

11.4 APPROACHES FOR INFECTIOUS DISEASE PREDICTION

As mentioned earlier, ML algorithms and modeling approaches are being used to study the various aspects of infectious diseases, such as the prediction of antigenic variation in pathogens, the design of vaccination policies, the role of disease determinants, and the impact of infectious disease on population growth (26).

An overview of research articles cataogued in PubMed from 2019 till the beginning of 2024 revealed that the SARS-CoV-2 virus was the most frequently studied for prediction-related objectives (Figure 11.4). A large spectrum of other diseases from different fields, such as zoonotic diseases (influenza, tuberculosis, tetanus, leishmaniasis, malaria, chikungunya, toxoplasmosis, lyssavirus, norovirus, calcivirus, anaplasmosis, chickenpox, chagas disease) and human medical diseases (HIV, syphilis, rubella, measles, cytomegalovirus, metapneumovirus, entivirus, leprosy, ebolavirus, polio, anellovirus, papillomavirus, chlamydia, arbovirus, rotavirus), were also studied.

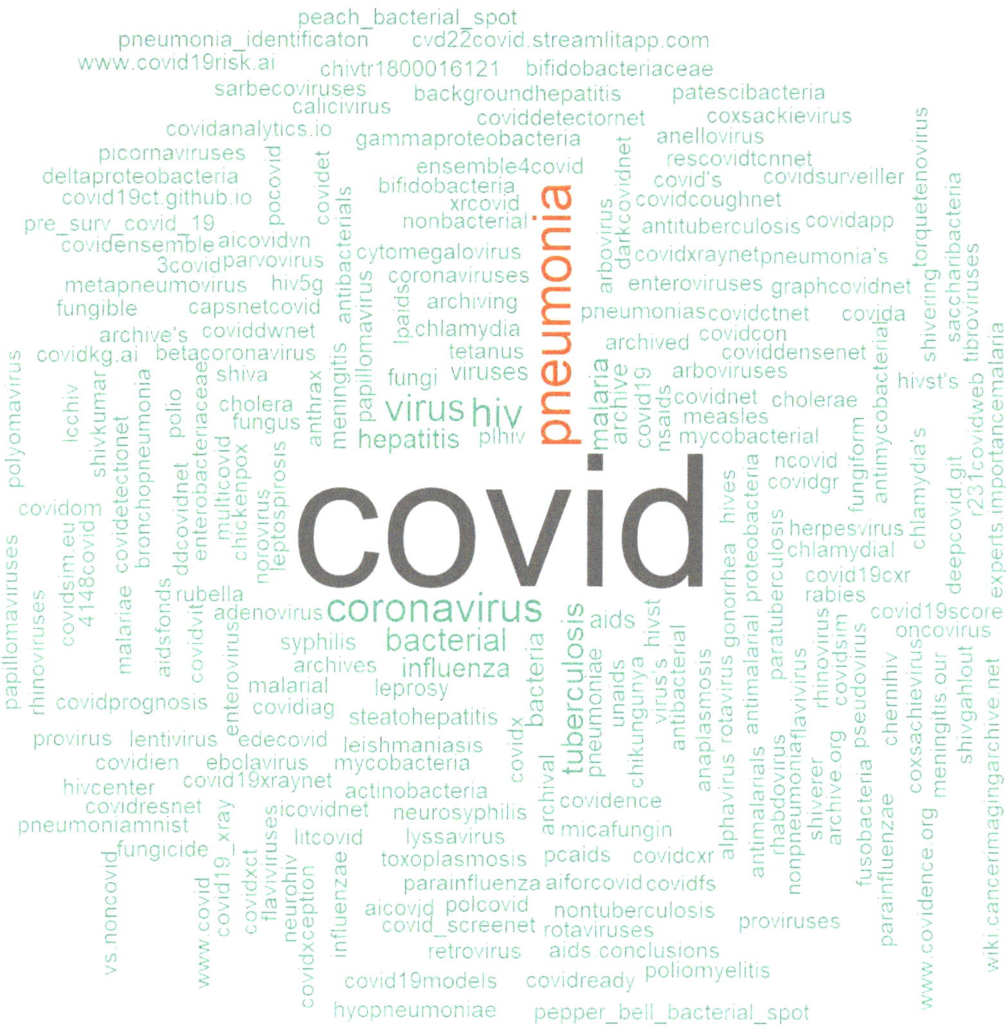

FIGURE 11.4 An overview of infectious-disease-related studies catalogued in PubMed during 2019–2023 (data not shown here).

Various types of approaches are being used in infectious disease predictions. These approaches have different pros and cons and data requirements are also different. Implementing such approaches requires expertise from various disciplines, such as mathematicians, statisticians, epidemiologists, computer scientists, and complex sciences experts. Based on this interdisciplinary expertise, approaches will be discussed in the following sections.

11.4.1 Epidemiological Approaches

Epidemiological approaches in infectious disease prediction involve integrating various data sources, such as structured and unstructured studies, social-media-based data, and data related to environmental parameters, to build robust predictions (27). Epidemiological methodologies such as outbreak investigations and disease surveillance systems play crucial roles in understanding the disease, collecting the data, and predicting the occurrence (28). Both methodologies provide valuable data for designing effective interventions. Classical outbreak investigations and conventional surveillance methods have been significantly improved with the advent of whole genome sequencing (29). Surveillance systems, especially those utilizing novel web-based data sources, contribute significantly to disease understanding and prediction. Event-based surveillance systems are usually classified into three categories, including news aggregators (e.g., Google Flu Trends), automatic systems (e.g., EpiSPIDER, HealthMap, EpiSimS, MedISys, and GETWELL), and moderated systems (e.g., ProMED-Mail, GPHIN, GOARN, and BioCaster) (30, 31). Event-based surveillance systems gather information from news or event reports and help in estimation of disease trends, distribution, and occurrences. Data obtained from these methodologies are usually used with statistical techniques for prediction purposes.

Recent utilization of computational algorithms has improved epidemiological approaches and associated outcomes. Several incidence prediction models of infectious diseases have been developed using these algorithms, for vector-borne diseases, typhus fevers, hemorrhagic fever, leptospirosis, malaria, *Vibrio vulnificus* infections, and bacterial dysentery. Most prediction models related to vector-borne diseases utilize algorithms to correlate disease occurrence and meteorological factors (32, 33). Advanced methods such as Bayesian regression neural networks, cubist regression, and support vector regression offer promising disease dynamic predictions (34). On the other hand, computational algorithms such as Bayesian latent point process are utilized to enhance the accuracy of spatial epidemiology (27). These advancements will ultimately guide public health interventions and the development of effective prevention and control strategies.

11.4.2 Simulation Modeling Approaches

Simulation modeling approaches address specific aspects of real-world systems. These approaches include system dynamic models, agent-based models (ABMs), discrete event models (DE), and Monte Carlo simulation models (35). SIM-D, COVASIM, ProMES, CoPE, PanSim, and CityCOVID are a few examples of simulation modeling approaches (8, 36). System dynamic models attempt to understand dynamic system behavior over time by utilizing feedback loops (7). On the other hand, ABMs utilize the actions and interactions of agents to understand the emerging behavior of individual-level decisions (37). DE models analyze systems with well-defined events and processes and explain the discrete event occurrence at specific points in time. Monte Carlo simulation models involve probability distributions in order to analyze system behavior under uncertainty (38). Among these approaches, ABMs are a highly used modeling approach, followed by DE models (39).

Interdisciplinary approaches involving infectious diseases, simulation models, and other methods have yielded exciting developments. Harmonic decomposition analysis is used to understand the role of underlying factors in disease occurrence (40), for instance, Fourier and wavelet analysis, a type of harmonic analysis, are utilized to understand the role of weather-related factors in spatio-temporal malaria data (41). The creation of digital organisms is another interesting approach in this regard (42). Digital microbes simulate microbial evolution and adaptation using evolutionary principles, offering insights into problems difficult to study in nature (43). Development of network models and graph theory have relatively better explained the network dynamics, genomic pathways, and co-occurrence of associated characteristics (44) (45). There are many examples of algorithms and software to run network models such as Gephi (46), Node2Vec (47), NetworkX (48), igraph R (49), etc. These approaches will provide valuable tools not only for informed decision-making processes but also in scenarios such as bioterrorism response.

11.4.3 Mathematical Modeling

Mathematical modeling approaches are based on creating a set of complex equations to describe certain phenomena or systems which enable the modelers to predict and explain certain mechanisms (50). Generally every model using mathematical generics is considered a mathematical model. On the other hand, mathematical models are based on specific model parameters such as optimization (linear vs. non-linear), time representation (static vs. dynamic), calculation mode (deterministic vs. stochastic), and nature (empirical vs. mechanistic) (51). Dynamical system models developed from case notification data are based on the Sparse Identification of Nonlinear Dynamics (SINDy) algorithm. Such disease dynamic models have been proved helpful in measles and rubella (52). Stochastic models, such as those for Ebola transmission, offer flexible frameworks for forecasting disease dynamics (53). Additionally, differential-equations-based compartmental models such as susceptible-infected-recovered (SIR) models and their related variants have been widely utilized (54). Population-based susceptible-exposed-infected-removed (SEIR) models have been developed to advise governments on COVID-19 responses, with challenges including accurate calibration to observed data (55). Furthermore, there is a call to explore predictive tools such as machine learning and integrated approaches to develop robust prediction systems.

11.4.4 Bioinformatics and Infectious Disease Prediction

With progress in the field of genomic sequencing, numerous infectious diseases have been studied at DNA and gene levels. Additionally, the genomic sequences available in databases have been utilized by bioinformaticians for their in-silico predictions of drugs as treatments against infectious diseases. Researchers utilize ML algorithms, such as random forests (RF), support vector machine (SVM), k-nearest neighbors (kNN) and others, to process genomic data, to predict genome sequences (56), to identify unconventional chromosomes, for domains of splice junction classification (57), for gene regulatory network reconstruction (58), and to analyze DNA datasets. Such approaches in bioinformatics have led to understanding the existing complexity of the pathogens (59). These approaches have been proved significantly less time-consuming and more cost-effective in infection biology and molecular medicine.

11.5 APPROACHES TO NON-COMMUNICABLE DISEASE PREDICTION

Among developed countries, non-communicable diseases (NCDs) are the leading cause of death and a higher disease burden than communicable disease. Cardiovascular diseases comprise a major share of deaths related to NCDs, followed by different types of cancer, chronic obstructive pulmonary disease (COPD), and diabetes (60). A quick review of papers catalogued in PubMed revealed that cancer of different types, benign diseases, liver diseases, cardiovascular diseases (heart conditions, stroke, angiodysplasia), eye conditions (glaucoma, eyelid, eyeball), thyroid conditions (parathyroid, hypothyroid), sleep issues, dementia, and skin conditions were highly studied during 2019–2023 (Figure 11.5). Cancer conditions are highly studied compared to all the other diseases.

Developing accurate prediction models for non-communicable diseases is crucial for early detection, effective management, and improved patient outcomes. Researchers have employed various methodologies to construct prediction models for non-communicable diseases, including traditional statistical approaches and modern ML techniques such as decision trees, random forests (61), and multivariable chronic disease risk assessment (62). There are similar approaches available for both infectious and non-communicable diseases. To avoid repetition, already discussed approaches will not be discussed here. However, aspects specific to non-communicable disease prediction models will be discussed in this section.

Various studies highlight the utilization of ML algorithms in predicting and classifying different diseases. For instance, a study employs non-invasive tools to associate tissue changes with cardiovascular health outcomes in elderly individuals (63). Another study compares different techniques for cardiovascular disease prediction, including Decision Tree, KNN, Naïve Bayes, SVM, XGBoost, and Random Forest (64). However, it has already been proved that ML-based approaches outperform non-ML clinical prediction models (65). With the advent of DL algorithms, various strategies such

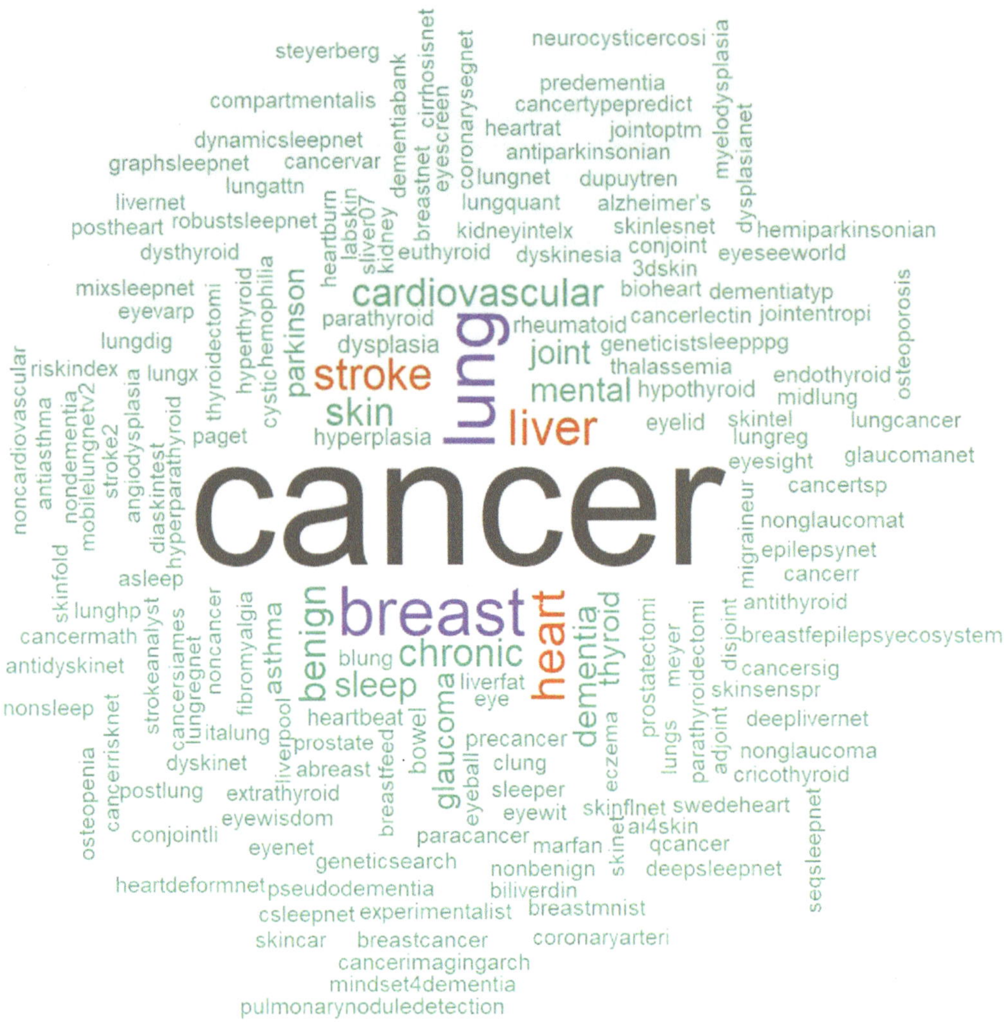

FIGURE 11.5 An overview of non-communicable-disease-related studies catalogued in PubMed during 2019–2023 (data not shown here).

as using Internet of Things (IoT) devices (66) and wearable sensors (67) have been used for cardiac disease prediction, showcasing the integration of technology into disease prediction frameworks.

Risk prediction models have been tailored to specific patient populations and disease contexts—such as cardiovascular diseases with COVID-19 (68), non-alcoholic fatty liver disease with diabetes (69), cardiovascular diseases in type 2 diabetes (70), HIV and cardiovascular diseases (71), hypertension and obstructive sleep apnea (72)—and innovative frameworks facilitate early risk prediction, as seen in diabetes (67). Similarly, treatment-related factors in risk assessment have been studied, such as adverse cardiovascular events after breast cancer therapy (73). Another study elaborates on hospital readmission risk in diabetic patients (74) while another model predicts dropout risk in outpatient treatment for alcohol use disorder (75). A few studies emphasize embedding biological knowledge in ML classifiers for predicting cancer phenotypes (76). To educate patients, the chronic prediction model aims to control risk factors affecting their diseases (77). These studies collectively demonstrate the versatility and efficacy of ML in disease prediction across various medical domains.

Ensemble ML models have improved accuracy of disease prediction as compared to single ML models (78), as studied in robust heart disease predictions (79). Ensemble models have also proved accurate in predicting perioperative complications after cervical fusion (80). Predicting disease severity, particularly in conditions like dementia, has also been studied (81). Additionally, researchers have modeled regional health outcomes from sociodemographic determinants using combined or ensemble models (82) and developed frameworks for complex health interventions (83).

11.6 STEPS AND METHODOLOGY

The methodology of disease prediction modeling includes multiple important steps such as data collection, data preprocessing, feature selection, feature extraction, model development, and evaluation (84). Data collection begins with the aim to collect data of appropriate quality and quantity, while preprocessing steps, including data-cleaning and imputation, are essential to ensure the reliability and representativeness of the data. Feature selection and extraction are the next crucial steps, where relevant features must be carefully chosen or extracted to enhance prediction accuracy (85, 86). On the other hand, selecting the appropriate ML algorithm or technique is vital in developing a model. It considers various factors like input data type, learning type, and data completeness (87). The last step in this methodology is to evaluate the performance of the developed model using evaluation measures derived from confusion matrices, like accuracy and precision, and metrics such as AUROC (84). These steps collectively ensure the development of robust disease prediction models suitable for real-world applications.

11.6.1 Data Collection and Preprocessing

All the modeling approaches rely highly on good data. Model performance and accuracy depend upon good quality and appropriate quantity of data (88). Data preprocessing also contributes significantly to the development of accurate disease forecasting models, along with data collection. Poor-quality data, such as data with numerous missing values or containing outliers, can lead to inaccurate and unreliable predictions.

While designing the study, data collection should be designed carefully. Data quality factors such as reliability, completeness, representativeness, and class imbalance significantly impact model performance (84). Thus, ensuring complete, reliable, representative, and balanced training datasets is essential for accurate predictions and generalization to different regions and populations (89). On the other hand, factors related to data quantity include large size and diversity that show underlying patterns effectively with no multicollinearity and dimensionality (90). In other words, a good dataset will have maximum factors of data quality and quantity.

After collecting data, the preprocessing step ensures data quality, consistency, and suitability for model development. It involves various techniques such as cleaning, normalization, imputation of missing values, and feature engineering. These steps help remove noise, handle outliers, and transform the data (2, 91). Moreover, preprocessing strategies designed for specific data types, such as sensor data or electronic health records, have been shown to have significant results (92).

11.6.2 Feature Selection and Extraction

The feature selection step identifies the most important elements that provide the most information about the problem. On the other hand, feature extraction is the step that is used to transform the original dataset into a newer dataset with a reduced number of features (4). Feature selection and feature extraction are used to reduce redundancy, multicollinearity, and effects of unimportant variables on the model performance. All these methods remove irrelevant features to produce more accurate models (93).

11.6.2.1 Feature Selection

Feature selection is crucial in disease prediction models for better accuracy of the models. Various feature selection techniques, such as filter, wrapper, and embedded methods, aim to identify relevant and most important features while ignoring the redundant ones, which improves accuracy and reduces computational costs (94). Filter methods rank features based on their correlation with the output variable. Some commonly used statistical measures include the chi-squared test, mutual-information, correlation coefficient, and ANOVA, while wrapper methods assess feature subsets for performance on a validation set such as sequential forward and backward selection, randomized hill climbing, and recursive feature elimination. Embedded methods integrate feature selection into the ML process, utilizing techniques like Lasso (L1 regularization) regression, elastic net regressions, and random forests (95, 96). Decision trees are commonly used to select the most informative features by selecting the Gini values that partition the nodes of trees (97). Notably, the necessity of feature selection varies based on the dataset and model type, with some models, like deep neural networks, capable of automatically learning relevant features (85).

Various studies have explored the strengths and weaknesses of these feature selection algorithms. Filter methods can be classified into univariate and multivariate based on their applicability. Univariate filter methods such as chi-square, Pearson correlation, t-test, and Whitney U test are fast and scalable with reduced overfitting and classifier independence but cannot model feature dependencies (96). Multivariate filter methods such as fast correlation-based filter (FCBF) and minimal redundancy maximal relevance (mRMR) are relatively slower and less scalable but can model feature dependencies with reduced overfitting (98, 99). Wrapper methods can model feature dependencies and model interaction with classifiers but are slower than other techniques and more prone to overfitting (100). On the other hand, embedded methods are better than wrapper methods but slower than filter methods. These methods can also model feature dependencies and interaction with classifiers (96, 101). Among all methods, the wrapper and embedded methods are found to achieve better prediction performance but are complex to implement and computation-intensive (102).

In medical data analysis, various innovative approaches and techniques for feature selection have been proposed. To diagnose diseases such as breast cancer and heart disease, novel approaches such as Group Method of Data Handling (GMDH) (103), and Particle Swarm Optimization (PSO) and its associated supervised feature selection methods such as PSO-relative Reduct and PSO-quick reduct (104, 105), have recently been proposed. Penalty-based regularizations involving Lasso regressions, Tree-Lasso model, and ridge regressions methods have recently addressed stability issues in feature selection in medical datasets, making these crucial for clinical decision-making processes (106). Overall these approaches demonstrate the improved disease diagnostics, risk assessment, and early detection that could lead to better health outcomes.

11.6.2.2 Feature Extraction

Data with a high number of features is known as high-dimensional data. The goal of feature extraction is to extract the essential feature information from the original features in the dataset and represent it in a lower dimension by reducing dimensionality (93). Feature extraction methods can be categorized into linear methods and nonlinear methods. Linear methods are easier as compared to nonlinear methods, and therefore were mostly preferred in early data dimensionality reduction. However, many of real-world problems follow nonlinear distribution and time-varying systems, therefore recent studies are focusing on nonlinear feature reduction methods.

Linear methods such as principal component analysis (PCA), linear discriminant analysis (LDA), independent component analysis (ICA), singular value decomposition (SVD), multi-dimensional scaling (MDS), and factor analysis (FA) consider that the data lies on a lower-dimensional linear subspace (93). It projects a dataset on this subspace using matrix factorization. On the other hand, nonlinear methods involve kernel-based method (kernel-PCA), nonnegative matrix factorization, wavelet transformation (Fourier transformation), and manifold learning based methods such as isometric map (ISOMAP), locally linear embedding (LLE), locality preserving projection (LPP) (107). These methods extract the nonlinear information present in the original dataset space and transform it into the linear problem in the feature space.

Several studies employ various feature extraction techniques to improve classification accuracy in medical applications such as Alzheimer's disease, end-stage renal disease patients, COVID-19, and cardiovascular and respiratory diseases. Hybrid feature extraction techniques have also been applied in medical images—ultrasound images, brain MRI, and breast cancer images—such as discrete wavelet transform (DWT), principal component analysis (PCA), and gray-level co-occurrence matrix (GLCM) (108). Similarly, a few studies report use of algorithms such as python-based TPOT genetic programming and Boruta to predict indicators and feature selection for medical diseases (109, 110). Additionally, feature extraction techniques like short-time Fourier transform (STFT) (111) and multichannel feature extraction modules (112) have been developed to enhance classification models.

11.6.3 Model Development and Evaluation

Disease prediction model development refers to the step that includes creating models using tailored algorithms. These computational algorithm-based models analyze medical datasets and make predictions. Different computational algorithms have different task types and output formats (94). Therefore, a heuristic selection of algorithms is considered essential in developing a good disease prediction model. This decision significantly influences the accuracy of the final model and should be decided on multiple factors such as type of input data (113), type of learning and prediction (94), targeted labels such as multi-label or binary classification problems (87), and requirements of large-high-quality data. Therefore, a good knowledge of the problem and related data is vital for selecting the most suitable algorithm for model development.

Various studies have discussed various computational algorithms, including supervised, unsupervised, and reinforcement learning. Each type of algorithm has its advantages and suitability. Supervised learning is effective when labeled data is abundant, unsupervised learning is useful for exploring data patterns in the absence of labels, and reinforcement learning excels in tasks involving sequential decision-making. A recent study has compared various algorithms of supervised and unsupervised learning. It found that supervised learning and unsupervised learning performed well, especially SVM among all the algorithms for clustering. Overall, supervised learning algorithms are suitable for classification tasks with labeled data, while unsupervised learning algorithms are useful for clustering tasks and identifying patterns in unlabeled data (114). There is still a gap in research when comparing different algorithms on different bases and conditions.

There is no one-size-fits-all answer to which machine learning algorithm is better, as the choice depends on the specific requirements and constraints of the problem. Experimentation, evaluation, and consideration of domain knowledge are considered basic requirements while selecting the best algorithm for the task. Additionally, ensembling techniques, which combine the predictions of multiple algorithms, can often lead to improved performance compared to using a single algorithm alone. For instance, a recently utilized approach for COVID-19 prediction based on population-based Harris hawks optimization (HHO) and SHAP (Shapley Additive exPlanations) features importance analysis (115).

11.6.3.1 Model Evaluation

Several metrics are usually used to evaluate the performance of computational models, particularly in medical applications. These metrics include accuracy, precision, F1-score, and area under the receiver operating curve (AUROC) derived from recall and specificity values. Accuracy measures the proportion of correct predictions out of the total predictions made by the model. However, accuracy alone may not provide sufficient insight, especially in cases of class imbalance. In such cases, the accuracy of the model is often misleading. Therefore, AUROC is preferable to use in this case

(116). AUROC assesses the model's performance and represents the model's ability to distinguish between different classes. Other performance matrices are derived from the confusion matrix, such as precision, recall, and specificity. Precision measures the proportion of true positive predictions out of all positive predictions made by the model, while recall (also known as sensitivity) measures the proportion of true positives correctly identified by the model. Specificity measures the proportion of true negatives correctly identified by the model. F1-score is the harmonic mean of precision and recall, providing a balanced measure of the model's performance. Overall, the evaluation of ML models should be thorough, considering multiple metrics to ensure their effectiveness and reliability, especially when deploying them in clinical settings. Validation with separate datasets not used during model training is crucial to validate the model's performance accurately.

11.7 TYPES AND APPLICATIONS OF DISEASE PREDICTION MODELS

Artificial intelligence (AI) is a field that encompasses techniques enabling computers to imitate human behavior and excel in decision-making. ML and DL as advanced AI techniques are two main subfields applied in the healthcare system to diagnose and predict diseases. There are many types of computational algorithms being used in disease prediction modeling.

11.7.1 Machine Learning Algorithms for Disease Prediction

Machine learning is a sort of artificial intelligence that enables a computer to learn and improve its performance without being explicitly programmed. ML approaches have been used in diverse dimensions of medical fields such as gene expression analysis, disease progression (Parkinson's disease), detecting diseases (Alzheimer's disease, epilepsy), detecting tumors, and many more (117, 118). ML has three main types: supervised learning, unsupervised learning, and reinforcement learning. (119). All of these types have different requirements and applications in the field of medical sciences.

Supervised learning algorithms rely on labeled data for training and are widely used in regression and data classification tasks such as breast cancer classification (120). These learning techniques include logistic regression, decision tree, K-nearest neighbors, and naive Bayes. On the other hand, unsupervised learning algorithms, including k-means clustering and hierarchical clustering, are used for clustering when the data is unlabeled or lacks clear categories. These techniques are valuable for tasks such as anomaly detection and association mining. Lastly, reinforcement learning algorithms learn to make decisions by interacting with an environment and receiving feedback in the form of rewards or penalties (121). These are commonly applied in tasks where sequential decision-making is required, such as game-playing and robotics (122). A comprehensive comparison of various ML approaches is needed in terms of computational time, data availability, and classification accuracy.

11.7.2 Deep Learning Techniques in Disease Prediction

Deep learning (DL) algorithm is an ML technique that does not rely on expert feature extraction, unlike classical neural network algorithms (121). Unlike traditional networks, deep neural networks are complex architectures consisting of multiple hidden layers and utilize advanced neurons that can perform intricate operations like convolutions and employing multiple activations within a single neuron. Moreover, deep learning excels in handling large and high-dimensional datasets common in domains such as medical records, medical images (123), and medical device data (124, 125). Despite its impressive performance, deep learning has its limitations, especially in tasks requiring strong artificial intelligence capabilities such as literal comprehension and intentionality (126). DL techniques are broadly divided into three major categories: (i) deep networks for supervised or discriminative learning; (ii) deep networks for unsupervised or generative learning; and (iii) deep networks for hybrid learning combining both and relevant others.

Discriminative learning techniques are typically designed to give discriminative power for pattern classification by describing the posterior distributions of classes conditioned on visible data. These architectures mainly include multi-layer perceptron (MLP), convolutional neural networks (CNN), recurrent neural networks (RNN), and related variants. On the other hand, generative learning methods are essentially applied for unsupervised learning and can be used as preprocessing for supervised learning tasks as well, which ensures the discriminative model accuracy. Commonly used deep neural network techniques for unsupervised or generative learning are generative adversarial network (GAN), autoencoder (AE), restricted Boltzmann machine (RBM), self-organizing map (SOM), and deep belief network (DBN) along with their variants. Lastly, hybrid deep learning models are typically composed of multiple (two or more) deep basic learning models, where the basic model is a discriminative or generative deep learning model discussed earlier. This type of DL model includes deep transfer learning (DTL) and deep reinforcement learning (DRL). In recent years, DL algorithms have emerged as a powerful tool in medical image analysis, disease prediction, and diagnosis showcasing its potential in various healthcare applications.

There are multiple DL algorithms popular for their applications in medical and healthcare data. One of the most popular is CNN for disease classification such as COVID-19, acute myocardial infarction using ECG dataset, heart failure patients' survival prediction, diabetes, cancer, and eye disease prediction (9, 127–131). For image classifications, recurrent neural networks with long short-term memory have also been utilized to predict lung diseases from X-ray images (5). Additionally, hybrid deep learning techniques Alexnet, GoogleNet, SegNet, and MultiResUNet are also popular for disease prediction models such as liver cancer and pap-smear image classification (132, 133). A recent approach, CanDiag, is employing transfer deep learning (TDL) with fog computing for cancer diagnosis, combining transfer learning techniques with convolutional neural networks (CNNs) (134).

11.7.3 Statistical Models for Disease Prediction

Statistical modeling is the process of using statistical techniques to create a model that describes a system or phenomenon and to make predictions about future events. There are many different types of statistical models, and they can be used to analyze a wide range of data types, including linear models, time series models, generalized linear models, Bayesian models, and survival analysis models.

Among the statistical models, linear models are the most popular models. These have been extensively utilized in medical disease prediction using parameters in medical lab records, addressing challenges and showcasing effectiveness across various domains such as breast cancer diagnosis, heart disease detection, etc. (135, 136). On the other hand, time series models such ARIMA and its variants are pivotal in offering personalized and reliable forecasts for healthcare services using time series data—for instance, COVID-19 trends forecasting (137), and Alzheimer's disease progression prediction (138). A recently popular statistical approach is the Bayesian models which offer a versatile framework for medical disease prediction, covering a wide range of applications in healthcare. These models can offer dynamic prognosis in clinical medicine (139), risk characterization from electronic medical records (140), disease topic discovery from medical records (141), and mortality prediction using fitness data (142). Lastly, survival analysis models are crucial for predicting disease outcomes and improving patient care. Various studies have focused on survival prediction in cancer (143), cardiovascular diseases (144), and infectious diseases like COVID-19 (145).

11.8 ROLE OF BIG DATA IN DISEASE PREDICTION

"Big data" is large amounts of information generated or produced that cannot be stored using traditional methodologies (10). Handling this big data has been a hot topic for the past two decades.

All public and private industries use this big data to improve their services. Like other sectors, the healthcare industry also generates a vast amount of data that cannot be handled and analyzed using traditional methods. Sources of healthcare big data include hospital admission records, patient medical records, medical examinations, laboratory examination records, and IoT devices being used by patients. The healthcare industry is also striving to properly manage and analyze big data to derive meaningful information. There is still a big gap and an urgent need for development in medical big-data-related issues.

11.9 BENEFITS FROM BIG DATA

The convergence of digitization and cloud computing has revolutionized healthcare by enabling the utilization of big data analytics and machine learning techniques. Several studies have explored the benefits of medical big data and big data analytics. The most common utilization is for prognostics and predicting medical conditions and diseases. Other utilizations involve analysis of general body-based illnesses, predicting novel diseases (146), disease heterogeneity (147), clinical big data analysis (148) including pediatrics data (149), disease risk prediction (150), and healthcare infrastructure investments (151). Various data-mining models and computational algorithms have been applied in developing intelligent medical systems (152). Overall, these studies underscore the pivotal role of big data in disease prediction and healthcare analysis.

11.10 CAUTION AND CONCERNS

The utilization of big data raises significant ethical, privacy, and security concerns across various domains. The risks associated with unethical use, data ownership, and privacy challenges are the big problems in this area (153). Furthermore, new frameworks such as Clinical Practice Research Datalink (CPRD) (154), and EULAR (155) points are being proposed to address privacy issues, emphasizing the need for foundational knowledge and ethical considerations when handling this medical big data. On the other hand, empirical ethical study methodologies have also been proposed to handle the ethical concerns regarding informed consent, privacy, and data ownership in big-data-driven personalized medicine (156). Similarly, many studies have stressed informed consent (157), transparent data-handling (155), and urgent need for policy development and ethical engagement (158). Lastly, the emerging discipline of cyberbiosecurity has categorized ethical issues related to big data and AI (159), emphasizing ongoing discussions and standardizations of ethical frameworks across sectors to ensure responsible technology use.

11.11 FUTURE PERSPECTIVES AND CONCLUSION

With advancements in computational technology and data analysis techniques, the future of disease prediction modeling is promising. Recently developed DL algorithms have exponentially increased the handling of medical big data. Additionally, the use of edge computing architecture, blockchain, and cloud technology can enhance the security and privacy of personal health data. The integration of these technologies can lead to the full mining of data value while protecting data-sharing security and privacy. Application of these techniques in medical data such as histopathological image analysis (160), pre-miRNA prediction (161), disease progression based on historical measurements (162), time series prediction (163), and many other healthcare areas still requires huge improvements. Overall, the future of disease prediction modeling lies in the continued advancement and integration of ML, AI, and data analysis techniques to improve healthcare outcomes.

REFERENCES

1. Toma M, Wei OC. Predictive modeling in medicine. *Encyclopedia*. 2023;3(2):590–601.
2. Yang Z, Mitra A, Liu W, Berlowitz D, Yu H. TransformEHR: transformer-based encoder-decoder generative model to enhance prediction of disease outcomes using electronic health records. *Nature Communications*. 2023;14(1):7857.
3. Li Y, Schoenfeld AH, diSessa AA, Graesser AC, Benson LC, English LD, et al. Computational thinking is more about thinking than computing. *Journal for STEM Education Research*. 2020;3(1):1–18.
4. Nogales RE, Benalcázar ME. Analysis and evaluation of feature selection and feature extraction methods. *International Journal of Computational Intelligence Systems*. 2023;16(1):153.
5. Goyal S, Singh R. Detection and classification of lung diseases for pneumonia and Covid-19 using machine and deep learning techniques. *Journal of Ambient Intelligence and Humanized Computing*. 2023;14(4):3239–59.
6. Dixon S, Keshavamurthy R, Farber DH, Stevens A, Pazdernik KT, Charles LE. A comparison of infectious disease forecasting methods across locations, diseases, and time. *Pathogens (Basel, Switzerland)*. 2022;11(2):185.
7. Sansone M, Holmstrom P, Hallberg S, Nordén R, Andersson L-M, Westin J. System dynamic modelling of healthcare associated influenza -a tool for infection control. *BMC Health Services Research*. 2022;22(1):709.
8. Reveil M, Chen Y-H. Predicting and preventing COVID-19 outbreaks in indoor environments: an agent-based modeling study. *Scientific Reports*. 2022;12(1):16076.
9. Umer M, Sadiq S, Karamti H, Karamti W, Majeed R, NAPPI M. IoT based smart monitoring of patients' with acute heart failure. *Sensors*. 2022;22(7):2431.
10. Batko K, Ślęzak A. The use of big data analytics in healthcare. *Journal of the Big Data*. 2022;9(1):3.
11. Barragán-Montero A, Javaid U, Valdés G, Nguyen D, Desbordes P, Macq B, et al. Artificial intelligence and machine learning for medical imaging: a technology review. *Physica Medica*. 2021;83:242–56.
12. van Boheemen L, van Schaardenburg D. Predicting rheumatoid arthritis in At-risk individuals. *Clinical Therapeutics*. 2019;41(7):1286–98.
13. Schreuder A, Prokop M, Scholten ET, Mets OM, Chung K, Mohamed Hoesein FAA, et al. CT-detected subsolid nodules: a predictor of lung cancer development at another location? *Cancers*. 2021;13(11):2812.
14. Mok Y, Ballew SH, Matsushita K. Chronic kidney disease measures for cardiovascular risk prediction. *Atherosclerosis*. 2021;335:110–8.
15. Ho DSW, Schierding W, Wake M, Saffery R, O'Sullivan J. Machine learning SNP based prediction for precision medicine. Frontiers in Genetics. 2019;10: 267.
16. Shah DR, Dhawan DA, editors. Disease Prediction Based on Symptoms Using Various Machine Learning Techniques. 2023; Singapore: Springer Nature Singapore.
17. Jenner AL, Aogo RA, Davis CL, Smith AM, Craig M. Leveraging computational modeling to understand infectious diseases. *Current Pathobiology Reports*. 2020;8(4):149–61.
18. Rabbani G, Baig MH, Ahmad K, Choi I. Protein-protein interactions and their role in various diseases and their prediction techniques. *Current Protein & Peptide Science*. 2018;19(10):948–57.
19. Muhammad Y, Tahir M, Hayat M, Chong KT. Early and accurate detection and diagnosis of heart disease using intelligent computational model. *Science Reports*. 2020;10(1):19747.
20. You Z-H, Huang Z-A, Zhu Z, Yan G-Y, Li Z-W, Wen Z, et al. PBMDA: a novel and effective path-based computational model for miRNA-disease association prediction. *PLOS Computational Biology*. 2017;13(3):e1005455.
21. Peng L-H, Zhou L-Q, Chen X, Piao X. A computational study of potential miRNA-disease association inference based on ensemble learning and kernel ridge regression. *Frontiers in Bioengineering and Biotechnology*. 2020;8:40.
22. Wang B, Zhang C, Du XX, Zheng XD, Li JY. lncRNA-disease association prediction based on the weight matrix and projection score. *PLoS One*. 2023;18(1):e0278817.
23. McInnes G, Sharo AG, Koleske ML, Brown JEH, Norstad M, Adhikari AN, et al. Opportunities and challenges for the computational interpretation of rare variation in clinically important genes. *The American Journal of Human Genetics*. 2021;108(4):535–48.
24. Dongarra J, Heroux MA, Luszczek P. A new metric for ranking high-performance computing systems. *National Science Review*. 2016;3(1):30–5.

25. Davis NA, Pandey A, McKinney BA. Real-world comparison of CPU and GPU implementations of SNPrank: a network analysis tool for GWAS. *Bioinformatics (Oxford, England)*. 2011;27(2):284–5.
26. Santangelo OE, Gentile V, Pizzo S, Giordano D, Cedrone F. Machine learning and prediction of infectious diseases: a systematic review. *Machine Learning and Knowledge Extraction*. 2023;5(1):175–98.
27. Nelli L, Ferguson HM, Matthiopoulos J. Achieving explanatory depth and spatial breadth in infectious disease modelling: integrating active and passive case surveillance. *Statistical Methods in Medical Research*. 2020;29(5):1273–87.
28. Tao D, Zhang D, Hu R, Rundensteiner E, Feng H. Epidemiological data mining for assisting with foodborne outbreak investigation. *Foods*. 2023;12(20):3825.
29. Revez J, Espinosa L, Albiger B, Leitmeyer KC, Struelens MJ, ENMFP, et al. Survey on the use of whole-genome sequencing for infectious diseases surveillance: rapid expansion of European national capacities, 2015–2016. *Frontiers in Public Health*. 2017;5:347.
30. Choi J, Cho Y, Shim E, Woo H. Web-based infectious disease surveillance systems and public health perspectives: a systematic review. *BMC Public Health*. 2016;16(1):1238.
31. Brett TS, Drake JM, Rohani P. Anticipating the emergence of infectious diseases. *Journal of The Royal Society Interface*. 2017;14(132):20170115.
32. Meng Q, Liu X, Xie J, Xiao D, Wang Y, Deng D. Epidemiological characteristics of bacillary dysentery from 2009 to 2016 and its incidence prediction model based on meteorological factors. *Environmental Health and Preventive Medicine*. 2019;24(1):82.
33. Kim SH, Jang JY. [Correlations between climate change-related infectious diseases and meteorological factors in Korea]. *Journal of Preventive Medicine and public health = Yebang Uihakhoe chi*. 2010;43(5):436–44.
34. da Silva RG, Ribeiro M, Mariani VC, Coelho LDS. Forecasting Brazilian and American COVID-19 cases based on artificial intelligence coupled with climatic exogenous variables. *Chaos, Solitons, and Fractals*. 2020;139:110027.
35. Nguyen LKN, Howick S, Megiddo I. A framework for conceptualising hybrid system dynamics and agent-based simulation models. *European Journal of Operational Research*. 2024;315(3):1153–66.
36. Latkowski R, Dunin-Kęplicz B. An agent-based Covid-19 simulator: extending Covasim to the polish context. *Procedia Computer Science*. 2021;192:3607–16.
37. Waleed M, Um T-W, Kamal T, Khan A, Zahid ZU. SIM-D: an agent-based simulator for modeling contagion in population. *Applied Sciences*. 2020;10(21):7745.
38. Stojanović O, Leugering J, Pipa G, Ghozzi S, Ullrich A. A Bayesian Monte Carlo approach for predicting the spread of infectious diseases. *PLOS ONE*. 2019;14(12):e0225838.
39. Zhang W, Liu S, Osgood N, Zhu H, Qian Y, Jia P. Using simulation modelling and systems science to help contain COVID-19: a systematic review. *Systems Research and Behavioral Science*. 2023;40(1):207–34.
40. Krantz SG, Polyakov P, Rao A. True epidemic growth construction through harmonic analysis. *Journal of Theoretical Biology*. 2020;494:110243.
41. Rogers DJ, Randolph SE, Snow RW, Hay SI. Satellite imagery in the study and forecast of malaria. *Nature*. 2002;415(6872):710–5.
42. Wilke CO, Adami C. The biology of digital organisms. *Trends in Ecology & Evolution*. 2002;17(11):528–32.
43. Egli A, Schrenzel J, Greub G. Digital microbiology. *Clinical Microbiology and Infection*. 2020;26(10):1324–31.
44. Zhang W, Chien J, Yong J, Kuang R. Network-based machine learning and graph theory algorithms for precision oncology. *npj Precision Oncology*. 2017;1(1):25.
45. Lloyd AL, May RM. Epidemiology. How viruses spread among computers and people. *Science (New York, NY)*. 2001;292(5520):1316–7.
46. Wolf FA, Angerer P, Theis FJ. SCANPY: large-scale single-cell gene expression data analysis. *Genome Biology*. 2018;19(1):15.
47. Grover A, Leskovec J. node2vec: scalable feature learning for networks. *KDD: proceedings International Conference on Knowledge Discovery & Data Mining*. 2016;2016:855–64.
48. Hagberg A, Swart P, Chult D. Exploring network structure, dynamics, and function using NetworkX. Python in Science Conference, 2008.
49. Csardi G, Nepusz T. The Igraph Software package for complex network research. *InterJournal: Complex Systems*. 2005;1695:1–9.

50. Mohamadou Y, Halidou A, Kapen PT. A review of mathematical modeling, artificial intelligence and datasets used in the study, prediction and management of COVID-19. *Applied Intelligence*. 2020;50(11):3913–25.
51. Tedeschi LO. Review: the prevailing mathematical modeling classifications and paradigms to support the advancement of sustainable animal production. *Animal*. 2023;17:100813.
52. Horrocks J, Bauch CT. Algorithmic discovery of dynamic models from infectious disease data. *Scientific Reports*. 2020;10(1):7061.
53. Asher J. Forecasting Ebola with a regression transmission model. *Epidemics*. 2018;22:50–5.
54. Yang Z, Zeng Z, Wang K, Wong SS, Liang W, Zanin M, et al. Modified SEIR and AI prediction of the epidemics trend of COVID-19 in China under public health interventions. *Journal of Thoracic Disease*. 2020;12(3):165–74.
55. Liu X-X, Fong SJ, Dey N, Crespo RG, Herrera-Viedma E. A new SEAIRD pandemic prediction model with clinical and epidemiological data analysis on COVID-19 outbreak. *Applied Intelligence*. 2021;51(7):4162–98.
56. Mahood EH, Kruse LH, Moghe GD. Machine learning: a powerful tool for gene function prediction in plants. *Applications in Plant Sciences*. 2020;8(7):e11376.
57. Zhang Y, Liu X, MacLeod J, Liu J. Discerning novel splice junctions derived from RNA-seq alignment: a deep learning approach. *BMC Genomics*. 2018;19(1):971.
58. Zhang Y, Zhang X, Lane AN, Fan TW, Liu J. Inferring gene regulatory networks of metabolic enzymes using gradient boosted trees. *IEEE Journal of Biomedical and Health Informatics*. 2020;24(5):1528–36.
59. Law GL, Tisoncik-Go J, Korth MJ, Katze MG. Drug repurposing: a better approach for infectious disease drug discovery? *Current Opinion in Immunology*. 2013;25(5):588–92.
60. Hasanzad M, Sarhangi N, Aghaei Meybodi HR, Nikfar S, Khatami F, Larijani B. Precision medicine in non communicable diseases. *International Journal of Molecular and Cellular Medicine*. 2019;8(Suppl1):1–18.
61. Kim J-K, Son E-J, Lee Y, Park D-K. Decision tree driven rule induction for heart disease prediction model Korean National Health and Nutrition Examinations Survey V-1. *Lecture Notes in Electrical Engineering*. 2012;215.
62. Oh SM, Stefani KM, Kim HC. Development and application of chronic disease risk prediction models. *Yonsei Medical Journal*. 2014;55(4):853–60.
63. Ricciardi C, Edmunds KJ, Recenti M, Sigurdsson S, Gudnason V, Carraro U, et al. Assessing cardiovascular risks from a mid-thigh CT image: a tree-based machine learning approach using radiodensitometric distributions. *Science Reports*. 2020;10(1):2863.
64. D D, Goel A. Cardio vascular diseases detection using ultrasonic image by retaining deep learning model. *International Journal of Electrical and Electronics Research*. 2022;10:639–43.
65. Allan S, Olaiya R, Burhan R. Reviewing the use and quality of machine learning in developing clinical prediction models for cardiovascular disease. *Postgraduate Medical Journal*. 2021;98(1161):551–8.
66. Tabassum S, Zaman MIU, Ullah MS, Rahaman A, Nahar S, Islam AKMM, editors. The cardiac disease predictor: IoT and ML driven healthcare system. *4th International Conference on Electrical Information and Communication Technology (EICT)*. 2019.
67. Ferdousi R, Hossain MA, Saddik AE. Early-stage risk prediction of non-communicable disease using machine learning in health CPS. *IEEE Access*. 2021;9:96823–37.
68. Azarpazhooh MR, Morovatdar N, Avan A, Phan TG, Divani AA, Yassi N, et al. COVID-19 pandemic and Burden of non-communicable diseases: an ecological study on data of 185 countries. *Journal of Stroke and Cerebrovascular Diseases: The Official Journal of National Stroke Association*. 2020;29(9):105089.
69. Stefan N, Cusi K. A global view of the interplay between non-alcoholic fatty liver disease and diabetes. *The Lancet Diabetes & Endocrinology*. 2022;10(4):284–96.
70. Ramírez-Prado D, Palazón-Bru A, Folgado-de la Rosa DM, Carbonell-Torregrosa M, Martínez-Díaz AM, Martínez-St John DR, et al. A four-year cardiovascular risk score for type 2 diabetic inpatients. *PeerJ*. 2015;3:e984.
71. Raghavan A, Rimmelin DE, Fitch KV, Zanni MV. Sex differences in select non-communicable HIV-associated comorbidities: exploring the role of systemic immune activation/inflammation. *Current HIV/AIDS Reports*. 2017;14(6):220–8.
72. Shi Y, Ma L, Chen X, Li W, Feng Y, Zhang Y, et al. Prediction model of obstructive sleep apnea-related hypertension: machine learning-based development and interpretation study. *Frontiers in Cardiovascular Medicine*. 2022;9:1042996.

73. Abdel-Qadir H, Thavendiranathan P, Austin PC, Lee DS, Amir E, Tu JV, et al. Development and validation of a multivariable prediction model for major adverse cardiovascular events after early stage breast cancer: a population-based cohort study. *European Heart Journal*. 2019;40(48):3913–20.
74. Shang Y, Jiang K, Wang L, Zhang Z, Zhou S, Liu Y, et al. The 30-days hospital readmission risk in diabetic patients: predictive modeling with machine learning classifiers. *BMC Medical Informatics and Decision Making*. 2021;21(2):57.
75. Park SJ, Lee SJ, Kim H, Kim JK, Chun J-W, Lee S-J, et al. Machine learning prediction of dropping out of outpatients with alcohol use disorders. *PLOS ONE*. 2021;16(8):e0255626.
76. Omar M, Dinalankara W, Mulder L, Coady T, Zanettini C, Imada EL, et al. Using biological constraints to improve prediction in precision oncology. *iScience*. 2023;26(3):106108.
77. Pittoli F, Vianna HD, Barbosa JLV. An education driven model for non-communicable diseases care. *Handbook of Research on Advancing Health Education through Technology*: IGI Global; 2016. p. 391–418.
78. Sultana N, Sharma N, Sharma K, Verma S. A sequential ensemble model for communicable disease forecasting. *Current Bioinformatics*. 2019;14:309–17.
79. Alim M, Habib S, Farooq Y, Jungsher A. Robust heart disease prediction: a novel approach based on significant feature and ensemble learning model. *3rd International Conference on Computing, Mathematics and Engineering Technologies (iCoMET)*. 2020.
80. Shah AA, Devana S, Lee C, Bugarin A, Upfill-Brown A, Lord EL, et al. P125. Development of a novel ensemble machine learning algorithm for prediction of complications and readmission after anterior cervical spinal fusion. *The Spine Journal*. 2021;21(9, Supplement):S201–S2.
81. Huang W. A novel disease severity prediction scheme via big pair-wise ranking and learning techniques using image-based personal clinical data. *Signal Processing*. 2016;124:233–45.
82. Luo L, Luo L, Zhang X, He X. Hospital daily outpatient visits forecasting using a combinatorial model based on ARIMA and SES models. *BMC Health Service Research*. 2017;17(1):469.
83. Mahomed OH, Asmall S. Development and implementation of an integrated chronic disease model in South Africa: lessons in the management of change through improving the quality of clinical practice. *International Journal of Integrated Care*. 2015;15:e038.
84. Anahtar MN, Yang JH, Kanjilal S. Applications of machine learning to the problem of antimicrobial resistance: an emerging model for translational research. *Journal Clinical Microbiology*. 2021;59(7):e0126020.
85. Najafabadi MM, Villanustre F, Khoshgoftaar TM, Seliya N, Wald R, Muharemagic E. Deep learning applications and challenges in big data analytics. *Journal of the Big Data*. 2015;2(1):1.
86. Aytan-Aktug D, Clausen PTLC, Bortolaia V, Aarestrup FM, Lund O. Prediction of acquired antimicrobial resistance for multiple bacterial species using neural networks. *mSystems*. 2020;5(1):e00774–19.
87. Vamathevan J, Clark D, Czodrowski P, Dunham I, Ferran E, Lee G, et al. Applications of machine learning in drug discovery and development. *Nature Reviews Drug Discovery*. 2019;18(6):463–77.
88. Obermeyer Z, Emanuel EJ. Predicting the future – big data, machine learning, and clinical medicine. *New England Journal of Medicine*. 2016;375(13):1216–9.
89. Jain A, Patel H, Nagalapatti L, Gupta N, Mehta S, Guttula S, et al. Overview and importance of data quality for machine learning tasks. *Proceedings of the 26th ACM SIGKDD International Conference on Knowledge Discovery & Data Mining*; Virtual Event, CA, USA: Association for Computing Machinery; 2020. p. 3561–2.
90. OBrien ARS, PiotrLuo, Oscar Junhong George, Andrew Dunne, Rob Bauer, Denis C., editor Breaking the curse of dimensionality for machine learning on genomic data. *International Joint Conference on Artificial Intelligence*; Melbourne, Australia: SCHLOSS DAGSTUHL; 2017.
91. Zhao M, Song C, Luo T, Huang T, Lin S. Fatty liver disease prediction model based on big data of electronic physical examination records. *Frontiers in Public Health*. 2021;9:668351.
92. Park JR, Kim S, Kim T, Jin SW, Kim JL, Shin J, et al. Data preprocessing and augmentation improved visual field prediction of recurrent neural network with multi-central datasets. *Ophthalmic Research*. 2023;66(1):978–91.
93. Hira ZM, Gillies DF. A review of feature selection and feature extraction methods applied on microarray data. *Advances in Bioinformatics*. 2015;2015:198363.
94. Kohavi R, editor. A study of cross-validation and bootstrap for accuracy estimation and model selection. *International Joint Conference on Artificial Intelligence*; Standorf University, USA: GCS Stanford EDU; 1995.

95. Cai J, Luo J, Wang S, Yang S. Feature selection in machine learning: a new perspective. *Neurocomputing*. 2018;300:70–9.
96. Pudjihartono N, Fadason T, Kempa-Liehr AW, O'Sullivan JM. A Review of feature selection methods for machine learning-based disease risk prediction. *Frontiers in Bioinformatics*. 2022;2:927312.
97. Pirgazi J, Alimoradi M, Esmaeili Abharian T, Olyaee MH. An efficient hybrid filter-wrapper metaheuristic-based gene selection method for high dimensional datasets. *Science Reports*. 2019;9(1):18580.
98. Peng H, Long F, Ding C. Feature selection based on mutual information: criteria of max-dependency, max-relevance, and min-redundancy. *IEEE Transactions on Pattern Analysis and Machine Intelligence*. 2005;27(8):1226–38.
99. Yu L, Liu H. Eficient feature selection via analysis of relevance and redundancy. *Journal of Machine Learning Research*. 2004;5:1205–24.
100. Skalak DB. Prototype and feature selection by sampling and random mutation hill climbing algorithms. Machine Learning Proceedings 1994, ed. WW Cohen and H Hirsh. Amsterdam: Elsevier; 1994. p. 293–301.
101. Breiman LJMl. Random forests. Machine Learning. 2001;45:5–32.
102. Qu L, He W, Li J, Zhang H, Yang C, Xie B. Explicit and size-adaptive PSO-based feature selection for classification. *Swarm and Evolutionary Computation*. 2023;77:101249.
103. Abdel-Aal RE. GMDH-based feature ranking and selection for improved classification of medical data. *Journal of Biomedical Informatics*. 2005;38(6):456–68.
104. Abbasi MS, Al-Sahaf H, Mansoori M, Welch I. Behavior-based ransomware classification: a particle swarm optimization wrapper-based approach for feature selection. *Applied Soft Computing*. 2022;121:108744.
105. Inbarani HH, Azar AT, Jothi G. Supervised hybrid feature selection based on PSO and rough sets for medical diagnosis. *Computer Methods and Programs in Biomedicine*. 2014;113(1):175–85.
106. Kamkar I, Gupta SK, Phung D, Venkatesh S. Stable feature selection for clinical prediction: exploiting ICD tree structure using Tree-Lasso. *Journal of Biomedical Informatics*. 2015;53:277–90.
107. Jia W, Sun M, Lian J, Hou S. Feature dimensionality reduction: a review. *Complex & Intelligent Systems*. 2022;8(3):2663–93.
108. Madhu, Kumar R. A hybrid feature extraction technique for content based medical image retrieval using segmentation and clustering techniques. *Multimedia Tools and Applications*. 2022;81(6):8871–904.
109. Ahmed U, Lin JC-W, Srivastava G. Multivariate time-series sensor vital sign forecasting of cardiovascular and chronic respiratory diseases. *Sustainable Computing: Informatics and Systems*. 2023;38:100868.
110. Zhou H, Xin Y, Li S. A diabetes prediction model based on Boruta feature selection and ensemble learning. *BMC Bioinformatics*. 2023;24(1):224.
111. MS K, Rajaguru H, Nair AR. Evaluation and exploration of machine learning and convolutional neural network classifiers in detection of lung cancer from microarray gene—a paradigm shift. *Bioengineering*. 2023;10(8):933.
112. Ma L, Zhang C, Gao J, Jiao X, Yu Z, Zhu Y, et al. Mortality prediction with adaptive feature importance recalibration for peritoneal dialysis patients. *Patterns (New York, NY)*. 2023;4(12):100892.
113. Shorten C, Khoshgoftaar TM. A survey on image data augmentation for deep learning. *Journal of Big Data*. 2019;6(1):60.
114. Azmat F, Chen Y, Stocks N. Analysis of spectrum occupancy using machine learning algorithms. *IEEE Transactions on Vehicular Technology*. 2015;65(9):6853–60.
115. Debjit K, Islam MS, Rahman MA, Pinki FT, Nath RD, Al-Ahmadi S, et al. An improved machine-learning approach for COVID-19 prediction using Harris hawks optimization and feature analysis using SHAP. *Diagnostics (Basel, Switzerland)*. 2022;12(5):1023.
116. Eskofier BM, Klucken J. Predictive models for health deterioration: understanding disease pathways for personalized medicine. *Annual Review of Biomedical Engineering*. 2023;25:131–56.
117. Khan P, Kader MF, Islam SMR, Rahman AB, Kamal MS, Toha MU, et al. Machine learning and deep learning approaches for brain disease diagnosis: principles and recent advances. *IEEE Access*. 2021;9:37622–55.
118. Maharana C, Mishra BB, Dash CSK. A topical survey: applications of machine learning in medical issues. *Journal of Computational and Theoretical Nanoscience*. 2020;17(11):5010–9.
119. Ramaswamy S, DeClerck N. Customer perception analysis using deep Learning and NLP. *Procedia Computer Science*. 2018;140:170–8.

120. Shah A. Prediction of malignant and benign tumor using machine learning. *International Journal of Computer Applications*. 2016;135:19–23.
121. Ghaffar Nia N, Kaplanoglu E, Nasab A. Evaluation of artificial intelligence techniques in disease diagnosis and prediction. *Discovery Artifical Intelligent*. 2023;3:5.
122. Sivamayil K, Rajasekar E, Aljafari B, Nikolovski S, Vairavasundaram S, Vairavasundaram I. A systematic study on reinforcement learning based applications. *Energies*. 2023;16(3):1512.
123. Shoeibi A, Khodatars M, Jafari M, Ghassemi N, Sadeghi D, Moridian P, et al. Automated detection and forecasting of covid-19 using deep learning techniques: a review. Neurocomputing. 2024;577:127317.
124. Yang Y, Li D. Medical data feature learning based on probability and depth learning mining: model development and validation. *JMIR Medical Informatics*. 2021;9(4):e19055.
125. Raj RJS, Shobana SJ, Pustokhina IV, Pustokhin DA, Gupta D, Shankar K. Optimal feature selection-based medical image classification using deep learning model in internet of medical things. *IEEE Access*. 2020;8:58006–17.
126. Hauser L. Searle's Chinese box: debunking the Chinese room argument. *Minds and Machines*. 1997;7(2):199–226.
127. Tsarapatsani K-H, Sakellarios A, Pezoulas V, Tsakanikas V, Matsopoulos G, Marz W, et al. Machine learning models to predict myocardial infarction within 10-years follow-up of cardiovascular disease progression. *IEEE-EMBS International Conference on Biomedical and Health Informatics*. 2022.
128. Toda R, Oda M, Hayashi Y, Otake Y, Hashimoto M, Akashi T, et al. *Improved Method for COVID-19 Classification of Complex-Architecture CNN from Chest CT Volumes Using Orthogonal Ensemble Networks*: SPIE; 2023.
129. Ibrahim L, Mesinovic M, Yang K-W, Eid MAJIA. Explainable prediction of acute myocardial infarction using machine learning and shapley values. *IEEE Access*. 2020;8:210410–7.
130. Sekar S, Palani U. An effective disease prediction system using incremental feature selection and temporal convolutional neural network. *Journal of Ambient Intelligence and Humanized Computing*. 2020;11:5547–60
131. Chelaramani S, Gupta M, Agarwal V, Gupta P, Habash R. Multi-task learning for fine-grained eye disease prediction. In Proceedings of the Asian Conference on Pattern Recognition. Berlin: Springer; 2020. p. 734–49.
132. Ramamurthy M, Krishnamurthi I, Shanmuganathan V, Robinson Y. Deep learning based genome analysis and NGS-RNA LL identification with a novel hybrid model. *Biosystems*. 2020;197:104211.
133. Kalbhor M, Shinde S, Popescu DE, Hemanth DJ. Hybridization of deep learning pre-trained models with machine learning classifiers and fuzzy min–max neural network for cervical cancer diagnosis. Diagnostics. 2023;13(7):1363.
134. Pati A, Parhi M, Pattanayak BK, Sahu B, Khasim S. CanDiag: fog empowered transfer deep learning based approach for cancer diagnosis. *Designs*. 2023;7(3):57.
135. Guo C, Zhang J, Liu Y, Xie Y, Han Z, Yu J. Recursion enhanced random forest with an improved linear model (RERF-ILM) for heart disease detection on the internet of medical things platform. *IEEE Access*. 2020;8:59247–56.
136. Adebiyi MO, Arowolo MO, Mshelia MD, Olugbara OO. A linear discriminant analysis and classification model for breast cancer diagnosis. *IEEE Access*. 2022;12(22):11455.
137. Tandon H, Ranjan P, Chakraborty T, Suhag V. Coronavirus (COVID-19): ARIMA-based time-series analysis to forecast near future and the effect of school reopening in India. *Journal of Health Management*. 2022;24(3):373–88.
138. El-Sappagh S, Saleh H, Sahal R, Abuhmed T, Islam SMR, Ali F, et al. Alzheimer's disease progression detection model based on an early fusion of cost-effective multimodal data. *Future Generation Computer Systems*. 2021;115:680–99.
139. Verduijn M, Peek N, Rosseel PMJ, de Jonge E, de Mol BAJM. Prognostic Bayesian networks: I: rationale, learning procedure, and clinical use. *Journal of Biomedical Informatics*. 2007;40(6):609–18.
140. Bouktif S, Hanna EM, Zaki N, Khousa EA. Ant colony optimization algorithm for interpretable Bayesian classifiers combination: application to medical predictions. *PLOS ONE*. 2014;9(2):e86456.
141. Li Y, Nair P, Lu XH, Wen Z, Wang Y, Dehaghi AAK, et al. Inferring multimodal latent topics from electronic health records. *Nature Communications*. 2020;11(1):2536.

142. Sakr S, Elshawi R, Ahmed AM, Qureshi WT, Brawner CA, Keteyian SJ, et al. Comparison of machine learning techniques to predict all-cause mortality using fitness data: the Henry ford exerclse testing (FIT) project. *BMC Medical Informatics and Decision Making*. 2017;17(1):174.
143. Numico G, Occelli M, Russi EG, Silvestris N, Pasero R, Fea E, et al. Survival prediction and frequency of anticancer treatment in cancer patients hospitalized due to acute conditions. Role of clinical parameters and PaP score. *Supportive Care in Cancer: Official Journal of the Multinational Association of Supportive Care in Cancer*. 2011;19(11):1823–30.
144. Guan H, Dai GH, Gao WL, Zhao X, Cai ZH, Zhang JZ, et al. A 5-year survival prediction model for chronic heart failure patients induced by coronary heart disease with traditional Chinese medicine intervention. *Evidence-Based Complementary and Alternative Medicine: eCAM*: 2021;2021:4381256.
145. Dong YM, Sun J, Li YX, Chen Q, Liu QQ, Sun Z, et al. Development and validation of a nomogram for assessing survival in patients with COVID-19 pneumonia. *Clinical Infectious Diseases: An Official Publication of the Infectious Diseases Society of America*. 2021;72(4):652–60.
146. Ahmed I, Ahmad M, Jeon G, Piccialli F. A framework for pandemic prediction using big data analytics. *Big Data Research*. 2021;25:100190.
147. Bhattarai S, Gupta A, Ali E, Ali M, Riad M, Adhikari P, et al. Can big data and machine learning improve our understanding of acute respiratory distress syndrome? *Cureus*. 2021;13(2):e13529.
148. Wu W-T, Li Y-J, Feng A-Z, Li L, Huang T, Xu A-D, et al. Data mining in clinical big data: the frequently used databases, steps, and methodological models. *Military Medical Research*. 2021;8(1):44.
149. Misra S, Mukhopadhyay K. Data harnessing to nurture the human mind for a tailored approach to the child. *Pediatric Research*. 2022;93:357–65.
150. Moro Visconti R, Morea D. Big data for the sustainability of healthcare project financing. *Sustainability*. 2019;11(13):3748.
151. Chen M, Hao Y, Hwang K, Wang L, Wang L. Disease prediction by machine learning over big data from healthcare communities. *IEEE Access*. 2017;5:1.
152. Salma Banu NK, Swamy S. Prediction of heart disease at early stage using data mining and big data analytics: a survey. *International Conference on Electrical, Electronics, Communication, Computer and Optimization Techniques*. 2016.
153. Löfgren K, Webster CW. The value of Big Data in government: the case of 'smart cities'. *Big Data & Society*. 2020;7(1):2053951720912775.
154. Car J, Sheikh A, Wicks P, Williams MS. Beyond the hype of big data and artificial intelligence: building foundations for knowledge and wisdom. *BMC Medicine*. 2019;17(1):143.
155. Gossec L, Kedra J, Servy H, Pandit A, Stones S, Berenbaum F, et al. EULAR points to consider for the use of big data in rheumatic and musculoskeletal diseases. *Annals of the Rheumatic Diseases*. 2020;79(1):69–76.
156. Rauter CM, Wöhlke S, Schicktanz S. My data, my choice? – German patient organizations' attitudes towards big data-driven approaches in personalized medicine. An empirical-ethical study. *Journal of Medical Systems*. 2021;45(4):43.
157. Fothergill BT, Knight W, Stahl BC, Ulnicane I. Responsible data governance of neuroscience big data. *Frontiers in Neuroinformatics*. 2019;13:28.
158. Braunack-Mayer AJ, Street JM, Tooher R, Feng X, Scharling-Gamba K. Student and staff perspectives on the use of big data in the tertiary education sector: a scoping review and reflection on the ethical issues. *Review of Educational Research*. 2020;90(6):788–823.
159. Murch RS, So WK, Buchholz WG, Raman S, Peccoud J. Cyberbiosecurity: An emerging new discipline to help safeguard the bioeconomy. *Frontiers in Bioengineering and Biotechnology*. 2018;6:39.
160. Izadi S, Shiri I, Uribe C, Geramifar P, Zaidi H, Rahmim A, et al. Enhanced direct joint attenuation and scatter correction of whole-body PET images via context-aware deep networks. *Zeitschrift für medizinische Physik*. 2024; online ahead of print.
161. Mohanty S, Mahapatra S, Swarnkar T, editors. Deep learning techniques for pre-miRNA prediction: current challenges and future directions. *2021 9th International Conference on Reliability, Infocom Technologies and Optimization (Trends and Future Directions) (ICRITO)*. Piscataway, NJ: IEEE; 2021. p. 1–4.

162. Liang W, Zhang K, Cao P, Liu X, Yang J, Zaiane O. Rethinking modeling Alzheimer's disease progression from a multi-task learning perspective with deep recurrent neural network. *Computers in Biology and Medicine*. 2021;138:104935.
163. Han Z, Zhao J, Leung H, Ma K. A review of deep learning models for time series prediction. *IEEE Sensors Journal*. 2019;21(6):7833–48.

12 Bioinformatics Applied to Microbiology

Maimoona Sabir, Anabia Sohail, Ayesha Saeed, Sidra Abbas, Kishmain Ahmad Khan, and Misbah Khalil

12.1 INTRODUCTION

Microorganisms, spanning bacteria, archaea, fungi, viruses, and protists, pervade every ecosystem, intricately intertwined with diverse multicellular organisms. The earth harbors an extensive array of predominantly uncharacterized bacteria and archaea (1). Despite their minute size, microorganisms exert profound influence over the biosphere, impacting human health, agriculture, industry, and the environment (2–4). Operating across multifaceted roles, they drive biogeochemical cycles, decompose organic matter, facilitate nitrogen fixation and oxygen production, and act as symbionts or pathogens (5, 6). Within the human body, microbial populations, numbering in the billions, coexist as both benign inhabitants and active participants in physiological processes, such as the breakdown of intestinal contents. Notably, the microbiota inhabiting niches such as the gut and skin profoundly influence metabolism, immunity, and disease susceptibility, while pathogenic microbes pose significant threats, causing infections ranging from mild discomfort to life-threatening illnesses (7).

Microbiology, the study of microorganisms, has long stood as a cornerstone of scientific inquiry, significantly shaping our comprehension of life on earth. From Antonie van Leeuwenhoek's 17th-century discovery of bacteria to contemporary advancements in genetic elucidation and molecular techniques, microbiology has witnessed remarkable progress. Early efforts, spearheaded by scientists like Robert Koch, focused on cultivating and isolating microbes to enumerate and visualize them using nutrient media like potato slices or gelatin. These pioneering isolation methods provided insights into microbial physiology. Subsequently, microscopy supplanted other instruments as the primary tool for microbial identification, complemented by culture techniques to elucidate physiology, genetic makeup, metabolism, and pathogenicity. Microscopy resolution was significantly enhanced by staining procedures pioneered by Schaeffer and Fulton, and by Ziehl and Neelsen.

For over three centuries, microbial study relied on morphological characteristics, growth patterns, and specific biochemical profiles for classification. However, these methods, although providing insight into the microbial world, possess limited resolution for broader applications and are labor-intensive. The late Carl Woese advocated for using ribosomal RNA genes as molecular markers for life classification in the 1970s, revolutionizing microbial research. The advent of the Sanger automated sequencing method further transformed microbial classification, paving the way for molecular techniques that unveiled the "new uncultured world" of microbial communities. Advanced techniques such as high-throughput sequencing and next-generation sequencing have generated vast genomic datasets, with initiatives like the Human Microbiome Project shedding light on diverse microbial communities' intricacies (8).

Understanding microbial physiology, genetics, and interactions is paramount for addressing global challenges, including infectious diseases, antimicrobial resistance, and climate change. The

DOI: 10.1201/9781003410164-12

staggering microbial diversity, estimated at approximately 1 trillion species on Earth, predominantly uncultivated and uncharacterized, spans various habitats, from the human gut to deep-sea hydrothermal vents, showcasing their adaptability and resilience. Characterizing this diversity necessitates interdisciplinary approaches integrating microbiology, genomics, ecology, and bioinformatics. Microbiologists grapple with summarizing genetic diversity, prompting debates on species definition, speciation mechanisms, and the existence of microbial species.

The wealth of data from microbial genomics domains has spawned novel diagnostic and genotyping tools, unraveled microbial genetic diversity, and elucidated virulence and resistance mechanisms. However, optimal analysis and interpretation of these intricate datasets pose significant challenges, exceeding human expertise, which is prone to errors, and necessitating advanced computational techniques capable of discerning meaningful patterns. In-silico methods aid in mining genetic information to identify therapeutic targets, probe host–pathogen interactions, and elucidate mechanisms of antibiotic resistance and virulence (9).

12.2 BIOINFORMATICS

Bioinformatics, an application of information technology, aids in the processing and analysis of data generated in biological research and experiments by employing computer-based algorithms. It serves as an integrative field spanning various scientific disciplines, with a primary focus on translating terms and concepts between these domains. The interdisciplinary nature of bioinformatics encompasses computer science, mathematics, statistics, molecular biology, genetics, and other related fields. Notably, it addresses large-scale, data-intensive biological challenges from a computational standpoint (10). Bioinformatics applies principles of computer science, mathematics, and statistics to the biological sciences, providing software, methodologies, and solutions for identifying patterns, constructing models, and comprehending complex biological systems (11).

Bioinformatics has emerged as a frontline applied science crucial for exploring new realms of biology, recognized as a prominent scientific pursuit of the twenty-first century. Advances in microarray technology, proteomics, structural genomics, and complete genomic sequencing would stagnate without the application of bioinformatics, highlighting its indispensable role in solving intricate biological problems. At its core, two important developments in biological study have come together to establish bioinformatics: molecular sequences being kept in computer databases, and the utilization of computational algorithms for analyzing DNA and protein sequences. Its applications span from DNA barcoding to the design of disease outbreak patterns and novel biological product development. Additionally, bioinformatics plays a central role in genomic exploration, particularly in sequencing and annotating microbial genomes.

The significance of bioinformatics is notably pronounced in microbiology, where the field has become integral to advancing our understanding of microbial life and exploiting its potential applications. Microbiology benefits from molecular investigations, particularly DNA sequencing, facilitated by bioinformatics tools and techniques. The synergy between microbiology and bioinformatics enables researchers to characterize microbial communities, annotate genomes, predict gene functions, and unravel microbial interactions. Bioinformatics provides indispensable tools and methodologies for analyzing and interpreting vast biological datasets, especially concerning microorganisms.

From the perspective of the application of bioinformatics in microbiology, microbiologists formulate biological questions and hypotheses. For example, a microbiologist seeking to utilize an enzyme from seawater for synthetic polymer production would collaborate with biochemists to identify relevant artificial polymers connected to the sample or organism. Computer programs, incorporating algorithms created by mathematicians and put into practice by computer scientists, are then utilized to evaluate the data and pinpoint significant protein sequences similar to known enzymes involved in polymerization. Further interpretation and annotation of enzyme information within databases require collaboration with molecular biologists and physicists. Bioinformaticians play a crucial role in integrating information across various scientific disciplines, often stemming from backgrounds in conventional fields, including microbiology, physics, biochemistry, and computer science (12).

Examples highlighting the significance and relevance of bioinformatics in microbiology include proteomics for studying protein structures and identifying protein–protein interaction sites, metabolomics for understanding cell dynamics and interactions, and genome sequencing for drawing phylogenetic relationships and detecting transcription factor-binding sites. Bioinformatics tools enable the analysis of microarray data, crucial for interpreting the exponentially growing biological datasets resulting from low-cost sequencing technologies. This influx of data has led to the development of databases housing nucleic acid and protein sequences, along with their respective structures (13).

12.3 BIOINFORMATICS DATABASES

Bioinformatics databases serve as repositories for biological information obtained from scientific experiments, primarily encompassing DNA and protein sequences, as well as protein structures. Protein sequences and structures, metabolites and reactions, genes, genomes, expression data, interactions, and pathways literature archives, computational analyses of primary data, and metadata are all included in these databases. Equipped with query facilities and data analysis tools, bioinformatics databases play a crucial role across various fields within bioinformatics. The significance of these facts cannot be overstated for microbiologists. Furthermore, the fields of microbiology have their databases covering disease, physiology, microbial diversity, and model microorganisms (14). These databases must exhibit high diversity in information and support various data formats compatible with most computer systems. Furthermore, reliability, regular updates, and easily accessible documentation are essential attributes expected from bioinformatics databases. The databases are typically maintained and curated by bioinformatics institutions, which allocate resources for their upkeep. Different categorizations of databases can be defined based on source, information, and format.

12.4 CATEGORIZATION BASED ON SOURCE

The group of biological databases defined based on the source may be subdivided into primary (archival), secondary (curated), or composite databases. Primary databases contain data generated experimentally like nucleotide sequences of DNA/RNA, amino acid sequences of proteins, and 3D structures. Primary databases include GenBank from the National Centre for Biotechnology Information (NCBI), DNA Databank of Japan (DDBJ), and European Molecular Biology Laboratory (EMBL) (15), for nucleic acids. Amino acid sequence databases include Protein Information Resource (PIR), Universal Protein Resource (UniProt), Translated European Molecular Biology Laboratory (TrEMBL), and SWISS-PROT. Protein Data Bank (PDB) and Nucleic Acid Database (NDB) are examples of structural databases (16).

A secondary database represents sequences that are widely accepted by scientists and hold data that is extracted from the primary databases. The prominent secondary databases are eMOTIF, PROSITE, The Protein Families Database (Pfam), Blocks, Prints, Structural Classification of Proteins database (SCOP), Class, Architecture, Topography and Homology (CATH) (databases of conserved features and motifs of a sequence), Online Mendelian Inheritance in Man (OMIM), Kyoto Encyclopedia of Genes and Genomes (KEGG), etc. Multiple primary databases are combined into a composite database, which offers many database search options (e.g., The National Center for Biotechnology Information (NCBI), or The Maize Genetics and Genomics Database (MaizeGDB).

12.5 CATEGORIZATION BASED ON TYPE

12.5.1 Nucleotide Sequence Databases

Protein and nucleic acid sequences can be found in sequence databases. Nucleotide sequence repositories contain raw nucleic acid sequences and are made available through Internet sources. GenBank, the European Molecular Biology Laboratory (EMBL), and the DNA Data Bank of Japan (DDBJ) are the three primary repositories of nucleotide sequences. While these databases operated separately at first, DDBJ, GenBank, and EMBL eventually formed the International Nucleotide Sequence Database Collaboration (INSDC). To obtain the sequences that are available in each of the three formats, these databases began transferring their data via continuous team communication at each cooperating organization.

12.5.2 GenBank

Protein and nucleotide data, both raw and annotated, are compiled in GenBank. The National Centre for Biotechnology Information is the organization that manages and provides access to GenBank (NCBI). There is a new release every two months. As a component of the INSDC, it is maintained by NCBI. Enter "insulin" in the search bar on the GenBank homepage to see a list of partial or complete insulin gene sequences from various taxa.

12.5.3 European Molecular Biology Laboratory (EMBL)

The European Molecular Biology Laboratory (EMBL) (www.embl.org) (17) is a molecular biology organization that has Australia as an associate member and is run by 20 European nations. It was established in 1974 as an intergovernmental organization. It creates and manages a sizable number of databases, which are freely accessible to scientists. The European Bioinformatics Institute (EBI), located in Heidelberg, Germany, is the primary location of this research laboratory. It is run by Rolf Apweiler and Ewan Birney and serve as a hub for bioinformatics research and services. It is a component of INSDC, along with GenBank and DDBJ.

12.5.4 DNA Data Bank of Japan (DDBJ)

With sponsorship from the Japanese Ministry of Education, Culture, Sports, Science, and Technology, the DDBJ component of INSDC was founded in 1986 at the National Institute of Genetics (NIG) in Japan.

12.6 PROTEIN SEQUENCE DATABASES

Protein Information Resource (PIR) and UniProt are two of the various protein sequence databases that are accessible.

12.6.1 Protein Information Resource (PIR)

In order to study the evolutionary links between proteins, Margaret Dayhoff created the Protein Information Resource (PIR) at the National Biomedical Research Foundation (NBRF) in the 1960s. PIR makes its analysis tools for the protein database available to scientists without charge.

12.6.2 UniProt

The National Institutes of Health (NIH) awarded Protein Information Resource and its global partners, the European Bioinformatics Institute (EBI) and Swiss Institute of Bioinformatics (SIB), funding in 2002 so that they could combine the databases of PIR-PSD, SWISS-PROT, and TrEMBL to create UniProt. UniProt is divided into two sections: Translated EMBL (TrEMBL) and SWISS-PROT. Established in 1896, the most popular protein sequence database, developed jointly by the EMBL and the University of Geneva, is called SWISS-PROT (www.uniprot.org). The partnership relocated to the EBI, EMBL's UK outpost, in 1994. Translations of every coding sequence (CDS) in EMBL are included in TrEMBL, which makes use of the SWISS-PROT format. Its two main divisions are called REM-TrEMBL and SWISSPROT-TrEMBL (18).

12.7 STRUCTURAL DATABASES

The structural databases include 2D and 3D structures of nucleic acid and amino acids. Protein Data Bank (PDB), Molecular Modelling Database (MMDB), Vector Alignment Search Tool (VAST), Conserved Domain Database (CDD), and Nucleic Acid Structure Database (NDB) are among them. The highly renowned databases are discussed in more detail below.

12.7.1 Protein Data Bank (PDB)

Proteins and nucleic acids can be found in the PDB (www.rcsb.org/pdb/home/home.do), a source for three-dimensional structural information on large biological molecules. The Research Collaborators for Structural Bioinformatics (RCSB) founded it in 1971. Through the Internet, scientists from all over the world can readily and freely access the data they have submitted. The Worldwide Protein Data Bank (wwPDB) oversees the PDB. Every structure has been given a PDB ID, which is made up of four alphabetic and numeric characters. The final three characters can be either letters or digits, with the first character being a number (19).

12.8 LITERATURE DATABASES

Literature databases give us access to a global collection of life science research. MEDLINE, CiteXplore, OMIM, Patent abstracts, FlyBase archives, and other literature databases are available.

12.9 STRUCTURAL CLASSIFICATION OF PROTEINS DATABASE

It offers a hierarchical classification of protein structures that indicates how proteins have evolved together. www.ebi.ac.uk/pdbe/scop is the website for Structural Classification of Proteins (SCOP). Scientists can access Class, Architecture, Topology, and Homologous Superfamily (CATH) at no cost (www.cathdb.info).

12.10 CATEGORIZATION BASED ON STRUCTURE

From a technical perspective, databases can be categorized based on their file structures, including flat file, object-oriented, and relational databases. Flat file databases consist of ordered collections of files in a standard format, such as ASCII text files, made searchable through indexing. Notably, many sequence databases, like GenBank, utilize this format. Relational databases store information in tables, facilitating the extraction of specific data types across the database content. Object-oriented databases, more intricate in design, manage data as objects, enabling storage in various formats, from text to multimedia (20).

12.11 RELEVANCE OF DATABASES IN MICROBIOLOGY

Comprehensive information about microorganisms focused on comparative genomics can be found in a number of sizable databases. The databases specific to microbiologists or storing information

TABLE 12.1
Bioinformatics Tools for Microbial Research

Category	Database	URL	Brief Description
Microbial genomic resources	IMG	img.jgi.doe.gov	All-inclusive platform for metagenome and microbial genome annotation and analysis
	MicrobesOnline	www.microbesonline.org	Gateway for functional and comparative microbial genomics
	SEED	www.theseed.org	Portal for automatically annotating microbial genomes and curating genomic data
	GOLD	https://gold.jgi-psf.org	A source of thorough data regarding genome and metagenome sequencing initiatives
	MBGD		Using a predetermined collection of organisms, MBGD enables users to generate their orthologous categorization table.
	MiMeDB		Information regarding human microorganisms, the metabolites they are linked to, and the effects these metabolites have on people may be found in MiMeDB.
Protein families	CDD	www.ncbi.nlm.nih.gov/cdd	Conserved domain database
	Pfam	http://pfam.xfam.org	Protein family database
Protein–protein interactions	STRING	http://string-db.org	Protein association network database
Microbial diversity	RDP	http://rdp.cme.msu.edu	Database of 16S rRNA genes
	SILVA	http://www.arb-silva.de	Database of rRNA genes
	GREENGENES	http://greengenes.lbl.gov	Database of 16S rRNA genes
	BIGSdb	http://pubmlst.org/software/database/bigsdb	Database of bacterial isolate genomic sequences
	EBI metagenomics	www.ebi.ac.uk/metagenomics	Portal for metagenomics data submission and analysis
Model organisms	EcoCyc	http://EcoCyc.org	Knowledge based on *E. Coli* genome and metabolism
	RegulonDB	http://regulondb.ccg.unam.mx	Transcriptional regulation resource for *E. coli*
	Pseudomonas	http://pseudomonas.com	Database of *pseudomonas* genomes
Pathogenesis	PATRIC	www.patricbrc.org	Gateway to many prokaryotic infections
	EuPathDB	http://eupathdb.org	Gateway to several eukaryotic infections
	TBDB	www.tbdb.org	An integrated platform for research on tuberculosis
Transport and metabolism	TCDB	www.tcdb.org	Database for transporter classification
	TransportDB	www.membranetransport.org	Database for study of transporter proteins
	MetaCyc	http://metacyc.org	Database of metabolic pathways
	KEGG	www.genome.jp/kegg	Database of genomes with a focus on metabolism
Signal transduction and gene regulation	MiST	http://mistdb.com	Database for Microbial Signal Transduction
	SwissRegulon	http://swissregulon.unibas.ch	Annotations of regulatory locations across the genome in model organisms
	RegPrecise	http://regprecise.lbl.gov/RegPrecise	Regulon database in prokaryotic genomes

relevant to microbiologists are listed in Table 12.1. Some of them are discussed as follows. One of the most extensive resources dedicated to microbes is the Integrated Microbial Genomes (IMG) data warehouse. It incorporates genomes and associated metadata from viruses, eukaryotic microorganisms, bacteria, and archaea. IMG offers a method for comparing and analyzing genome data. It provides superior annotation over a vast array of microbial genomes using a broad range of bioinformatics tools and curation by subject experts. Furthermore, IMG integrates recently made proteomics and high-throughput RNA sequencing (RNA-seq) datasets and includes details on biosynthetic clusters and groups of genes that code for pathways leading to the manufacture of secondary metabolites in specific bacterial genomes. Additional resources, including Microbial Genome Database (MBGD), xBase, and MicrobesOnline, are designed to give the microbiology community the tools they need to analyze comparative genomic data. MicrobesOnline offers access to fitness and gene expression data in addition to thousands of bacterial, archaeal, and fungal genomes.

The world's most extensive suites of genomic and molecular biology data collections are provided by the unquestionable leaders, the European Molecular Biology Laboratory/European Bioinformatics Institute (EMBL-EBI) and the National Centre for Biotechnology Information (NCBI) at the National Institutes of Health in the United States.

Databases offer several utilities to microbiologists. All genomes of microorganisms that have been deposited to GenBank, EMBL Bank, or DNA Data Bank of Japan (DDBJ) are made available to the public through a range of text- and sequence-based search engines as essential components of the NCBI and EMBL-EBI collection of databases. Resources for NCBI literature are essential to microbiologists. The complete text of peer-reviewed journal articles is accessible via PubMed and PubMed Central, as well as NCBI Bookshelf, which provides free access to books and reports in their entirety (21). All sequenced microbial genomes are included in the NCBI Genome database, which also aggregates genome sequencing projects. Corresponding records in nonredundant (NR) databases of protein and nucleotide sequences and BioProject, a central repository for primary data from sequencing projects, are linked to these records. Archive for Sequence Reads (SRA) (22), at GEO, the archive for functional genomics datasets, and NCBI, a public repository for next-generation sequencing data, both offer an R-based web tool to assist users in analyzing their data (23). The most widely used sequence database search engine, BLAST, now allows users to search for sequence similarity against any taxonomic group directly from the NCBI website.

A user-centered design is another feature that EMBL-EBI offers for its cross-linked databases and tools. Numerous databases are arranged according to numerous categories: systems (reactions, interactions, and pathways), metabolites (chemogenomics and metabolomics), proteins (sequences, families, and structures), and DNA and RNA (genes, genomes, and expression data). The Metagenomics portal at EMBL-EBI is among the most fascinating tools for microbiologists (24). Researchers can submit, archive, and analyze genomic data from a variety of locations using this tool. Additionally, EMBL-EBI helps to establish several databases with specific uses for microbiologists.

Many hundreds to over ten thousand proteins are encoded by the genomes of different bacteria. Their biological role can be defined by grouping them into different protein families that are distinguished by conserved domains and evolutionary links. Many resources try to categorize proteins based on their structure or sequence. Pfam (25), SMART (26), and TIGRFAM (27) are carefully curated collections of models that identify conserved proteins and protein domains. The phylogenetic classification of proteins is provided by the Clusters of Orthologous Groups of Proteins (COGs) database. Increased microbial genome coverage significantly improves the most recent version of COGs (28, 29). A Conserved Domain Database (CDD) at NCBI combined information from all four of the aforementioned databases with its models to create a single, cohesive collection of these models. The EMBL-EBI InterPro database for protein sequence analysis and classification (30) has a similar function and, by combining data from multiple sources, offers thorough information on protein domains, regions, motifs, and functional locations.

A deeper comprehension of biological macromolecules' functions results from an understanding of their structure. The Protein Data Bank (PDB), a crucial central repository of atomic coordinates and pertinent information on the three-dimensional (3D) structures of proteins, nucleic acids, and their complexes, is in addition to the Research Collaboratory for Structural Bioinformatics

(RCSB) (31) the SCOP (Structural Classification of Proteins) (32) and CATH (Class, Architecture, Topology, and Homology) (33) databases, which provide extensive resources for protein structure classification. Predicting protein function also involves identifying orthologs, or genes and products in various species that diverged from a common ancestor gene through speciation, and paralogs, or genes and products originating from gene duplication. The OrthoDB database offers thorough details on gene/protein orthology in addition to COGs (34).

One of the most important processes in modern biology is the reconstruction of metabolism from the genome sequence, which combines the capacity of comparative genome analysis with the abundance of information gained from experimental biochemistry. Kyoto Genome and Gene Encyclopaedia (KEGG) (35) is among the pioneers in this domain. The annotation platform for almost all bacterial and archaeal genome sequencing projects includes KEGG metabolic maps. A sizable BioCyc collection's metabolic arm is called MetaCyc (36). Protein–protein interactions are essential for biological processes such as signal transduction, metabolism, and transport. For functional protein networks, the STRING database is the most comprehensive source (37). Information regarding genetic and protein interactions in a variety of model organisms, such as viruses, eukaryotic microorganisms, *B. subtilis*, and *E. coli*, is available in the BioGRID database (38).

12.12 SEQUENCE ALIGNMENT

Sequence alignment stands as a vibrant research domain within bioinformatics. It serves as a cornerstone for numerous critical tasks. These tasks include phylogenetic analysis and prediction of biological macromolecule functions and structures. They encompass DNA, RNA, and proteins that are the fundamental building blocks of all living organisms. In its essence, sequence alignment entails comparing and identifying similarities between biological sequences, with the nature of these similarities contingent upon the specific objectives of the alignment process. Given its versatility, sequence alignment finds widespread utility across various bioinformatics applications (39).

Typically, sequence alignment serves as the initial step in understanding the molecular phylogeny of unknown sequences. This involves aligning the unknown sequence with known database sequences to infer common regions, as residues/sequence-block crucial for function and structure are typically conserved through natural selection. Alignment aims to increase the quantity of identical or similar residues matched between two or more sequences, which can be nucleotide sequences (DNA or RNA) or amino acid sequences (proteins). The alignment process may introduce gaps, signifying possible evolutionary events such as insertions or deletions (indels), translocations, and inversions. Alignment is also integral to sequencing techniques, particularly in next-generation sequencing (NGS), where short reads need to be aligned to reference sequences for gene annotation and transcriptome analysis.

12.12.1 Sequence Alignment Types

Sequence alignment is carried out in two main ways: pairwise sequence alignment (PSA) and multiple sequence alignment (MSA). PSA compares two sequences at a time, while MSA aligns multiple related sequences. Multiple sequence alignment, or MSA, is the alignment of three or more biological sequences (proteins or nucleic acids) of similar length (see Table 12.2). On the other hand, regions of similarity that might point to structural, functional, or evolutionary links between two biological sequences are found using pairwise sequence alignment techniques. MSA, being more comprehensive, offers advantages in capturing biological information from sequence families and is essential for comparative genomic analyses, evolutionary relationship estimation, homology inference, and ancestral sequence profiling (40). MSA is useful for deriving motifs and is frequently used to find conserved subsequences in gene sequences that are quite similar. Proteins are characterized by motifs, which are distinct subsequences that have been discovered to be highly helpful in identifying genes that share the same functionality. Once the sequences have been aligned,

TABLE 12.2
Bioinformatics Tools for Multiple Sequence Alignment

Tool	Description
Clustal Omega	A novel MSA tool that generates alignments through the use of HMM profile-profile algorithms and seeded guide trees. Ideal for alignments of medium to large sizes.
Cons (EMBOSS)	A consensus sequence is produced by EMBOSS Cons from a protein or nucleotide multiple alignment.
Kalign	Extremely quick MSA tool focused on nearby areas. Ideal for extensive alignments.
MAFFT	Fast Fourier transforms are used by the MSA tool. Ideal for alignments of medium to large sizes.
MUSCLE	Precise MSA instrument that works particularly well with proteins. In line with middle alignments.
MView	Use the MView program to convert a Sequence Similarity Search result into a Multiple Sequence Alignment or to reformat a Multiple Sequence Alignment.
T-Coffee	A consistency-based MSA tool designed to lessen the drawbacks of progressive alignment techniques. Ideal for minor alignments.
WebPRANK	A novel phylogeny-aware multiple sequence alignment program developed by the EBI uses evolutionary data to assist in determining the locations of insertions and deletions.

the conserved subsequences of the functionally equivalent genes from other organisms are identified to generate motif sequences.

12.12.2 Amino Acid-based Sequence Alignment

Alignment at the amino acid level is particularly relevant due to proteins' central role in biological function, making sequence alignment a crucial tool in structural biology as well. Different scoring methods, such as nucleotide scoring for identifying identical bases and similarity scoring for proteins based on physicochemical properties, are employed to assess identity or similarity levels between sequences. Notably, substitution matrices like point accepted mutation (PAM) (41) and BLOcked SUbstitution Matrix (BLOSUM) (42) are frequently used for protein sequence alignment (43).

Different biochemical and physical characteristics of amino acids affect how replaceable they are compared to one another during evolution. The possibility of being substituted depends upon the physiochemical characteristics of the amino acids. For example, hydrophobic amino acids may easily be substituted by other hydrophobic amino acids as compared to charged amino acids. These attributes are reflected in the substitution matrices. Alignments of closely similar proteins worldwide serve as the foundation for the PAM (percent acceptable mutations) matrices. The number of steps needed for a specific percent change is indicated after a PAM matrix. PAM1 denotes 99% identity with 1% variation among the matched sequences. This indicates that PAM112 will identify sequences with 40% identity that are more distantly related than PAM23 (80% identity). Due to its global alignment foundation, PAM matrices are most appropriate for global pairwise alignments found in evolutionary research. The foundation of BLOSUM (BLOcked SUbstitution Matrix) matrices is local alignment. The minimal percent identity permitted in the collection of aligned sequences used to construct a BLOSUM matrix is indicated by the number at the end of the matrix. As a result, BLOSUM50 will be able to identify sequences that are more distantly related than BLOSUM80.

12.12.3 Nucleotide-based Sequence Alignment

Nucleotide sequence alignment is a valuable technique for matching and coordinating DNA sequencing data with either a parental DNA template or the sequence of a known gene. Nucleotide substitution matrices assign greater weight to transversions (G to C, G to T, A to C, A to T, and vice versa) than to transitions (G to A or C to T, and vice versa). This is mostly because purine nucleotides are larger than pyrimidine nucleotides. The Jukes and Cantor model, which assumes equal odds for the substitution of each of the four nucleotides, is the most straightforward model. The general reversible model, HKY, Kimura, Tamura-Nei, and Tamura models all have the transversion/transition bias. Models of Tamura-Nei, equal input, and generic reversible models all include an extra account for the frequency of nucleotides. The nucleotides under comparison have these measured frequencies. Different probabilities of modifications are allowed for each of the 12 nucleotide pairs in the unrestricted model.

12.12.4 Sequence Alignment Methods

Sequence alignment methods can be broadly categorized into global and local alignment. Global alignment (44) assesses similarity across the entire length of sequences, while local alignment, focuses on identifying similar regions within sequences. To identify sequence sections that are comparable or identical between two DNA, RNA, or protein sequences, pairwise local sequence alignment techniques are utilized. Aligning pairings of mutant DNA or protein sequences is a common use.

The choice of whether to establish a pairwise alignment locally or globally must be made before proceeding. A pairwise alignment is seen as global if it is known that the sequences are homologous throughout. It makes sense to align both sequences to their complete length in this case. An entire gene sequence can be sequenced and a global alignment can be created if early data indicates that the genes encode the same proteins. A local alignment is necessary when it is known that one sequence is shorter than the other and that their lengths cannot be coupled. This may occur when one gene's entire DNA sequence is unknown or when the domain structures of the proteins that the genes encode for differ significantly.

The pairwise alignment can be computed using dynamic programming. If all possible scenarios for the pairwise alignment of two sequences were to be explored, there would be around 1,059 choices for two nucleotide sequences of length 100. This is almost equivalent to the amount of molecules in the Milky Way, and the challenge would be too big even for the biggest computer. Additionally, it is not necessary to take into account irrelevant scenarios, such as the idea that only gaps in one sequence would allow a single nucleotide to pair with another.

Dynamic programming is used to minimize the computational effort while maintaining a thorough analysis. Only pertinent comparisons (high scores) are tested by the computer, which retains the highest scores it has already computed in memory. The most pertinent dynamic programming algorithms were developed by Smith and Waterman and by Wunsch and Needleman, respectively, for comparing local and global pairwise alignments. The Needleman and Wunsch approach is used to create global pairwise alignments once it is established that the sequences exactly symbolize a gene or protein in its entirety (a small number of amino acid or nucleotide variations are accepted). To create local alignments, the Smith and Waterman algorithm is applied to incomplete sequences or sequences of undetermined length. The gap penalty function adjustments for both algorithms rely on prior knowledge about domain structures, repeats, and other features. A combination of methods and computer programs implementing them can be applied to nucleotide and amino acid sequences.

Dynamic programming (DP) (45) is commonly utilized for PSA to achieve optimal alignment, but its complexity poses challenges for MSA, where exact methods become computationally infeasible for large datasets. As a result, most MSA methods rely on heuristic or approximate approaches, with three main categories: exact, progressive, and iterative alignment. While exact algorithms strive for high-quality alignments close to optimal, they are limited by computational complexity, leading to ongoing research efforts to develop efficient MSA methods capable of handling large datasets and complex sequence relationships.

12.12.5 Tools for Sequence Alignment

Common software tools used in bioinformatics for sequence analysis include BLAST, Clustal Omega, BioPython, and MEGA. BLAST, or Basic Local Alignment Search Tool, is a widely used software tool in bioinformatics.

12.12.6 Sequence Alignment in Microbiology

The pairwise gene alignment and common sequence search methods are used to annotate genes with correct structure and function. The four most widely used algorithms for functional annotation of genes are BLAST and its variants; Smith-Waterman alignment and its variations, a dynamic programming technique; index-based scheme FASTA and its variations; and BLOCKS, which identifies protein patterns by multiple sequence alignment of conserved domains to find motifs.

The most widely used sequence database search engine, BLAST, can now compare sequences to any taxonomic group. To find the greatest matching nonrandom segment, a BLAST search expands numerous plausible seed points (greater than four nucleotides) that match. This is done with the use of scoring matrices like BLOSUM or PAM. This facility is available on the NCBI web page. An individual can opt to look for resemblance solely in *Proteobacteria* or *Firmicutes*, or even in a specific organism like *Bacillus subtilis* or *Escherichia coli.* Alternatively, an organism or any taxonomic group might be left out of the search. The NCBI BLAST tool facilitates the exploration of extensive metagenomics data by enabling users to search genetic data from environmental samples. Bench microbiologists can create and evaluate PCR primers with the use of NCBI Primer-BLAST. Microbiologists can also benefit from the NCBI Taxonomy database, which has data for almost all formally documented prokaryotic species and roughly 10% of eukaryotic species at every taxonomic level, from super kingdoms to subspecies. The NCBI Virus Variation database connects West Nile, dengue, and influenza virus genome sequence data to related literature, structures, and population studies.

12.13 PHYLOGENETIC ANALYSIS

The term "phylogenetics" comes from the Greek *phyle*, meaning "tribe" or "race," while "genetics" means "relative to birth" from *genesis* or "birth." The term refers to the investigation of evolutionary relationships between various organismal groups, such as species or populations. Essentially, phylogenetic analysis elucidates how a group of organisms, such as a family, may have evolved. Understanding phylogenetic relationships among species is fundamental to various studies in biology. An accurate phylogenetic tree serves as a cornerstone for comprehending major evolutionary transitions, It is crucial for deducing the formation of novel genes, identifying molecular adaptability, comprehending the evolution of physical characteristics, and reconstructing demographic shifts in recently separated species, among other things. Examples of these include the emergence of new body plans or metabolic pathways (46).

Phylogenetic analysis is widely regarded as a highly reliable and important bioinformatics tool due to its straightforward representation and ease of data-handling. The simplicity of the tree representation simplifies comprehension and facilitates its representation across different fields of biology, making phylogenetic analysis indispensable. A phylogenetic tree serves as a two-dimensional representation of relatedness among various biological species, depicting their evolutionary relationships and indicating their chronological origins. Phylogenetic trees can be represented in three forms: phylogram, dendrogram, and cladogram (47).

12.13.1 Phylogenetic Techniques

Phylogenetic trees are constructed primarily using distance-based or character-based methods. Distance-based methods, including UPGMA (unweighted paired group method with arithmetic mean), NJ (neighbor-joining), ME (minimum evolution method), and FM (Fitch-Margoliash method), focus on computing the distance between sequences and arranging them accordingly. Character-based methods maximize the way that real data patterns are distributed for each character, with the most used methods being maximum parsimony (MP) and maximum likelihood (ML).

Comparing different tree-building techniques requires considering a number of factors, including processing speed, consistency of estimated topology, statistical consistency of phylogenetic trees, probability of achieving the proper topology, and reliability of estimated branch length. The NJ method stands out for its computational speed, especially in handling large datasets with bootstrap tests, while MP, ME, and ML methods exhaustively explore all possible topologies, making them less efficient for large sequence sets. Advanced algorithms have been developed to enhance the efficiency of ME and MP methods, such as the branch and bound method and algorithms proposed by Rzhetsky and Nei. NJ and ME methods exhibit consistency in estimating trees, while MP methods may be inconsistent. ML methods offer flexibility in choosing evolutionary models but are computationally intensive and time-consuming. Overall, the choice of tree-building method depends on factors such as dataset size, computational resources, and the desired balance between speed and accuracy (48).

12.14 TRANSCRIPTOMICS

Novel methodologies that redefine what is possible and surpass previous technologies have defined the area of transcriptomics every ten years or so. In 1991, 609 human brain mRNA sequences were revealed, marking the first attempt to partially sequence the human transcriptome (49). Two human transcriptomes, including millions of transcript-derived sequences spanning 16,000 genes, were released in 2008 (50), and transcriptomes for hundreds of people had been made public by 2015 (51). These days, it is common practice to create transcriptomes of various disease states, organs, or even individual cells (52, 53). The rapid creation of new technologies with increased sensitivity and efficiency has propelled the expansion of transcriptomics.

In the 1990s, the term "transcriptome" was coined (54). Among the most sophisticated disciplines in the post-genomic age is transcriptomics. The transcriptome is the whole set of messenger, transfer, ribosomal, and other non-coding RNA transcripts in a certain cell type or tissue at a given developmental stage and/or physiological condition. With an emphasis on RNA-level gene expression, transcriptomics provides genome-wide details on gene structure and function, enabling the identification of the molecular mechanisms underlying certain biological processes. Our comprehension of the RNA-based gene regulation network has gradually improved thanks to transcriptome research and the advancement of next-generation high-throughput sequencing technology (55).

Serial analysis of gene expression (SAGE), created in 1995, was among the initial transcriptomic methods based on sequencing. Sanger sequencing of concatenated random transcript segments was how it worked (56). The fragments were compared to known genes to quantify the transcripts. A high-throughput sequencing form of SAGE known as digital gene expression analysis was also briefly employed. High-throughput sequencing of whole transcripts, however, largely surpassed previous techniques and offered more details on transcript structure, such as splice variants (57). The two most widely used contemporary methods, RNA-Seq and microarrays, were developed in the latter part of the 1990s and the early 2000s. The first microarrays were released in 1995, and they are used to assess the surpluses of a certain set of transcripts by hybridizing them to a variety of complementary probes (58). During this time, many microarrays covering known genes in models or economically significant organisms were developed. Additional genes may be examined on a single array thanks to developments in array design and manufacturing that increased probe specificity. The precision of the measurement and sensitivity for transcripts with low abundance were improved by developments in fluorescence detection (59).

The term "RNA-Seq" describes the sequencing of cDNA transcripts, where the quantity of each transcript is determined by counting it. As such, the advancement of high-throughput sequencing technologies has had a significant impact on the approach. An early form, known as "massively parallel signature sequencing" (MPSS), relied on producing 16–20 bp sequences through a convoluted series of hybridizations. It was employed in 2004 to confirm the expression of 104 genes in *Arabidopsis thaliana* (60). A lot of raw sequence reads are produced by RNA-Seq experiments; these reads must be processed to produce meaningful data. Depending on the objectives and design of the experiment, a variety of bioinformatics software packages are typically needed for data analysis. The following four steps comprise the process breakdown: differential expression, quantification, alignment, and quality control (61).

12.14.1 Quality Control

Because sequence readings aren't always accurate, it's necessary to assess each base's accuracy for further analysis. The raw data are analyzed to look for things like a surprisingly elevated read duplication rate, overrepresentation of specific short sequence motifs (k-mers), high-quality scores for base calls, and guanine-cytosine content that matches the expected distribution (62). For sequence quality analysis, there are several choices available, such as the software packages FastQC and FaQCs. Any abnormalities found can be trimmed away or marked for additional care at a later step.

12.14.2 Alignment

Transcript sequences are matched to a reference genome, or de novo aligned to each other in the absence of a reference, to establish a connection between sequence read abundance and gene expression. The main difficulties facing alignment software are having enough speed to align over 109 short sequences in a reasonable amount of time, flexibility to handle eukaryotic mRNA intron splicing, and accurate read assignment that maps to numerous sites. Software developments have significantly resolved these problems, and multimapping reads are being further reduced by longer sequencing reads. The EBI maintains a list of high-throughput sequence aligners that are currently accessible (63). Since mature mRNA lacks intron sequences, which need to be linked to a reference genome, primary transcript mRNA sequences from eukaryotes must be handled cautiously. Using information from known intron splice sites and canonical splice site sequences, short-read aligners carry out a second cycle of alignments exclusively intended to find splice junctions. By identifying intron splice junctions, reads can be more accurately matched to the reference genome and gene expression estimations can be made more accurately since they are not misaligned across splice junctions or incorrectly discarded. Splice-aware alignments also allow for the discovery of isoform abundance changes that would otherwise be missed in a bulked analysis, as gene regulation may take place at the mRNA isoform level (64). Without the need for a reference genome, de novo assembly can be used to match reads to one another to create full-length transcript sequences. Better measures of assembly completeness are annotation-based metrics, including contig reciprocal best hit count. De novo assembly can serve for quantitative gene expression analysis and subsequent sequence alignment techniques.

12.14.3 Quantification

Sequence alignments can be quantified at the transcript, exon, or gene levels. A table of read counts for each feature that the software received, such as genes in a standard feature format file, is one of the typical outputs. HTSeq software makes it simple to calculate exon and gene read counts (65). Transcript-level quantification is more complex and involves probabilistic techniques, such as the use of Cufflinks software, to estimate transcript isoform abundance from short read data. Certain quantification techniques can completely avoid the requirement for a precise alignment of a read to a reference sequence. The Kallisto approach runs two orders of magnitude faster and requires less computing power than related methods like Tophat/Cufflinks because it integrates quantification and pseudo alignment into a single step (66).

12.14.4 Differential Expression

Differential gene expression is then assessed by normalizing, modeling, and statistically evaluating the data when quantitative counts of each transcript are obtained. Next, differential gene expression is evaluated by data normalization, modeling, and statistical analysis when numerical totals for every transcript are obtained.

12.14.5 Validation

An independent method that can be recognized and evaluated statistically, such as quantitative PCR (qPCR), can be used to validate transcriptome findings (67). Gene expression for both the target gene and control genes is quantified in relation to predetermined benchmarks. Generally, qPCR validation of RNA-Seq data has demonstrated a strong correlation between various RNA-Seq techniques.

12.14.6 Technology for Transcriptome Analysis

For nearly two decades, methods have changed: from the original expression sequencing tag (EST) scheme to gene chips and, more recently, RNA-seq. Transcriptome analysis becomes more affordable with increased throughput, improved sensitivity, and reduced starting RNA (68). In the early days of genome research, information for genome annotation was obtained via Sanger sequencing of EST or cDNA libraries. EST approaches are not able to provide transcriptome quantitative analysis due to cost and throughput restrictions. Using CAGE and SAGE, sequential study of gene expression, many 3′ and 5′ cDNA ends were joined to form a single clone. Therefore, such limitations are removed, and quantitative analysis is made possible by the fact that many sequence tags can be recovered from a single Sanger sequencing procedure. However, DNA microarray quickly supplanted CAGE and SAGE because of the high cost of Sanger sequencing and the challenge of mapping the short sequence (~20 bp) tags to the genome (69).

Nucleic acid hybridization forms the basis of the DNA microarray or chip technique. To determine the amount of RNA present, the fluorescence density is evaluated following the incubation of fluorescently labeled cDNAs with oligonucleotide probes on the chip. With a precision of a few nucleotides, the genome tiling array is an unbiased design that detects expression of the entire genome through a series of overlapping oligonucleotide probes. This is accomplished without requiring an earlier understanding of genome transcription data. For large genomes, tiling arrays are expensive. Another limitation of the hybridization process is a high background, which prevents the hybridization process from distinguishing between RNA molecules with high sequence similarity (70).

Our understanding of how genes are expressed has undergone a fundamental shift as a result of the study of transcriptomics. Over the last three decades, new technologies have increased the range of possible studies, and when these technologies are combined with other omics technologies, the cellular world's complex workings become more fully understood. Small laboratories can now conduct transcriptomics studies thanks to their falling costs, and large-scale transcriptomics consortia can conduct studies that compare the transcriptomes of thousands of different organisms, tissues, or environmental situations. With advancements in sequencing technologies, this trend is probably here to stay.

12.15 PROTEOMICS

Since the earliest days of biological inquiry, the dynamic role that molecules play in supporting life has been observed. The term "protein," which comes from the Greek word *proteios*, which means "the first rank," was given to these molecules by Berzelius in 1838 to emphasize their significance (71). The total amount of proteins in a cell that are identified based on their location, interactions, post-translational modifications, and turnover during a specific period is known as the "proteome." Marc Wilkins coined the word "proteomics" in 1996 to refer to the "PROTein complement of a genOME" (72). Proteomics is crucial for prognosis, early disease identification, and monitoring of the course of illness. In addition, it plays a crucial part in the creation of drugs as target molecules. Proteomics is the study of proteins, including their expression, structure, activities, interactions, and changes throughout time. As biological effectors, proteins are regulated and controlled by the host translation machinery in addition to the matching mRNA levels. Proteomics is therefore thought to be the most pertinent dataset for characterizing biological systems (73).

Size exclusion chromatography (SEC), affinity chromatography, and ion exchange chromatography (IEC) are chromatography-based conventional methods for protein purification. Western blotting and the enzyme-linked immunosorbent assay (ELISA) can be used to analyze specific proteins. These techniques might only be able to analyze a limited variety of unique proteins, and they can't tell you how much a protein is expressed. Complex protein samples are separated using two-dimensional gel electrophoresis (2-DE), two-dimensional differential gel electrophoresis (2D-DIGE), and sodium dodecyl sulphate polyacrylamide gel electrophoresis (SDS-PAGE). Although high-throughput and speedy expression analysis has been made possible by the development of protein microarrays or chips, it is still challenging to properly progress a protein microarray to study the function of a full genome. Different proteomics techniques, such as mass spectrometry (MS), have been developed to more sensitively analyze complicated protein mixtures. Furthermore, the technique of Edman degradation has been established to ascertain a protein's amino acid sequence. Techniques for quantitative proteomics have recently evolved, including an isobaric tag for relative and absolute quantification (iTRAQ), stable isotope labeling with amino acids in cell culture (SILAC), and isotope-coded affinity tag labeling (ICAT). Nuclear magnetic resonance (NMR) spectroscopy and X-ray crystallography are two important high-throughput methods that offer the three-dimensional (3D) structure of proteins, which may be useful to understand their biological function (74).

For 3D structure prediction, protein domain and motif analysis, quick protein–protein interaction analysis, and MS data interpretation, several bioinformatics tools have been created. For sequence and structural alignment to determine the evolutionary link, the alignment tools are useful (75). Proteomics, the high-throughput analysis of proteins, is seeing rapid technological advancements that lead to better scalability, higher-resolution spectra capture, and more effective protein extraction. Proteomics has developed into a stand-alone discovery platform thanks to these. One of the most notable instances is the identification of the initial human proteomes by high-resolution mass spectrometry (MS). These developments produce more data, but the quality may degrade, posing new computing difficulties. It is important to think about how bioinformatics will change considering these advancements to address these new difficulties. Among the significant accomplishments is the widespread adoption of the mzML common data standard for proteomics (76), while Mega open-access software, such as OpenMS, offers unparalleled analytical flexibility and cross-hardware compatibility.

Because bioinformatics is a crucial part of proteomics, its ramifications have grown over time with the introduction of high-throughput techniques that rely on robust data analysis. This is a new and developing field that is advancing the discovery process and offering innovative techniques to handle massive and diverse proteome data (77). A class of antibacterial enzymes called endolysins is finding use as a beneficial tool to stop the proliferation of microorganisms resistant to many antibiotics. Mutagenesis, domain shifting, and gene shuffles can all be used to expand or change the antibacterial property. In-silico analysis for protein domains found in prophage and phage endolysins has shown difficulty in building specific endolysins. Multiple sequence alignment has been used to determine the sequence type with domain layout and conserved amino acids after the combination of domains has been examined. Additionally, research has been done on the quantity, kinds, and presence of binding domains in the endolysin sequence. The distribution of plant food allergens into protein families and the identification of the conserved surface necessary for IgE cross-reactivity were computed using an in-silico analytical approach. Four families were identified from the plant food allergen sequences, indicating the importance of biological activities and conserved structures in promoting allergic qualities (78).

To acquire a particular collection of peptides for MS, the proteins are removed from the sample and digested using one or more proteases. When it's necessary to analyze a particular group of proteins or to reduce the complexity of the sample, further procedures like fractionation and enrichment might be implemented at the protein or peptide level. Mass spectrometry combined with liquid chromatography (LC-MS) is used to analyze the extracted peptides. Typical methods involve either quantitative analysis for a specific group of proteins using targeted MS or study of the proteome with deep coverage using shotgun MS (79). The sequence information inferred from the resultant spectra is crucial for protein identification. The acquired data may be shown as a three-dimensional map that includes fragmentation spectra, retention times (RT), and peptide intensities in addition to the mass-to-charge (m/z) ratio. The most extensive proteome archives, such as the PeptideAtlas project, Proteome Commons, and PRIDE proteomics identification database, offer direct access to the majority of recorded data and are useful resources for data mining (80).

Proteomics has seen enormously beneficial advancements during the past few years. Increased proteome coverage, speed, and sensitivity are provided by the technologies. The combination of these methods has allowed for the successful purification, characterization, quantification, sequencing, and structural and bioinformatic analysis of a vast array of proteins present in both prokaryotic and eukaryotic species. The application of proteomics technology has proved beneficial to all biological science-related sectors. To increase the performance and reproducibility of well-known proteomics techniques, more work is still needed.

12.16 BIOINFORMATICS METHODS FOR DECIPHERING ANTIMICROBIAL RESISTANCE MOLECULAR MECHANISMS

Bioinformatics methods are vital for understanding the molecular mechanisms underlying antimicrobial resistance. By analyzing vast datasets, these techniques enable researchers to identify genetic mutations, pathways, and mechanisms contributing to antimicrobial resistance. Through computational analysis and modeling, bioinformatics provides insights into how resistance develops and spreads, informing strategies for combating antimicrobial resistance.

12.16.1 Innovative Bioinformatics Strategies to Prevent Antibiotic Resistance

Antibiotic resistance poses a grave global health threat, demanding the exploration of innovative strategies to mitigate its impact. The rise of resistant strains, such as methicillin-resistant *Staphylococcus aureus* and vancomycin-resistant *Enterococcus faecium*, has led to complications in intensive care units, escalating medical costs and jeopardizing patient lives. The emergence of these resistant strains, coupled with the challenges in finding new antimicrobials, has raised alarms within the scientific community. The majority of the current procedures used to create new antibiotics emphasize innovative methods for drug design that are based on prodrugs or logical compound design (81). In this context, the use of computational methods for analyzing biological data emerges as a potent ally with immense promise. Bioinformatics has a significant part in the multifaceted and data-driven fight against antibiotic resistance, expanding on the broad application of genomic methods in diagnosing and managing viral, bacterial, and fungal infections. Applications of bioinformatics span pathogen identification, virulence factor detection, resistance analysis, and strain typing (82).

12.16.2 Target Identification and Validation

Conventional methods for identifying drug–target interactions through biological experiments were both costly and labor-intensive, prompting a shift toward computational approaches (83). Computational methods have proven highly efficient and cost-effective in predicting drug–target interactions. The process of drug discovery starts with the identification of a possible biological target and understanding its role in the disease. A target, typically a biochemical entity like a protein, RNA, or gene, serves as the binding site for a drug, inducing a shift in physiology. For a potential medication to bind, a target needs to have an active site or binding pocket. A target that meets commercial criteria and is able to be securely and effectively drugged is desirable (84).

While drug–target interaction (DTI) forms the foundation of drug discovery, experimental determination of DTIs is a time-consuming and expensive endeavor (85). In the last ten years, various computational methods have emerged to predict potential DTIs with high efficiency and low costs. These techniques can be divided into a number of groups, including network-based, pharmacophore-based, molecular docking-based, similarity-based, and machine learning-based techniques (86). Especially, network-based techniques show clear advantages over other approaches since they do not depend on the three-dimensional architectures of targets and negative samples (87).

12.16.3 Antimicrobial Drug Repurposing

The ongoing need for novel antimicrobial medications that are effective against bacteria resistant to several antibiotics has compelled researchers to investigate a range of drug discovery approaches, such as drug repurposing, which involves finding new uses for already approved medications (88). Although relatively young in the antimicrobial field, drug repurposing is gaining traction in both the public and private sectors (89). Antimicrobial drug repurposing, sometimes referred to as drug repositioning or drug reprofiling, is the process of developing new therapeutic applications for antimicrobial drugs that are already on the market. This strategy addresses rising antimicrobial resistance and the limited supply of innovative antimicrobial medicines, contributing to the global health challenge. In drug repurposing, identifying drugs with established safety records and mechanisms of action could be applied to the treatment of different microbiological targets or diseases. By using existing information and research on existing medications, repurposing accelerates the discovery of new treatment options, bypassing many early stages of drug development. Furthermore, repurposing may be able to combat drug resistance by employing drugs with unique modes of action. There are frequent checkpoints in the process, such as comprehending the drug's mode of action, carrying out high-throughput screening, preclinical studies, and clinical trials to assess efficacy and safety. The potential of repurposing antimicrobial drugs to address unmet requirements in infectious diseases is demonstrated by successful cases of this approach. Overall, antimicrobial drug repurposing, offering faster and more cost-effective alternatives to conventional drug development methods, presents a promising approach to combat antimicrobial resistance (90).

12.16.4 Culture-independent Gene Surveillance for Antibiotic Resistance Made Possible by Metagenomics

The application of bioinformatics, artificial intelligence (AI), and metagenomic profiling—the untargeted sequencing of bacterial communities—in antibiotic resistance research has garnered more interest recently. Metagenomics facilitates the observation of microbial populations without regard to culture, allowing for the study of all bacteria potentially harboring antibiotic resistance genes (ARGs) (91). With gene identification in metagenomic data, along with using sequence similarity-based grouping, gene catalogs can be produced (92). To ascertain whether or not ARGs are present in the microbial population and how abundant they are, these catalogs can then be compared with ARG databases. Furthermore, metagenomics research clarifies how ARGs move horizontally across bacterial genomes (93). Essentially, these methods take advantage of differences in phylogenetic and genetic information between genomic sequences that are inherited vertically and horizontally and have different evolutionary histories. The addition of techniques that are dependent on culture can reinforce metagenomic approaches. The examination of bacterial ARG profiles linked to hospitals, for example, has been made possible by long-read metagenomic sequencing of hospital samples after culture-based enrichment, revealing unique ARG combinations. Culture-based methods also make it easier to distinguish between ARGs that come from viable and nonviable sources, which affects how and to what extent they spread (94).

12.16.5 Artificial Intelligence Facilitates the Discovery of New Antibiotic Compounds

The discovery and development of antibiotics that can treat illnesses caused by bacteria, especially those that are resistant to current antibiotics, is a critical component of the reaction of humans to antibiotic resistance. The conventional methods for novel drug discovery typically revolve around two approaches: target-based screening, which concentrates on creating medications to target a specific molecular target like a gene or protein; and phenotypic-based methods, which assess chemical compounds for their ability to induce desired phenotypic changes (95). Target-based screening techniques have been a common tool used in the antibiotic drug discovery era to find broad-spectrum antibiotics that are effective against a variety of possible infections (96). However, by having an impact on other non-pathogenic bacteria, broad-spectrum antibiotics can aid in the development of antibiotic resistance (97). Furthermore, the ongoing emphasis on ubiquitous targets would not be sufficient to combat the increasing incidence of multidrug-resistant bacteria with diverse resistance mechanisms. Drug repurposing screens are a relatively recent method of identifying novel antibiotics; in comparison to established methods, they are quick and affordable (98). Machine learning algorithms play a pivotal role in guiding the repurposing of existing medications as antibiotics, even when the tested drugs have structural differences or were originally intended for distinct objectives. This is exemplified by the discovery of the unique antibacterial molecule Halicin. Utilizing algorithms trained on a wide range of molecular datasets, the researchers were able to forecast molecular features from the Drug Repurposing Hub, including antibiotic activity. Among other high-priority pathogens, halicin showed activity both in vitro and in vivo against *Acinetobacter baumannii* (99, 100). With the increasing use of machine learning models in antibiotic and antimicrobial peptide discovery, the Halicin breakthrough may signify a broader trend where future developments in AI techniques will help in the discovery of new antibiotics by forecasting antibacterial characteristics from molecular or metagenomic data (101).

12.16.6 Pathogenic Interaction and Drug Targeting (Drug Development and Antimicrobial Agents Targeting)

The hardest and most frustrating part of the drug research and discovery process is finding appropriate and bioactive pharmaceutical compounds among the around 10^{60} compounds in the large chemical space. The process of finding and developing new drugs is also seen to be a time- and money-consuming undertaking. The most frustrating thing is that, usually, nine out of ten therapeutic compounds are not approved by regulators or do not complete phase II clinical studies (102).

To address these challenges, the integration of AI-based and bioinformatics tools and techniques has emerged as a promising solution. Small molecule design, medication dosage and related efficacy, bioactive agent prediction, interactions between proteins, the discovery of protein folding and misfolding, structure and ligand-based virtual screening, QSAR modelling, drug repurposing, bioactive property and toxicity prediction, and mode of action identification of drug compounds are just a few of the stages in which these tools and techniques are used in the drug development process. These are explained in more detail below.

12.16.7 Design of Small Molecules and Peptide Synthesis

Peptides, which are tiny, physiologically active chains made up of roughly two to fifty amino acids, are being studied more and more for potential medicinal uses because of their capacity to penetrate cellular membranes and specifically target locations (103). Researchers have found new peptides in recent years by using AI and other bioinformatic methods. Yan created the Deep-AmPEP30 platform, which uses DL to detect short AMPs, or short antimicrobial peptides (104). Using Deep-AmPEP30, Yan was able to identify novel AMPs from the genome sequence of *Candida glabrate*, a fungal disease that affects the gastrointestinal tract. This CNN-driven technique predicts short AMPs from DNA sequence data. Similarly, Plisson (105) combined machine learning algorithms with an outlier detection technique to discover AMPs with non-hemolytic profiles. Furthermore, Kavousi et al. created IAMPE, a web server for AMP identification that integrates physicochemical and 13CNMR-based peptide properties as input to machine learning algorithms (106). Furthermore, employing AI-based technologies, tiny compounds with extremely low molecular weights, such as peptides, are being investigated for potential medicinal uses. Generative tensorial reinforcement learning (GENTRL), a de novo small molecule design method based on generative reinforcement learning, was created by Zhavoronkov (107).

12.17 NGS IN CLINICAL MICROBIOLOGY AND INFECTIOUS DISEASE PREVENTION

In clinical microbiology, bioinformatics tools play an important role in leveraging genome sequence data for diagnostic, therapeutic, and preventive purposes. These tools help in comprehensive analyses of microbial content within clinical samples, enabling rapid identification of pathogens and understanding of outbreak structures (108). The integration of bioinformatics algorithms focusing on resistome-based prevention is essential, offering insights into strategies for discovering new antibiotics. Further, applications of next-generation sequencing (NGS) in clinical microbiology provide advanced capabilities for detailed genomic analyses, aiding in the early detection and management of infectious diseases (109).

12.17.1 Bioinformatics Tools in Clinical Microbiology

One important application of bioinformatics resources is the use of genomic sequence data for preventive, therapeutic, and diagnostic objectives. Clinical genomics is a vital tool in the field of infectious diseases, with applications ranging from accurately characterizing the composition of epidemic communities to detecting pathogens in clinical samples. Furthermore, one of the biggest challenges facing syndromic monitoring systems is the early diagnosis of infectious diseases during outbreaks. The central concept behind clinical genomics involves producing extensive sequence datasets that comprehensively sample a specimen for microbial content. The sequences are then resolved into distinct species, genes, pathways, or other properties using computer techniques. By facilitating a prompt epidemiological response, this strategy eventually lowers morbidity and death. Thus, pathogen transmission and resistome-based prevention are important bioinformatics methods to consider. Future studies on the resistome offer insights into strategies applicable to the discovery of new antibiotics (110).

12.7.2 A Resistome Algorithm Model

The tiny, crosshatched boxes stand for clinically relevant drugs and resistance genes. These represent tiny bioactive compounds and the resistome, a universe of putative resistance determinants, and make up a subset of the parvome. The resistome comprises genes with the potential to encode antibiotic resistance, while the mobilome includes the mobile segment of bacterial genomes. The resistome and mobilome intersect because a large number of resistance genes are found on mobile elements like transposons. Subsets of the pangenome, or complete coding capacity of prokaryotic cells, are represented by the resistome and mobilome, which together make up the pan-proteome (111).

It is significant to remember that humans only use a small portion of parvome for antibiotic purposes and that commercial antibiotic manufacture is probably produced on a larger scale than the global microbiota's natural production of these compounds. A recent study compared the use of an RNA sequencing-based metagenomics methodology with a more traditional molecular method for the detection of viruses in previously unanalyzed nasopharyngeal swabs and known respiratory virus-positive samples. Taxonomer, a quick and interactive web-based tool for metagenomics data analysis, was used to perform the data analysis. Overall, the metagenomics methodology showed strong conformity with the molecular approach, identified viruses that were not targeted by the molecular approach, and produced sequencing information that was useful to epidemiology and clinical practice (112, 113).

12.17.3 Predicting and Preventing Outbreaks with Mathematical Models and Bioinformatics

A mathematical model constitutes an abstract microcosm comprising entities adhering to precisely defined rules. Mathematics serves as a language to articulate these behavioral rules concisely and unambiguously, compelling us to express our assumptions clearly. Following the creation of a mathematical model, mathematical analysis—often in conjunction with computer simulations—facilitates investigating the global behavior of the model, elucidating the implications of our underlying assumptions. As a result, we can forecast the future of our hypothetical world inside the framework of the model and also examine how these forecasts alter as the laws controlling the entities the model describes change (114, 115).

Therefore, based on factors such as patterns of contact between susceptible and infectious individuals, the amount of time it takes for an infection to become infectious, the duration of infectiousness, the degree of immunity developed after infection, and so forth, the transmission of the pathogen among hosts is described by a mathematical model for the spread of an infectious disease in a population of hosts (116). Once these factors are integrated into a model, predictions can be made regarding the anticipated number of infections during an epidemic, the epidemic's duration, peak incidence, and the entire epidemic curve, furnishing expected case numbers at each time point.

In conclusion, antibiotic resistance presents a severe global health threat, necessitating innovative strategies for mitigation. The emergence of resistant strains, escalating complications in healthcare settings, and challenges in finding new antimicrobials underscore the urgency of addressing this issue. Bioinformatics, particularly in genomic approaches, plays a pivotal role in identifying, understanding, and combating antibiotic resistance. Computational methods streamline drug–target interaction identification, offering cost-effective and efficient alternatives to traditional experimental approaches. Antimicrobial drug repurposing provides rapid and economical solutions, capitalizing on existing medications. Metagenomics, coupled with artificial intelligence, enhances antibiotic resistance gene surveillance, while mathematical models, informed by bioinformatics, predict and prevent infectious disease outbreaks. Overall, the integration of AI and bioinformatics offers a multifaceted approach to tackling antibiotic resistance, promising advancements in drug discovery, surveillance, and outbreak management.

12.18 APPLICATIONS OF BIOINFORMATICS IN MICROBIAL RESEARCH

Bioinformatics plays a vital role in microbial research, especially in industries for the production of food, enzymes, and probiotics (117). It also provides advanced tools and methods for integration of computational approaches that have significantly enhanced the identification of pathogenic microbes, contributing to more precise and timely diagnoses. Computational tools, such as sequence alignment algorithms and comparative genomics, play a crucial role in distinguishing pathogenic strains from non-pathogenic ones. This aids in the development of molecular assays and next-generation sequencing techniques, allowing for rapid and accurate detection of microbial infections. Furthermore, in the field of environmental microbiology, bioinformatics tools are important in metagenomics for the exploration of microbial populations and bioresources (118).

12.8.1 Bioinformatics' Role in Drug Development and Discovery

Bioinformatics is used for analyzing DNA, microbial genomes, and small organic molecules to understand disease and design drugs with desirable properties. Bioinformatics streamline the drug discovery process by making it possible to investigate and understand vast amounts of biological data efficiently, which helps with target identification, lead compound optimization, and drug–target interaction prediction in the drug development process. In-silico screening of chemical libraries against these targets expedites the identification of promising drug candidates. This not only saves time and resources but also increases the efficiency of the drug discovery process.

In bioinformatics, the integration of data from diverse origins, including genomics, proteomics, and clinical data, emerges as a powerful approach. This incorporation facilitates a general understanding of disease mechanisms, disease progression, and drug responses.

The Biological Database identifies novel molecular objects and has successfully assembled excellent design libraries and datasets that could be improved for molecular similarity or diversity. For a drug design, traditionally a natural active compound is compulsory for treatment and then its chemical modification is required. A desirable change for the production of better medicine is an experimental approach.

The Swiss Institute of Bioinformatics is actively involved in the development and maintenance of software, databases, and web facilities associated with drug design and discovery. Valuable software includes ChEMBL, Swiss Similarity, and Swiss Target Prediction.

12.18.2 A Roadmap for the Design of Screening Drugs

Utilizing bioinformatics methods for drug creation, certain features are chosen based on availability and efficacy. The roadmap for drug development involves a series of interconnected stages, each supported by bioinformatics tools and software. These tools are essential in various aspects of drug development, from the early stages of drug design to clinical trials and pharmacovigilance. Some key stages and the corresponding use of bioinformatics tools include:

12.18.2.1 Target Identification and Validation

By using interfering target protein binding sites and a bioinformatics tool, target selection sequencing of protein can find all members of gene families and uncover phylogenetic analysis between sequencing, creating a 3D structure of the protein or target sequences and screening the archive preparation via a web-based service (119).

12.18.2.2 Drug Design and Development

Bioinformatics tools are essential for drug design and development, including molecular modeling, homology modeling, and molecular docking. These tools enable the analysis of protein expression, dynamics, and interactions, as well as the prediction of ligand–receptor interactions crucial for drug design (120).

12.18.2.3 Clinical Trials

Bioinformatics tools are used to predict drug response or toxicity, find biomarkers, and analyze data from clinical trials. Through the integration of data from several sources, including proteomics, genomes, and clinical data, these tools aid in the development of a thorough understanding of medication response and disease causes (121).

12.19 BIOINFORMATICS SOFTWARE INVOLVED IN THE DRUG DEVELOPMENT PROCESS

In the dynamic pathway of drug development, bioinformatics tools emerge as indispensable instruments, offering comprehensive solutions at each stage of the complicated process. Drug discovery is a multistep process which involves target identification and verification, drug target interaction, and its stability. The majority of the time, bioinformatics software is utilized for many aspects of drug design characterization, including behavioral research, data analysis, image analysis, visualization, structural activity relationship, the interaction between ligand molecules, pharmacokinetics molecular modeling, and molecular dynamics (122).

12.19.1 Sequence Retrieval Databases

In the initial phases, sequence databases such as GenBank, DDBJ, PDP, SwissProt, and UniProt facilitate sequence retrieval for thousands of nucleotides, mRNA, and proteins (123).

12.19.2 Web Services

Numerous web services and tools are available. BLAST and Graph Align can be used for analyzing similarities between DNA and protein and sequences. BLAST has diverse variants like BLASTN, BLASTX, BLASTP, TBLASTX, and TBLASTN. The results of BLAST include an alignment text, a match list, and a graphical summary box (124).

12.19.3 Molecular Docking Software

For molecular docking and screening, AutoDock and DOCK are used. AutoDock is a user-friendly server that predicts the interactions between ligand molecules and their biological targets, aiding in the rational design of drugs. Developments in bimolecular X-ray crystallography help to make structural data on intricate biomolecules like proteins and nucleic acids accessible. These structures may be used as targets for novel drugs to treat illnesses and diseases in humans, animals, and plants, as well as to comprehend basic biological concepts (125).

12.19.4 Ligand Design and Interaction Analysis Software

Software such as LigPlot and FlexX, which are essential for ligand design and the analysis of ligand–receptor interactions, contribute to the rational design of drug candidates (120). PocKdrug-Server is utilized to forecast a drug's logical potential for two purposes: (a) ligand binding proximity, and (b) protein structural information (126).

12.19.5 Databases for Gene Expressions

An array of databases and tools has been devised for the exploration of gene expression in both cancerous and normal tissues. The International Database of Cancer Gene Expression, established by Strausberg in 2002, encompasses 6 million gene tags derived from diverse cancerous and normal tissue samples. "Gene logic" is involved in the expression of various gene data (123).

12.19.6 Molecular Modeling Programs

Molecular modeling software such as SwissParam, BioAdviser, BioJava, and poseView are tools for analysis of protein expression and dynamics in interaction studies. SwissParam is used for ligands to expand swift binding free energy and for proteins (127).

12.19.7 Homology Modeling Programs

SWISS-MODEL, Modeller, and LOMESTS are used in homology modeling. With the help of global interference, users can customize the SWISS-MODEL server's computerized comparative modeling of protein structures in three dimensions. The "First approach mode" allows users to construct a protein's three-dimensional structure using only its amino acids (128).

12.20 MICROBIOME ANALYSIS AND HUMAN HEALTH

The human microbiome is made up of a variety of vibrant microbial communities that live in different bodily parts. As a result, these communities have a significant impact on supporting human health due to the coevolution of the microbiome and the host. Bioinformatics plays an important role in microbiome analysis. Microbiome studies generate vast amounts of sequencing data due to the diversity and abundance of microbial species. Bioinformatics tools help manage, process, and analyze these large datasets efficiently. Bioinformatics tools aid in assigning taxonomic classifications to microbial sequences, enabling researchers to identify and categorize the different microorganisms present in a sample (129). Bioinformatics tools help predict the functional capabilities of the microbial community by annotating genes and identifying potential pathways. This information is crucial for understanding the role of microbiota in human health. This is particularly important in understanding the role of microbiota in diseases. Microbiota and major diseases and ecological shifts are increasingly being linked through the use of metagenomics and metatranscriptomic sequencing techniques. For example, variations in carbohydrate utilization profiles within the *Bifidobacterium* genus are due to genetic variances between strains of the same species (130). This includes understanding how microbial communities respond to environmental factors or therapeutic interventions.

Gut microbiome data is recognized by computational tools as one of the inputs and outputs of evidence-based, automated, and personalized dietary recommendations for blood sugar homeostasis (131). Microbiome analysis for blood sugar levels needs bioinformatics software along with data mining and machine learning (132). Some investigations concentrate on evaluating the capacity of microbial communities to generate certain substances and require the extraction of bacterial genomes from intricate (such as soil) microbiomes (131). Phenotypes can differ from one another in vivo despite otherwise being genetically identical bacterial strains, due to slight variations in their gene composition. Therefore, to prove that the probiotic candidate itself has the intended health impact, well-conducted clinical trials are required. In order to ensure that the benefits observed are solely attributed to the probiotic that was consumed, it is required to use metagenomic and bioinformatics approaches that enable the identification and tracking of the probiotic under study at the strain level.

Because even identical microbial strains can exhibit wildly disparate behaviors, strain-level recognition is essential within the procedure for quality control to identify potential contaminants (133). To identify single-strain probiotic contamination, conventional typing methods such as PCR (Random Amplification of Polymorphic DNA) can be employed; however, these methods necessitate cultivation (134), thereby rendering them inappropriate for high-throughput examination of complex community goods (dairy products, probiotics, etc.). In addition to avoiding these drawn-out procedures to provide strain-level information, whole-metagenome sequencing and analysis may make it possible to screen for undesirable features like (spore) heat resistance based on the presence of related genes (135).

Research on the skin microbiota with possible medicinal objectives for conditions like acne, eczema, and Malassezia folliculitis is gaining traction in the cosmetics sector (136).

12.21 APPLICATIONS OF BIOINFORMATICS IN INDUSTRIAL PRODUCTS

Microorganisms play vital roles in the manufacturing process of food, probiotics, and cosmetics and in the production of enzymes inside industries (137). The improvement of bioinformatics enables study of the microbiome genomes as well as taxonomic composition by using 16sRNA. For this purpose, studying the taxonomic makeup and functional traits of these bacteria with marker genes—such as 16S rRNA—is crucial, and metaproteomics promises to share functional information of genes for the development of products, their optimization, and quality control (138). Utilizing shotgun metagenome sequencing and marker genes (such as 16S rRNA) to investigate the taxonomic makeup and functional properties of these microbes for product development, optimization, and quality control has become industry standard practice (138).

12.21.1 Dairy Products

Microbial strain-specific enzymes have a significant impact on the food- and beverage-producing process, including for cheese, meat, yogurt, and wine. Bioinformatics applications such as metagenomics, sequencing, assembly, and annotation steps are used for the optimization of end products instead of studying such enzymes by isolation of strain which is a time-consuming process (139). Metagenomics is also important for the examination of bacteriophages in the context of diversity, abundance, and development (140). Phage studies are important for the prevention of infection in industrial processes and it is helpful for the discovery of new phages against foodborne illnesses (140).

12.21.2 Probiotics

Probiotics are live microorganisms that provide benefits to the host's health, infection control, and disease when ingested in reasonable quantities. Finding new probiotics by using the culturomics approach is a challenging process (141). In recent years, there has been a rise in the wide characterization of probiotics along with the development of high-throughput multiomics tools. Multiomics techniques, including metagenomic, transcriptomic, proteomic, and metabolomic data, have been useful in shedding light on the molecular alterations brought about by probiotic treatment (142). These alterations include immune system modulation, interactions with mucin and the intestinal epithelium, bacteriocin-mediated protection against infections, and competitive exclusion. The potential use of probiotics in medical nutrition therapy for various conditions, including diarrhea, irritable bowel syndrome, and cancer, has also been explored. Our understanding of probiotic activity and the unique interactions between probiotics and their human hosts has improved with the development of omics tools and sequencing techniques. Furthermore, information from more omics sources, such as metabolomics and meta transcriptomics, can be included in integrative research to produce leads, as in the case of enzyme discovery (143).

12.21.3 Cosmetics

Research on the skin microbiota as a possible therapeutic target for conditions like acne, eczema, and Malassezia folliculitis is gaining traction in the cosmetics sector (136). Sadly, the low biomass of skin samples sometimes makes these investigations difficult to conduct and makes it easy for tiny contaminations (from nearby skin or reagents, for example) to produce inaccurate results. The role of bioinformatics in the cosmetics industry is significant, as it enables the exploration of microbial communities that contribute to product development (117). The cosmetic industry is also discovering the benefits of bioinformatics for formulation optimization. Bioinformatics analyses allow scientists to study gene expression patterns under various conditions. This then enables the design of personalized formulations that cater to diverse skin types and genetic profiles. Such an individualized approach to cosmetic development contributes to greater product effectiveness and customer satisfaction, which positively impacts revenue. Another way bioinformatics tools benefit cosmetic development is by verifying the authenticity of cosmetic ingredients, such as plant extracts, by analyzing their genetic signatures. This ensures that products are accurately labeled and free from contaminants.

12.21.4 Enzymes

Various industrial enzymes that are used to produce cleaning agents, detergents, textiles, and paper need to become cheaper, greener, and more efficient. Bioinformatics plays an important role in this regard by providing online databases. BRENDA is an online database that provides enzyme chemistry, substrate specificity, and kinetics. KEGG provides metabolic pathways and an understanding of enzymes as well. Phylogenetic studies are involved in the evolution of enzymes.

In addition, bioinformatics tools can also be used to predict how a particular gene or protein will behave in different environments, which can help to optimize production processes (117).

12.22 BIOINFORMATICS IN CLINICAL MICROBIOLOGY

The identification and characterization of the microbes that cause infections is essential for effective treatment, patient recovery, and safety. Nevertheless, not all types of bacteria can be effectively cultivated in a diagnostic laboratory, and current molecular tests cannot identify newly developing genetic traits in diseases that have evolved successfully and spread to humans, animals, and the environment. Hospital outbreaks can be quickly caused by unidentified infections, placing patients in danger while they are being admitted (112). The analysis of next-generation sequencing (NGS) data using bioinformatics tools and methodologies is becoming more and more common in the diagnosis and surveillance of infectious illnesses. Interest among clinical microbiologists is piqued by its emphasis on molecular identification, genotyping, microbiome research, analysis of antibiotic resistance, and identification of pathogens associated with unknown diseases in clinical specimens. Additionally, a wide range of pathogen types are identified, described, and typed using bioinformatics methods. This came after the identification and treatment of bacterial, fungal, and viral illnesses was widely achieved using genomic techniques. Bioinformatics has been applied to resistance analysis, strain type, pathogen identification, and virulence factor detection. With next-generation sequencing (NGS), the entire bacterial genome's DNA sequence may be found in a single sequencing run. This information can be used to type, resist, and evaluate virulence, all of which are important for investigating outbreaks. An outbreak-specific screening test can be developed using the available genomic data. An overview of NGS is provided in this article, along with information on library preparation and the salient features of the most used NGS platforms. NGS has great potential for testing infectious diseases in clinical settings, but there are still several obstacles to overcome before NGS is widely used in clinical microbiology labs. These obstacles include automation, standardizing technical procedures and bioinformatics pipelines, enhancing referencing databases, putting in place quality control and competency testing procedures, and cutting costs and turnaround times (144).

Sequence analyses can be used to find answers to a variety of diagnostic questions, including the genetic relationships between bacteria and viruses, the identification of fungi through sequence analyses of the internal transcribed spacer (ITS) region of the 18S ribosomal deoxyribonucleic acid (rDNA), and the identification of alterations in the genomes of bacteria or viruses that result in resistance to certain antibiotics or antivirals (145). This is usually done with Sanger sequencing, which is preceded by primer-specific amplification of every gene or region of the genome. The pathogens in clinical material can be identified using the same methodology. However, when dealing with more complex clinical material containing many species, such as faeces samples, this approach becomes challenging. With NGS, it is possible to sequence the entire genome of many pathogens in a single sequence run, whether they are bacterial isolates from various patients or multiple species found in patient material from a single individual (metagenomics). Over the past ten years, NGS's investment and operating costs have dropped significantly (146). Numerous medical microbiology laboratories, including the one at the UMCG, already use next-generation sequencing (NGS) for a variety of purposes. These include managing outbreaks, identifying molecular cases, characterizing and monitoring pathogens, quickly identifying bacteria using the 16S-23S rRNA region, taxonomy, metagenomics techniques on clinical samples, and determining the transmission of zoonotic microorganisms from animals to humans.

The data analysis presents the largest obstacle to the use of NGS in the clinical microbiology lab. However, with the many user-friendly software programs available, NGS data analysis for diagnostic purposes can be carried out even by those with minimal background expertise in bioinformatics (147). In the lab, genome assembly is done using software programs such as Velvet, SPAdes, and CLC Genomic Workbench (Qiagen). Using a multi-locus sequencing typing (MLST) approach, the genetic relationships between isolates can be examined gene-by-gene through the investigation of either the conserved core genome (cgMLST) or the entire genome (wgMLST), which includes a set of variable accessory genes. This method can be applied to a number of software programs, including BioNumerics (Applied Maths, Biomérieux) and SeqSphere (Ridom), as well as web resources, like EnteroBase and BIGSdb (Bacterial Isolate Genome Sequence Database) (148).

A number of options for analyzing NGS data are available on the Centre for Genomic Epidemiology website. VirulenceFinder and ResFinder can be utilized to find genes associated with virulence and resistance. Alternatively, the Virulence Factor Database (VFDB) and the Comprehensive Antibiotic Resistance Database (CARD) can be used to find information on virulence and resistance genes. The assembled genome and the non-assembled sequencing data can both be submitted via these web resources. Nevertheless, further measures are required to confirm the data received from these websites. Regarding *S. aureus*, reports indicate a strong association between the phenotypic resistance pattern and the presence of resistance genes (149).

In the coming years, NGS will play a bigger role in medical microbiology labs. Research will make greater use of it, possibly most significantly for molecular diagnostics, infection control, investigating outbreaks using a special outbreak marker approach, characterizing and monitoring pathogens, finding new resistance genes, and applying a metagenomics approach to clinical specimens.

Further investigation is required to optimize the NGS workflow, particularly to shorten turnaround times for NGS platform runs and library preparation while simultaneously reducing expenses even further. Subsequently, user-friendly metagenomics software and automated procedures for data analysis must be created.

The clinical microbiology laboratory's responsibilities include everything from determining the bacteria causing an infection in a patient to helping to spot global disease epidemics. Informatics and all these procedures are becoming closer and closer partners. The efficient use of informatics technologies can reduce laboratory burden and improve productivity and efficiency in the laboratory by improving the completeness, accuracy, and timeliness of microbiology testing. With the introduction of fully automated laboratories, sophisticated instrument interfaces, electronic health records, clinical decision support systems, and the clinical use of microbial genome sequencing, informatics is anticipated to play an increasingly important role in clinical microbiology (108). There is no denying that whole genome sequencing can be used in research to improve our knowledge of many aspects of microbiology and infectious diseases. The goal of research into these areas is to improve patient care by indirectly influencing microbiology and clinical infectious disease practices. These fields of research include the evolution of pathogens, epidemiology, virulence factors, and the creation and dissemination of mechanisms behind antibiotic resistance. Our comprehension of the human microbiome and other microbial ecosystems is beginning to expand with the growing application of metagenomic studies and genomics in the identification of possible targets for drugs and vaccines. WGS is currently not widely used in public health and clinical microbiology as a real-time test that directly informs clinical procedures because of some limitations. WGS is unlikely to replace present techniques for regular detection of microorganisms and antibiotic susceptibility testing very soon, despite rapidly increasing sequencing efficiency. It would appear to play an increasingly important role in infection control, public health, and reference laboratories for thorough isolate characterization, outbreak investigation, and disease transmission detection. Metagenomic analysis of polymicrobial samples for pathogen and "resistome" identification, as well as routine use in diagnostic facilities for identifying pathogens in samples that do not culture, may be adopted when sequencing becomes more accessible and useful. With more widespread bioinformatics skills, it is evident that user-friendly bioinformatic pipelines are essential to enabling more broad adoption of WGS. The most practical implementation approach, at least until this bottleneck is removed, is for regional facilities to outsource WGS and data analysis to centralized state reference laboratories as needed. A more potent system of epidemiological surveillance might be created with sequencing carried out at peripheral nodes, and bioinformatic analysis and supervision of the sequencing at the centralized reference centre, with WGS adoption improving and costs declining. We need to maximize the few but effective means at our disposal to keep an eye on and stop the spread of infectious diseases in an era of rising drug resistance worldwide, easy access to foreign countries, and minimal investment in the development of antimicrobial drugs (150).

12.23 ETHICAL ISSUES IN BIOINFORMATICS RESEARCH

Bioinformatics is an important tool for the analysis of biological data by using computational and statistical methods. It has several ethical, legal, and social concerns leading to serious consequences. Here are some challenges associated with this dominant tool.

12.23.1 Privacy Concerns

Personal privacy is the greatest concern, as most bioinformatics applications deal with genes, which are specific and private to each individual (151). Computational analysis of genetic data can expose personal hidden information, resulting in psychological stress and tensions in families because the result may reveal some unpleasant or socially unacceptable truths. Germline genomic data of a person gives evidence allowing identification of an individual and remains unchanged throughout life (152).

Obtaining informed consent from participants is essential, especially when dealing with human subjects. Researchers must communicate the purpose of the study, potential risks, and how the data will be used.

12.23.2 Data Security

Ensuring the security of genomic and other biological data is critical to prevent unauthorized access, data breaches, or misuse. As precision medication comprises many kinds of data, for example genomic data and personal information, it is currently a significant challenge and cause for concern. Specifically, personal information may be exploited for commercial purposes and utilized as proof to deny someone access to insurance or a job. Furthermore, those who confess to drug abuse are reluctant to participate in precision medicine research initiatives due to concerns that their data could be utilized to solve unsolved crimes (153).

Traditional privacy regulations were designed for health data, and they only provide partial protection for genomic data. Utilizing the connection between genetic data and other information that is available to the public, such as family name, demographics, and observable traits like eye and hair colour, an attacker may discover private information about a target individual (154).

12.23.3 Misuse

Bioinformatics research has the potential to be misused or abused, leading to serious consequences. Misuse can arise from various sources, including the manipulation of human cells via genetic engineering, cloning, and the potential revelation of previously unknown relationships with intellectual property. The use of certain bioinformatics tools in a context of structural rather than functional genomics can also cast doubts on the authenticity of a large dataset. The misuse and malicious uses of new biotechnologies, such as directed drug design and synthetic biology, can lead to public health risks for bioterrorism. By using bioinformatics techniques, it is possible to develop more effective aerosols for biowarfare (155). To address these issues, it is important to reflect carefully on the ethical consequences of bioinformatics technology and to develop regulations and guidelines that ensure the responsible use of this technology. Scientists, scholars, and healthcare authorities should collaborate to resolve this issue (156).

12.23.4 Legal Issues

Legal issues are also associated with bioinformatics tools, for example, software development licensing. Some other legal issues are database security, information protection, consent, etc. Human genetic material legal issues are also of important concern in terms of confidentiality and privacy (155). Pharmacogenomics studies are accessible but risks occur when a single gene change affects the behavior of a whole generation. Legal considerations for research environments, including biopharma collaborations and university R&D, are also important. The ongoing discussion requires cooperation from social scientists and biologists, medical experts, historians, legal scholars, ethicists, social workers, and others (157).

12.23.5 Data Protection Law

Precision medicine includes information that is beyond privacy and it is an important topic of discussion. Specifically, personal data has sometimes been commercially misused. Furthermore, patients who disclosed substance addiction indicated concerns that their information would be utilized to conceal unresolved corruption and, as a result, they declined to take part in research projects related to precision medicine.

The economic impact of precision medicine is also important in terms of treatment actual cost and its consequences for insurance. Concerns about the true cost of medical testing, insurance companies' coverage of such costs, and losing insurance because of a genetic condition are legitimate, and as a result, the US recently passed the Genetic Information Nondiscrimination Act (GINA) (158).

12.23.6 Regulatory Frameworks

Several regulatory frameworks and laws are relevant to data privacy in bioinformatics research. These include:

1. **EU General Data Protection Regulation (GDPR):** The GDPR offers strong privacy protection and allows individuals better data control, including the right to delete data. It applies to EU members and has a significant impact on data privacy in bioinformatics research. It also provides rules for processing genetic data for research purposes, including the conditions under which further use of genetic data for scientific research purposes is allowed (158).
2. **Health Insurance Portability and Accountability Act (HIPAA):** HIPAA regulates the use and disclosure of protected health information in the United States, including genetic and genomic data. However, there is no clear controlling direction on how to apply HIPAA's minimum essential standard when allowing for the sharing of facts in the data-rich environment of genomic testing. Genomic testing laboratories should collaborate with legislators to support sensible, informed regulations and proper data-sharing activity characterization to reduce negative effects on daily operations (159).
3. **Sociotechnical Safeguards for Genomic Data Privacy:** This framework requires companies handling genetic data to notify of security breaches, and grants data subjects access and deletion rights. Additionally, it discusses the intricate legal, ethical, and technological difficulties that come with maintaining privacy in genetic research (160). These regulatory frameworks and laws play a crucial role in ensuring data privacy and security in bioinformatics research, and researchers and organizations involved in this field need to comply with these regulations to protect individuals' privacy and data.

REFERENCES

1. Ciuffreda L, Rodríguez-Pérez H, Flores C. Nanopore sequencing and its application to the study of microbial communities. *Computational and Structural Biotechnology Journal*. 2021;19:1497–511.
2. Daniel R. The metagenomics of soil. *Nature Reviews Microbiology*. 2005;3(6):470–8.
3. DeLong EF. Microbial community genomics in the ocean. *Nature Reviews Microbiology*. 2005;3(6):459–69.
4. Peterson J, Garges S, Giovanni M, McInnes P, Wang L, Schloss JA, et al. The NIH human microbiome project. *Genome Research*. 2009;19(12):2317–23.
5. Handelsman J. Metagenomics: application of genomics to uncultured microorganisms. *Microbiology and Molecular Biology Reviews*. 2004;68(4):669–685.
6. Wooley JC, Godzik A, Friedberg I. A primer on metagenomics. *PLoS Computational Biology*. 2010;6(2):e1000667.
7. Kolde R, Franzosa EA, Rahnavard G, Hall AB, Vlamakis H, Stevens C, et al. Host genetic variation and its microbiome interactions within the Human Microbiome Project. *Genome Medicine*. 2018;10(1):1–13.
8. Kobras CM, Fenton AK, Sheppard SK. Next-generation microbiology: from comparative genomics to gene function. *Genome Biology*. 2021;22(1):1–16.
9. Zhang X, Wu F, Yang N, Zhan X, Liao J, Mai S, et al. In silico methods for identification of potential therapeutic targets. *Interdisciplinary Sciences: Computational Life Sciences*. 2022;14:1–26.
10. Can T. Introduction to bioinformatics. In *miRNomics: MicroRNA Biology and Computational Analysis*, ed. M Yousef and J Allmer. Springer; 2014. p. 51–71.
11. Carriço JA, Rossi M, Moran-Gilad J, Van Domselaar G, Ramirez M. A primer on microbial bioinformatics for nonbioinformaticians. *Clinical Microbiology and Infection*. 2018;24(4):342–9.
12. Christensen H, Andersson AJ, Jørgensen SL, Vogt JK. 16S rRNA amplicon sequencing for metagenomics. In *Introduction to Bioinformatics in Microbiology*, ed. H Christensen. Springer; 2018. p. 135–61.
13. Peter SC, Dhanjal JK, Malik V, Radhakrishnan N, Jayakanthan M, Sundar D, et al. *Encyclopedia of Bioinformatics and Computational Biology*. Elsevier. Ranganathan, S, Grib-skov, M, Nakai, K, Schönbach, C, Eds. 2019:661–76.
14. Zhulin IB. Databases for microbiologists. *Journal of Bacteriology*. 2015;197(15):2458–67.
15. Sayers EW, Bolton EE, Brister JR, Canese K, Chan J, Comeau DC, et al. Database resources of the National Center for Biotechnology Information in 2023. *Nucleic Acids Research*. 2023;51(D1):D29–D38.

16. Bairoch A, Apweiler R, Wu CH, Barker WC, Boeckmann B, Ferro S, et al. The universal protein resource (UniProt). *Nucleic Acids Research*. 2005;33(suppl_1):D154–D9.
17. Brooksbank C, Bergman MT, Apweiler R, Birney E, Thornton J. The european bioinformatics institute's data resources 2014. *Nucleic Acids Research*. 2014;42(D1):D18–D25.
18. Bairoch A, Apweiler R. The SWISS-PROT protein sequence data bank and its supplement TrEMBL. *Nucleic Acids Research*. 1997;25(1):31–6.
19. Taubig H, Buchner A, Griebsch J. PAST: fast structure-based searching in the PDB. *Nucleic Acids Research*. 2006;34(Web Server issue):W20–3.
20. Gonnella G. TextFormats: simplifying the definition and parsing of text formats in bioinformatics. *Plos One*. 2022;17(5):e0268910.
21. White J. PubMed 2.0. *Medical Reference Services Quarterly*. 2020;39(4):382–7.
22. Kodama Y, Shumway M, Leinonen R. The sequence read archive: explosive growth of sequencing data. *Nucleic Acids Research*. 2012;40(D1):D54–D6.
23. Barrett T, Wilhite SE, Ledoux P, Evangelista C, Kim IF, Tomashevsky M, et al. NCBI GEO: archive for functional genomics data sets—update. *Nucleic Acids Research*. 2012;41(D1):D991–D5.
24. Mitchell AL, Almeida A, Beracochea M, Boland M, Burgin J, Cochrane G, et al. MGnify: the microbiome analysis resource in 2020. *Nucleic Acids Research*. 2020;48(D1):D570–D8.
25. Mistry J, Chuguransky S, Williams L, Qureshi M, Salazar GA, Sonnhammer EL, et al. Pfam: the protein families database in 2021. *Nucleic Acids Research*. 2021;49(D1):D412–D9.
26. Letunic I, Doerks T, Bork P. SMART: recent updates, new developments and status in 2015. *Nucleic Acids Research*. 2015;43(D1):D257–D60.
27. Haft DH, Selengut JD, Richter RA, Harkins D, Basu MK, Beck E. TIGRFAMs and genome properties in 2013. *Nucleic Acids Research*. 2012;41(D1):D387–D95.
28. Tatusov RL, Fedorova ND, Jackson JD, Jacobs AR, Kiryutin B, Koonin EV, et al. The COG database: an updated version includes eukaryotes. *BMC Bioinformatics*. 2003;4(1):1–14.
29. Galperin MY, Makarova KS, Wolf YI, Koonin EV. Expanded microbial genome coverage and improved protein family annotation in the COG database. *Nucleic Acids Research*. 2015;43(D1):D261–D9.
30. Mitchell A, Chang H-Y, Daugherty L, Fraser M, Hunter S, Lopez R, et al. The InterPro protein families database: the classification resource after 15 years. *Nucleic Acids Research*. 2015;43(D1):D213–D21.
31. Burley SK, Berman HM, Christie C, Duarte JM, Feng Z, Westbrook J, et al. RCSB Protein Data Bank: sustaining a living digital data resource that enables breakthroughs in scientific research and biomedical education. *Protein Science*. 2018;27(1):316–30.
32. Andreeva A, Howorth D, Chothia C, Kulesha E, Murzin AG. SCOP2 prototype: a new approach to protein structure mining. *Nucleic Acids Research*. 2014;42(D1):D310–D4.
33. Sillitoe I, Lewis TE, Cuff A, Das S, Ashford P, Dawson NL, et al. CATH: comprehensive structural and functional annotations for genome sequences. *Nucleic Acids Research*. 2015;43(D1):D376–D81.
34. Kriventseva EV, Tegenfeldt F, Petty TJ, Waterhouse RM, Simao FA, Pozdnyakov IA, et al. OrthoDB v8: update of the hierarchical catalog of orthologs and the underlying free software. *Nucleic Acids Research*. 2015;43(D1):D250–D6.
35. Kanehisa M, Goto S, Sato Y, Kawashima M, Furumichi M, Tanabe M. Data, information, knowledge and principle: back to metabolism in KEGG. *Nucleic Acids Research*. 2014;42(D1):D199–D205.
36. Caspi R, Billington R, Ferrer L, Foerster H, Fulcher CA, Keseler IM, et al. The MetaCyc database of metabolic pathways and enzymes and the BioCyc collection of pathway/genome databases. *Nucleic Acids Research*. 2016;44(D1):D471–D80.
37. Franceschini A, Szklarczyk D, Frankild S, Kuhn M, Simonovic M, Roth A, et al. STRING v9. 1: protein-protein interaction networks, with increased coverage and integration. *Nucleic Acids Research*. 2012;41(D1):D808–D15.
38. Chatr-Aryamontri A, Oughtred R, Boucher L, Rust J, Chang C, Kolas NK, et al. The BioGRID interaction database: 2017 update. *Nucleic Acids Research*. 2017;45(D1):D369–D79.
39. Chowdhury B, Garai G. A review on multiple sequence alignment from the perspective of genetic algorithm. *Genomics*. 2017;109(5–6):419–31.
40. Chao J, Tang F, Xu L. Developments in algorithms for sequence alignment: a review. *Biomolecules*. 2022;12(4):546.
41. Dayhoff M. *Atlas of Protein Sequence and Structure*. National Biomedical Research Foundation; 1978. p. 345.

42. Henikoff S, Henikoff JG. Amino acid substitution matrices from protein blocks. *Proceedings of the National Academy of Sciences*. 1992;89(22):10915–9.
43. Madhavan G. *Bioinformatics: A Practical Guide to the Analysis of Genes and Proteins*. Edited by Andreas D. Baxevanis and BF Francis Ouellette. Hardcover, October 2004, 540 pages. US $79.95, ISBN: 0-471-47878-4. Springer; 2006.
44. Needleman SB, Wunsch CD. A general method applicable to the search for similarities in the amino acid sequence of two proteins. *Journal of Molecular Biology*. 1970;48(3):443–53.
45. Shyu C, Sheneman L, Foster JA. Multiple sequence alignment with evolutionary computation. *Genetic Programming and Evolvable Machines*. 2004;5:121–44.
46. Kapli P, Yang Z, Telford MJ. Phylogenetic tree building in the genomic age. *Nature Reviews Genetics*. 2020;21(7):428–44.
47. Dion MB, Oechslin F, Moineau S. Phage diversity, genomics and phylogeny. *Nature Reviews Microbiology*. 2020;18(3):125–38.
48. Lees JA, Kendall M, Parkhill J, Colijn C, Bentley SD, Harris SR. Evaluation of phylogenetic reconstruction methods using bacterial whole genomes: a simulation based study. *Wellcome Open Research*. 2018;3.
49. Adams MD, Kelley JM, Gocayne JD, Dubnick M, Polymeropoulos MH, Xiao H, et al. Complementary DNA sequencing: expressed sequence tags and human genome project. *Science*. 1991;252(5013):1651–6.
50. Pan Q, Shai O, Lee LJ, Frey BJ, Blencowe BJ. Deep surveying of alternative splicing complexity in the human transcriptome by high-throughput sequencing. *Nature Genetics*. 2008;40(12):1413–5.
51. Lappalainen T, Sammeth M, Friedländer MR, 't Hoen PA, Monlong J, Rivas MA, et al. Transcriptome and genome sequencing uncovers functional variation in humans. *Nature*. 2013;501(7468):506–11.
52. Melé M, Ferreira PG, Reverter F, DeLuca DS, Monlong J, Sammeth M, et al. The human transcriptome across tissues and individuals. *Science*. 2015;348(6235):660–5.
53. Sandberg R. Entering the era of single-cell transcriptomics in biology and medicine. *Nature Methods*. 2014;11(1):22–4.
54. Piétu G, Mariage-Samson R, Fayein N-A, Matingou C, Eveno E, Houlgatte R, et al. The Genexpress IMAGE knowledge base of the human brain transcriptome: a prototype integrated resource for functional and computational genomics. *Genome Research*. 1999;9(2):195–209.
55. Kukurba KR, Montgomery SB. RNA sequencing and analysis. *Cold Spring Harbor Protocols*. 2015;2015(11):pdb.top084970.
56. Velculescu VE, Zhang L, Vogelstein B, Kinzler KW. Serial analysis of gene expression. *Science*. 1995;270(5235):484–7.
57. Wang Z, Gerstein M, Snyder M. RNA-Seq: a revolutionary tool for transcriptomics. *Nature Reviews Genetics*. 2009;10(1):57–63.
58. Schena M, Shalon D, Davis RW, Brown PO. Quantitative monitoring of gene expression patterns with a complementary DNA microarray. *Science*. 1995;270(5235):467–70.
59. Pozhitkov AE, Tautz D, Noble PA. Oligonucleotide microarrays: widely applied—poorly understood. *Briefings in Functional Genomics and Proteomics*. 2007;6(2):141–8.
60. Meyers BC, Vu TH, Tej SS, Ghazal H, Matvienko M, Agrawal V, et al. Analysis of the transcriptional complexity of Arabidopsis thaliana by massively parallel signature sequencing. *Nature Biotechnology*. 2004;22(8):1006–11.
61. Van Verk MC, Hickman R, Pieterse CM, Van Wees SC. RNA-Seq: revelation of the messengers. *Trends in Plant Science*. 2013;18(4):175–9.
62. Conesa A, Madrigal P, Tarazona S, Gomez-Cabrero D, Cervera A, McPherson A, et al. A survey of best practices for RNA-seq data analysis. *Genome Biology*. 2016;17(1):1–19.
63. Fonseca NA, Rung J, Brazma A, Marioni JC. Tools for mapping high-throughput sequencing data. *Bioinformatics*. 2012;28(24):3169–77.
64. Trapnell C, Williams BA, Pertea G, Mortazavi A, Kwan G, Van Baren MJ, et al. Transcript assembly and quantification by RNA-Seq reveals unannotated transcripts and isoform switching during cell differentiation. *Nature Biotechnology*. 2010;28(5):511–5.
65. Anders S, Pyl PT, Huber W. HTSeq—a Python framework to work with high-throughput sequencing data. *Bioinformatics*. 2015;31(2):166–9.
66. Bray NL, Pimentel H, Melsted P, Pachter L. Near-optimal probabilistic RNA-seq quantification. *Nature Biotechnology*. 2016;34(5):525–7.

67. Fang Z, Cui X. Design and validation issues in RNA-seq experiments. *Briefings in Bioinformatics*. 2011;12(3):280–7.
68. Mardis ER. Next-generation DNA sequencing methods. *Annual Review Genomics Human Genetics*. 2008;9:387–402.
69. Dong Z, Chen Y. Transcriptomics: advances and approaches. *Science China Life Sciences*. 2013;56:960–7.
70. Draghici S, Khatri P, Eklund AC, Szallasi Z. Reliability and reproducibility issues in DNA microarray measurements. *TRENDS in Genetics*. 2006;22(2):101–9.
71. Cristea IM, Gaskell SJ, Whetton AD. Proteomics techniques and their application to hematology. *Blood*. 2004;103(10):3624–34.
72. Wilkins MR, Sanchez J-C, Gooley AA, Appel RD, Humphery-Smith I, Hochstrasser DF, et al. Progress with proteome projects: why all proteins expressed by a genome should be identified and how to do it. *Biotechnology and Genetic Engineering Reviews*. 1996;13(1):19–50.
73. Cox J, Mann M. Is proteomics the new genomics? *Cell*. 2007;130(3):395–8.
74. Aslam B, Basit M, Nisar MA, Khurshid M, Rasool MH. Proteomics: technologies and their applications. *Journal of Chromatographic Science*. 2016:1–15.
75. Perez-Riverol Y, Alpi E, Wang R, Hermjakob H, Vizcaíno JA. Making proteomics data accessible and reusable: current state of proteomics databases and repositories. *Proteomics*. 2015;15(5-6):930–50.
76. Deutsch EW. Mass spectrometer output file format mzML. In Proteome Bioinformatics, ed. SJ Hubbard and AR Jones. Springer; 2010. p. 319–31.
77. Vihinen M. Bioinformatics in proteomics. *Biomolecular Engineering*. 2001;18(5):241–8.
78. Jenkins JA, Griffiths-Jones S, Shewry PR, Breiteneder H, Mills EC. Structural relatedness of plant food allergens with specific reference to cross-reactive allergens: an in silico analysis. *Journal of Allergy and Clinical Immunology*. 2005;115(1):163–70.
79. Schmidt A, Forne I, Imhof A. Bioinformatic analysis of proteomics data. *BMC Systems Biology*. 2014;8(2):1–7.
80. Vizcaíno JA, Côté RG, Csordas A, Dianes JA, Fabregat A, Foster JM, et al. The PRoteomics IDEntifications (PRIDE) database and associated tools: status in 2013. *Nucleic Acids Research*. 2012;41(D1):D1063–D9.
81. Monserrat-Martinez A, Gambin Y, Sierecki E. Thinking outside the bug: molecular targets and strategies to overcome antibiotic resistance. *International Journal of Molecular Sciences*. 2019;20(6):1255.
82. Seoane A, Bou G. Bioinformatics approaches to the study of antimicrobial resistance. *Revista Española de Quimioterapia*. 2021;34(Suppl1):15.
83. Moll J, Colombo R. *Target Identification and Validation in Drug Discovery*: Springer; 2019.
84. Bull SC, Doig AJ. Properties of protein drug target classes. *PloS One*. 2015;10(3):e0117955.
85. Mahmud SH, Chen W, Meng H, Jahan H, Liu Y, Hasan SM. Prediction of drug-target interaction based on protein features using undersampling and feature selection techniques with boosting. *Analytical Biochemistry*. 2020;589:113507.
86. Chen P, Zheng H. Drug-target interaction prediction based on spatial consistency constraint and graph convolutional autoencoder. *BMC Bioinformatics*. 2023;24(1):1–21.
87. Wu Z, Li W, Liu G, Tang Y. Network-based methods for prediction of drug-target interactions. *Frontiers in Pharmacology*. 2018;9:1134.
88. Ahmad S, Qazi S, Raza K. Translational bioinformatics methods for drug discovery and drug repurposing. In *Translational Bioinformatics in Healthcare and Medicine*, ed. K Rasa and N Dey. Academic Press; 2021. p. 127–39.
89. Farha MA, Brown ED. Drug repurposing for antimicrobial discovery. *Nature Microbiology*. 2019;4(4):565–77.
90. Gurajala S. Antimicrobial drug repurposing. *Journal of Population Therapeutics and Clinical Pharmacology*. 2023;30(17):14–39.
91. Boolchandani M, D'Souza AW, Dantas G. Sequencing-based methods and resources to study antimicrobial resistance. *Nature Reviews Genetics*. 2019;20(6):356–70.
92. Commichaux S, Shah N, Ghurye J, Stoppel A, Goodheart JA, Luque GG, et al. A critical assessment of gene catalogs for metagenomic analysis. *Bioinformatics*. 2021;37(18):2848–57.
93. Douglas GM, Langille MG. Current and promising approaches to identify horizontal gene transfer events in metagenomes. *Genome Biology and Evolution*. 2019;11(10):2750–66.
94. Chng KR, Li C, Bertrand D, Ng AHQ, Kwah JS, Low HM, et al. Cartography of opportunistic pathogens and antibiotic resistance genes in a tertiary hospital environment. *Nature Medicine*. 2020;26(6):941–51.

95. Swinney D. Phenotypic vs. target-based drug discovery for first-in-class medicines. *Clinical Pharmacology & Therapeutics*. 2013;93(4):299–301.
96. Brown ED, Wright GD. Antibacterial drug discovery in the resistance era. *Nature*. 2016;529(7586):336–43.
97. Cižman M, Srovin TP. Antibiotic consumption and resistance of gram-negative pathogens (collateral damage). *GMS Infectious Diseases*. 2018;6:Doc05.
98. Matlock A, Garcia JA, Moussavi K, Long B, Liang SY-T. Advances in novel antibiotics to treat multidrug-resistant gram-negative bacterial infections. *Internal and Emergency Medicine*. 2021;16(8):2231–41.
99. Perez F, Hujer AM, Hujer KM, Decker BK, Rather PN, Bonomo RA. Global challenge of multidrug-resistant Acinetobacter baumannii. *Antimicrobial Agents and Chemotherapy*. 2007;51(10):3471–84.
100. Stokes JM, Yang K, Swanson K, Jin W, Cubillos-Ruiz A, Donghia NM, et al. A deep learning approach to antibiotic discovery. *Cell*. 2020;180(4):688–702. e13.
101. Wang Y, Li F, Bharathwaj M, Rosas NC, Leier A, Akutsu T, et al. DeepBL: a deep learning-based approach for in silico discovery of beta-lactamases. *Briefings in Bioinformatics*. 2021;22(4):bbaa301.
102. Segler MH, Kogej T, Tyrchan C, Waller MP. Generating focused molecule libraries for drug discovery with recurrent neural networks. *ACS Central Science*. 2018;4(1):120–31.
103. Bruno BJ, Miller GD, Lim CS. Basics and recent advances in peptide and protein drug delivery. *Therapeutic Delivery*. 2013;4(11):1443–67.
104. Yan J, Bhadra P, Li A, Sethiya P, Qin L, Tai HK, et al. Deep-AmPEP30: improve short antimicrobial peptides prediction with deep learning. *Molecular Therapy-Nucleic Acids*. 2020;20:882–94.
105. Plisson F, Ramírez-Sánchez O, Martínez-Hernández C. Machine learning-guided discovery and design of non-hemolytic peptides. *Scientific Reports*. 2020;10(1):16581.
106. Kavousi K, Bagheri M, Behrouzi S, Vafadar S, Atanaki FF, Lotfabadi BT, et al. IAMPE: NMR-assisted computational prediction of antimicrobial peptides. *Journal of Chemical Information and Modeling*. 2020;60(10):4691–701.
107. Zhavoronkov A, Ivanenkov YA, Aliper A, Veselov MS, Aladinskiy VA, Aladinskaya AV, et al. Deep learning enables rapid identification of potent DDR1 kinase inhibitors. *Nature Biotechnology*. 2019;37(9):1038–40.
108. Kim H. Bioinformatics technology in clinical and public health microbiology applying computational methods. *Journal of Medical Biomedical and Applied Science*. 2018;6(9):136–40.
109. d'Humières C, Salmona M, Dellière S, Leo S, Rodriguez C, Angebault C, et al. The potential role of clinical metagenomics in infectious diseases: therapeutic perspectives. *Drugs*. 2021;81(13):1453–66.
110. Bianconi I, Aschbacher R, Pagani E. Current uses and future perspectives of genomic technologies in clinical microbiology. *Antibiotics*. 2023;12(11):1580.
111. Gillings MR. Evolutionary consequences of antibiotic use for the resistome, mobilome and microbial pangenome. *Frontiers in Microbiology*. 2013;4:4.
112. Deurenberg RH, Bathoorn E, Chlebowicz MA, Couto N, Ferdous M, García-Cobos S, et al. Application of next generation sequencing in clinical microbiology and infection prevention. *Journal of Biotechnology*. 2017;243:16–24.
113. Wang X, Stelzer-Braid S, Scotch M, Rawlinson WD. Detection of respiratory viruses directly from clinical samples using next-generation sequencing: a literature review of recent advances and potential for routine clinical use. *Reviews in Medical Virology*. 2022;32(5):e2375.
114. Huppert A, Katriel G. Mathematical modelling and prediction in infectious disease epidemiology. *Clinical Microbiology and Infection*. 2013;19(11):999–1005.
115. Yanchevskaya EY, Mesnyankina O. Mathematical modelling and prediction in infectious disease epidemiology. *RUDN Journal of Medicine*. 2019;23(3):328–34.
116. Lange A. A mathematical framework for predicting lifestyles of viral pathogens. *Bulletin of Mathematical Biology*. 2020;82(5):54.
117. Van den Bogert B, Boekhorst J, Pirovano W, May A. On the role of bioinformatics and data science in industrial microbiome applications. *Frontiers in Genetics*. 2019;10:721.
118. Nwachukwu BC, Babalola OO. Metagenomics: a tool for exploring key microbiome with the potentials for improving sustainable agriculture. *Frontiers in Sustainable Food Systems*. 2022;6:886987.
119. Song CM, Lim SJ, Tong JC. Recent advances in computer-aided drug design. *Briefings in Bioinformatics*. 2009;10(5):579–91.

120. Browne RB, Thomas SC, Roy JD. Bioinfomatics as a tool in drug designing. In *Computation in Bioinformatics: Multidisciplinary Applications*, ed. S Balamurugan et al. Scrivener; 2021. p. 1–24.
121. Celi LA, Moseley E, Moses C, Ryan P, Somai M, Stone D, et al. From pharmacovigilance to clinical care optimization. *Big Data*. 2014;2(3):134–41.
122. Jamkhande PG, Ghante MH, Ajgunde BR. Software based approaches for drug designing and development: a systematic review on commonly used software and its applications. *Bulletin of Faculty of Pharmacy, Cairo University*. 2017;55(2):203–10.
123. Khan N. Bioinformatics in drug development. *Bioceramics Devlopment and Applications*. 2017;7(104):2.
124. Wheeler D, Bhagwat M. *BLAST QuickStart: Example-driven Web-based BLAST Tutorial*: Springer; 2008.
125. Rizvi SMD, Shakil S, Haneef M. A simple click by click protocol to perform docking: AutoDock 4.2 made easy for non-bioinformaticians. *EXCLI Journal*. 2013;12:831.
126. Hussein Z, Caleb OJ, Opara UL. Perforation-mediated modified atmosphere packaging of fresh and minimally processed produce—A review. *Food Packaging and Shelf Life*. 2015;6:7–20.
127. Zoete V, Cuendet MA, Grosdidier A, Michielin O. SwissParam: a fast force field generation tool for small organic molecules. *Journal of Computational Chemistry*. 2011;32(11):2359–68.
128. Schwede T, Kopp J, Guex N, Peitsch MC. SWISS-MODEL: an automated protein homology-modeling server. *Nucleic Acids Research*. 2003;31(13):3381–5.
129. Nnachi RC, Sui N, Ke B, Luo Z, Bhalla N, He D, et al. Biosensors for rapid detection of bacterial pathogens in water, food and environment. *Environment International*. 2022;166:107357.
130. Arboleya S, Bottacini F, O'Connell-Motherway M, Ryan CA, Ross RP, Van Sinderen D, et al. Gene-trait matching across the Bifidobacterium longum pan-genome reveals considerable diversity in carbohydrate catabolism among human infant strains. *BMC Genomics*. 2018;19(1):1–16.
131. Howe AC, Jansson JK, Malfatti SA, Tringe SG, Tiedje JM, Brown CT. Tackling soil diversity with the assembly of large, complex metagenomes. *Proceedings of the National Academy of Sciences*. 2014;111(13):4904–9.
132. Zeevi D, Korem T, Zmora N, Israeli D, Rothschild D, Weinberger A, et al. Personalized nutrition by prediction of glycemic responses. *Cell*. 2015;163(5):1079–94.
133. Huys G, Botteldoorn N, Delvigne F, De Vuyst L, Heyndrickx M, Pot B, et al. Microbial characterization of probiotics–Advisory report of the Working Group "8651 Probiotics" of the Belgian Superior Health Council (SHC). *Molecular Nutrition & Food Research*. 2013;57(8):1479–504.
134. Mohkam M, Nezafat N, Berenjian A, Mobasher MA, Ghasemi Y. Identification of Bacillus probiotics isolated from soil rhizosphere using 16S rRNA, recA, rpoB gene sequencing and RAPD-PCR. *Probiotics and Antimicrobial Proteins*. 2016;8:8–18.
135. Berendsen EM, Boekhorst J, Kuipers OP, Wells-Bennik MH. A mobile genetic element profoundly increases heat resistance of bacterial spores. *The ISME Journal*. 2016;10(11):2633–42.
136. Wallen-Russell C. The role of every-day cosmetics in altering the skin microbiome: a study using biodiversity. *Cosmetics*. 2018;6(1):2.
137. Singh R, Kumar M, Mittal A, Mehta PK. Microbial enzymes: industrial progress in 21st century. *3 Biotech*. 2016;6:1–15.
138. Costessi A, van den Bogert B, May A, Ver Loren van Themaat E, Roubos JA, Kolkman MA, et al. Novel sequencing technologies to support industrial biotechnology. *FEMS Microbiology Letters*. 2018;365(16):fny103.
139. De Filippis F, Parente E, Ercolini D. Metagenomics insights into food fermentations. *Microbial Biotechnology*. 2017;10(1):91–102.
140. Muhammed MK, Kot W, Neve H, Mahony J, Castro-Mejía JL, Krych L, et al. Metagenomic analysis of dairy bacteriophages: extraction method and pilot study on whey samples derived from using undefined and defined mesophilic starter cultures. *Applied and Environmental Microbiology*. 2017;83(19):e00888–17.
141. Lagier J-C, Khelaifia S, Alou MT, Ndongo S, Dione N, Hugon P, et al. Culture of previously uncultured members of the human gut microbiota by culturomics. *Nature Microbiology*. 2016;1(12):16203.
142. Kiousi DE, Rathosi M, Tsifintaris M, Chondrou P, Galanis A. Pro-biomics: omics technologies to unravel the role of probiotics in health and disease. *Advances in Nutrition*. 2021;12(5):1802–20.
143. McFarland LV, Evans CT, Goldstein EJ. Strain-specificity and disease-specificity of probiotic efficacy: a systematic review and meta-analysis. *Frontiers in Medicine*. 2018;5:124.
144. Kim H. Bioinformatics tools in clinical microbiology and infectious disease prevention algorithms. *Journal of Medical Biomedical and Applied Science*. 2018;6(9):159–65.

145. Bush K. Proliferation and significance of clinically relevant β-lactamases. *Annals of the New York Academy of Sciences*. 2013;1277(1):84–90.
146. Dark MJ. Whole-genome sequencing in bacteriology: state of the art. *Infection and Drug Resistance*. 2013;6:115–23.
147. Edwards DJ, Holt KE. Beginner's guide to comparative bacterial genome analysis using next-generation sequence data. *Microbial Informatics and Experimentation*. 2013;3:1–9.
148. Jolley KA, Maiden MC. BIGSdb: scalable analysis of bacterial genome variation at the population level. *BMC Bioinformatics*. 2010;11:1–11.
149. Aanensen DM, Feil EJ, Holden MT, Dordel J, Yeats CA, Fedosejev A, et al. Whole-genome sequencing for routine pathogen surveillance in public health: a population snapshot of invasive Staphylococcus aureus in Europe. *MBio*. 2016;7(3):10.1128/mbio. 00444-16.
150. Kwong JC, McCallum N, Sintchenko V, Howden BP. Whole genome sequencing in clinical and public health microbiology. *Pathology*. 2015;47(3):199–210.
151. Rahnama H, Pentland A. The new rules of data and privacy. *Harvard Business Review Retrieved*. 2022;9(6):23.
152. Lin Z, Owen AB, Altman RB. *Genomic Research and Human Subject privacy*: American Association for the Advancement of Science; 2004. p. 183.
153. Perlman DC, Gelpí-Acosta C, Friedman SR, Jordan AE, Hagan H. Perceptions of genetic testing and genomic medicine among drug users. *International Journal of Drug Policy*. 2015;26(1):100–6.
154. Bonomi L, Huang Y, Ohno-Machado L. Privacy challenges and research opportunities for genomic data sharing. *Nature Genetics*. 2020;52(7):646–54.
155. Galatas I, editor. The misuse and malicious uses of the new biotechnologies. In *Annales des Mines-Réalités industrielles*; 2017: Cairn/Softwin.
156. Munoz-Fambuena N, Mesejo C, González-Mas MC, Primo-Millo E, Agustí M, Iglesias DJ. Use or abuse of bioinformatic tools: a response to Samach. *Annals of Botany*. 2013;111(3):335–6.
157. Kesh S, Raghupathi W. Critical issues in bioinformatics and computing. *Perspectives in health information management/AHIMA, American Health Information Management Association*. 2004;1.
158. Shabani M, Chassang G, Marelli L. The Impact of the GDPR on the Governance of Biobank Research. In *GDPR and Biobanking: Individual Rights, Public Interest and Research Regulation Across Europe*, ed. S Slokenberga, O Tzortzatou, and J Reichel. Springer; 2021. p. 45–60.
159. Evans BJ, Jarvik GP. Impact of HIPAA's minimum necessary standard on genomic data sharing. *Genetics in Medicine*. 2018;20(5):531–5.
160. Shabani M, Borry P. Rules for processing genetic data for research purposes in view of the new EU General Data Protection Regulation. *European Journal of Human Genetics*. 2018;26(2):149–56.

13 Whole Genome Sequencing and Making Sense of Molecular Data

Arif Ullah, Rafiq Ahmad, Muhammad Imran, and Muhammad Ali Syed

13.1 INTRODUCTION

Whole genome sequencing (WGS) enables a researcher to read the entire genetic blueprint of an organism in one stretch. This modern holistic approach allows scientists to appreciate how genes guide the growth, development, and maintenance of a whole organism (1). WGS technique has changed our overall understanding of organisms like bacteria, viruses, fungi, and archaea in the field of microbiology. By sequencing the genomes of these organisms, scientists can analyze microbial diversity in great detail for the first time (2).

The ability of WGS to unveil the functional and structural traits of microbial genomes is among the key advantages of the technique (3). To this list belong genes which represent the lowest level of heredity and are regulatory regions of DNA responsible for transcription initiation within cells. Similarly, WGS can also detect regulatory elements that control gene expression, repeats which are regions of DNA that are repeated, and mobile elements, which are segments of DNA that can move around within the genome (4). In addition, WGS can also detect genetic mutations, which are alterations in the DNA sequence responsible for variation in characteristics and even prototypes of new species (5).

WGS can also reveal the dynamics of microbial interactions and relations provided by temporal sequences and information on differential expression (6). For instance, it is able to follow horizontal gene transfer, a way for organisms to exchange genetic material. This is the main force creating genetic variability and evolution. Moreover, WGS can give a clearer picture of phylogenetic trees, depicting the evolutionary interconnections of different species (7). It may also give insights into co-evolution, the phenomenon where multiple species mutually influence each other's evolution. Other interesting phenomena that WGS may aid in investigating are symbiosis (interaction between two species), pathogenicity (ability to cause disease), and microbial resistance to drugs like antibiotics, one of the major global health concerns (8).

Shotgun sequencing and targeted sequencing are two differing techniques of WGS. Shotgun sequencing is random shattering of DNA sequences into a lot of minuscule portions and then joining of the sequence by focusing on sections of overlap. This method works well for new genome sequencing and for studying the genetic diversity within a species or the corresponding community of organisms (metagenomics) (9). It is also especially useful for detecting large structural variations in the genome. It utilizes next-generation sequencing (NGS) platforms such as Illumina, PacBio, Oxford Nanopore, and Ion Torrent that can produce a huge quantity of short or long reads. These reads span the entire genome from end to end, capturing variable depths and coverages, depending on the platform (1, 10).

DOI: 10.1201/9781003410164-13

In contrast to shotgun sequencing, targeted sequencing enriches a subset of desired sequences, but excludes undesired sequences (11). It is a cost-effective method for obtaining sequence reads of a targeted region of the genome. This technique utilizes nucleic acid probes that hybridize to the regions of interest such as genes, exons, or single nucleotide polymorphisms (SNPs). As a result of hybridization, one may get high quality and accurate reads that cover the regions of interest in high depth and coverage. Targeted sequencing is particularly optimal for resequencing of known genomes, genotyping of variants, and functional analysis of genes (1, 12). Both of these techniques have their own merits, and thus are selected depending on the research question involved. Nevertheless, it should be considered that the quality and precision of the WGS data may be influenced by a number of factors, such as the DNA sample quality and quantity, the performance of the sequencing platform, and the design and performance of the sequencing protocol (2).

The DNA sample used for sequencing must be of good quality and in adequate quantity to run the test. It must be pure, whole, and representative of the organism or population under study. Contamination or degradation of the DNA sample may lead to errors in the sequencing data (13).

Another important factor to be considered is the choice of sequencing platform. Different platforms, such as Illumina, PacBio, Oxford Nanopore, and Ion Torrent, have distinct strengths and weaknesses in terms of read length, error rate, throughput, and cost. The platform should be selected depending on the particular necessities of the sequencing project like the genome complexity, project size, and budget (1).

The design and execution of the sequencing protocol can also influence the quality and accuracy of the data. The protocol should be optimized for the corresponding DNA sample and sequencing platform. It will comprise steps for quality control, such as verifying the quality and the quantity of the DNA, fabricating the sequencing library, and testing the sequencing results (2). The quality and accuracy of the sequencing reads are equality important. They should be evaluated for possible sources of error, biases, contaminants, and duplicates. Any minor issue with the reads may result in inaccuracy of the assembled genome or detected variants (13).

Ultimately, the use of data analysis helps in interpreting the sequencing data. The analysis should be in accordance with the type of data, the quality of data, and the goal of the analysis. It should have sections for read mapping, variant calling, genome assembly, annotation, and interpretation. Likewise, analysis should be validated and verified by independent methods, for example, polymerase chain reaction (PCR), Sanger sequencing, or other sequencing platforms (2). Taken together, the quality and precision of WGS data are an outgrowth of a mix of several parameters. Adequate evaluation and optimization of these components contribute greatly to the robustness of WGS in microbiology and other biological disciplines.

13.2 CHALLENGES AND LIMITATIONS OF WGS DATA ANALYSIS

Analysis of WGS data may be a complex and technically challenging job. One of the major challenges is storage of a vast amount of data generated through sequencing. One run of sequencing may generate gigabytes to terabytes of data depending on the genome size and sequencing depth, demanding large storage capabilities, which are costly, particularly when working with multiple genomes or large-scale projects (14). Furthermore, the process of transferring data from the sequencing platform to the analysis platform may also be challenging and problematic. The speed and width of transmission may be restricted by the network infrastructure and the data format. Efficient data compression and transfer methods are required to handle large pieces of data.

Data integration is combining and comparing WGS data between multiple sources, platforms, and formats. It may be obstructed by the absence of standards and interoperability. Standardization and formatting of genomic data toward universal unification is being worked upon to achieve data integration. The process called data annotation, the manual and automatic labeling of the particular features and roles of the WGS data, is difficult due to the complexity and diversity of the microbial genomes. The presence of incompleteness and inconsistency in existing databases and tools necessitates novel methods to accurately annotate genomic datasets (15).

Data visualization is indispensable for representing and analyzing the WGS data effectively. Nevertheless, the great dimensionality and variability of the data and the balance between straightforwardness and details make it a difficult thing to visualize (16). At the end, data sharing also comprises releasing the WGS data for use by the scientific community and the public. This is limited by ethical, legal, and social consequences, including privacy issues and data ownership problems, for instance (17).

The ethical, legal, and social implications (ELSI) of WGS data may become a serious challenge and must be considered when analyzing data. These entail the rights and obligations of the data producers, users, and the human subjects. The key areas to be covered are privacy, security, consent, and ownership. Confidentiality is an utmost issue while performing WGS data analysis. It requires regulation of the information collected from WGS data about individuals. This information may comprise identity, health status, disease predisposition, ancestry, kinship, etc. The potential for reidentification, linkage, or likely inference from the WGS data violates privacy, posing major ethical and legal problems (18, 19).

Protection of the data, i.e., data security, is also important. It has to do with the prevention and detection of unauthorized access and use, or misuse, of WGS data. The WGS data may be vulnerable to hacking, leaking, or disclosure, which compromises the confidentiality of the subjects (20).

Consent is a main ethical principle in any research involving human subjects, including WGS. It is about the consent provided by data subjects to be captured in the WGS data generation process, without any coercion or compelled engagement in tasks. Subjects' awareness, cognizance, and perception of data protection can heavily influence the process of obtaining consent. Ownership here denotes the rights and interests that the data subjects and the data producers have over the WGS data. Such rights and interests may include access, control, benefits, and recognition of the WGS data. Issues with data ownership arise from conflicts, disputes, or even agreements between data subjects and data producers (21). Given ELSI of WGS data analysis are intrinsic and multilevel, they demand meticulous discussion and constant conversation among all the stakeholders—scientists, ethicists, legal experts, and the public—to guarantee the appropriate and ethical utilization of WGS data (17).

13.3 COMMON STEPS AND WORKFLOW OF WGS DATA ANALYSIS

WGS data analysis is a step-by-step process with various stages, each of which helps bring a deeper understanding of genomic data. Data preprocessing constitutes the first stage in the WGS data analysis. It consists of enhancing the quality and precision of the sequencing reads through base trimming and removal of low-quality reads, correction of sequencing errors, and removal of biases, contaminants, and duplicates (22). This step is an important step of data analysis, as it can greatly improve the performance and robustness of the following analysis steps, which in turn contribute to the accuracy of the outcomes (23) (Figure 13.1).

In assembly, the genome sequence is reconstituted from the sequencing reads. This can be attained either by working with a reference genome (reference-based assembly) or by employing the overlaps between the reads (de novo assembly). Genome assembly can produce contigs, scaffolds, or whole genomes, depending on coverage and quality of the reads, and on the complexity and size of the genome. Alignment implies that a sequencing read or an assembled genome is mapped to a reference genome or a reference database. This procedure employs advanced algorithms that consider variations, e.g., mismatches, insertions, deletions, and rearrangements. The matching process in alignment identifies the similarities and differences among the query and reference, such as single nucleotide polymorphisms (SNPs), insertions and deletions (indels), structural variations (SVs), and copy number variations (CNVs) (24). Genome comparison is the job of comparing multiple genomes of different species or samples. This includes methods calculating the distances/similarities/divergences between the genomes, such as phylogenetic trees, clustering, principal component

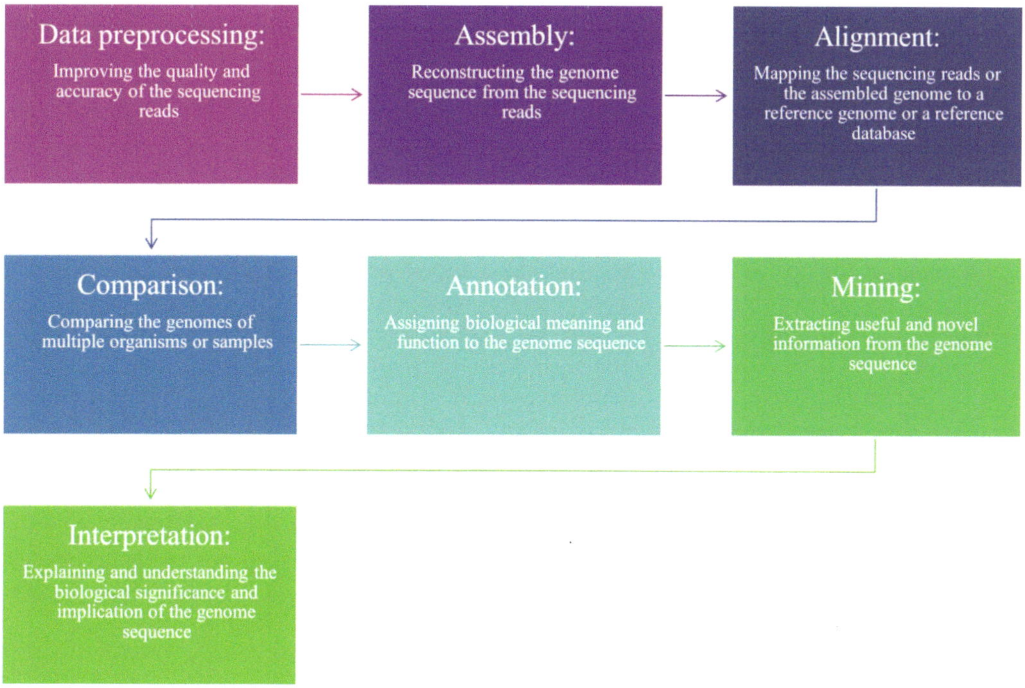

FIGURE 13.1 Common workflow of whole genome sequencing data analysis.

analysis, network analysis, etc. Comparison can unveil the phylogenetic relationships, population structure, genetic diversity, and genomic variation of the organisms or samples (25). As sequencing technologies are developing at a rapid pace, so will methods and tools in WGS data analysis.

Annotation, mining, and interpretation are the three crucial points of the analysis of whole genome sequencing (WGS) data. Each of these steps has its distinct function in eliciting biologically significant information from a raw genome. Annotation, viewed as the assignment of biological meaning to genome sequence, is the procedure. Here such genomic features as genes, operons, promoters, regulatory elements, repeats, mobile elements, and mutations are to be defined. In carrying out this process, researchers employ the methods that forecast, deduce, or determine these attributes through an experimental approach. The annotation step may bring forth informative findings concerning the functional and structural traits of the genome, i.e., gene expression, gene regulation, gene interaction, gene ontology, pathway analysis, and metabolic modeling (15).

Mining stands for the process where we extract useful and novel information out of the gene sequence. This encompasses pattern, motif, signature, or association search using bioinformatics tools. Genome mining enables the discovery of new and interesting aspects, for example, novel genes, novel operons, novel promoters, novel regulatory elements, novel repeats, novel mobile elements, and novel mutations (26, 27). Interpretation is the final step and is the process where the biological importance and implication of the genome sequence are revealed and understood. It involves using tools that integrate, synthesize, or model the genome, providing biological answers or solutions to problems. Interpretation offers, as regards all aspects and at all levels of the organism or the sample, insights into origin, evolution, adaptation, diversity, and function. It also enables the understanding of various aspects, including the phylogeny, taxonomy, ecology, epidemiology, pathogenicity, virulence, and antibiotic resistance of such an organism or sample (28). Equipped with suitable tools and methods, abundant information can be retrieved from WGS data, which can aid in exploring the biology of the organism being examined.

WGS data analysis is a multifaceted process that can follow different pathways dependent on the goal and analysis type. The workflows consist of phylogenetic analysis, comparative genomics, functional genomics, metagenomics, epidemiology, and diagnostics. Phylogenetic analysis is a process that comprises building a tree showing the evolutionary history and relationship of organisms or samples. This is done through the inference of phylogenetic trees or networks from the genome sequence or genomic variation. Phylogenetic analysis gives signals about the origin, diversity and taxonomy of organisms or samples. It can help trace the ancestry of particular species, explain how species have evolved, and classify organisms based on their evolutionary relationships (29).

The more common workflow when using comparative genomics is comparing the genomes of different organisms or different samples. It shows how the genomes differ and resemble; such parameters as the core and accessory genome and pan-genome, synteny and collinearity, homology and orthology are implicated. Comparative genomics unveils the genomic diversity, variation, and also adaptation of the organisms or samples. It could help to detect genes conserved across the species (core genome), genes specific to certain species (accessory genome), and also the total gene set of a group of closely related species (pan-genome). Comparative genomics pinpoints synteny (blocks of order conserved in two sets of chromosomes joined for comparison) and also collinearity (linear sequence of genes of two species being similar) (30, 31).

Functional genomics is a workflow which studies the correlation between the genome of an organism and its phenotype. The eventual goal of functional genomics is to unravel the biological role of all genes in a genome, determining how they interact with each other and with the environment. This is accomplished by employing a range of experimental and computational approaches such as gene expression profiling, protein–protein interaction mapping. and mutagenesis studies (32).

Metagenomics is a process that involves the study of genetic material, directly from environmental samples. Metagenomics allows the study of communities of organisms directly in their natural environments, eliminating the need for the isolation and also cultivation of the individual species in the lab. This enables an understanding of the richness and the performance of the microbial communities in different environments, for example, soil, water, and also the human body (33, 34).

Epidemiology and diagnostics are workflows that mandate the use of WGS data to trace disease epidemics and diagnose diseases. In epidemiology, whole genome sequencing can be employed as a means of tracking the propagation of infectious diseases and a way of pinpointing the origin of outbreaks. WGS can be used in diagnostics to identify the cause of an infection and decide on the proper treatment (35, 36). Every workflow unveils peculiar facets of genome structure, function, diversity, and evolution.

13.4 TOOLS AND RESOURCES FOR WGS DATA ANALYSIS

To facilitate and ease the WGS data analysis process and outcome, a number of databases, softwares, pipelines, platforms, and standards have been developed. These resources constitute the foundation for WGS data analysis, furnishing scientists with the appropriate tools and a conducive environment where they can systematically and conveniently analyze the genomic information.

Databases are vital to WGS data analysis as they function as the repositories for the genomic data and also the accompanying metadata. They house a variety of information such as DNA sequences, genome annotation, genetic variations, functional annotations, and also phylogenetic associations. Some popular databases are widely used in WGS data analysis, such as NCBI GenBank, NCBI RefSeq, NCBI SRA, EMBL-EBI ENA, EMBL-EBI, EMBL, EMBL-EBI MGnify, IMG, PATRIC, and UniProt. These databases are a vital source of extensive genomic data for researchers, allowing the examination and comparison of genomes across several organisms (Table 13.1).

Software tools are non-negotiable for performing various stages in the WGS data analysis (46). They cover a wide range of functionalities: data preprocessing, assembly, alignment, comparison, annotation, mining, and interpretation. Researchers utilize several software programs to accomplish

TABLE 13.1
Some Examples of Databases for WGS Data Analysis

Name of Database	Resource Description
NCBI GenBank	A comprehensive archive of all publicly available DNA sequences, including whole genomes and their annotations (37)
NCBI RefSeq	A curated and non-redundant set of reference sequences that serve as a foundation for comparative genomics and functional annotation (38)
NCBI SRA	A public archive of high-throughput sequencing data, including raw sequence reads, alignment files, and variant calls (39)
EMBL-EBI ENA	A comprehensive archive of the world's nucleotide sequences, including raw sequencing data, sequence assemblies, and functional annotations (40)
EMBL-EBI EMBL	A nucleotide sequence database that provides a comprehensive view of the genetic information for various organisms, including whole genomes and their functional annotations (41)
EMBL-EBI MGnify	A large-scale metagenomics analysis platform that provides a comprehensive view of the microbial and viral communities in different environments and hosts (40)
PATRIC	A comprehensive bioinformatics resource for studying bacterial pathogens, including whole genomes, genome annotations, and comparative genomics tools (42)
IMG	A comprehensive genome-scale metabolic network reconstruction platform that integrates genomic, metabolic, and phenotypic data for various organisms (43)
KEGG	A database of biological pathways, functions, and networks (44)
UniProt	A database of protein sequences, structures, and functions (45)

TABLE 13.2
Some Examples of Tools for WGS Data Analysis

Tool Name	Resource Description
Trimmomatic	A tool for trimming and filtering low-quality reads and adapters (47)
FastQC	A tool for quality control and assessment of sequencing reads (48)
SPAdes	A tool for de novo assembly of short reads from Illumina, Ion Torrent, and PacBio (49)
MEGAHIT	A tool for de novo assembly of metagenomic data from short reads (50)
BWA	A tool for aligning short reads to a reference genome using Burrows-Wheeler transform (51)
Bowtie	A tool for aligning short reads to a reference genome using FM-index (52)
SAMtools	A tool for manipulating and analyzing SAM/BAM files (53)
GATK	A tool for variant discovery and genotyping from high-throughput sequencing data (54)
Mauve	A tool for multiple genome alignment and visualization (55)
MUMmer	A tool for rapid alignment of large genomes and detection of structural variation (56)
Prokka	A tool for rapid annotation of prokaryotic genomes (57)
RAST	A tool for annotation and comparative analysis of bacterial genomes (58)
BLAST	A tool for finding regions of similarity between biological sequences (59)
HMMER	A tool for searching and aligning protein sequences using profile hidden Markov models (60)
PhyML	A tool for inferring phylogenetic trees from DNA or protein sequences (61)

this well. Some well-known software tools that are utilized in WGS data analysis are Trimmomatic, FastQC, SPAdes, MEGAHIT, BWA, Bowtie, SAMtools, GATK, Mauve, MUMmer, Prokka, RAST, BLAST, HMMER, and PhyML (Table 13.2). They incorporate complex algorithms and deliver strong efficiency to garner vital inferences from WGS.

Pipelines are an essential tool in rendering the WGS data analysis workflow more efficient and also in automating it (62). They unite different software tools and analytical stages to allow for convenient handling of large-scale genomics data. Data preprocessing, assembly, alignment, comparison, annotation, mining, and interpretation are facilitated by the pipelines, thus ensuring a systematic and reproducible analysis. Well-known WGS data analysis pipelines are Galaxy, Snakemake, CGE, Nullarbor, and IRIDA. These pipelines offer researchers preconfigured workflows and scripts, hence making the analytical process straightforward, and the reproducibility of outcomes is enhanced.

Platforms constitute the base for computational infrastructure and WGS data analysis. They provide storage, computational resources, data transfer, data integration, annotation, visualization, and data sharing capabilities (69). Platforms allow researchers to manipulate and perform a sophisticated analysis of the big WGS data. The key platforms employed in WGS data analysis are AWS, Google Cloud, Microsoft Azure, BaseSpace, CLC Genomics, and DNAnexus. These platforms provide researchers with a flexible and scalable environment in which they can perform many computational analyses, storage, and collaboration.

Standards play a key role in ensuring the interoperability as well as compatibility of WGS data analysis (76). They specify the formats and protocols for data representation, exchange, annotation, visualization, and also distribution. Adhering to the common standards, researchers will enable a smooth sharing and integration of genomic information, enhancing collaboration and comparability across various research projects. The basic standards in WGS data analysis are FASTA, FASTQ, SAM, BAM, VCF, GFF, GBK, Newick, Nexus, and also BioPAX. These standards guarantee conformity and let researchers exactly understand and process WGS data via various software and also databases.

The wide range of databases, software, pipelines, platforms, and standards for whole genome sequencing (WGS) data processing sufficiently supports genome research. These resources equip researchers with the required instruments, systems, and rules to delve into and investigate WGS data widely (86). With the use of these resources, researchers can discover new aspects about the structure, function, and evolution of genomes, which will help in improving fields such as medicine, agriculture, and evolutionary biology. The constantly manifest development and improvement of these resources will unquestionably speed up further the advancement of WGS data analysis and genomics research.

Some commonly used criteria to measure the performance and usability of the available tools and resources for WGS data analysis are accuracy, sensitivity, specificity, precision, recall, F1-score, ROC curve, AUC, speed, memory, scalability, robustness, reproducibility, user-friendliness, documentation, and also support. Accuracy is a core metric of any computational tool. It relates to the percentage of the correct results or predictions produced by the tool or resource. A high accuracy rate signifies a very trustworthy and also dependable instrument. Sensitivity and specificity are two very important indicators in the case of binary classification problems. Sensitivity (it is also referred to as the true positive rate) is the proportion of the actual positives that are correctly identified. Specificity, however, represents the ratio of the negative examples that are correctly identified by the model. Precision and recall are also vital metrics. Precision is the fraction of the true positives of the tool or resource for all the positives produced. Also known as the sensitivity, it is the ratio of true positives to all the relevant results. F1-scores are the harmonic mean of the precision and recall, giving a single measure that equally considers both. ROC curve is a graphic plot that demonstrates the diagnostic capability of a binary classification system as its discrimination threshold is varied. The AUC, also known as the Area Under the ROC Curve, gives us a measure of performance, across all classification thresholds. When it comes to dealing with datasets, two crucial factors are speed (how long it takes to run or complete a tool or resource) and memory (the space needed to store or load the tool or resource). Scalability refers to how a tool or resource can handle complex datasets. Robustness relates to its ability to handle errors or uncertainties in the data. Reproducibility is producing reliable results or outputs. User-friendliness describes how easy it is to use and install the tool or resource. Documentation refers to the availability and quality of instructions and manuals for using the tool or resource. Lastly, support refers to being able to access help and feedback for the tool or resource.

These metrics offer a framework for assessing the effectiveness and user-friendliness of tools and resources for analyzing whole genome sequencing (WGS) data. By taking these factors into

TABLE 13.3
Some Examples of Pipelines for WGS Data Analysis

Pipeline Name	Resource Description
Galaxy	A web-based platform for data integration and analysis of genomic data (63)
Nextflow	A framework for scalable and reproducible data analysis pipelines (64)
Snakemake	A workflow management system for data analysis pipelines (65)
CGE	A web-based platform for microbial genomics analysis (66)
Nullarbor	A pipeline for public health microbiology and outbreak investigation (67)
IRIDA	A platform for genomic epidemiology and outbreak investigation (68)

TABLE 13.4
Platforms Available for WGS Data Analysis

Platform Name	Resource Description
AWS	A cloud computing service that provides infrastructure and resources for data storage and analysis (70)
Google Cloud	A cloud computing service that provides infrastructure and resources for data storage and analysis (71)
Microsoft Azure	A cloud computing service that provides infrastructure and resources for data storage and analysis (72)
BaseSpace	A cloud-based platform for genomic data management and analysis (73)
CLC Genomics	A platform for genomic data analysis and visualization (74)
DNAnexus	A platform for genomic data management, analysis, and collaboration (75)

TABLE 13.5
Some Examples of Common Formats and Standards for WGS Data Analysis

Format Name	Resource Description
FASTA	A format for representing nucleotide or protein sequences (77)
FASTQ	A format for representing nucleotide sequences and their quality scores (78)
SAM	A format for representing aligned sequencing reads (79)
BAM	A binary version of SAM format (80)
VCF	A format for representing genomic variants (81)
GFF	A format for representing genomic features and annotations (82)
GBK	A format for representing genomic sequences and annotations (82)
Newick	A format for representing phylogenetic trees (83)
Nexus	A format for representing phylogenetic data and metadata (84)
BioPAX	A format for representing biological pathways and networks (85)

account, researchers can make informed choices regarding the tools and resources they employ in their work (Tables 13.3, 13.4, and 13.5).

13.5 FUTURE TRENDS OF WGS DATA ANALYSIS

WGS data analysis is a rapidly growing area of biomedical science, with constant changes and improvements in sequencing technologies, analysis methods, and applications, as well as the emergence of new challenges (87).

With the passage of time, novel sequencing technologies are being introduced for WGS data generation and analysis. New platforms and methods may provide more accurate, faster, cheaper, portable WGS data. Examples include nanopore sequencing, single-cell sequencing, single-molecule sequencing, and in situ sequencing. These technologies have brought revolution in the field of genomics by enabling rapid and cost-effective WGS data analysis.

Moreover, sophisticated algorithms as well as models which can analyze more complicated, diverse and dynamic WGS data are being introduced. Techniques such as machine learning, deep learning, artificial intelligence, and network analysis are being increasingly employed for interpreting the huge volumes of data generated through WGS. By so doing they can uncover patterns or insights that may not be found through traditional ways of analyzing.

Different fields are now adopting different aspects of WGS data analysis for different domains, leading to new perspectives on how it is used. Some examples include synthetic biology, personalized medicine, precision agriculture, and biotechnology. In these areas, WGS data analysis can bring unfamiliar ideas and answers to fundamental as well as applied questions and problems.

Existing and emerging issues and problems with WGS data analysis include data quality, data integration, data annotation, data visualization, data sharing, the privacy of the subjects' information, security of research findings and materials at hand, consent in the sample collection processes, as well as ownership of information (88).

WGS has extraordinary potential for transforming life sciences, microbiology, biotechnology, as well as the medical sciences (89). It is hoped that through WGS, we shall be able to get deep insights into basic and applied questions regarding organisms and their interactions with other organisms or the environment (89). WGS has been able to play a key role in fingerprinting bacterial pathogens for epidemiological surveillance purposes, outbreak detection, and infection control. In addition, it has been used to identify genetic disorders, characterize mutations driving cancer progression, and trace disease outbreaks.

REFERENCES

1. Goodwin S, McPherson JD, McCombie WR. Coming of age: ten years of next-generation sequencing technologies. *Nature Reviews Genetics*. 2016;17(6):333–51.
2. Didelot X, Walker AS, Peto TE, Crook DW, Wilson DJ. Within-host evolution of bacterial pathogens. *Nature Reviews Microbiology*. 2016;14(3):150–62.
3. Pérez-Cobas AE, Gomez-Valero L, Buchrieser C. Metagenomic approaches in microbial ecology: an update on whole-genome and marker gene sequencing analyses. *Microbial Genomics*. 2020;6(8):e000409.
4. Oniciuc EA, Likotrafiti E, Alvarez-Molina A, Prieto M, Santos JA, Alvarez-Ordóñez A. The present and future of whole genome sequencing (WGS) and whole metagenome sequencing (WMS) for surveillance of antimicrobial resistant microorganisms and antimicrobial resistance genes across the food chain. *Genes*. 2018;9(5):268.
5. Zankari E, Allesøe R, Joensen KG, Cavaco LM, Lund O, Aarestrup FM. PointFinder: a novel web tool for WGS-based detection of antimicrobial resistance associated with chromosomal point mutations in bacterial pathogens. *Journal of Antimicrobial Chemotherapy*. 2017;72(10):2764–8.
6. Hazards EPoB, Koutsoumanis K, Allende A, Alvarez-Ordóñez A, Bolton D, Bover-Cid S, et al. Whole genome sequencing and metagenomics for outbreak investigation, source attribution and risk assessment of food-borne microorganisms. *EFSA Journal*. 2019;17(12):e05898.
7. Schürch A, Arredondo-Alonso S, Willems R, Goering R. Whole genome sequencing options for bacterial strain typing and epidemiologic analysis based on single nucleotide polymorphism versus gene-by-gene–based approaches. *Clinical Microbiology and Infection*. 2018;24(4):350–4.
8. Fraser C, Alm EJ, Polz MF, Spratt BG, Hanage WP. The bacterial species challenge: making sense of genetic and ecological diversity. *Science*. 2009;323(5915):741–6.
9. Bell KL, Petit III RA, Cutler A, Dobbs EK, Macpherson JM, Read TD, et al. Comparing whole-genome shotgun sequencing and DNA metabarcoding approaches for species identification and quantification of pollen species mixtures. *Ecology Evolution*. 2021;11(22):16082–98.

10. Loman NJ, Pallen MJ. Twenty years of bacterial genome sequencing. *Nature Reviews Microbiology*. 2015;13(12):787–94.
11. De Koning TJ, Jongbloed JD, Sikkema-Raddatz B, Sinke RJ. Targeted next-generation sequencing panels for monogenetic disorders in clinical diagnostics: the opportunities and challenges. *Expert Review of Molecular Diagnostics*. 2015;15(1):61–70.
12. Mertes F, Elsharawy A, Sauer S, van Helvoort JMLM, van der Zaag PJ, Franke A, et al. Targeted enrichment of genomic DNA regions for next-generation sequencing. *Briefings in Functional Genomics*. 2011;10(6):374–86.
13. Quail MA, Smith M, Coupland P, Otto TD, Harris SR, Connor TR, et al. A tale of three next generation sequencing platforms: comparison of Ion Torrent, Pacific Biosciences and Illumina MiSeq sequencers. *BMC Genomics*. 2012;13:341.
14. Stein LD. The case for cloud computing in genome informatics. *Genome Biology*. 2010;11(5):207-.
15. Yandell M, Ence D. A beginner's guide to eukaryotic genome annotation. *Nature Reviews Genetics*. 2012;13(5):329–42.
16. Pavlopoulos GA, Soldatos TG, Barbosa-Silva A, Schneider R. A reference guide for tree analysis and visualization. *BioData Mining*. 2010;3(1):1.
17. Knoppers BM, Harris JR, Tassé AM, Budin-Ljøsne I, Kaye J, Deschênes M, et al. Towards a data sharing code of conduct for international genomic research. *Genome Medicine*. 2011;3(7):46.
18. Kaye J, Boddington P, de Vries J, Hawkins N, Melham K. Ethical implications of the use of whole genome methods in medical research. *European Journal of Human Genetics: EJHG*. 2010;18(4):398–403.
19. Gymrek M, McGuire AL, Golan D, Halperin E, Erlich Y. Identifying personal genomes by surname inference. *Science*. 2013;339(6117):321–4.
20. Malin B, Sweeney L. How (not) to protect genomic data privacy in a distributed network: using trail re-identification to evaluate and design anonymity protection systems. *Journal of Biomedical Informatics*. 2004;37(3):179–92.
21. Bovenberg J, Gevers S, Meulenkamp T, Smets E. Biobank research: reporting results to individual participants. *European Journal of Health Law*. 2009;16(3):229–47.
22. Schmieder R, Edwards R. Quality control and preprocessing of metagenomic datasets. *Bioinformatics (Oxford, England)*. 2011;27(6):863–4.
23. Zhou Q, Su X, Wang A, Xu J, Ning K. QC-Chain: fast and holistic quality control method for next-generation sequencing data. *PLoS One*. 2013;8(4):e60234–e.
24. Li H, Durbin R. Fast and accurate short read alignment with Burrows-Wheeler transform. *Bioinformatics (Oxford, England)*. 2009;25(14):1754–60.
25. Huson DH, Bryant D. Application of phylogenetic networks in evolutionary studies. *Molecular Biology and Evolution*. 2005;23(2):254–67.
26. Bailey TL, Boden M, Buske FA, Frith M, Grant CE, Clementi L, et al. MEME SUITE: tools for motif discovery and searching. *Nucleic Acids Research*. 2009;37(Web Server issue):W202–W8.
27. Nguyen NTT, Contreras-Moreira B, Castro-Mondragon JA, Santana-Garcia W, Ossio R, Robles-Espinoza CD, et al. RSAT 2018: regulatory sequence analysis tools 20th anniversary. *Nucleic Acids Research*. 2018;46(W1):W209–W14.
28. Storz JF. Causes of molecular convergence and parallelism in protein evolution. *Nature Reviews Genetics*. 2016;17(4):239–50.
29. Kumar S, Stecher G, Tamura K. MEGA7: Molecular Evolutionary Genetics Analysis version 7.0 for bigger datasets. *Molecular Biology and Evolution*. 2016;33(7):1870–4.
30. Tettelin H, Masignani V, Cieslewicz MJ, Donati C, Medini D, Ward NL, et al. Genome analysis of multiple pathogenic isolates of Streptococcus agalactiae: implications for the microbial "pan-genome". *Proceedings of the National Academy of Sciences of the United States of America*. 2005;102(39):13950–5.
31. Vernikos G, Medini D, Riley DR, Tettelin H. Ten years of pan-genome analyses. *Current Opinion Microbiology*. 2015;23:148–54.
32. Hieter P, Boguski M. Functional genomics: it's all how you read it. *Science*. 1997;278(5338):601–2.
33. Handelsman J. Metagenomics: application of genomics to uncultured microorganisms. *Microbiology and Molecular Biology Reviews: MMBR*. 2004;68(4):669–85.
34. Wooley JC, Godzik A, Friedberg I. A primer on metagenomics. *PLoS Computational Biology*. 2010;6(2):e1000667–e.

35. Köser CU, Ellington MJ, Peacock SJ. Whole-genome sequencing to control antimicrobial resistance. *Trends in Genetics: TIG*. 2014;30(9):401–7.
36. Loman NJ, Constantinidou C, Christner M, Rohde H, Chan JZM, Quick J, et al. A culture-independent sequence-based metagenomics approach to the investigation of an outbreak of Shiga-Toxigenic Escherichia coli O104:H4. *JAMA*. 2013;309(14):1502.
37. Sayers EW, Cavanaugh M, Clark K, Pruitt KD, Schoch CL, Sherry ST, et al. GenBank. *Nucleic Acids Research*. 2022;50(D1):D161–D4.
38. Farrell CM, Goldfarb T, Rangwala SH, Astashyn A, Ermolaeva OD, Hem V, et al. RefSeq functional elements as experimentally assayed nongenic reference standards and functional interactions in human and mouse. *Genome Research*. 2022;32(1):175–88.
39. Sayers EW, O'Sullivan C, Karsch-Mizrachi I. *Using GenBank and SRA. Plant Bioinformatics: Methods and Protocols*: Springer; 2022. p. 1–25.
40. Thakur M, Bateman A, Brooksbank C, Freeberg M, Harrison M, Hartley M, et al. Embl's European Bioinformatics Institute (embl-ebi) in 2022. *Nucleic Acids Research*. 2023;51(D1):D9–D17.
41. Madeira F, Pearce M, Tivey AR, Basutkar P, Lee J, Edbali O, et al. Search and sequence analysis tools services from EMBL-EBI in 2022. *Nucleic Acids Research*. 2022;50(W1):W276–W9.
42. Davis JJ, Wattam AR, Aziz RK, Brettin T, Butler R, Butler RM, et al. The PATRIC Bioinformatics Resource Center: expanding data and analysis capabilities. *Nucleic Acids Research*. 2020;48(D1):D606–D12.
43. Chen I-MA, Chu K, Palaniappan K, Ratner A, Huang J, Huntemann M, et al. The IMG/M data management and analysis system v. 6.0: new tools and advanced capabilities. *Nucleic Acids Research*. 2021;49(D1):D751–D63.
44. Kanehisa M, Sato Y, Kawashima M. KEGG mapping tools for uncovering hidden features in biological data. *Protein Science*. 2022;31(1):47–53.
45. Bateman A, Martin M-J, Orchard S, Magrane M, Ahmad S, Alpi E, et al. UniProt: the universal protein knowledgebase in 2023. *Nucleic Acids Research*. 2022;51(D1):D523–D531.
46. Meggendorfer M, Jobanputra V, Wrzeszczynski KO, Roepman P, de Bruijn E, Cuppen E, et al. Analytical demands to use whole-genome sequencing in precision oncology. *Seminars in Cancer Biology*. 2022;84:16–22.
47. Sewe SO, Silva G, Sicat P, Seal SE, Visendi P. Trimming and validation of illumina short reads using Trimmomatic, Trinity assembly, and assessment of RNA-seq data. In *Methods in Molecular Biology*: 2022;2443:211–32.
48. Ward CM, To T-H, Pederson SM. ngsReports: a Bioconductor package for managing FastQC reports and other NGS related log files. *Bioinformatics*. 2020;36(8):2587–8.
49. Prjibelski A, Antipov D, Meleshko D, Lapidus A, Korobeynikov A. Using SPAdes de novo assembler. *Current Protocols in Bioinformatics*. 2020;70(1):e102.
50. Xie H, Yang C, Sun Y, Igarashi Y, Jin T, Luo F. PacBio long reads improve metagenomic assemblies, gene catalogs, and genome binning. *Frontiers in Genetics*. 2020;11:516269.
51. Jung Y, Han D. BWA-MEME: BWA-MEM emulated with a machine learning approach. *Bioinformatics*. 2022;38(9):2404–13.
52. Kumar P, Vo T, Cha M, Visheratina A, Kim J-Y, Xu W, et al. Photonically active bowtie nanoassemblies with chirality continuum. *Nature*. 2023;615(7952):418–24.
53. Danecek P, Bonfield JK, Liddle J, Marshall J, Ohan V, Pollard MO, et al. Twelve years of SAMtools and BCFtools. *Gigascience*. 2021;10(2):giab008.
54. Babadi M, Fu JM, Lee SK, Smirnov AN, Gauthier LD, Walker M, et al. GATK-gCNV enables the discovery of rare copy number variants from exome sequencing data. *Nature Genetics*. 2023;55(9):1589–97.
55. Darling AC, Mau B, Blattner FR, Perna NT. Mauve: multiple alignment of conserved genomic sequence with rearrangements. *Genome Research*. 2004;14(7):1394–403.
56. Cirkoska E, Kishore C, Bose SC, Mubarakali A, Reshmi T, Marina N, editors. A study on suffix trees and their applications in genome sequences using MUMmer. *Computer Networks and Inventive Communication Technologies: Proceedings of Third ICCNCT 2020*; 2021: Springer.
57. Seemann T. Prokka: rapid prokaryotic genome annotation. *Bioinformatics*. 2014;30(14):2068–9.
58. Aziz RK, Bartels D, Best AA, DeJongh M, Disz T, Edwards RA, et al. The RAST Server: rapid annotations using subsystems technology. *BMC Genomics*. 2008;9(1):1–15.
59. Madden T. The BLAST sequence analysis tool. *The NCBI Handbook*. 2013;2:425–36.

60. Potter SC, Luciani A, Eddy SR, Park Y, Lopez R, Finn RD. HMMER web server: 2018 update. *Nucleic Acids Research*. 2018;46(W1):W200–W4.
61. Guindon S, Dufayard J-F, Lefort V, Anisimova M, Hordijk W, Gascuel O. New algorithms and methods to estimate maximum-likelihood phylogenies: assessing the performance of PhyML 3.0. *Systematic Biology*. 2010;59(3):307–21.
62. Ahmed Z, Renart EG, Zeeshan S. Genomics pipelines to investigate susceptibility in whole genome and exome sequenced data for variant discovery, annotation, prediction and genotyping. *PeerJ*. 2021;9:e11724.
63. Chappell K, Francou B, Habib C, Huby T, Leoni M, Cottin A, et al. Galaxy is a suitable bioinformatics platform for the molecular diagnosis of human genetic disorders using high-throughput sequencing data analysis: five years of experience in a clinical laboratory. *Clinical Chemistry*. 2022;68(2):313–21.
64. Di Tommaso P, Chatzou M, Floden EW, Barja PP, Palumbo E, Notredame C. Nextflow enables reproducible computational workflows. *Nature Biotechnology*. 2017;35(4):316–9.
65. Mölder F, Jablonski KP, Letcher B, Hall MB, Tomkins-Tinch CH, Sochat V, et al. Sustainable data analysis with Snakemake. *F1000Research*. 2021;10:33.
66. Larsen MV, Joensen KG, Zankari E, Ahrenfeldt J, Lukjancenko O, Kaas RS, et al. The CGE tool box. In *Applied Genomics of Foodborne Pathogens*, ed. X. Deng, H den Bakker, and RE Hendriksen. Springer; 2017. p. 65–90.
67. Olawoye IB, Frost SD, Happi CT. The Bacteria Genome Pipeline (BAGEP): an automated, scalable workflow for bacteria genomes with Snakemake. *PeerJ*. 2020;8:e10121.
68. Matthews TC, Bristow FR, Griffiths EJ, Petkau A, Adam J, Dooley D, et al. The integrated rapid infectious disease analysis (IRIDA) platform. BioRxiv. 2018:381830.
69. Brancato V, Esposito G, Coppola L, Cavaliere C, Mirabelli P, Scapicchio C, et al. Standardizing digital biobanks: integrating imaging, genomic, and clinical data for precision medicine. *Journal of Translational Medicine*. 2024;22(1):136.
70. Wittig A, Wittig M. *Amazon Web Services in Action: An in-depth Guide to AWS*: Simon and Schuster; 2023.
71. Bisong E. *Building Machine Learning and Deep Learning Models on Google Cloud Platform: A Comprehensive Guide for Beginners*. Apress; 2019. p. 7–10.
72. Collier M, Shahan R. *Microsoft Azure Essentials-Fundamentals of Azure*: Microsoft Press; 2015.
73. Start HDI, Interface BU. *BaseSpace User Guide*. 2014.
74. Zeden MS, Gründling A. Bacterial whole-genome-resequencing analysis: basic steps using the CLC Genomics Workbench software. *Cold Spring Harbor Protocols*. 2023;2023(11):pdb. prot107901.
75. Clode C, editor. DNANexus-The Data Backbone for the Vertebrate Genome Consortia. *Plant and Animal Genome XXVI Conference (January 13-17, 2018)*; 2018: PAG.
76. Hazra A, Adhikari M, Amgoth T, Srirama SN. A comprehensive survey on interoperability for IIoT: taxonomy, standards, and future directions. *ACM Computing Surveys (CSUR)*. 2021;55(1):1–35.
77. Tang T, Li J. Transformation of FASTA files into feature vectors for unsupervised compression of short reads databases. *Journal of Bioinformatics and Computational Biology*. 2021;19(01):2050048.
78. Kumar S, Singh MP, Nayak SR, Khan AU, Jain AK, Singh P, et al. A new efficient referential genome compression technique for FastQ files. *Functional Integrative Genomics*. 2023;23(4):333.
79. Renaud A, Walsh I, Kalika M. Is SAM still alive? A bibliometric and interpretive mapping of the strategic alignment research field. *The Journal of Strategic Information Systems*. 2016;25(2):75–103.
80. Miller CA, Qiao Y, DiSera T, D'astous B, Marth GT. Bam. Iobio: a Web-based, real-time, sequence alignment file inspector. *Natural Methods*. 2014;11(12):1189–.
81. Tollefson GA, Schuster J, Gelin F, Agudelo A, Ragavendran A, Restrepo I, et al. VIVA (VIsualization of VAriants): a VCF file visualization tool. *Science Report*. 2019;9(1):12648.
82. Pertea G, Pertea M. GFF utilities: GffRead and GffCompare. *F1000Research*. 2020;9:304.
83. Rezaee MJ, Eshkevari M, Saberi M, Hussain O. GBK-means clustering algorithm: an improvement to the K-means algorithm based on the bargaining game. *Knowledge-Based Systems*. 2021;213:106672.
84. Dereeper A, Guignon V, Blanc G, Audic S, Buffet S, Chevenet F, et al. Phylogeny. fr: robust phylogenetic analysis for the non-specialist. *Nucleic Acids Research*. 2008;36(suppl_2):W465–W9.
85. Agapito G, Pastrello C, Guzzi PH, Jurisica I, Cannataro M. BioPAX-Parser: parsing and enrichment analysis of BioPAX pathways. *Bioinformatics*. 2020;36(15):4377–8.
86. Meehan CJ, Goig GA, Kohl TA, Verboven L, Dippenaar A, Ezewudo M, et al. Whole genome sequencing of Mycobacterium tuberculosis: current standards and open issues. *Nature Reviews Microbiology*. 2019;17(9):533–45.

87. Souche E, Beltran S, Brosens E, Belmont JW, Fossum M, Riess O, et al. Recommendations for whole genome sequencing in diagnostics for rare diseases. *European Journal of Human Genetics*. 2022;30(9):1017–21.
88. Hwang KB, Lee IH, Li H, Won DG, Hernandez-Ferrer C, Negron JA, et al. Comparative analysis of whole-genome sequencing pipelines to minimize false negative findings. *Scientific Reports*. 2019;9(1):3219.
89. Wyres KL, Conway TC, Garg S, Queiroz C, Reumann M, Holt K, et al. WGS analysis and interpretation in clinical and public health microbiology laboratories: what are the requirements and how do existing tools compare? *Pathogens*. 2014;3(2):437–58.

14 Types of Whole Genome Sequencing Techniques and Their Usefulness in Microbiology

Javed Muhammad, Hafsah Muhammad, Anza Abbas, and Nayyer Siddique

14.1 INTRODUCTION

Whole genome sequencing (WGS) is a laboratory technique used to analyze the complete genetic makeup of organisms. It involves the identification of the entire base sequence of the genome at a single time. Alongside chromosomal DNA, sequencing technologies also provide the base sequence of chloroplasts or mitochondrial DNA. The main purpose of whole genome sequencing is to identify inherited diseases, locate mutations responsible for disorders such as cancer, and track disease outbreaks. With the advancements of sequencing technologies, the cost has been significantly reduced, which makes the use of sequencers affordable and widespread for genomics research. Initially, WGS was used as a research tool, but now WGS data provides crucial information regarding therapeutic options for personalized medicine. It helps to understand the data of association studies and functional variants in evolutionary biology and provides estimations of disease susceptibility and drug response. Below are some of the advantages of WGS:

- High resolution and base-by-base identification of genome sequence
- Identification of all the large and small mutations which might be crucial for disease identification, gene expression, and regulation mechanisms
- Rapid assembly of novel genomes using large amounts of data

In 1953, Sanger used the chromatography technique to sequence the two polypeptide chains of the insulin protein (1, 2). He deduced the complete protein sequence by the random fragmentation of the protein, separation of fragments by chromatography, and reading the sequences from the overlapping fragments. Similarly, RNA sequencing was performed based on the same principle and a 76 bases alanine tRNA from *Saccharomyces cerevisiae* was successfully sequenced in 1965 (3). In 1968, the first DNA molecule sequenced was a 12 bases-long part of phage λ cos-site (4). Furthermore, the 510 bp gene of the coat protein of bacteriophage MS2 was sequenced in 1972, and the complete RNA genome of the virus was sequenced in 1976 (5, 6). In 1977, Sanger developed a method for sequencing which consisted of four different polymerization reactions, each containing tritium-radiolabeled primers, and one chain-terminating 2,3-dideoxynucleoside triphosphate (ddNTP) resulting in the production of DNA fragments with different lengths. The analysis of these fragments on gel electrophoresis then produces the complete sequence of the DNA template. In the same year, Maxam and Gilbert presented a chemical cleavage method which relies on the

DOI: 10.1201/9781003410164-14

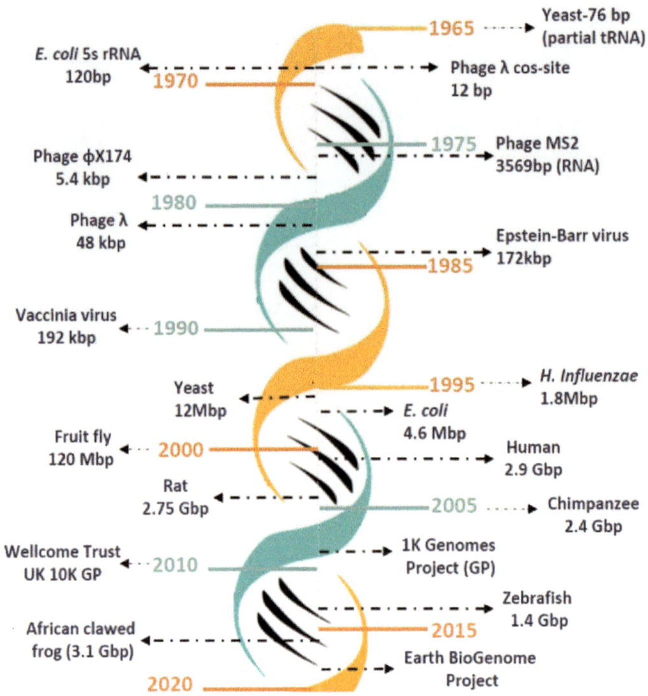

FIGURE 14.1 Landmarks in the assembly of the genome: a timeline showing many of the significant accomplishments in genome assembly, from the start of the sequencing era to the ongoing large-scale genome projects.

radioactive labeling of the DNA fragments and sequencing of fragments cleaved at specific bases (7). However, further improvements in the chain termination (Sanger) method led to its widespread use in sequencing studies. In the 1990s, rapid and automated sequencing methods facilitated the whole genome sequencing of complete bacterial and eukaryotic genomes.

In 1999, human chromosome 22 was sequenced, and subsequently, the complete sequence of another animal genome (fruit fly *Drosophila melanogaster*), and a plant genome (*Arabidopsis thaliana*) were published by the year 2000. A draft of the complete human genome sequencing was released in 2001. Furthermore, researchers published an incomplete version of the human genome in 2004 which consisted of about 3 billion bases found in a precise order (Figure 14.1). Currently, advances in sequencing technologies have led to the sequencing of the complete human genome, and thousands of other genomes have been sequenced completely (8).

14.2 AN OVERVIEW OF WHOLE GENOME SEQUENCING TECHNOLOGIES

Whole genome sequencing has emerged as a powerful tool in modern microbiology, particularly in the extensive sequencing of bacterial genomes. It is not only used in research projects but also has significant applications in surveillance projects and outbreak investigations. Below, we present a condensed overview of the most commonly utilized sequencing technologies in whole genome sequencing.

14.2.1 Sanger Sequencing

Sanger sequencing is a technique used to identify the order of nucleotides in a DNA fragment. It was developed by Frederick Sanger and colleagues in 1977 and is recognized as the gold standard of sequencing technologies based on its diverse applications in different research fields. It is also

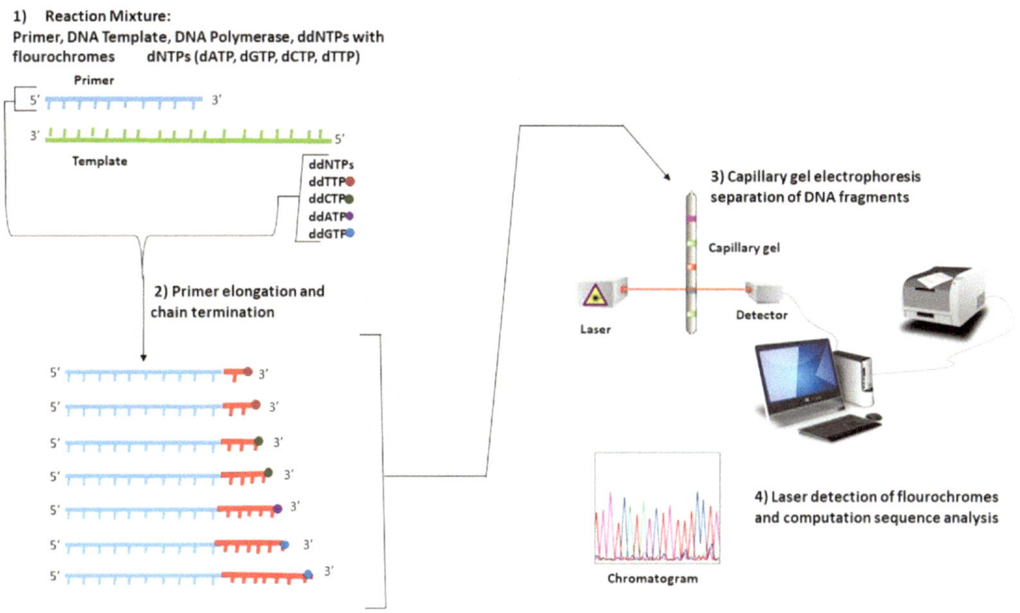

FIGURE 14.2 Schematic representation of workflow of Sanger sequencing.

called the chain termination method because DNA polymerase incorporates chain-terminating dideoxynucleotides (ddNTPs) during the process of *in vitro* DNA replication. The incorporation of ddNTPs will terminate the synthesis of new DNA strands due to the absence of hydroxyl group at the 3'-C of the deoxyribose. This results in DNA fragments of various lengths, which can be further used to determine the order of nucleotides (Figure 14.2).

14.2.1.1 Workflow

It works by *in vitro* replication of target DNA using primers complementary to the sequences adjacent to the region of interest. In PCR, primers bind to the DNA after denaturation, and DNA polymerase will carry out the synthesis of new DNA strands. DNA polymerase adds thymine (T) to the adenine (T) on the template strand and guanine (G) to the cytosine (C) and vice versa. In the extension step, DNA polymerase will add one of the four fluorescently labeled chain-terminating ddNTPs to the growing polynucleotide chain leading to the termination of DNA strand. Next, DNA fragments are subjected to electrophoresis and passed through a gel-filled glass capillary in which they migrate according to their particular length. A laser is directed to the migrating DNA fragments which generates fluorescence. As all the ddNTPs are labeled with different colors, each signal emitted by the ddNTPs' fluorescence indicates the presence of a particular base at the end of the DNA fragment. Software is used to detect the particular fluorescence and generate a chromatogram indicating different fluorescence peaks for the labeled fragments (9, 10).

14.2.1.2 Advantages of Sanger Sequencing

Sanger sequencing has numerous benefits over other sequencing techniques, some of which are listed below.

a. Since the approach is validated on one or a small number of sequences of interest, extensive validation is not necessary.
b. It is less reliant on computational tools for data processing.
c. It is cost-effective for the sequencing of single genes.

d. It provides higher specificity in the analysis of the variants of the same families.
e. It also offers high accuracy, particularly for sequencing of cloned inserts or site-directed mutagenesis.

14.2.1.3 Limitations of Sanger Sequencing

Despite being the best method for sequencing DNA, Sanger sequencing has certain drawbacks. These are as follows:

a. It has limited capacity to sequence long DNA fragments, usually sequences 300-1kb bases long fragments.
b. It does not detect different genes at one time.
c. It usually requires a higher concentration of DNA as input.
d. It results in poor quality of sequence in the first 15–40 bases due to the binding of primers in this region.
e. It is more time-consuming than other sequencing technologies.

14.2.1.4 Applications of Sanger Sequencing:

a. **Diagnostics:** Sanger sequencing is employed to check mutations in genes linked to inherited disorders, for instance, sickle cell anemia and cystic fibrosis. It is also used to sequence particular HLA gene sections in order to identify an individual's unique human leukocyte antigen (HLA) type.
b. **Genetic mapping:** It can be used to find the positions of particular genes on chromosomes, which may shed light on genetic features of inherited traits and complex diseases.
c. **Gene cloning:** It can be applied in the confirmation of cloned gene sequences, which ensures that genes have been copied completely and accurately.
d. **Bacterial and viral identification:** It can be used to sequence conserved regions of microbial genomes, which helps to identify specific viral or bacterial strains.
e. **Protein engineering:** It helps to determine the sequences of proteins, crucial to understanding the protein function. Sequence of proteins is also used to design new proteins with desired traits.
f. **Identifying genetic differences in population studies:** It helps to understand genetic variations in the case of population studies, giving insights into the genetic basis of disease and population history.

14.2.2 Next-Generation Sequencing

Since the era of Sanger sequencing, advancements in sequencing technologies have brought about significant improvements in data output, efficiency, and versatility. These advancements can be broadly classified into two categories based on read length: short-read, often referred to as "second-generation" sequencing technologies, and long-read, known as "third-generation" sequencing technologies. Short-read sequencing methods are exemplified by widely used platforms such as Illumina and Ion Torrent, while long-read sequencing technologies include Pacific Biosciences and Oxford Nanopore technologies. Each category offers distinct advantages and applications in genomic research and analysis and is discussed in detail below.

14.2.2.1 Illumina Sequencing

Illumina dye sequencing determines the sequence of nucleotides in DNA fragments using reversible dye terminators. The concept of reversible terminated chemistry was invented by Simon Sarfati and Bruno Canard at the Pasteur Institute in Paris. This sequencing technique is widely used, and

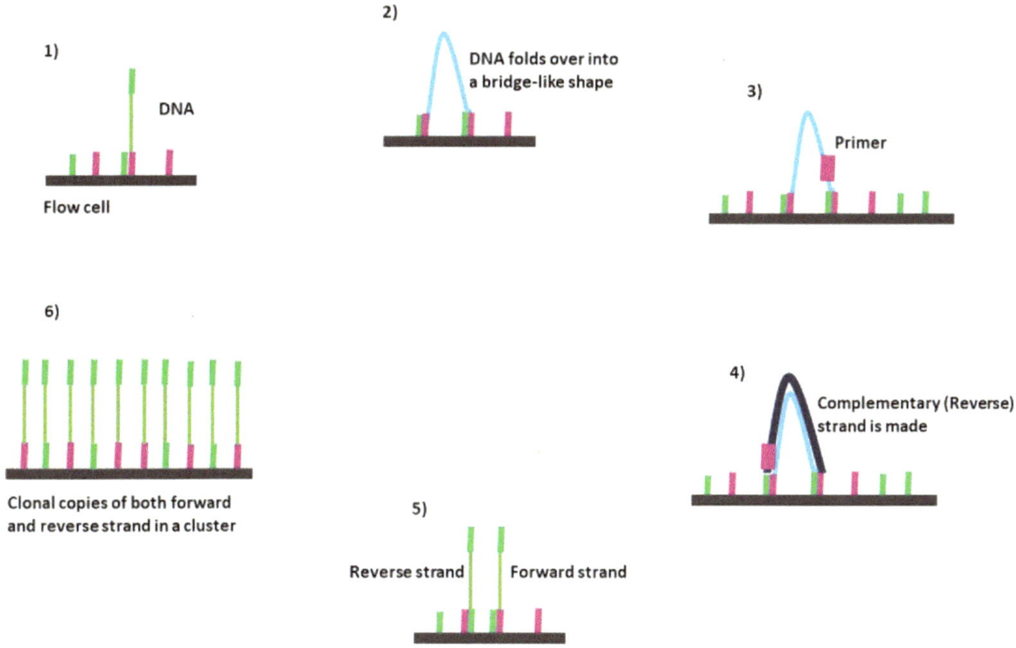

FIGURE 14.3 Schematic representation of Illumina dye sequencing workflow.

it works by converting DNA into fragments, linking them to a flow cell, amplifying them with PCR, and determining the sequence of the complementary strand one base at a time. Modified nucleotides are used in the synthesis of the complementary strands, emitting a light signal when they are incorporated into the growing nucleotide chain. The process records the emitted color of each base as it is added to the complementary strand until all bases have been incorporated (Figure 14.3). The basic workflow of Illumina dye sequencing involves three steps: amplification, sequencing, and analysis. Firstly, adapters containing reference points for amplification are added to the DNA fragments. The DNA is then loaded onto a flow cell containing nanowells with specific oligonucleotide sequences complementary to the adapter sequences. After the DNA fragments bind to the flow cell, they are amplified through bridge amplification PCR. Next, primers and fluorescently labeled nucleotides are added to the chip, and DNA polymerase adds one nucleotide at a time due to the presence of reversible fluorescent blockers. After each base addition, the fluorescent signal is recorded, and a computer determines the presence of a specific nucleotide based on the wavelength of the fluorescent tag. Subsequently, unbound nucleotides are washed away, and a deblocking agent is used to remove the 3' fluorescent terminal blocking group. This process repeats until the full DNA strand is synthesized (11).

14.2.2.2 454 Sequencing

454 Life Sciences Corp gained a reputation for its innovative technology that revolutionized the field of genetic sequencing. The technology, commonly referred to as "454-sequencing," enabled scientists to generate massive amounts of DNA sequence data. This breakthrough technology has since become synonymous with the company's name and has been widely adopted by researchers worldwide (12). It works on the basis of light detection during DNA synthesis. Initially, emulsion PCR is performed to increase the number of library molecules on DNA capture beads. Next, these beads are allowed to attach to the microwells of a PicoTiterPlate (PTP). During sequencing, the addition of nucleotides releases molecular byproducts such as

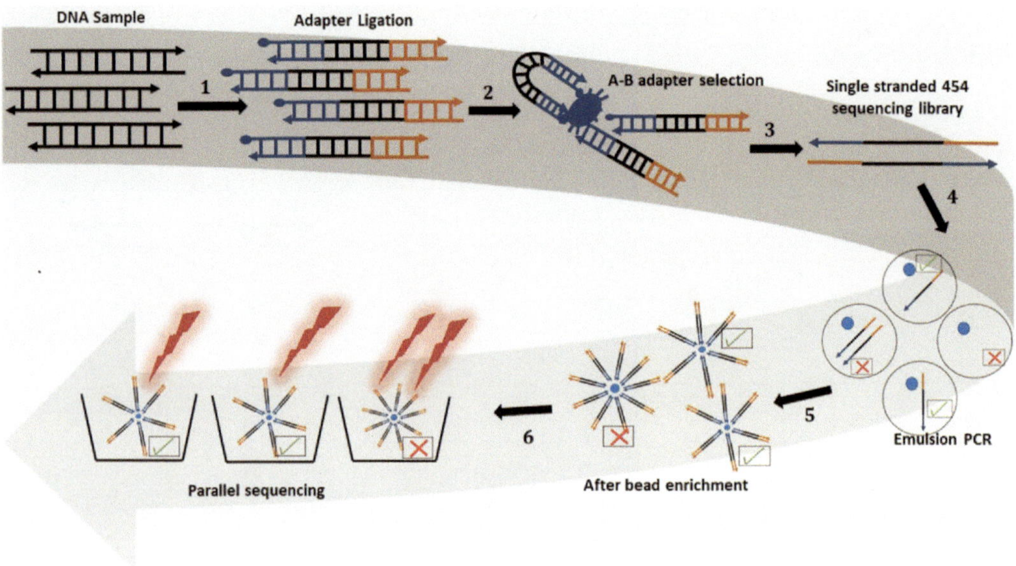

FIGURE 14.4 Schematic representation of 454 sequencing workflow.

ATP which is used by 454 chemistry-specific enzymes as substrates. The enzymatic reaction generates a particular fluorescence, which is recorded and processed for every base added to the polynucleotide chain.

In 454 sequencing, special adapters called A and B are attached to the ends of DNA templates. These DNA molecules, now linked with the adapters, are separated and isolated using Streptavidin beads. After this, a process called alkaline treatment helps to separate the single-stranded sequencing library from the beads. Next, the library molecules are mixed with PCR reaction mix and 600,000 oligonucleotide-coated sepharose beads, creating tiny droplets. These droplets act as individual reaction vessels. Once the DNA is amplified, the emulsion is removed while the PCR products remain attached to the beads. Through a process called bead enrichment, only the amplified beads are isolated. A total of 50,000 beads are then loaded onto the wells of a 16th 454 FLX picotitre sequencing plate, and nucleotides are added. This addition of nucleotides causes the emission of light signals, which are captured and recorded by computer software. The sequencing reaction is conducted by flowing nucleotides over the plate and measuring the emitted light (see Figure 14.4). Any mixed signals produced by beads carrying multiple DNA fragments are identified and filtered out by specialized software (13).

14.2.2.3 SOLiD Sequencing

This next-generation sequencing technology was introduced by Life Technologies in 2006. It stands out because it relies on primer ligation, a unique approach among sequencing methods. It ensures high accuracy by incorporating internal checkpoints: each ligation reaction is marked by two bases, meaning every nucleotide is sequenced twice, providing the highest overall accuracy. Its basic principle involves the detection of fluorophores. Initially, emulsion PCR is performed to increase the number of single library molecules onto beads. Then, oligonucleotides flow over a chip surface, binding to the library beads. Each sequencing oligo carries a distinct fluorophore. When the oligo binds to its complementary sequence on the DNA template, an enzymatic reaction removes the fluorophore, emitting a specific color (see Figure 14.5). A camera captures images, and specialized software deciphers the sequence of bases (14).

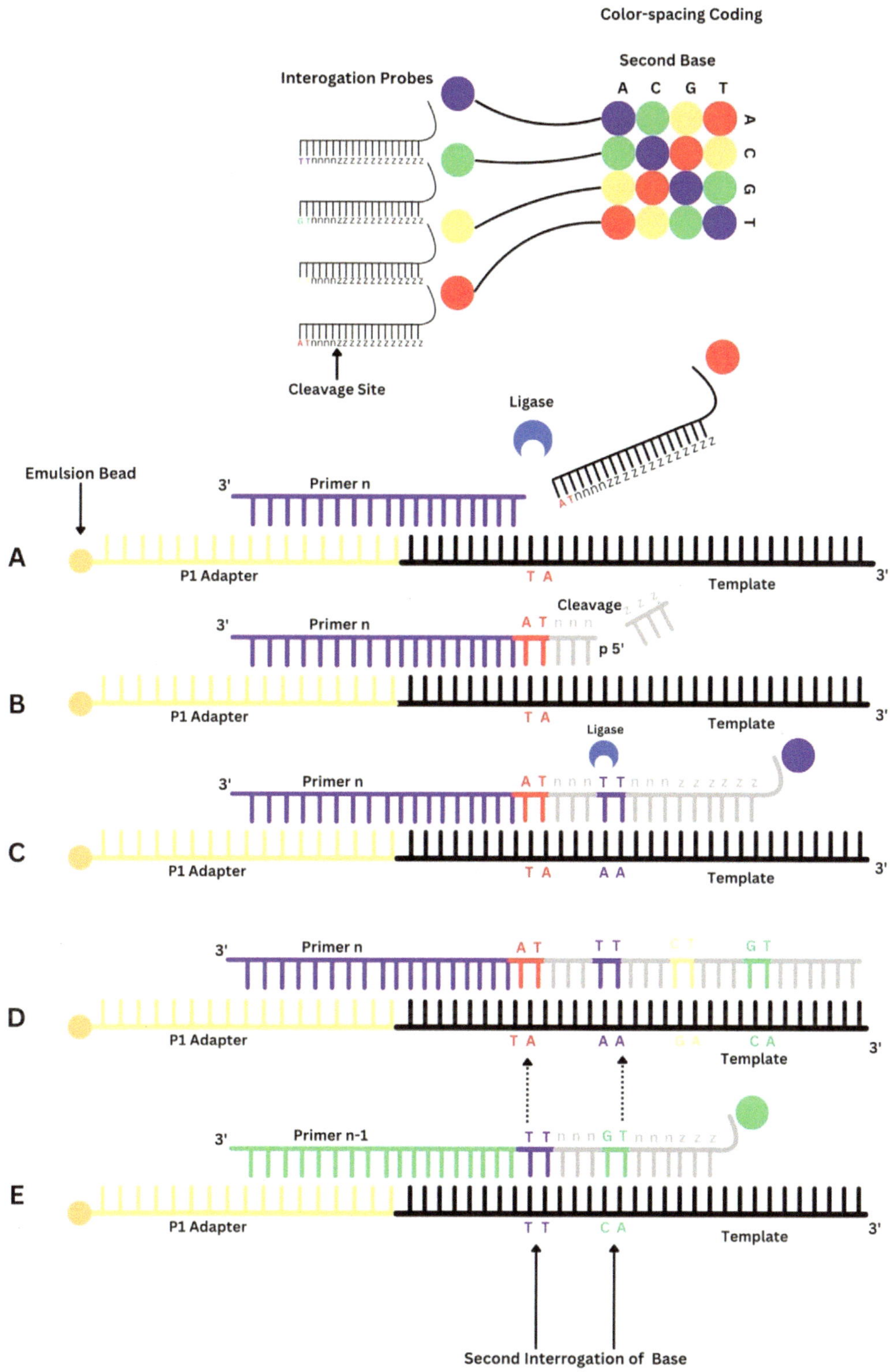

FIGURE 14.5 Schematic representation of SOLiD sequencing workflow.

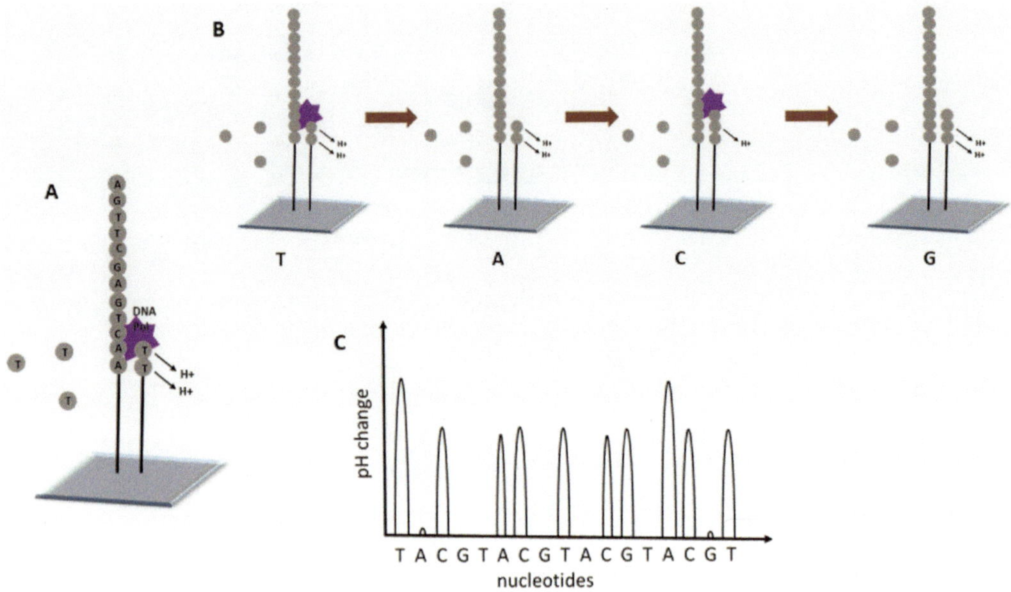

FIGURE 14.6 Schematic representation of Ion Torrent sequencing workflow.

14.2.2.4 Ion Torrent Sequencing

Ion Torrent sequencing operates on the principle that adding dNTPs to a nucleotide chain releases an H+ ion. This process causes fluctuations in pH levels, which can be analyzed to identify the specific base and the total number of bases added to the template (see Figure 14.6). Unlike other methods that rely on fluorescence signals, this technology converts chemical information directly into digital data (15).

14.2.3 Third-Generation Sequencing (TGS)

The advent of next-generation sequencing (NGS) technologies was a significant revolution in genomics, but it came with its own set of challenges, primarily the generation of short reads. This characteristic, known as "short-read sequencing," posed difficulties in bioinformatics analysis and extended the time required for data manipulation (16). These limitations of NGS paved the way for third-generation sequencing techniques (TGSTs), introducing a new era in genomic studies with unprecedented capabilities. The TGSTs were developed in 2011 and are discussed below in detail. These technologies work by single-molecule sequencing and provide the following advantages:

a. Long reads with no amplification.
b. Direct detection of epigenetic modifications on native DNA.
c. Direct sequencing through regions of the genome is inaccessible or difficult to analyze by short-read platforms.
d. Uniform coverage of the genome, as they are not as sensitive to GC content as short-read platforms are.

14.2.3.1 Single-Molecule Sequencing in Real Time (SMRT)

Single-Molecule Sequencing in Real Time (SMRT), developed by PacBio in 2011, revolutionized genomic studies by enabling the sequencing of DNA from individual cells with remarkable accuracy. Over the years, SMRT evolved from reading fragments of 1,000 bases to impressive 30,000-base long reads. Its unique approach eliminates the need for PCR during DNA library preparation, allowing for automatic replication of closed and circular single-stranded DNA templates. The incorporation of labeled dNTPs into DNA triggers fluorescence signals, recorded in real time by a camera system equipped with zero mode waveguides (17). SMRT's capabilities extend to direct RNA sequencing, making it a versatile and powerful TGS technology (18).

14.2.3.2 Oxford Nanopore Technology (ONT)

Introduced in 2015, Oxford Nanopore Technology (ONT) offers a distinct approach to sequencing by utilizing α-hemolysin, a nanopore protein, to measure characteristic disruptions in the electrical current as DNA passes through (19). This method eliminates the need for DNA polymerases and allows for the sequencing of ultra-long reads, some reaching close to 1 million base pairs. ONT's pocket-sized sequencers provide portability and have been utilized in real-world scenarios, such as the screening of Ebola and Lassa virus outbreaks in Africa (20–22).

14.2.3.3 Applications and Advancements of TGS

14.2.3.3.1 Drug Discovery

TGSTs have a promising future in drug discovery by enhancing sequencing capabilities, reducing biases, and minimizing costs. Epigenomics studies benefit from TGS, offering a comprehensive understanding of epigenetic mechanisms crucial in disease etiology. TGSTs contribute to aptamer-based drug discovery by improving the SELEX process and enabling accurate single-cell measurements of DNA and RNA molecules (23). TGS has the ability to yield long reads for whole genome sequencing applications, especially in repeat region analysis. This can help with full de novo genome assemblies and microbial sequencing investigations (24, 25).

14.2.3.3.2 Genetic Disease Identification

TGS has played a pivotal role in the discovery and reevaluation of genetic diseases. SMRT sequencing, with its high accuracy, excels in detecting structural variants (SVs) associated with genetic disorders. TGS is instrumental in characterizing complex genomic regions related to diseases such as autosomal-dominant polycystic kidney disease. By offering real-time detection of epigenetic modifications and full-length mRNA transcripts, TGS provides a holistic view of the genome, revolutionizing genetic disease research (26).

14.2.3.3.3 TGS Technologies for Cancer

The application of TGS in cancer studies is rapidly progressing, with a focus on decoding allele-resolution mutation statuses and complex cancer genome structures. Long-read sequencing, either alone or combined with short-read sequencing, enables the detection of low-frequency cancer mutations. For instance, PacBio has been applied to screen for actionable mutations in tumor suppressor TP53 (27), providing insights into allele-specific mutations in diseases like acute myeloid leukemia (28).

14.2.3.3.4 Challenges of TGT

TGS technologies present challenges despite their revolutionary potential. Practical challenges include the requirement for new material and intact cells for the production of DNA libraries. Problems arise from higher rates of sequencing error and systematic errors, and TGS is still more expensive than NGS. Another important point is the lack of database systems in bioinformatics analysis that are capable of interpreting complex changes in structure (20, 29).

14.3 COMPARATIVE ANALYSIS OF TECHNIQUES

The choice between Sanger sequencing, NGS, or TGS depends on the specific needs of the study, considering factors such as sequencing volume, read length, error rates, and cost. Advances in bioinformatics and machine learning contribute to improving the accuracy of variant calling across

TABLE 14.1
Comparative Analysis of All Three WGS Techniques

Criteria	Sanger Sequencing	Next-generation Sequencing (NGS)	Third-generation Sequencing (TGS)
Principle	Adds fluorescent nucleotides one by one onto a growing DNA template strand.	Adds fluorescent nucleotides onto multiple DNA fragments simultaneously.	Sequences DNA in real time using single-molecule or nanopore technologies.
Sequencing volume	Sequences a single DNA fragment at a time.	Massively parallel, sequencing millions of fragments simultaneously.	High-throughput, real-time sequencing of individual molecules.
Discovery power	Limited to sequencing one fragment at a time.	Greater discovery power, detecting novel or rare variants with deep sequencing.	Offers haplotype-phasing for variant discovery on homologous genes and CNV calling.
Error rates	Generally lower error rates.	NGS has lower error rates compared to TGS.	TGS may have higher error rates, but advancements in bioinformatics tools enhance accuracy.
Read length	Shorter read lengths.	Varied read lengths, depending on the NGS platform.	Longer read lengths, allowing for haplotype-phasing and comprehensive genome analysis.
Variant calling accuracy	High accuracy for individual fragments.	High accuracy, and lower error rates for variant calling.	Lower accuracy in variant calling, but improved accuracy with machine learning-based tools.

all sequencing techniques. The ongoing transition from conventional DNA analysis to NGS or TGS is shaping the landscape of genomics, enabling more comprehensive and efficient diagnostics approaches. The overall comparison between these techniques is presented in Table 14.1.

14.4 APPLICATIONS OF WHOLE GENOME SEQUENCING TECHNIQUES IN MICROBIOLOGY

Whole genome sequencing (WGS) has emerged as a pivotal technology in microbiology, facilitating the sequencing of entire bacterial, viral, and other microbial genomes. There are diverse applications of WGS in the fields of food, agriculture, and clinical microbiology.

14.4.1 Food Microbiology

The whole genome sequencing of food microorganisms and bioinformatics data analysis can help predict genes that are important for the fermentation process or the quality of the final product, giving insights into the metabolic pathways of the microorganisms and how they interact with their surroundings. Microbial whole genome sequencing provides the option for modifying microorganisms to make them more efficient in the manufacture of vinegar, liquor, yogurt, and many other fermentation processes (30).

14.4.2 Agricultural Microbiology

In agriculture, microorganisms are used for planting, breeding, product processing, agricultural biotechnology, agricultural ecology, and other research and applications. People may develop a deeper understanding of agricultural microbes at the genomic level through microbial whole genome sequencing, and the ensuing research on the structure and function of genomes lays a crucial foundation for agromicrobiology in the following fields:

- establishing agricultural microbial gene bank
- soil microorganisms (including root microorganism)
- plant nutrition
- biological nitrogen fixation
- microbial pesticide
- microbial fertilizer
- feed additive
- biogas fermentation

14.4.3 Clinical Microbiology

WGS has gained attention as a technology that could completely transform clinical microbiology in recent years. The goal of clinical microbiology is to quickly detect and identify microorganisms, whether or not antibiotic susceptibility testing (AST) is used. Recent advancements in sequencing technologies have resulted in faster sequencing rates and improved output-to-cost ratios, making whole genome sequencing (WGS) useful for numerous clinical microbiology applications, such as pathogen epidemiology and infectious disease control (31).

14.4.3.1 Surveillance of Outbreaks and Antimicrobial Resistance

WGS is a relatively recent widely available method that has proven successful in several application contexts for monitoring disease epidemics. The full or almost complete genomes of bacterial isolates can now be examined using massively parallel (or "next-generation") DNA sequencing technology. WGS has used many online tools to decrease the time required for strain identification. To classify the bacteria into serotypes, biotypes, pathotypes, and sequence types, all that is needed is the sequence that was obtained by sequencing.

The molecular aspects of antibiotic resistance have been made public by WGS, which has also uncovered surprising patterns of transmission in healthcare environments. Furthermore, WGS research has revealed insights into the relationship between humans and animals, challenging earlier ideas regarding the sources of infection. In general, WGS is essential for monitoring and comprehending the formation and spread of antibiotic resistance, which helps to inform public health policies and infection control strategies (32).

14.5 CHALLENGES AND FUTURE PERSPECTIVES

The lack of funding and the usage of chemicals, ethical issues like discrimination based on genetics, and scientific difficulties in storing human genome sequences are among the greatest challenges to genome sequencing. The following section is an overview of a few of the challenges one faces while doing WGS.

14.5.1 Genome Sequence Analysis

The maturation of sequencing technology has shifted the bottleneck to data analysis, requiring faster and more accessible methods. The challenge lies in developing analysis techniques that can be readily applied at scale, addressing the current complexity and potential cost implications.

14.5.2 Education of Health Care Professionals

Healthcare workers need to be educated on genomic medicine because of the difficulties posed by the deterministic perspective on genetics. The lack of genetic counselors renders incorporating WGS into standard medical procedures even more challenging.

14.5.3 Cost and Security of Data Storage

The computational data generated by WGS demands substantial storage capabilities, leading to increased costs for both sequencing and storage. Privacy concerns regarding the security of patient data also need careful consideration.

14.5.4 Ethical Issues

There are various ethical and practical considerations surrounding the use of WGS in medical settings, including concerns about the accuracy of genetic mutation databases, with an estimated 25% of gene–disease associations being incorrect (33). There are potential psychological harms to patients and legal consequences for healthcare professionals due to inaccurate information. In the context of incidental findings during WGS, there is a debate on whether to disclose information about disease-associated mutations to patients, considering the vast number of genes and the potential for unexpected findings. The ethical dilemma involves balancing patient autonomy with the need to protect individuals from unnecessary psychological distress. Some suggest disclosing only actionable findings that can be prevented or treated.

Research on the preferences of individuals for knowing about their genetic makeup reveals a need for knowledge, particularly when it comes to disorders that are treatable or preventable. However, concerns arise as studies show that awareness of genetic risks may not lead to behavioral modifications. There is a need to address the use of WGS in newborn screening programs, where the practicality and cost-effectiveness of validating and treating incidental findings are questioned.

Lastly, the contentious issue of testing fetuses through WGS is important, with potential implications for parental decisions, including the termination of pregnancies based on desired genetic traits. The applications of WGS should be limited to diagnosing diseases rather than predicting complex traits like intelligence or athletic ability until more is known about its benefits and harms (34).

14.6 FUTURE OF WGS

In the near future, technological advancements will usher in faster and more accurate sequencing methods. Future sequencing platforms will require less input DNA and reagents and become more portable for diagnostic use in medical, agricultural, ecological, and other fields. Technological developments in liquid handling, robotics, bioinformatics, and nucleic acid production will all contribute to its facilitation.

14.7 CONCLUSION

In conclusion, genome sequencing has undergone tremendous advancements over the years, with the development of second-generation sequencing and more recently, third-generation sequencing technologies. These developments have created huge amounts of genetic data and created a wide range of opportunities for applications across various fields. Genome sequencing has made it possible for scientists to investigate the genetic causes of diseases, find novel targets for drugs, increase crop productivity, and learn more about the evolutionary history of different species. Genome sequencing has the potential to continue making significant contributions to various fields, leading to discoveries that are improving our perception of the world around us, as a result of ongoing technological developments and reducing costs.

REFERENCES

1. Sanger F, Thompson E. The amino-acid sequence in the glycyl chain of insulin. 2. The investigation of peptides from enzymic hydrolysates. *Biochemical Journal*. 1953;53(3):366.
2. Sanger F, Thompson E. The amino-acid sequence in the glycyl chain of insulin. 1. The identification of lower peptides from partial hydrolysates. *Biochemical Journal*. 1953;53(3):353.
3. Holley RW, Apgar J, Everett GA, Madison JT, Marquisee M, Merrill SH, et al. Structure of a ribonucleic acid. *Science*. 1965;147(3664):1462–5.
4. Wu R, Kaiser A. Structure and base sequence in the cohesive ends of bacteriophage lambda DNA. *Journal of Molecular Biology*. 1968;35(3):523–37.
5. Jou WM, Haegeman G, Ysebaert M, Fiers W. Nucleotide sequence of the gene coding for the bacteriophage MS2 coat protein. *Nature*. 1972;237(5350):82–8.
6. Fiers W, Contreras R, Duerinck F, Haegeman G, Iserentant D, Merregaert J, et al. Complete nucleotide sequence of bacteriophage MS2 RNA: primary and secondary structure of the replicase gene. *Nature*. 1976;260(5551):500–7.
7. Sanger F, Nicklen S, Coulson AR. DNA sequencing with chain-terminating inhibitors. *Proceedings of the National Academy of Sciences*. 1977;74(12):5463–7.
8. Giani AM, Gallo GR, Gianfranceschi L, Formenti G. Long walk to genomics: history and current approaches to genome sequencing and assembly. *Computational and Structural Biotechnology Journal*. 2020;18:9–19.
9. Valencia CA, Pervaiz MA, Husami A, Qian Y, Zhang K, Valencia CA, et al. Sanger sequencing principles, history, and landmarks. In *Next Generation Sequencing Technologies in Medical Genetics*, ed. C Alexander Valencia et al. Springer; 2013. p. 3–11.
10. Qiang-long Z, Shi L, Peng G, Fei-shi L. High-throughput sequencing technology and its application. *Journal of Northeast Agricultural University (English Edition)*. 2014;21(3):84–96.
11. Lu Y, Shen Y, Warren W, Walter R. Next generation sequencing in aquatic models. *Next Generation Sequencing-Advances, Applications and Challenges*. 2016;1:13.
12. Goldberg SM, Johnson J, Busam D, Feldblyum T, Ferriera S, Friedman R, et al. A Sanger/pyrosequencing hybrid approach for the generation of high-quality draft assemblies of marine microbial genomes. *Proceedings of the National Academy of Sciences*. 2006;103(30):11240–5.
13. Meyer M, Briggs AW, Maricic T, Höber B, Höffner B, Krause J, et al. From micrograms to picograms: quantitative PCR reduces the material demands of high-throughput sequencing. *Nucleic Acids Research*. 2008;36(1):e5–e.
14. Voelkerding KV, Dames SA, Durtschi JD. Next-generation sequencing: from basic research to diagnostics. *Clinical Chemistry*. 2009;55(4):641–58.
15. Flatt PM. *Biochemistry – Defining Life at the Molecular Level*. Western Oregon University, Monmouth; 2019.
16. Van Dijk EL, Jaszczyszyn Y, Naquin D, Thermes C. The third revolution in sequencing technology. *Trends in Genetics*. 2018;34(9):666–81.
17. Flusberg BA, Webster DR, Lee JH, Travers KJ, Olivares EC, Clark TA, et al. Direct detection of DNA methylation during single-molecule, real-time sequencing. *Nature Methods*. 2010;7(6):461–5.
18. Eid J, Fehr A, Gray J, Luong K, Lyle J, Otto G, et al. Real-time DNA sequencing from single polymerase molecules. *Science*. 2009;323(5910):133–8.
19. Schatz MC. Nanopore sequencing meets epigenetics. *Nature Methods*. 2017;14(4):347–8.
20. Midha MK, Wu M, Chiu K-P. Long-read sequencing in deciphering human genetics to a greater depth. *Human Genetics*. 2019;138(11–12):1201–15.
21. Kafetzopoulou L, Pullan S, Lemey P, Suchard M, Ehichioya D, Pahlmann M, et al. Metagenomic sequencing at the epicenter of the Nigeria 2018 Lassa fever outbreak. *Science*. 2019;363(6422):74–7.
22. Hoenen T, Groseth A, Rosenke K, Fischer RJ, Hoenen A, Judson SD, et al. Nanopore sequencing as a rapidly deployable Ebola outbreak tool. *Emerging Infectious Diseases*. 2016;22(2):331.
23. Zimmermann B, Gesell T, Chen D, Lorenz C, Schroeder R. Monitoring genomic sequences during SELEX using high-throughput sequencing: neutral SELEX. *PloS One*. 2010;5(2):e9169.
24. Comas I, Borrell S, Roetzer A, Rose G, Malla B, Kato-Maeda M, et al. Whole-genome sequencing of rifampicin-resistant Mycobacterium tuberculosis strains identifies compensatory mutations in RNA polymerase genes. *Nature Genetics*. 2012;44(1):106–10.

25. Toprak E, Veres A, Michel J-B, Chait R, Hartl DL, Kishony R. Evolutionary paths to antibiotic resistance under dynamically sustained drug selection. *Nature Genetics*. 2012;44(1):101–5.
26. Xiao T, Zhou W. The third generation sequencing: the advanced approach to genetic diseases. *Translational Pediatrics*. 2020;9(2):163.
27. Lodé L, Ameur A, Coste T, Ménard A, Richebourg S, Gaillard J-B, et al. Single-molecule DNA sequencing of acute myeloid leukemia and myelodysplastic syndromes with multiple TP53 alterations. *Haematologica*. 2018;103(1):e13.
28. Suzuki A, Suzuki M, Mizushima-Sugano J, Frith MC, Makałowski W, Kohno T, et al. Sequencing and phasing cancer mutations in lung cancers using a long-read portable sequencer. *DNA Research*. 2017;24(6):585–96.
29. Mohandas T, Sparkes R, Shapiro L. Reactivation of an inactive human X chromosome: evidence for X inactivation by DNA methylation. *Science*. 1981;211(4480):393–6.
30. Abriouel H, Lerma LL, Casado Muñoz MdC, Montoro BP, Kabisch J, Pichner R, et al. The controversial nature of the Weissella genus: technological and functional aspects versus whole genome analysis-based pathogenic potential for their application in food and health. *Frontiers in Microbiology*. 2015;6:1197.
31. Bianconi I, Aschbacher R, Pagani E. Current uses and future perspectives of genomic technologies in clinical microbiology. *Antibiotics*. 2023;12(11):1580.
32. Salipante SJ, SenGupta DJ, Cummings LA, Land TA, Hoogestraat DR, Cookson BT. Application of whole-genome sequencing for bacterial strain typing in molecular epidemiology. *Journal of Clinical Microbiology*. 2015;53(4):1072–9.
33. Bell CJ, Dinwiddie DL, Miller NA, Hateley SL, Ganusova EE, Mudge J, et al. Carrier testing for severe childhood recessive diseases by next-generation sequencing. *Science Translational Medicine*. 2011;3(65):65ra4–ra4.
34. Witte S. *Challenges to Whole Genome Sequencing in National Healthcare Systems*. Cambridge University Science and Policy Exchange Commonications; Summer 2016. www.repository.cam.ac.uk/handle/1810/278277 (Accessed on 11.09.2024).

15 Artificial Intelligence and Machine Learning Techniques Applied to Microbiology

Zaara Ishaq, Fiza Faiz, Uzair Haider, and Amjad Ali

15.1 INTRODUCTION

The scientific world is constantly evolving, and the integration of cutting-edge technologies like artificial intelligence (AI) and machine learning (ML) with traditional disciplines like microbiology is leading to groundbreaking discoveries (1, 2). Microbiology, traditionally relying on well-established methods, is currently undergoing a transformative shift driven by AI and ML techniques (2). This chapter discusses how AI and ML are gaining popularity in microbiology and reshaping our perception of microbes, their functions, and their roles in various ecosystems.

Microbiology is the science concerned with studying microscopic organisms like bacteria, viruses, fungi, and protozoa. It has always played an important role in helping us understand biology (3). However, microbes themselves are very complex. They have multiple forms of energy acquisition, well-developed control systems, and can adapt quickly. This complexity, and the large amount of detailed data collected using modern research methods, make it difficult to analyze them using traditional techniques. AI and ML have emerged as powerful tools for scientists to decipher the intricate world of microbes (Figure 15.1) (4). These techniques are used in many areas of microbiology, from studying individual genes and entire sets of genes (genomes and metagenomes) to investigating microbes in their environment and fighting infectious diseases (5, 6). AI algorithms can analyze massive amounts of data rapidly and accurately, which has completely changed how we identify new types of microbes, predict what they will do, and understand their role in complex ecosystems (6, 7).

With the decreasing cost of DNA sequencing, there has been a rapid and significant increase in the size of genomic data (7, 8). AI algorithms are essential for analyzing this enormous collection of genetic information. With AI, scientists can pinpoint important genetic markers associated with specific traits or functions in microbes. AI can also make good predictions about what a microbe will look or act like based on its genes. This allows us to retrieve the hidden information contained in microbial DNA and thus gain a new perspective on the complexity and versatility of microbes (9).

Furthermore, we will explore the significance of AI and ML in metagenomics (10). Conventional methods often struggle to extract meaningful patterns from complex data, underlining the need for AI and ML to reveal hidden connections, identify major microbial roles, and predict community dynamics in diverse environments (11, 12).

AI and ML continue to be a driving force for enhancing diagnostic technologies in infectious diseases (13). These technologies are continuously improving capability to deal with microbial challenges, such as rapid and accurate identification of pathogenic organisms, modeling their growth, and predicting the patterns of antibiotic resistance (14).

DOI: 10.1201/9781003410164-15

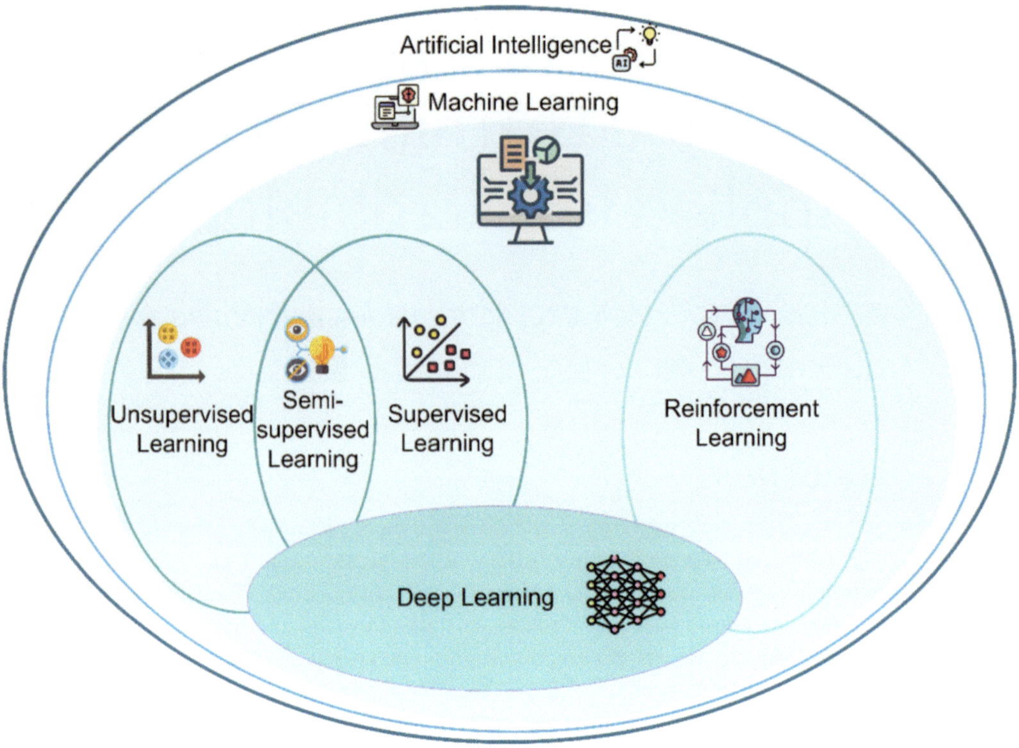

FIGURE 15.1 A visual representation of the hierarchical relationship between artificial intelligence (AI), machine learning (ML), and deep learning (DL). AI is the broad domain, while ML is a specialized subset that falls under its umbrella. DL falls under the field of ML, and it is distinguished by the utilization of multilayered deep neural networks (Adapted from reference 5).

15.2 GENERAL ARTIFICIAL INTELLIGENCE AND MACHINE LEARNING TECHNIQUES APPLIED TO LIFE SCIENCES

This chapter explores various AI and ML techniques that empower computer algorithms to learn, adapt, and draw insights independently. These techniques have become the foundation of a technical revolution, providing microbiologists with a clear lens to explore the complex intricacies of microbial life (Figure 15.2) (3).

One approach in this regard is supervised learning, where an algorithm is trained with a labeled dataset to identify the relationship between inputs and outputs (15). Applications predicting microbial behaviors, identification of their species, and assessing the genomic complexity within the microorganisms can be pursued under this model since it has been powerfully trained for both classification and regression tasks (2,4,16).

In contrast, unsupervised learning allows algorithms to reveal hidden patterns within unlabeled data, acting like skilled analysts (17). It performs like expert analysts, where latent patterns are developed in microbiological information. It may group similar microbes, for example, showing the pattern of microbial communities and environmental dynamics that would otherwise be hard to determine with traditional methods (5).

Deep learning (DL) is a revolutionary concept in AI, which harnesses the full potential of multilayer neural networks (19). To understand microbial genetic blueprints for making exceptionally accurate predictions of their behaviors, researchers are digging into DL methods to decode the secrets contained in massive genomic databases (8, 20).

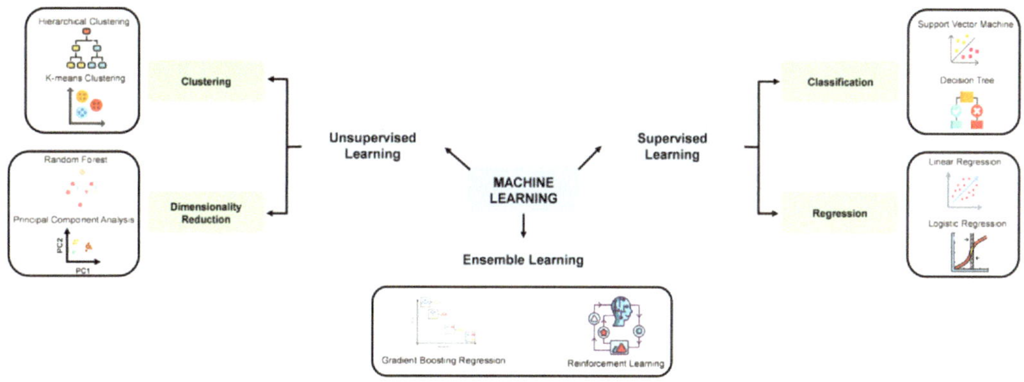

FIGURE 15.2 Examples of general machine learning algorithms. Examples are divided into three categories: supervised learning, unsupervised learning, and ensemble learning. Supervised algorithms: regression and classification, like linear regression, logistic regression, decision trees, and support vector machines, learn from labeled data. Unsupervised algorithms: clustering and dimensionality reduction, such as hierarchical clustering, k-means clustering, random forest, and principal component analysis, analyze unlabeled data to identify patterns. Methods of ensemble learning, such as gradient boosting and reinforcement learning, combine multiple models to achieve significant improvements in predicting performance.

Reinforcement learning introduces the concept of autonomous decision-making within a microbial context (18). This technique operates based on the premise of reinforcement in which the algorithm is rewarded for performing certain actions in a manner that is analogous to how microbes might behave in their environment at any given moment. Studying these interactions through reinforcement learning reflects on how microbes strategically navigate their environments.

Applications in microbiology also arise from NLP, a sub-area in AI uniquely tailored for understanding even the most intricate forms of languages (20, 21). It is through the use of NLP techniques that scientists decipher molecular conversations to understand microbial communication and find secrets hidden in the linguistic subtleties of microbial life (21, 22).

15.3 APPLICATIONS OF ARTIFICIAL INTELLIGENCE AND MACHINE LEARNING TECHNIQUES IN MICROBIOLOGY

The recent integration of AI and ML in microbiology has accelerated scientific discoveries about understanding and manipulating the microbial world. In the past, extensive research on invisible living organisms had to be carried out by laborious cultivation and was thoroughly examined over many years. AI and ML are changing research, diagnosis, and the fight against microbial menaces at a speed not achieved till now (Figure 15.3) (2). The future holds promise, and the collaboration of AI and human ingenuity, coupled with ML, may lead to the rapid identification of causative agents in numerous infectious diseases (13). The future appears promising.

15.3.1 Antimicrobial Resistance (AMR) Prediction

AI and ML are changing the game in the battle against AMR, one of the critical global health challenges (23, 24). AI and ML excel in analyzing complex datasets regarding understanding and managing AMR. One of the most important fields of influence of AI and ML is the identification of AMR markers (specific traits or genetic elements that may indicate the susceptibility of a micro-organism to antibiotics). This information is crucial to healthcare practitioners in choosing effective

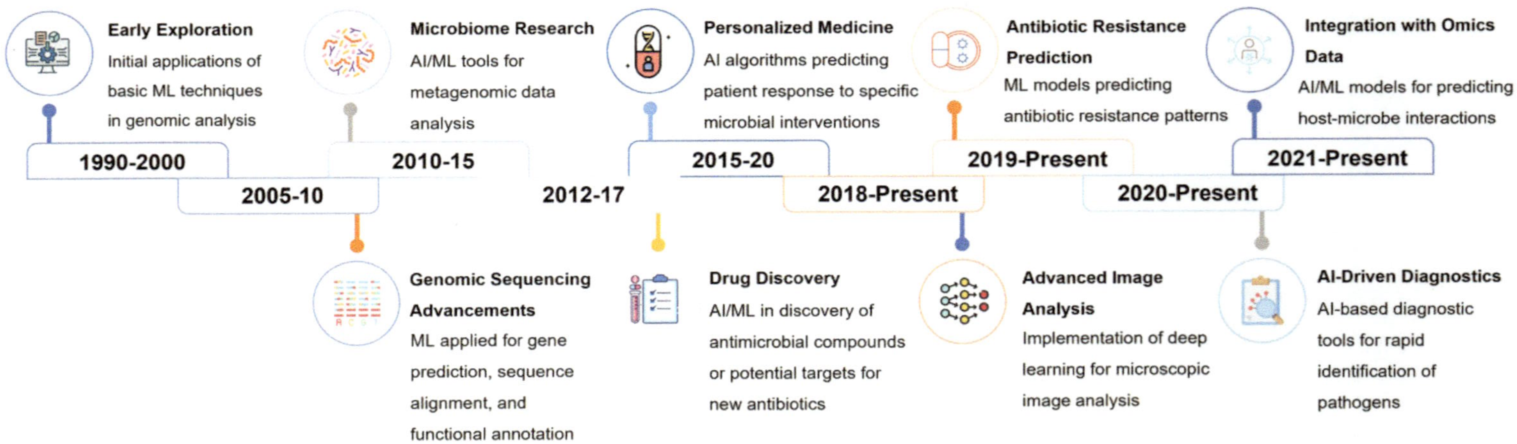

FIGURE 15.3 *S*equential timeline outlining important milestones in the utilization of artificial intelligence (AI) and machine learning (ML) in the discipline of microbiology. This visual depiction gives a thorough illustration of the historical development of AI/ML applications in microbiology. It provides valuable insights into the advancement of these technologies and their influence on different facets of microbiological research.

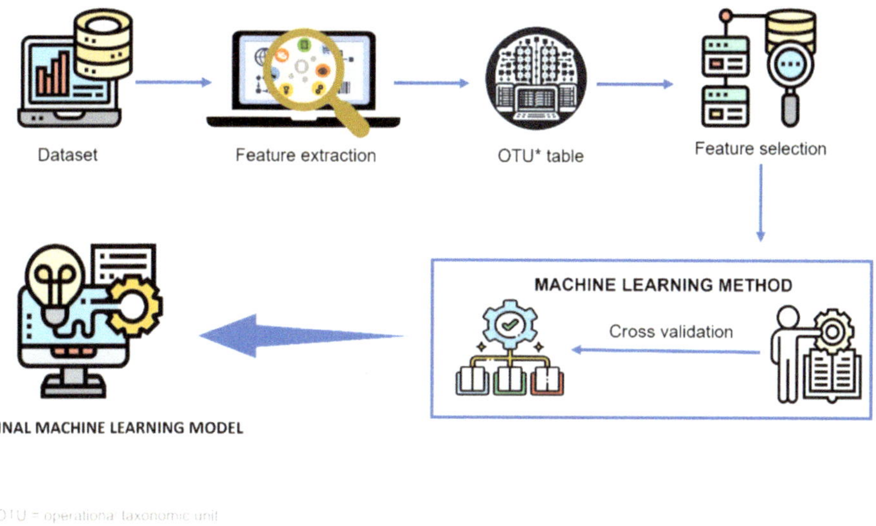

FIGURE 15.4 General procedural steps involved in the application of a machine learning model within the domain of microbiology. These steps typically include data collection and preprocessing, feature selection or extraction, model training, validation, and testing, as well as the final deployment of the trained model for practical use.

antibiotics, which improves the prognosis of patients while minimizing further resistance in the future (25).

Traditionally, the search for AMR markers used cultivation-based methodologies. However, there are limitations, given the impossibility of growing unculturable bacteria and the time it takes to grow others (26). AI and ML decision trees, DL, and neural networks can work more effectively, using large-scale genomic data and utilization patterns of antibiotics to predict resistance development (27).

Decision trees have been developed to differentiate between resistant and susceptible bacterial strains to inform optimal antibiotic choice. In one such model, a decision tree model described an approach that predicted antibiotic resistance for *E. coli* isolates with an average held-out accuracy of about 0.91 across 11 different compounds and 4 classes of antibiotics (28). Similarly, DL approaches such as DeepARG have proven better in the detection of antibiotic resistance genes in metagenomic data than traditional methods (29).

Recent AI and ML approaches employ homology modeling and molecular docking to discover resistance genes with high precision in various bacteria (30). For *Mycobacterium tuberculosis*, SVM- and DCNN-based models attained as high as 93% accuracy in identifying potential AMR genes. The deep neural pursuit-average activation potential (DNP-AAP) method has recently been applied to the *Neisseria gonorrhoeae* dataset, associating whole genome sequence data with resistance profiles to common antibiotics (31). An overview of the major steps involved in applying an ML model to microbiology is given in Figure 15.4.

15.3.2 Microbial Identification and Classification

The identification and classification of microorganisms have drastically changed with the emergence of AI and ML. These innovative tools offer unmatched accuracy, efficiency, and speed compared to traditional methods. For example, certain classical ML algorithms have gained a reputation as robust tools for the classification of microbes, hence being able to find patterns and relations within complex datasets, which has been of significant importance in this regard (32). SVM combined with Artificial Immune Recognition Systems has reached as high as 100% diagnostic accuracy in diagnosing *Mycobacterium tuberculosis* (33). RFs, also, have demonstrated unparalleled performance in microbial quantification and differentiation, achieving up to 99.7% accuracy (31). Especially, deep learning methodologies in neural networks are quickly improving the capacity of AI in microbiology. These advanced algorithms excel in feature extraction and processing autonomously from intricate datasets. Therefore, such developments allow one to go further into the details of precise microbial analysis. As an example, CNNs brought a revolution to the classification of harmful bacteria with the help of Raman spectroscopy. On average, CNNs perform with an accuracy of about 82%, outperforming the majority of the traditional ML methodologies (4, 34). For example, the classification of bacteria using FTIR spectra by MLP networks yielded excellent results with accuracy of more than 98%. These networks can also perform tasks of automatic spectral quality tests and classify bacterial spectra, hence presenting a quantum leap in microbial identification.

Application areas of AI and ML in microbial diagnostics are not restricted to enhancing accuracy. These technologies enable faster diagnosis with increased efficiency (4). They are fully capable of working through large datasets, which reveal patterns not visible to the human eye, and their ability to identify microbial threats quickly is vital in clinical settings and infectious disease control.

15.3.3 Predictive Modeling of Microbial Growth

Predictive modeling of microbial growth is intrinsic both to microbiological research and to food safety management. ML technologies handle such complex data and multifactor datasets, where pure mechanistic models often struggle. Applications of ML algorithms in predictive microbiology have focused on regression techniques and time series analysis (36): for instance, SVM and neural networks, which are specific applications in modeling growth curves of various microorganisms. Such models can take into consideration most of the factors that involve microbial growth, like temperature, pH, or humidity. In the food industry, predictive models like these can become very important for ensuring product safety and extension of shelf life (35, 37).

ML has been used to predict the growth of a dairy pathogen called *Listeria monocytogenes* (38). This is done by first training algorithms from historical data regarding temperature fluctuations and bacterial growth rates. Then, predictive models can predict quite accurately the risk of contamination. Similarly, in the healthcare industry, AI-driven predictive models are being used to forecast the growth pattern of hospital-acquired infections for effective infection control strategies (39).

15.3.4 Outbreak Detection and Surveillance

Outbreak detection and surveillance are key components of managing public health, in particular, finding and containing infectious disease outbreaks (40). Artificial intelligence technologies have greatly improved these systems by allowing more accurate, timely, and efficient means to identify potential outbreaks (41). Traditional approaches to outbreak detection involve the manual collection and screening of data, leading to labor-intensive tasks with incomplete dimensions of the emergent health threat.

AI and ML are a revolutionary new path to investigating all sorts of data from medical records to social networks, sometimes even flagging unusual disease patterns indicative of an outbreak well in advance (42). The application of AI and ML in the surveillance of influenza-like illnesses has been extremely promising (43). These algorithms analyze trends in visits to healthcare facilities, sales of medications, and even weather to make more accurate guesses about the onset and pattern of dispersion than traditional models can. Neural networks and time series forecasting models have been used to monitor flu trends to assist public health officials in better preparation and response (44).

Another application is the use of NLP to track symptoms of diseases and outbreaks through online platforms and news sources. It is a real-time surveillance system that can quickly detect any kind of outbreak, as in the COVID-19 pandemic where, using this facility, at the outset, it was detected with the help of NLP tools on online forums and news sites reporting unusual pneumonia cases (45).

Meanwhile, AI integrated with GIS has enabled much more advanced representative analytics of the spatial spread of diseases (46). ML models can, while analyzing data on population density, mobility patterns, and environmental factors, trace the patterns and hotspots of infection, which helps while devising targeted intervention strategies (47).

15.3.5 Drug Discovery and Development

The pharmaceutical industry has been revolutionized due to the inclusion of ML technologies which significantly speed up the discovery and development of therapeutic molecules, leading to cost reductions (48, 49). AI and ML algorithms have been applied in the early stage of drug discovery to analyze big biological datasets, which helped in target identification and candidate molecule identification (50).

For instance, CNNs analyzed protein structure to predict interactions with other molecules (51). This approach has been widely used in the search for potential inhibitors of the SARS-CoV-2 virus, whereby AI algorithms screened millions of compounds in a very short period to suggest potential drugs (52).

ML models have also assumed a key role in optimizing drug design by learning from comprehensive datasets of the chemical compounds and their biological activity. Such models can predict efficacy and toxicity for new compounds (53) and this predictive capability has been realized for drug development against Ebola virus infection, where ML algorithms were at the forefront in identifying high therapeutic potential compounds with reduced toxicity risks (54).

In recent advancements, drug development employs AI and ML to transform the clinical trial stage (55). Predictive models are built to identify appropriate candidates for clinical trials, hence improving the efficiency of patient recruitment with an increase in the success rate of clinical trials. AI-driven tools also track real-time data from clinical trials, thus providing data on drug efficacy and safety. This real-time analysis played a very important role during the accelerated development and approval of COVID-19 vaccines (56–58).

AI and ML, in turn, are opening routes toward personalized medicine. Based on the analysis of patient data at the genetic level, AI models can predict how a particular patient will react to a certain drug, which helps personalize that patient's treatment plan (59). This is exclusively being applied to treat cancers. AI for analyzing tumor genomics has led to the development of targeted cancer therapies (60).

15.3.6 Vaccine Antigen Prediction

Antigen prediction is one of the most important first steps in vaccine development, where parts of a pathogen are identified that can safely induce a protective immune response (61). Conventionally, antigen prediction has been a long and laborious process based on biological intuition and experimentation (62). AI and ML technologies have significantly enhanced this process, allowing better and faster identification of potential vaccine candidates and their targets (63–65).

ML algorithms, especially sequence analysis and pattern recognition, have played a significant role in epitope (specific regions of an antigen molecule that bind to antibodies). This is especially important in the design of peptide-based vaccines (63, 65). The DL models have been quite successful in complex biological data analysis. For example, during COVID-19 vaccine development, AI algorithms have successfully predicted viral epitopes of SARS-Co-V-2 virus that hold high potential for eliciting an immune response (66). Such predictions have been of paramount importance in rapid development and testing of effective vaccines (67).

Besides epitope prediction, ML algorithms help in understanding the evolution of antigens, which is important in diseases like influenza, where viruses mutate at a fast pace (68). For this, ML models analyze past data to predict future mutations to help in vaccine design that gives protection to an individual for a longer period (58, 69).

AI and ML also have applications in reverse vaccinology, where the genetics of a pathogen are used as a starting point to predict antigens (70). Analyzing the whole genome of a pathogen may lead an ML algorithm to identify novel candidate antigens that might have been missed previously by traditional methods (71). Such an approach has been followed for developing successful vaccines against Group B *Streptococcus* and meningococcus (72).

15.3.7 Microbiome Analysis and Functional Insights

High-throughput sequencing technologies have resulted in extensive genomic datasets from microbial communities. However, interpreting this data for functional insights is indeed a challenge. AI and ML algorithms have been built to tackle this challenge by offering efficient ways to analyze and interpret complex microbiome data (11).

ML has also been used to identify patterns and correlations in microbiome datasets, more precisely in supervised and unsupervised learning models (5). RF and Support Vector Machine are some of the algorithms used in associated studies that focused on specific microbial composition with health conditions, such as inflammatory bowel disease or obesity (12). This correlation helps in understanding the microbiome's role in maintaining pathogenesis and possible therapeutic interventions. Deep learning models further expanded the power of microbiome analysis. Such models will likely find reasons behind highly complicated patterns in the sequence data to provide better predictions of gene functions and metabolic pathways of the microbiome. In one such study, DL was used to predict functional capability in gut microbiota that allowed insights into their role in nutrient metabolic and immune-modulating processes (7).

AI/ML also play an imperative role in the emerging field of metagenomics (direct genetic analysis of environmental samples) (10, 11). The use of ML algorithms on metagenomic data allows for genome reconstruction even from noncultivable microorganisms, giving insights into their functional roles within the ecosystem (7). AI-driven network analysis has also been used to describe complex interactions between microbial populations, where the modeling of AI by the co-occurrence and co-exclusion relationships explains the symbiotic and antagonist relationships among these organisms. The latter is important for gaining insights into the stability and function of an ecosystem (73).

15.3.8 Improved Diagnostics and Treatment Strategies

AI and ML algorithms are used in diagnostic procedures, mainly in medical imaging. DL techniques are being utilized for analyzing radiological images of conditions such as tumors, fractures, and vascular diseases (74). These models are specifically helpful in discovering subtle patterns in imaging data which may be hardly noticeable by human radiologists. Another application is AI in mammography to enhance the detection of breast cancer. These approaches continue to show comparable or even superior accuracy to expert radiologists (75).

AI algorithms in treatment planning make predictions of the responses of patients to various therapies. Predictive capability is necessary for personalized medicine so that patients are subjected to effective therapy with minor side effects. Application of AI models has been made to predict the response of cancer patients to chemotherapy and thus assist the oncologist in framing a treatment plan based on the profile of individual patients (76).

AI and ML also tend to revolutioize patient monitoring and telemedicine. Wearables with embedded ML algorithms continuously monitor vital signs; once an abnormality is detected, an alarm is triggered to enable timely interference in conditions related to the cardiovascular system and diabetes (77). This technology has been found useful in the management of chronic diseases where continuous monitoring dramatically impacts the treatment results and quality of life.

15.3.9 Other Applications

One of the significant applications of AI and ML in microbiology is in the field of AMR management (78, 79). ML algorithms can analyze patterns of antibiotic prescriptions and subsequent resistance patterns, which can help develop more effective antibiotic stewardship programs (80–82).

Environmental microbiology benefits from AI in assessing microbial communities in ecosystems such as soil and water. This is vital for ecological studies, pollution control, and water quality management (83). For example, predicting pathogenic bacteria in water sources enhances water safety measures (84).

In industrial microbiology, AI and ML optimize fermentation processes and bioproduct synthesis, crucial for the efficient production of pharmaceuticals, biofuels, and food additives (85). These technologies optimize production conditions by analyzing data like temperature, pH, and microbial growth (86).

Bioinformatics, aided by AI, rapidly annotates microbial genomes, predicts gene functions, and unravels metabolic pathways, facilitating the discovery of novel biocatalysts and metabolic products (6, 7, 50).

In the food industry, ML improves food safety and quality control. Algorithms predict the shelf life of products by analyzing microbial growth and environmental factors, aiding in food preservation and waste reduction (73).

15.4 DEEP LEARNING IN MICROBIOLOGY

With the help of the powerful beam of DL, the world of microorganisms, which was formerly considered a mystery, is now being brought into the light. Many parts of microbiological research are being revolutionized by this subfield of artificial intelligence (87). Its multilayered architecture and ability to understand complicated patterns from data are two of the reasons for this revolution (19). In this section, we dig into three major DL techniques and their applications within the realm of microbes. Figure 15.5 indicates general steps followed by a DL algorithm in microbiological applications. We will see how these techniques are addressing some of the most serious concerns in the fields of health, environment, and other areas.

15.4.1 Convolutional Neural Networks (CNNs) for Image Analysis

With microscopy, a window into the world of microbes was opened, and large amounts of picture data were generated. On the other hand, manually evaluating these photos for identification, classification, and quantification continues to be a lengthy and error-prone task. Convolutional neural networks (CNNs) are highly effective computer image analysts that can extract important information and carry out reliable categorization (88).

CNNs are multilayered feature detectors, like experts that operate on a microscopic scale. Each layer functions as a filter, gradually extracting detailed patterns from the pixel data that is being processed (19). Deeper layers integrate these properties into sophisticated representations, which ultimately enables distinctions to be made between, for example, rod-shaped *Bacillus* bacteria and spherical *Staphylococcus* bacteria. Early layers are responsible for detecting fundamental edges and forms (19).

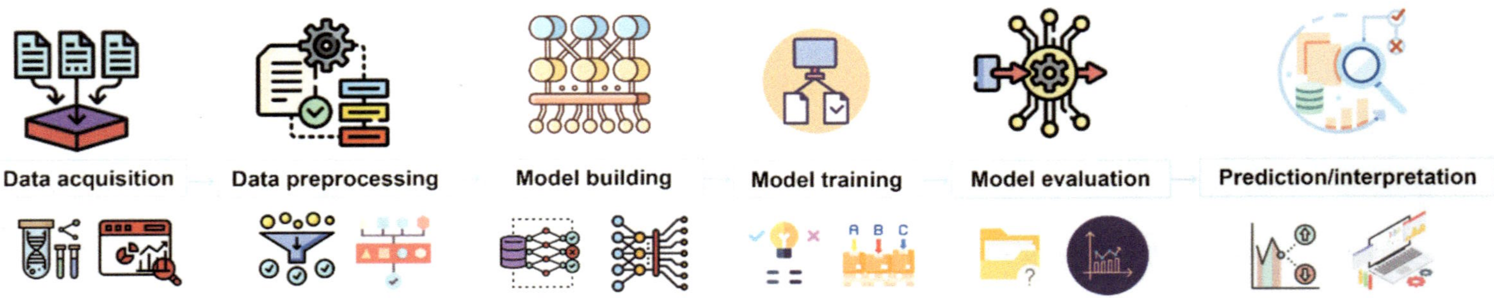

FIGURE 15.5 General steps of applying deep learning in microbiology. Starting with data collection and preprocessing, this illustration shows the overall flow of interconnected steps of deep learning. Each step is critical for making accurate predictions and interpretations.

Microbial CSI: Rapid identification of unknown bacteria in clinical samples, such as *Mycobacterium tuberculosis* in sputum smears, can be achieved by CNNs trained on labeled datasets of various pathogens (87, 89). This expedites diagnoses and guides focused antibiotic therapy. In the battle against infectious diseases like tuberculosis, this can prove to be pivotal. In a similar manner, they can quickly identify *Pneumocystis carinii* pneumonia, which is a fungal illness that is frequent in HIV-positive persons, from chest X-rays, which results in rapid antifungal therapy and improved patient outcomes (90, 91).

Drug discovery detectives: To predict the effectiveness of potential medications against infections like methicillin-resistant *Staphylococcus aureus* (MRSA), CNNs can examine microscope pictures of microbial interactions with these treatments (93, 94). Combating antibiotic resistance and speeding up the discovery of new life-saving medications are both achieved by this (81).

COVID-19 diagnosis: When it comes to the fight against COVID-19, CNNs that have been trained on chest X-ray and CT scan pictures are able to discriminate COVID-19 from other types of pneumonia with a high degree of accuracy, which helps with speedy diagnosis and early management (94, 95).

15.4.2 Recurrent Neural Networks (RNNs) for Sequence Data

The extensive reservoir of microbial genomic data, comprising DNA, RNA, and protein sequences, offers tremendous potential for comprehending the complicated mechanisms of life (92, 96). Nevertheless, converting these seemingly infinite sequences of nucleotides and amino acids into meaningful biological understandings poses a substantial difficulty (97). Recurrent neural networks (RNNs) are strong tools that are specifically built to decode the sequential language of life (98).

RNNs are highly skilled investigators, with an architecture that resembles a molecular Ferris wheel. By employing iterative loops, they navigate the sequence, capturing contextual dependencies between neighboring pieces (99). Initial iterations may detect fundamental patterns, while subsequent iterations expand on them, ultimately uncovering intricate connections and functionality embedded in the sequence (100).

Genome assembly experts: In the age of advanced sequencing methods, studying microbial genomes typically yields fragmented glimpses. After being trained on large genomic databases, RNNs demonstrate an impressive capacity to assemble these pieces into comprehensive blueprints, revealing the whole genetic structure of unknown microorganisms (101). This ability has facilitated the development of tailored therapy using individual microbial profiles and provided opportunities to comprehend previously mysterious species. RNNs are being used to construct the genetic sequence of a recently identified gut bacterium associated with inflammatory bowel disease (IBD). This has the capacity to transform the management of IBD by facilitating the creation of customized probiotics or dietary therapies designed for specific microbial makeup (102, 103).

Metabolic pathways prediction: If we want to understand the roles bacteria play in health and illness, the secret biochemical factories that power these processes (their metabolic pathways) must be unveiled. When trained on large datasets of known routes and gene sequences, RNNs have an extraordinary ability for accurately predicting new pathways (104). For instance, finding bacteria that can bioremediate environmental contaminants like oil spills or produce new antibiotics are two examples of where this predictive power could be quite useful (84).

15.4.3 Generative Adversarial Networks (GANs) for Microbial Simulation

Imagine taking part in the creation of the microbiome rather than just studying it. Generative adversarial networks (GANs) stand out in the DL field for their exceptional ability to creatively generate synthetic data that remarkably mimic actual microbiological processes (105). Two neural networks, the generator and the discriminator, compete with one another in GANs' operating framework. The goal of the generator is to produce real microbiological data, whereas the goal of the discriminator is to distinguish between real and fake cases. In the end, very complex and realistic microbiological simulations are produced by this antagonistic dynamic-driven iterative refinement within the generator (105, 106).

Think of GANs as digital artists that mimic the painstaking technique of Michelangelo. Nevertheless, GANs demonstrate an extraordinary ability to quickly produce large amounts of these microbial simulations, which is a sharp departure from their creative predecessors (107). With this capability, a wealth of potential for rapid and extensive study in microbiological research opens up.

Microbial community modelers: Predicting the effects of various microbial communities on agriculture, the environment, and human health requires an understanding of how these communities interact in real-world settings. Synthetic communities with tailored properties, such as certain species composition, metabolic relationships, and even spatial distribution, can be created by GANs (108, 109). By functioning as tiny ecosystems, these virtual communities let researchers test theories, forecast ecological dynamics, and create management plans for intricate microbial systems (110). Imagine creating and simulating distinct reef microbial communities under varying temperatures and ocean acidification scenarios to investigate the possible effects of climate change on the microbiome of coral reefs. This could help safeguard these important ecosystems and guide conservation efforts (106).

COVID-19 variant prediction: Using GANs, we can predict the emergence of new COVID-19 variants, which is useful information in our continuing fight against the virus. The use of

TABLE 15.1
Different Machine Learning Algorithms with Their Respective Applications in the Field of Microbiology

No.	Category	Machine Learning Algorithms	Use Cases in Microbiology	References
1.	Classification algorithms	Decision tree, SVM, naive Bayes, random forest, KNN	Identifying and categorizing microbial data into predefined classes.	(6, 112)
2.	Clustering algorithms	K-means, hierarchical clustering, DBSCAN	Grouping similar microbial entities without predefined categories.	(16, 113–115)
3.	Regression algorithms	Linear regression, lasso regression, logistic regression	Predicting continuous outcomes like growth rates or antibiotic resistance levels.	(16, 116, 117)
4.	Dimensionality reduction algorithms	PCA, t-distributed stochastic neighbor embedding (t-SNE)	Reducing the number of variables in complex microbial datasets.	(118, 119)
5.	Sequence analysis algorithms	Hidden Markov models (HMM), BLAST (Basic Local Alignment Search Tool)	Analyzing microbial DNA/RNA sequences.	(28, 120, 121)
6.	Deep learning algorithms	CNNs, RNNs, GANs	Complex tasks like image-based identification or genomic sequence analysis.	(118, 122–125)
7.	Reinforcement learning algorithms	Q-learning, Deep Q-Networks (DQN)	Used in experimental setups like adaptive laboratory evolution.	(18, 126)

GANs in evaluating viral sequences and modeling mutations could drive public health response measures and provide vital insights into the evolution of the virus (111).

ML and DL have many applications in microbiology beyond the ones given in Table 15.1. These techniques will transform many industries, from vaccine development to microbiome research. As researchers refine and apply these powerful approaches, we should expect additional groundbreaking discoveries and advances in microscopic comprehension and manipulation (25, 128).

15.5 CHALLENGES AND FUTURE DIRECTIONS

Despite the undeniable benefits to microbiology from the introduction of AI and ML, this paradigm change has not been without its share of difficulties. Data quantity and quality are major issues in applying AI and ML tools in microbiology. Microbial systems are complex and require big datasets to train. Microbes that have been barely investigated are hard to annotate (127). Building robust AI and ML models is difficult. Another issue is that the sophisticated models are confusing. Because of their complexity, especially deep learning neural networks, models are termed "black boxes" (54). Scientific knowledge and real-world use require comprehending model prediction reasoning. AI and ML are becoming more significant in microbiology research, making understandable models crucial. Combining multimodal data is complicated. Microbiology uses genetic sequences and environmental factors (128). Putting various data kinds together without affecting the original data is tricky. Models showing how different microorganisms interact require this. Understanding how microbial systems react to different environments requires proper integration (129).

Although the integration of AI and ML in microbiology may present hurdles, a promising road forward leads us toward future approaches that could transform our understanding of microbial life. The future of microbiological research depends on data collection and interpretation. High-throughput technologies and automated annotation methods improve dataset quality and AI model training, enabling more accurate microbial dynamics depiction. Explainable AI models are essential for the future. To understand microbial interactions, models must anticipate outcomes and reveal their decision-making processes. Explainable AI models will build confidence among researchers and practitioners, boosting scientific collaboration. More insights into microbial systems are possible with multi-omics data integration. Future research should create methods to integrate genomics, transcriptomics, proteomics, and metabolomics data. This holistic approach will improve microbial function understanding, enabling medical, biotech, and environmental research advances. Another future goal is to solve model generalization problems, especially in different microbial habitats (17). Transfer learning methods adapted for microbiological contexts can lead to models that adapt and perform well across environments, making AI and ML in microbiology more versatile and applicable (4). Finally, microbiological AI and ML ethics must be established. Gene editing and synthetic biology ethics must be thoroughly addressed as these technologies evolve. An ethical framework will ensure appropriate and transparent use of AI and ML, integrating them with microbiological research.

15.6 CONCLUSION

In conclusion, the integration of AI/ML with microbiology has sparked a significant change, enabling researchers to analyze extensive microbiological datasets with exceptional accuracy. DL techniques enable the exploration of novel approaches for automated microbial identification, predictive growth modelling, and prediction of AMR. This enables the proactive containment of outbreaks and the creation of precise therapeutic approaches to combat drug-resistant diseases. Metagenomic analysis, driven by AI and ML, reveals the complex structure of the microbiome, providing valuable knowledge about its impact on health and disease. This knowledge opens possibilities for tailored therapies. Despite the presence of data challenges, ethical concerns, and evolutionary obstacles, the future of microbiology appears bright with the potential of AI/ML. This advancement is expected to bring about a new era of scientific exploration, disease prevention, and medical treatment in our forever-evolving interaction with the microbial world.

REFERENCES

1. Peiffer-Smadja N, Dellière S, Rodriguez C, Birgand G, Lescure FX., Fourati S, et al. Machine learning in the clinical microbiology laboratory: has the time come for routine practice? *Clinical Microbiology and Infection*. 2020 Oct 1;26(10):1300–9. Available from: https://doi.org/10.1016/j.cmi.2020.02.006
2. Asnicar F, Thomas AM, Passerini A, Waldron L, Segata N. Machine learning for microbiologists. *Nature Reviews Microbiology*. 2024 Apr;22(4):191–205. Available from: www.nature.com/articles/s41579-023-00984-1
3. Egli A. Digitalization, clinical microbiology and infectious diseases. *Clinical Microbiology and Infection*. 2020 Oct 1;26(10):1289–90. Available from: https://doi.org/10.1016/j.cmi.2020.06.031
4. Jiang Y, Luo J, Huang D, Liu Y, Li DD. Machine learning advances in microbiology: a review of methods and applications. *Frontiers in Microbiology*. 2022 May 26;13:925454. Available from: https://doi.org/10.3389/fmicb.2022.925454
5. Ghannam RB, Techtmann SM. Machine learning applications in microbial ecology, human microbiome studies, and environmental monitoring. *Computational and Structural Biotechnology Journal*. 2021 Jan 1;19:1092–107. Available from: www.sciencedirect.com/science/article/pii/S2001037021000325
6. Goodswen SJ, Barratt JLN, Kennedy PJ, Kaufer A, Calarco L, Ellis JT. Machine learning and applications in microbiology. *FEMS Microbiology Reviews*. 2021 Mar 16;45(5). Available from: https://academic.oup.com/femsre/articleabstract/45/5/fuab015/6174022
7. Talukder A, Barham C, Li X, Hu H. Interpretation of deep learning in genomics and epigenomics. *Briefings in Bioinformatics*. 2021 May;22(3):bbaa177. Available from: https://academic.oup.com/bib/article/22/3/bbaa177/5894987
8. Auffray C, Imbeaud S, Roux-Rouquié M, Hood L. From functional genomics to systems biology: concepts and practices. *Comptes rendus. Biologies*. 2003;326(10–11):879–92. Available from: https://doi.org/10.1016/j.crvi.2003.09.033
9. Bailey AL, Ledeboer N, Burnham CAD. Clinical microbiology is growing up: the total Laboratory Automation revolution. *Clinical Chemistry*. 2019 May 1;65(5):634–43. Available from: https://academic.oup.com/clinchem/article/65/5/634/5608051
10. Reis T, Krause T, Bornschlegl MX, Hemmje ML. A conceptual architecture for AI-based big data analysis and visualization supporting metagenomics research. In *CERC* 2020 Sep (pp. 264–272). Available from: www.researchgate.net/publication/349462837
11. Wani AK, Roy P, Kumar V. Metagenomics and artificial intelligence in the context of human health. *Infection, Genetics and Evolution*. 2022 Jun 1;100:105267. Available from: https://doi.org/10.1016/j.meegid.2022.105267
12. Lin Y, Wang G, Yu J, Sung JJY. Artificial intelligence and metagenomics in intestinal diseases. *Journal of Gastroenterology and Hepatology*. 2021 Apr 1;36(4):841–7 Available from: https://doi.org/10.1111/jgh.15501
13. Han C, Rundo L, Murao K, Nemoto T, Nakayama H. Bridging the gap between AI and healthcare sides: towards developing clinically relevant AI-powered diagnosis systems. In *Artificial Intelligence Applications and Innovations: 16th IFIP WG 12.5 International Conference, AIAI 2020, Neos Marmaras, Greece, June 5–7, 2020, Proceedings, Part II* 16 2020 (pp. 320–333). Springer International Publishing. Available from: https://doi.org/10.1007/978-3-030-49186-4_27
14. Kricka LJ, Polevikov S, Park JY, Fortina P, Bernardini S, Satchkov D et al. Artificial intelligence-powered search tools and resources in the fight against covid-19. *Electronic Journal of the International Federation of Clinical Chemistry and Laboratory Medicine*. 2020;31(2):106–116.
15. Nasteski V. An overview of the supervised machine learning methods. *Horizons*. 2017 Dec 1;4(51–62):56. Available from: https://doi.org/10.20544/horizons.b.04.1.17.p05
16. Usama M, Qadir J, Raza A, Arif H, Yau KL, Elkhatib Y, Hussain A, Al-Fuqaha A. Unsupervised machine learning for networking: techniques, applications and research challenges. *IEEE Access*. 2019 May 14;7:65579–615. Available from: https://doi.org/10.1109/access.2019.2916648

17. Li Y. Deep reinforcement learning: an overview. arXiv preprint arXiv:1701.07274. 2017 Jan 25. Available from: www.semanticscholar.org/paper/Deep-Reinforcement-Learning%3A-An-Overview-Li/9f1e9e56d8014 6766bc2316efbc54d8b770a23df
18. Popescu MC, Balas VE, Perescu-Popescu L, Mastorakis N. Multilayer perceptron and neural networks. *WSEAS Transactions on Circuits and Systems*. 2009 Jul 1;8(7):579–88.
19. Zieliński B, Plichta A, Misztal K, Spurek P, Brzychczy-Włoch M, Ochońska D. Deep learning approach to bacterial colony classification. *PloS One*. 2017 Sep 14;12(9):e0184554.
20. Miller D, Stern A, Burstein D. Deciphering microbial gene function using natural language processing. *Nature Communications*. 2022 Sep 29;13(1):5731. Available from: https://doi.org/10.1038/s41 467-022-33397-4
21. Matougui B, Boukelia A, Belhadef H, Galiez C, Batouche M. NLP-metaxa: a natural language processing approach for metagenomic taxonomic binning based on deep learning. *Current Bioinformatics*. 2021 Aug 1;16(7):992–1003.
22. Qu K, Guo F, Liu X, Lin Y, Zou Q. Application of machine learning in microbiology. *Frontiers in Microbiology*. 2019 Apr 18;10:451710. Available from: www.frontiersin.org/articles/10.3389/fmicb.2019.00827/full
23. Rabaan AA, Alhumaid S, Mutair AA, Garout M, Abulhamayel Y, Halwani MA, Alestad JH, Bshabshe AA, Sulaiman T, AlFonaisan MK, Almusawi T. Application of artificial intelligence in combating high antimicrobial resistance rates. *Antibiotics*. 2022 Jun 8;11(6):784.
24. Masud RI, Fahim NA, Rana ML, Islam MS, Tanvir M. Artificial intelligence, a powerful tool to combat antimicrobial resistance: an update. *Journal Advanced Biotechnology Experimental Therapeutics*. 2023;6:711–27.
25. Ali T, Ahmed S, Aslam M. Artificial intelligence for antimicrobial resistance prediction: challenges and opportunities towards practical implementation. *Antibiotics*. 2023 Mar 6;12(3):523.
26. Sakagianni A, Koufopoulou C, Feretzakis G, Kalles D, Verykios VS, Myrianthefs P. Using machine learning to predict antimicrobial resistance—A literature review. *Antibiotics*. 2023 Feb 24;12(3):452.
27. Moradigaravand D, Palm M, Farewell A, Mustonen V, Warringer J, Parts L. Prediction of antibiotic resistance in Escherichia coli from large-scale pan-genome data. *PLoS Computational Biology*. 2018 Dec 14;14(12):e1006258.
28. Arango-Argoty G, Garner E, Pruden A, Heath LS, Vikesland P, Zhang L. DeepARG: a deep learning approach for predicting antibiotic resistance genes from metagenomic data. *Microbiome*. 2018 Dec;6:1–5.
29. Kim JI, Maguire F, Tsang KK, Gouliouris T, Peacock SJ, McAllister TA, McArthur AG, Beiko RG. Machine learning for antimicrobial resistance prediction: current practice, limitations, and clinical perspective. *Clinical Microbiology Reviews*. 2022 Sep 21;35(3):e00179–21.
30. Shi J, Yan Y, Links MG, Li L, Dillon JA, Horsch M, Kusalik A. Antimicrobial resistance genetic factor identification from whole-genome sequence data using deep feature selection. *BMC Bioinformatics*. 2019 Dec;20:1–4.
31. Wu Y, Gadsden SA. Machine learning algorithms in microbial classification: a comparative analysis. *Frontiers in Artificial Intelligence*. 2023;6.
32. Saybani MR, Shamshirband S, Hormozi SG, Wah TY, Aghabozorgi S, Pourhoseingholi MA, Olariu T. Diagnosing tuberculosis with a novel support vector machine-based artificial immune recognition system. *Iranian Red Crescent Medical Journal*. 2015 Apr;17(4):e24557.
33. Croxatto A, Marcelpoil R, Orny C, Morel D, Prod'hom G, Greub G. Towards automated detection, semi-quantification and identification of microbial growth in clinical bacteriology: a proof of concept. B*iomedical Journal*. 2017 Dec 1;40(6):317–28.
34. Ho CS, Jean N, Hogan CA, Blackmon L, Jeffrey SS, Holodniy M, Banaei N, Saleh AA, Ermon S, Dionne J. Rapid identification of pathogenic bacteria using Raman spectroscopy and deep learning. *Nature Communications*. 2019 Oct 30;10(1):4927.
35. Tarlak F. The use of predictive microbiology for the prediction of the shelf life of food products. *Foods*. 2023 Dec 13;12(24):4461. Available from: www.mdpi.com/2304-8158/12/24/4461
36. García-Gimeno RM, Hervás-Martínez C, Zurera-Cosano G. Application of artificial intelligence to predictive microbiology. In *Novel Food Processing Technologies*. CRC Press; 2005. p. 609–27.
37. Hiura S, Koseki S, Koyama K. Prediction of population behavior of Listeria monocytogenes in food using machine learning and a microbial growth and survival database. *Scientific Reports*. 2021 May 19;11(1):10613.

38. Santangelo OE, Gentile V, Pizzo S, Giordano D, Cedrone F. Machine learning and prediction of infectious diseases: a systematic review. *Machine Learning and Knowledge Extraction*. 2023 Feb 1;5(1):175–98.
39. Buckeridge DL, Burkom H, Campbell M, Hogan WR, Moore AW. Algorithms for rapid outbreak detection: a research synthesis. *Journal of Biomedical Informatics*. 2005 Apr 1;38(2):99–113.
40. Sabat AJ, Budimir A, Nashev D, Sá-Leão R, Van Dijl JM, Laurent F, Grundmann H, Friedrich AW, ESCMID Study Group of Epidemiological Markers (ESGEM). Overview of molecular typing methods for outbreak detection and epidemiological surveillance. *Eurosurveillance*. 2013 Jan 24;18(4):20380. Available from: www.eurosurveillance.org/ViewArticle.aspx?ArticleId=20380
41. Umapathy VR, Raj RD, Yadav S, Munavarah SA, Anandapandian PA, Mary AV, Padmavathy K, Akshay R, Rajkumar DS, Anandapandian IV PA, Mary V. Perspective of artificial intelligence in disease diagnosis: a review of current and future endeavours in the medical field. *Cureus*. 2023 Sep 21;15(9):e45684.
42. López Pineda A, Ye Y, Visweswaran S, Cooper GF, Wagner MM, Rich Tsui F. Comparison of machine learning classifiers for influenza detection from emergency department free-text reports. *Journal of Biomedical Informatics*. 2015 Dec 1;58:60–9.
43. Morris M, Hayes P, Cox IJ, Lampos V. Neural network models for influenza forecasting with associated uncertainty using Web search activity trends. *PLoS Computational Biology*. 2023 Aug 1;19(8):e1011392.
44. Homburg M, Meijer E, Berends M, Kupers T, Olde Hartman T, Muris J, de Schepper E, Velek P, Kuiper J, Berger M, Peters L. A natural language processing model for COVID-19 detection based on Dutch general practice electronic health records by using bidirectional encoder representations from transformers: development and validation study. *Journal of Medical Internet Research*. 2023 Oct 4;25:e49944.
45. Lewis MD, Pavlin JA, Mansfield JL, O'Brien S, Boomsma LG, Elbert Y, et al. Disease outbreak detection system using syndromic data in the greater Washington DC area. *American Journal of Preventive Medicine*. 2002 Oct 1;23(3):180–6.
46. Watkins RE, Eagleson S, Beckett S, Garner G, Veenendaal B, Wright G, Plant AJ. Using GIS to create synthetic disease outbreaks. *BMC Medical Informatics and Decision Making*. 2007 Dec;7:1–4. Available from: https://bmcmedinformdecismak.biomedcentral.com/articles/10.1186/1472-6947-7-4
47. Vemula D, Jayasurya P, Sushmitha V, Kumar YN, Bhandari V. CADD, AI and ML in drug discovery: a comprehensive review. *European Journal of Pharmaceutical Sciences*. 2023 Feb 1;181:106324.
48. Mohs RC, Greig NH. Drug discovery and development: role of basic biological research. *Alzheimer's & Dementia: Translational Research & Clinical Interventions*. 2017 Nov 1;3(4):651–7.
49. Khadela A, Popat S, Ajabiya J, Valu D, Savale S, Chavda VP. AI, ML and other bioinformatics tools for preclinical and clinical development of drug products. In *Bioinformatics Tools for Pharmaceutical Drug Product Development*, ed. V Chavda, K Anand, and V Apostolopoulos. Wiley; 2023. p. 255–84.
50. Bui-Thi D, Rivière E, Meysman P, Laukens K. Predicting compound-protein interaction using hierarchical graph convolutional networks. *Plos One*. 2022 Jul 21;17(7):e0258628.
51. Jamshidi MB, Lalbakhsh A, Talla J, Peroutka Z, Roshani S, Matousek V, Roshani S, Mirmozafari M, Malek Z, La Spada L, Sabet A. Deep learning techniques and COVID-19 drug discovery: Fundamentals, state-of-the-art and future directions. In *Emerging Technologies during the Era of COVID-19 Pandemic*, ed. I Arpaci et al. Springer; 2021. p. 9–31.
52. Aittokallio T. What are the current challenges for machine learning in drug discovery and repurposing? *Expert Opinion on Drug Discovery*. 2022 May 4;17(5):423–5.
53. Kwofie SK, Adams J, Broni E, Enninful KS, Agoni C, Soliman ME, Wilson MD. Artificial Intelligence, machine learning, and big data for Ebola virus drug discovery. *Pharmaceuticals*. 2023 Feb 21;16(3):332.
54. Askin S, Burkhalter D, Calado G, El Dakrouni S. Artificial intelligence applied to clinical trials: opportunities and challenges. *Health and Technology*. 2023 Mar;13(2):203–13.
55. Abdulkareem NM, Abdulazeez AM, Zeebaree DQ, Hasan DA. COVID-19 world vaccination progress using machine learning classification algorithms. *Qubahan Academic Journal*. 2021 May 3;1(2):100–5.
56. Abubaker Bagabir S, Ibrahim NK, Abubaker Bagabir H, Hashem Ateeq R. Covid-19 and artificial intelligence: genome sequencing, drug development and vaccine discovery. *Journal of Infection and Public Health*. 2022;15(2):289–96. Available from: www.sciencedirect.com/science/article/pii/S1876034122000144
57. Oyewola DO, Dada EG, Misra S. Machine learning for optimizing daily COVID-19 vaccine dissemination to combat the pandemic. *Health and Technology*. 2022 Nov;12(6):1277–93. Available from: https://doi.org/10.1007/s12553-022-00712-4

58. Schork NJ. Artificial intelligence and personalized medicine. In Cancer Treatment and Research. 2019;178:265–83.
59. Liao J, Li X, Gan Y, Han S, Rong P, Wang W, Li W, Zhou L. Artificial intelligence assists precision medicine in cancer treatment. *Frontiers in Oncology*. 2023 Jan 4;12:998222.
60. Thomas S, editor. *Vaccine Design: Methods and Protocols, Volume 1. Vaccines for Human Diseases*. Springer Nature; 2021 Dec 16.
61. Saleh A, Qamar S, Tekin A, Singh R, Kashyap R. Vaccine development throughout history. *Cureus*. 2021 Jul 26;13(7).
62. Thomas S, Abraham A, Baldwin J, Piplani S, Petrovsky N. Artificial intelligence in vaccine and drug design. In *Vaccine Design: Methods and Protocols, Volume 1. Vaccines for Human Diseases*, ed. S Thomas. Humana Press; 2022. p. 131–46.
63. Lee EK, Nakaya HI, Yuan F, Querec TD, Burel G, Pietz FH, Benecke BA, Pulendran B. Machine learning for predicting vaccine immunogenicity. *Interfaces*. 2016 Oct;46(5):368–90.
64. Liu G, Carter B, Bricken T, Jain S, Viard M, Carrington M, Gifford DK. Robust computational design and evaluation of peptide vaccines for cellular immunity with application to SARS-CoV-2. *bioRxiv*. 2020 May 17:2020-05. Available from: http://biorxiv.org/content/early/2020/05/17/2020.05.16.088989.
65. Hederman AP, Ackerman ME. Leveraging deep learning to improve vaccine design. *Trends in Immunology*. 2023 Mar 30;44;333–344.
66. Kaushal K, Sarma P, Rana SV, Medhi B, Naithani M. Emerging role of artificial intelligence in therapeutics for COVID-19: a systematic review. *Journal of Biomolecular Structure and Dynamics*. 2022 Jun 21;40(10):4750–65.
67. Parvizpour S, Pourseif MM, Razmara J, Rafi MA, Omidi Y. Epitope-based vaccine design: a comprehensive overview of bioinformatics approaches. *Drug Discovery Today*. 2020 Jun 1;25(6):1034–42.
68. Sharma A, Virmani T, Pathak V, Sharma A, Pathak K, Kumar G, Pathak D. Artificial intelligence-based data-driven strategy to accelerate research, development, and clinical trials of COVID vaccine. *BioMed Research International*. 2022 Jul 6;2022: 7205241.
69. Ong E, He Y. Vaccine design by reverse vaccinology and machine learning. *Methods in Molecular Biology*. 2022;2414:1–16.
70. Bhakta S, Choudhury S, Paul J, Bhattacharya A. Vaccine development through reverse vaccinology using artificial intelligence and machine learning approach. In *COVID-19: Tackling Global Pandemics through Scientific and Social Tools*, ed. S Chatterjee. Academic Press; 2022. p. 33–49. Academic Press.
71. Rappuoli R. Reverse vaccinology. *Current Opinion in Microbiology*. 2000 Oct 1;3(5):445–50.
72. Soni A, Dixit Y, Reis MM, Brightwell G. Hyperspectral imaging and machine learning in food microbiology: developments and challenges in detection of bacterial, fungal, and viral contaminants. *Comprehensive Reviews in Food Science and Food Safety*. 2022 Jul;21(4):3717–45. Available from: https://ift.onlinelibrary.wiley.com/doi/abs/10.1111/1541-4337.12983
73. Soffer S, Ben-Cohen A, Shimon O, Amitai MM, Greenspan H, Klang E. Convolutional neural networks for radiologic images: a radiologist's guide. *Radiology*. 2019 Mar;290(3):590–606.
74. Chen Y, Taib AG, Darker IT, James JJ. Performance of a breast cancer detection AI algorithm using the personal performance in mammographic screening scheme. *Radiology*. 2023 Sep 5;308(3):e223299.
75. Russo V, Lallo E, Munnia A, Spedicato M, Messerini L, D'Aurizio R, Ceroni EG, Brunelli G, Galvano A, Russo A, Landini I. Artificial intelligence predictive models of response to cytotoxic chemotherapy alone or combined to targeted therapy for metastatic colorectal cancer patients: a systematic review and meta-analysis. *Cancers*. 2022 Aug 19;14(16):4012.
76. Oikonomou EK, Khera R. Machine learning in precision diabetes care and cardiovascular risk prediction. *Cardiovascular Diabetology*. 2023 Sep 25;22(1):259.
77. Lv J, Deng S, Zhang L. A review of artificial intelligence applications for antimicrobial resistance. *Biosafety and Health*. 2021;3:22–31.
78. Burri RR, Pallu R, Burra PVLS. *Knowledge report on Integrated Strategies for containment of AMR for AP Govt*. 1st ed. APAPACR Support Consortium; August 2023. p.36.
79. Fitzpatrick F, Doherty A, Lacey G. Using artificial intelligence in infection prevention. *Current Treatment Options in Infectious Diseases*. 2020 Jun;12:135–44. Available from: https://link.springer.com/article/10.1007/s40506-020-00216-7
80. Lv J, Deng S, Zhang L. A review of artificial intelligence applications for antimicrobial resistance. *Biosafety and Health*. 2021 Feb 1;3(01):22–31.

81. Chakriswaran P, Vincent DR, Kadry S. Ensemble of artificial intelligence techniques for bacterial Antimicrobial Resistance (AMR) estimation using topic modeling and similarity measure. *International Journal of Uncertainty, Fuzziness and Knowledge-Based Systems*. 2022 Jun;30(03):541–65. Available from: www.worldscientific.com/doi/abs/10.1142/S0218488522400207
82. Sarrafzadeh MH, Mansouri SS, Zahiri J, Mussatto SI, Asgharnejad H. Artificial intelligence in environmental microbiology. *Frontiers in Microbiology*. 2022 Jun 13;13:944242.
83. Hallin S. Environmental microbiology going computational-Predictive ecology and unpredicted discoveries. *Environmental Microbiology*. 2023 Jan;25(1):111. Available from: https://ami-journals.onlinelibrary.wiley.com/doi/10.1111/1462-2920.16232
84. Signoroni A, Ferrari A, Lombardi S, Savardi M, Fontana S, Culbreath K. Hierarchical AI enables global interpretation of culture plates in the era of digital microbiology. *Nature Communications*. 2023 Oct 28;14(1):6874. Available from: https://doi.org/10.1038/s41467-023-42563-1
85. Vandenberg O, Durand G, Hallin M, Diefenbach A, Gant V, Murray P, Kozlakidis Z, van Belkum A. Consolidation of clinical microbiology laboratories and introduction of transformative technologies. *Clinical Microbiology Reviews*. 2020 Mar 18;33(2):10–128.
86. Zhang Y, Jiang H, Ye T, Juhas M. Deep learning for imaging and detection of microorganisms. *Trends in Microbiology*. 2021 Jul 1;29(7):569–72.
87. Shin HC, Roth HR, Gao M, Lu L, Xu Z, Nogues I, Yao J, Mollura D, Summers RM. Deep convolutional neural networks for computer-aided detection: CNN architectures, dataset characteristics and transfer learning. *IEEE Transactions on Medical Imaging*. 2016 Feb 11;35(5):1285–98.
88. Kermany DS, Goldbaum M, Cai W, Valentim CCS, Liang H, Baxter SL, et al. Identifying medical diagnoses and treatable diseases by image-based deep learning. *Cell*. 2018 Feb 22;172(5):1122–31.
89. Parveen N, Sathik MM. Detection of pneumonia in chest X-ray images. *Journal of X-ray Science and Technology*. 2011 Jan 1;19(4):423–8.
90. Vianna VP. Study and development of a computer-aided diagnosis system for classification of chest X-ray images using convolutional neural networks pre-trained for imagenet and data augmentation. arXiv preprint arXiv:1806.00839. 2018 Jun 3. Available from: http://arxiv.org/abs/1806.00839
91. Peng J, Li J, Shang X. A learning-based method for drug-target interaction prediction based on feature representation learning and deep neural network. *BMC Bioinformatics*. 2020 Sep 17;21(Suppl 13):394. Available from: https://bmcbioinformatics.biomedcentral.com/articles/10.1186/s12859-020-03677-1
92. Yamanishi Y, Kotera M, Kanehisa M, Goto S. Drug-target interaction prediction from chemical, genomic and pharmacological data in an integrated framework. *Bioinformatics*. 2010 Jun 1;26(12):246–54.
93. Reshi AA, Rustam F, Mehmood A, Alhossan A, Alrabiah Z, Ahmad A, Alsuwailem H, Choi GS. An efficient CNN model for COVID-19 disease detection based on X-ray image classification. *Complexity*. 2021 May 17;2021:1–2.
94. Dumakude A, Ezugwu AE. Automated COVID-19 detection with convolutional neural networks. *Scientific Reports*. 2023 Jun 30;13(1):10607. Available from: www.nature.com/articles/s41598-023-37743-4
95. Horgan RP, Kenny LC. 'Omic'technologies: genomics, transcriptomics, proteomics and metabolomics. *Obstetrician & Gynaecologist*. 2011 Jul 1;13(3). Available from: https://onlinelibrary.wiley.com/doi/full/10.1576/toag.13.3.189.27672
96. Kiechle FL, Holland-Staley CA. Genomics, transcriptomics, proteomics, and numbers. *Archives of Pathology & Laboratory Medicine*. 2003 Sep 1;127(9):1089–97. Available from: https://dx.doi.org/10.5858/2003-127-1089-GTPAN
97. Elhassani ME, Maisonnasse L, Olgiati A, Jerome R, Rehali M, Duroux P, Giudicelli V, Kossida S. Deep learning concepts for genomics: an overview. *EMBnet Journal*. 2022 Mar 21;27. Available from: https://hal.science/hal-04075722
98. Pouladi F, Salehinejad H, Gilani AM. Recurrent Neural Networks for Sequential Phenotype Prediction in Genomics. In *2015 International Conference on Developments of E-Systems Engineering (DeSE)* 2015 Dec 13 (pp. 225–230). IEEE.
99. Yue T, Wang Y, Zhang L, Gu C, Xue H, Wang W, Lyu Q, Dun Y. Deep learning for genomics: a concise overview. arXiv preprint arXiv:1802.00810. 2018 Feb 2. Available from: http://arxiv.org/abs/1802.00810

100. Lipton ZC, Berkowitz J, Elkan C. A critical review of recurrent neural networks for sequence learning. arXiv preprint arXiv:1506.00019. 2015 May 29.
101. Lugo L, Hernández EB. A Recurrent Neural Network approach for whole genome bacteria identification. *Applied Artificial Intelligence*. 2021 Jul 29 [cited 2024 Jan 7];35(9):642–56. Available from: www.tandfonline.com/doi/abs/10.1080/08839514.2021.1922842
102. Bahdanau D, Cho K, Bengio Y. Neural machine translation by jointly learning to align and translate. arXiv preprint arXiv:1409.0473. 2014 Sep 1.
103. Shen Z, Bao W, Huang DS. Recurrent neural network for predicting transcription factor binding sites. *Scientific Reports*. 2018 Oct 15;8(1):15270. Available from: www.nature.com/articles/s41598-018-33321-1
104. Frid-Adar M, Diamant I, Klang E, Amitai M, Goldberger J, Greenspan H. GAN-based synthetic medical image augmentation for increased CNN performance in liver lesion classification. *Neurocomputing*. 2018 Dec 10;321:321–31.
105. Creswell A, White T, Dumoulin V, Arulkumaran K, Sengupta B, Bharath AA. Generative adversarial networks: an overview. *IEEE Signal Processing Magazine*. 2018 Jan 10;35(1):53–65.
106. Chhibbar P, Joshi A. Generating protein sequences from antibiotic resistance genes data using generative adversarial networks. arXiv preprint arXiv:1904.13240. 2019 Apr 28. Available from: http://arxiv.org/abs/1904.13240
107. Yu S, Li H, Li X, Fu YV, Liu F. Classification of pathogens by Raman spectroscopy combined with generative adversarial networks. *Science of the Total Environment*. 2020 Jul 15;726:138477.
108. Reiman D, Dai Y. Using conditional generative adversarial networks to boost the performance of machine learning in microbiome datasets. bioRxiv. 2020 May 21:2020–05. Available from: www.biorxiv.org/content/10.1101/2020.05.18.102814v2
109. Rong R, Jiang S, Xu L, Xiao G, Xie Y, Liu DJ, Li Q, Zhan X. MB-GAN: microbiome simulation via generative adversarial network. *GigaScience*. 2021 Feb;10(2):giab005. Available from: https://dx.doi.org/10.1093/gigascience/giab005
110. Bej A, Maulik U, Sarkar A. Time-Series prediction for the epidemic trends of COVID-19 using Conditional Generative adversarial Networks Regression on country-wise case studies. *SN Computer Science*. 2022 Jun 29;3(5):352. Available from: https://link.springer.com/article/10.1007/s42979-022-01225-7
111. Wang Q, Garrity GM, Tiedje JM, Cole JR. Naive Bayesian classifier for rapid assignment of rRNA sequences into the new bacterial taxonomy. *Applied and Environmental Microbiology*. 2007 Aug 15;73(16):5261–7. Available from: https://journals.asm.org/doi/10.1128/AEM.00062-07
112. Lee JH, Yi H, Jeon YS, Won S, Chun J. TBC: a clustering algorithm based on prokaryotic taxonomy. *The Journal of Microbiology*. 2012 Apr;50:181–5. Available from: https://link.springer.com/article/10.1007/s12275-012-1214-6
113. Preheim SP, Perrotta AR, Martin-Platero AM, Gupta A, Alm EJ. Distribution-based clustering: using ecology to refine the operational taxonomic unit. *Applied and Environmental Microbiology*. 2013 Nov 1;79(21):6593–603. Available from: https://journals.asm.org/doi/10.1128/AEM.00342-13
114. Hallinan JS. Data mining for microbiologists. *Methods in Microbiology*. 2012 Jan 1;39:27–79.
115. Radu D, Ovidiu F, Coculescu C. The Usage of Linear Regression Models in Predictive Microbiology. In *Proceedings of the 14th WSEAS International Conference on Computers: Part of the 14th WSEAS CSCC Multiconference-Volume II* 2010 Jul 23 (pp. 665–668). Available from: https://dl.acm.org/doi/10.5555/1984366.1984408
116. Hajmeer M, Basheer I. Comparison of logistic regression and neural network-based classifiers for bacterial growth. *Food Microbiology*. 2003 Feb 1;20(1):43–55.
117. Oh M, Zhang L. DeepMicro: deep representation learning for disease prediction based on microbiome data. *Scientific Reports*. 2020 Apr 7;10(1):6026. Available from: www.nature.com/articles/s41598-020-63159-5
118. Treloar NJ, Fedorec AJ, Ingalls B, Barnes CP. Deep reinforcement learning for the control of microbial co-cultures in bioreactors. *PLoS Computational Biology*. 2020 Apr 10;16(4):e1007783. Available from: https://journals.plos.org/ploscompbiol/article?id=10.1371/journal.pcbi.1007783
119. Park H, Lim SJ, Cosme J, O'Connell K, Sandeep J, Gayanilo F, Cutter Jr GR, Montes E, Nitikitpaiboon C, Fisher S, Moustahfid H. Investigation of machine learning algorithms for taxonomic classification of marine metagenomes. *Microbiology Spectrum*. 2023 Oct 17;11(5):e05237–22.

120. Mathieu A, Leclercq M, Sanabria M, Perin O, Droit A. Machine learning and deep learning applications in metagenomic taxonomy and functional annotation. *Frontiers in Microbiology*. 2022 Mar 14;13:811495.
121. Vervier K, Mahé P, Vert JP. MetaVW: Large-scale machine learning for metagenomics sequence classification. *Methods in Molecular Biology*. 2018;1807:9–20.
122. Wang H, Zhang L, Han J, Weinan E. DeePMD-kit: a deep learning package for many-body potential energy representation and molecular dynamics. *Computer Physics Communications*. 2018 Jul 1;228:178–84. Available from: www.sciencedirect.com/science/article/pii/S0010465518300882
123. Torku TK, Khaliq AQ, Furati KM. Deep-data-driven neural networks for COVID-19 vaccine efficacy. *Epidemiologia*. 2021 Nov 30;2(4):564–86. Available from: www.mdpi.com/2673-3986/2/4/39
124. Min S, Lee B, Yoon S. Deep learning in bioinformatics. *Briefings in Bioinformatics*. 2017 Sep 1;18(5):851–69.
125. Liu J, Li J, Wang H, Yan J. Application of deep learning in genomics. *Science China Life Sciences*. 2020 Dec;63:1860–78. Available from: https://link.springer.com/article/10.1007/s11427-020-1804-5
126. Franco-Duarte R, Černáková L, Kadam S, S. Kaushik K, Salehi B, Bevilacqua A, Corbo MR, Antolak H, Dybka-Stępień K, Leszczewicz M, Relison Tintino S. Advances in chemical and biological methods to identify microorganisms-from past to present. *Microorganisms*. 2019 May 13;7(5):130.
127. Sapoval N, Aghazadeh A, Nute MG, Antunes DA, Balaji A, Baraniuk R, Barberan CJ, Dannenfelser R, Dun C, Edrisi M, Elworth RL. Current progress and open challenges for applying deep learning across the biosciences. *Nature Communications*. 2022 Apr 1;13(1):1728. Available from: www.nature.com/articles/s41467-022-29268-7
128. Rani P, Kotwal S, Manhas J, Sharma V, Sharma S. Machine learning and deep learning based computational approaches in automatic microorganisms image recognition: methodologies, challenges, and developments. *Archives of Computational Methods in Engineering*. 2022 May;29(3):1801–37. Available from: https://link.springer.com/article/10.1007/s11831-021-09639-x
129. Alzubaidi L, Zhang J, Humaidi AJ, Al-Dujaili A, Duan Y, Al-Shamma O, Santamaría J, Fadhel MA, Al-Amidie M, Farhan L. Review of deep learning: concepts, CNN architectures, challenges, applications, future directions. *Journal of Big Data*. 2021 Dec;8:1–74. Available from: https://journalofbigdata.springeropen.com/articles/10.1186/s40537-021-00444-8we

16 Future Microbiology Education

Mohammad Ejaz, Laiba Hassan, Usama Imtiaz, Maria Rukan, and Shahzad Ali

16.1 INTRODUCTION

Microbiology is the scientific discipline that focuses on investigating microbes and their interactions with other species and the environment. The field of microbiology, which is comparatively a new scientific discipline, has been taught as a separate branch of biology for about the last two centuries. Microbiology education has experienced significant changes and reforms over time, from the teachings of Mahavira in the sixth century to advanced microbiological techniques (1, 2). The field of microbiology received significant recognition after the observations made by Antonie van Leeuwenhoek in 1676, who observed microorganisms using the microscope. Afterward, the foundational knowledge for microbiology was laid in the second half of the nineteenth century when renowned and influential microbiologists Louis Pasteur, Martinus Beijerinck, and Sergei Winogradsky used interdisciplinary methods to study the wide variety of microbial processes occurring on Earth (3). The contributions of Robert Koch (1843–1910) were the primary impetus for the impending shift in perspective. Koch was a physician who lived during Pasteur's time and was particularly interested in infectious diseases. Medical applications of microbiology were thrust into the spotlight following Koch's compelling proof that bacteria were the reason behind significant infectious diseases. Koch's postulates and methods for obtaining pure cultures had a great impact on the development and progress of the field of microbiology (4). The continuous progress in technologies and the identification of microorganisms that cause diseases, as well as their beneficial aspects, have contributed to the constant evolution of this discipline.

The science of microbiology has recently gained significant attention from the general population and certain areas that have previously received significant attention are now more pertinent than ever. The shift stems from various factors, including major disease outbreaks and advancements in industries. Events like Ebola, Zika, influenza, dengue outbreaks, and the COVID-19 pandemic have propelled microbiology into the spotlight, showing the vital role of microbes in causing infectious diseases and the significance of research in understanding, combating, and preventing such events (5). Similarly, revolutions in the food and pharmaceutical industries were triggered by the utilization of microorganisms. Biotechnological advancements have broadened the range of microbiological uses, resulting in breakthroughs like the creation of biofuels, the removal of environmental contaminants through bioremediation, and the creation of new biopharmaceuticals. This development emphasizes how multidisciplinary microbiological study is and how much it affects industry, public health, and society at large.

Moreover, microorganisms are essential in influencing climate change. Throughout billions of years, they have altered our planet, guaranteeing the continuity of life by controlling the mixture of gases in Earth's atmosphere. Microorganisms are susceptible to the impacts of climate change,

DOI: 10.1201/9781003410164-16

including elevated temperatures, drought, and declining pH and oxygen concentrations in the world's oceans. Certain alterations may result in the redistribution of current pathogens or promote the establishment of novel infectious illnesses, necessitating the attention of the public health sector. This is particularly relevant as human populations become increasingly concentrated in urban settings. The intricate nature of the microbial biosphere and its interactions with plant and animal hosts presents significant difficulties in forecasting and alleviating the impacts of climate change. Nevertheless, there are possibilities to meticulously control microorganisms in a manner that could potentially address issues related to climate change (6, 7).

Furthermore, the improvements in molecular biology, next-generation sequencing, and bioinformatics have greatly revitalized various disciplines of research linked with microbiology. These advancements have led to the emergence of several new subfields in microbiology. Subsequently, microbiology education has become a significant component of both medical and biological sciences programs, at both the undergraduate and graduate levels (2, 8). Microbiology serves as a crucial link between fundamental, practical, and medical sciences. It has found practical applications in various disciplines including clinical medicine, environmental science, pharmaceutical science, molecular biology, and biotechnology (9). This chapter is focused on the comprehensive analysis of current challenges in microbiology education, emerging fields in microbiology, and the implementation of emerging technologies to improve microbiology education in the future.

16.2 CURRENT CHALLENGES IN MICROBIOLOGY EDUCATION

Microbiologists are currently experiencing a period in which they are gathering biological data in large quantities using high-throughput technology. A growing demand for skills beyond traditional laboratory and field research is the ability to utilize computational tools to extract valuable insights from the collected data (10). Next-generation sequencing (NGS), also known as high-throughput sequencing (HTS), offers novel approaches for detecting and characterizing microbial features, such as serotyping and genotyping, that go beyond traditional culture-based methods. The utilization of NGS in the fields of microbiology and medical research has become a prevalent and quickly expanding practice. Its profound influence on research, diagnostics, and clinical microbiology has been evident in recent years. Among the various forms of NGS, whole genome sequencing (WGS) has emerged as the most often employed method. The utilization of NGS for the identification of microorganisms is increasingly becoming widely accepted, particularly for bacteria, viruses, algae, and fungi (11, 12). The utilization of high-throughput technology necessitates the creation of novel approaches to handle the data produced by researchers. Hence, microbiologists and other life scientists must enhance their proficiency in handling the data produced by the recently adopted and evolving technologies (13).

Similarly, clinical microbiology covers a broad range of pathogenic microorganisms and some of the microorganisms cannot be identified using traditional microbiological methods. Clinical microbiology heavily relies on traditional culture-based and phenotypic methods. Traditional methods like broth dilution methods, and the Kirby-Baur disk diffusion method determining MIC and clinical breakpoints for antibiotic susceptibility profiling, remain gold standard for the prediction of antibiotic resistance, but many factors such as pre-analytical, analytical, or post-analytical errors can influence the results, leading to false negative or false positive results (14). Furthermore, it is not easily manageable through a limited number of inexpensive rapid diagnostic procedures or contained inside highly effective vertical control programs.

Over the past few decades, there has been a focused endeavor to tackle the current obstacles faced by clinical microbiology laboratories and ever-increasing antimicrobial resistance (AMR) in pathogenic microorganisms (15, 16). So, there is a need to update traditional microbiological approaches and couple them with advanced molecular biology techniques that can rapidly detect pathogenic microorganisms with utmost precision. Moreover, the detection of antimicrobial genes in resistant isolates using molecular biology techniques can improve molecular epidemiology of resistant strains.

A significant obstacle faced by many microbiological laboratories, especially in resource-constrained countries, is the limited capability of laboratory information systems (LIS) to support intricate data analytics (17). It would be optimal if these technologies could be incorporated into the electronic medical record, allowing for a two-way interchange of information. This integration has the potential to facilitate the analysis of clinical and laboratory data, resulting in a more comprehensive understanding of test utilization and the influence of laboratory results on patient treatment. It is anticipated that the experiences and investments made during the COVID-19 pandemic will stimulate the establishment of infrastructure capable of connecting to global disease burden and AMR surveillance networks (18). So, there is a need to go beyond the traditional curriculum to incorporate evolving, innovative, and emerging technologies in educating microbiology graduates. In addition, educators must regularly revise content to reflect the changing and evolving field of microbiology and global health concerns. The COVID-19 pandemic had a profound impact on higher education, necessitating the use of online learning as the sole method of instruction during the lockdowns (19). Moreover, COVID-19 has expedited the implementation of online learning, and translating hands-on laboratory experiences to virtual formats while preserving instructional quality emerged as a serious issue for educators.

The teaching of microbiology in many underdeveloped nations is characterized by unique aspects that are shaped by limitations in resources, socioeconomic issues, and educational infrastructure (20, 21). The prevalence of lecture-based instruction is attributed to the constraints of large class sizes and limited resources, which prioritize the teaching of theoretical concepts such as microbial taxonomy, morphology, and diseases. Practical sessions are essential; however, the focus may be on fundamental procedures due to difficulties in accessing well-equipped laboratories. Faculty frequently depend on textbooks as their main sources of information, because they have limited availability of online materials and scientific publications. Technological constraints are a major factor, as the absence of access to state-of-the-art equipment hampers the integration of sophisticated microbiological methods (22, 23). The curriculum frequently focuses on infectious diseases that are pertinent to the region, with particular emphasis on issues such as epidemiology and antibiotic resistance. The presence of insufficient laboratory facilities and obsolete equipment presents significant barriers to offering students comprehensive practical experiences (24).

Limited resources, including financial support for research initiatives and academic resources, have a direct impact on the overall standard of education (25). The field of faculty development encounters obstacles due to the scarcity of chances for educators to attend global conferences or engage in cooperative research endeavors. Despite these restrictions, educators strive to adapt traditional methods by incorporating local context into examples and case studies. Collaboration and networking initiatives in the region are essential for the exchange of knowledge and sharing of resources, which in turn enhance the quality of microbiological education by making it more relevant to the local environment. Moreover, there is a disparity between the microbiology curricula offered at institutions and the qualifications needed for employment in the job market. Employers are currently seeking and hiring graduates who possess the ability to solve human problems, streamline processes, optimize space and time, and deliver precise solutions to modern challenges. Upon examining the undergraduate microbiology curricula of most universities, especially in developing countries, it becomes evident that there is a pressing need to change the curricula and implement intervention courses aimed at enhancing competency (26). Consequently, a university graduate who possesses commendable microbiology achievements but lacks supplementary expertise may find himself unemployed or working in a position with lower compensation compared to a non-graduate who possesses the essential abilities required by the hiring firm (27). The persistent discrepancy between the qualifications and skills that graduates possess and the demands of the labor market has a detrimental impact year after year and necessitates immediate response.

16.3 EMERGING FIELDS SHAPING FUTURE MICROBIOLOGY EDUCATION

The field of microbiology is currently undergoing a period of captivating developments. It is becoming more evident that microorganisms are the fundamental building blocks of the biosphere. They serve as the progenitors of all extant organisms and provide the foundation for the existence of all other life forms. Paradoxically, specific microorganisms are potentially threatening to both human health and the well-being of plants and animals. Microbes have a crucial and essential function in life on earth, serving as the foundation of the biosphere and exerting significant influence on human health. Therefore, the study of microorganisms is crucial to systemic scientific studies of all organisms, making the field of microbiology indispensable for the investigation and comprehension of all life forms on this planet. Microbiology has been an ever-evolving field and, due to advancements in science and technology, new and advanced areas of microbiology are being studied, including interdisciplinary subjects. The science of microbiology is seeing a resurgence, propelled by technical advancements and a more profound comprehension of microbial processes. These nascent disciplines provide the potential for revolutionary advancements in the domains of medicine, industry, and environmental management. As scientists further explore the complexities of the microbial realm, the possibilities for revolutionary findings and advancements in microbiology seem limitless. The future of microbiology is a fascinating exploration of the microscopic domains that profoundly influence our world in ways that we are just beginning to comprehend.

16.3.1 Microbiome Studies

The human body harbors approximately 300 billion microorganisms that collectively form the microbiome (28). These organisms play a crucial role in coordinating many physiological activities and influencing susceptibility to diseases. Although the ratio of bacterial cells to human cells is similar, bacteria exhibit significantly higher genetic diversity. Bacteria possess a wide range of characteristics that enable them to excel in metabolism. These attributes not only influence their immediate surroundings but also have a significant impact on the immune cells of the host, affecting certain tissues and functions (29). The human gut microbiota comprises not just bacteria but also fungi, protozoa, and viruses (30). Typically, when the host and its microbiome are in good health, they maintain a mutually advantageous relationship called symbiosis. The host offers a conducive milieu abundant in nutrients, while the microbiome assists in digestion and metabolism. The idea that the host and microbiome work together as a unified entity is known as a meta-organism (31, 32). These microorganisms have a significant impact on overall health and balance between pathogenic and beneficial microbes within the body. The oral cavity and skin harbor a diverse and abundant assemblage of bacteria, while vaginal microbiota has been known for having lower diversity and being dominated by a few specific microbial species (33). Furthermore, the application of next-generation sequencing (NGS) has uncovered that organs and tissues such as the lung, prostate, bladder, breast, liver, and pancreas, traditionally thought to be devoid of microbes, may in fact harbor small communities of microorganisms (34–36). However, there is a continuing dispute about whether these microbiomes originate from separate populations at commensal sites or if they are only transitory migrants from surrounding areas.

Exploration of the microbiota has resulted in the creation of innovative treatments and diagnostic instruments (37, 38). Nevertheless, the outcomes of various investigations frequently diverge from prior discoveries (39). The presence of both random and systemic bias in sequence-based human microbiome investigations can contribute to the inconsistency of results observed across different studies. Over the past decade, there has been a significant increase in studies focusing on the microbiome. Notably, large-scale studies such as the Human Microbiome Project (40), the Metagenomics of the Human Intestinal Tract (41), the Molecular Metabolites Database, and the American Gut Project (42) have greatly expanded the amount of data available on the composition and function of the human microbiome. The data obtained through microbiome studies is typically stored in specialized databases that were designed for genomics, metagenomics, proteomics, and multi-omics data.

Enhanced data-analytical methods are required to effectively utilize the information contained within these biological datasets, taking into account the unique characteristics of microbiome data, such as its compositional nature and the heterogeneity and sparsity of the datasets. Utilizing taxonomy-informed feature selection to predict host phenotypes and build a correlation with the microbiome can have significant advantages for personalized treatment. This approach can help anticipate different disease states and enhance human health. The gut microbiome has become an essential component of personalized medicine. It not only plays a significant role in the differences between individuals' health and disease, but it also presents a potentially changeable factor that can be specifically addressed by therapeutics (43).

16.3.2 Synthetic Biology and Genetic Engineering

Synthetic biologists strive to engineer living systems that fulfill immediate human requirements, bypassing the need to rely on natural evolution for the emergence of beneficial biological functions. Organisms that have been reverse-engineered are presently employed in enclosed industrial environments for fuel, the creation of commercially valuable renewable chemicals, and the reduction of pharmaceutical production costs (44). Microorganisms have been widely employed in several industrial processes due to their ability to perform numerous complex and unique chemical reactions. In addition, synthetic biology is currently being employed to create customized microbes that may effectively tackle major social problems, in addition to reaping the benefits of naturally occurring traits. Research conducted in the past decade has resulted in new and enlarged insights into fundamental biological processes due to the rapid expansion of bioinformatics data and the development of increasingly sophisticated analysis methods. The application of this knowledge is now beginning to unlock the extraordinary promise of synthetic biology in the realm of biotechnology, as well as its significant capacity to address hitherto unresolved global challenges. The field of industrial microbiology has experienced significant advantages as a result of progress in synthetic biology (45). These technological developments have enabled the development of novel biosynthetic pathways for manufacturing sustainable fuels and chemicals, programmable logic controls for regulating and optimizing intricate biological operations, and robust microorganisms for eradicating detrimental environmental toxins (46).

Synthetic biology has greatly expanded the variety of chemical compounds that can be produced using biological methods by employing advanced tools and procedures to construct new enzymes and pathways. These hitherto elusive compounds have crucial functions in various sectors. They serve as monomers for the production of bioplastics as well as for bulk and fine chemicals (47). In addition, they are utilized for the synthesis of long-chain alcohols and other biofuels that exhibit improved compatibility for direct application (48, 49). Microbes can create a diverse array of chemicals known as isoprenoids. The microbial generation of isoprenoids has been substantially eased by recent developments in comprehending complicated regulatory and metabolic networks, as well as the engineering of biological devices utilizing standardized genetic pieces (50). Researchers did an innovative investigation to showcase the augmentation of dihydroartemisinic acid synthesis in yeast by discerning intricate metabolic engineering objectives through a combination of computer-based analysis (in silico) and live creature tests (in vivo) (51, 52).

With the advent of recombinant DNA technology, chimeric genes can be obtained that contain the coding region of one gene and are incorporated into the regulatory region of the target gene. Thus, chimeric genes upon expression result in the production of specific proteins. Several different types of proteins can be synthesized by genetic engineering and are used for medication. Some of the common proteins synthesized by recombinant DNA technology include human insulin, human growth hormone, factor VIII, and follicle-stimulating hormone (FSH). Recombinant DNA technology has brought many revolutionary changes in the field of medicine and introduced novel methods in the diagnosis of diseases, drug delivery, and treatment of diseases. Genetic factors are of great importance in adult-onset diseases such as cancer, neurodegenerative diseases, and atherosclerosis. Advances in genetic engineering have led to the development of prognostic approaches to medicine for such diseases (53).

Industry is also keen on employing genetic engineering approaches to optimize food fermentations and enzymatic processes. This is due to the large-scale production of food, which presents enormous potential advantages in terms of increased productivity. By ensuring fair competition, not only may the industry's profit margins be enhanced but customers may also benefit from lower prices. Moreover, due to the limitations of organic chemistry in achieving high yields, purity, and specificity in the synthesis of complex proteins, genetic engineering offers a valuable solution. Techniques involving the use of genetically modified microorganisms, particularly those with a safe track record in food production like yeasts and lactic acid bacteria and well-established nonpathogenic bacteria such as *Escherichia coli* K12, can also be utilized for this purpose (54).

16.3.3 Bioinformatics and Computational Microbiology

The advancements in computer technology and memory storage capacity in the last decade have facilitated the simulation of intricate tasks, such as sequencing large genomes and handling vast amounts of information on the internet. The substantial increase in computational capacity has led to the production of an enormous volume of genomic and proteomic data that is readily available to researchers across the globe. The availability of data has also facilitated numerous groundbreaking discoveries and allowed for the creation of procedures that were previously unachievable using conventional laboratory research methods.

The availability of genomic and proteomics data, along with progress in bioinformatics and biochemical procedures, has increased humanity's expectation of being able to alter existing bacteria to impact genetics. The advantages are significant, such as improved disease diagnosis by utilizing protein biomarkers, protection against diseases through affordable vaccines, rational drug design, increased agricultural productivity and quality, and the development of techniques that allow us to visualize and understand the complex microbial machinery at the systemic level (55–57).

Advancements in genomics have significantly impacted the study of microorganisms that can be cultivated. This encompasses the elucidation of microbial taxonomies and evolution using DNA/RNA sequences instead of physical or biochemical characteristics, the emergence of nucleotide-based diagnostic methods, molecular epidemiology, and the deciphering of the genetic foundation for pathogenic virulence, beneficial attributes in probiotics, or safety in starter cultures for fermentations (58). Although this knowledge has been highly valuable in terms of its foundational significance, it is increasingly being utilized in ways that will bring benefits to doctors, industry, and the general public for many years to come (59).

The vast amount of data generated by genome sequencing programs would be overwhelming and incomprehensible without the use of bioinformatics methods. Over the past decade, bioinformatics has discreetly taken up the responsibility of delivering cost-efficient data analysis. This has expedited the pace of discoveries, the advancement of pharmaceuticals and immunizations, and the formulation of antibacterial agents (60, 61). Moreover, bioinformatics analysis has enhanced our understanding of the genomic structure and the mechanism of microbial rearrangement. The utilization of bioinformatics analysis will accelerate the investigation of the overall behavior of cellular processes, facilitating the understanding of these processes to effectively control and regulate microbial cells as production facilities. Over the past decade, bioinformatics techniques have been developed to identify and analyze various components of cells, such as gene and protein activity, interactions, metabolic and regulatory processes, and generation of recombinant proteins. The next decade may focus on understanding cellular mechanisms and influencing cells by combining bioinformatics, wet lab investigations, and cell simulation technologies (62, 63). It is anticipated that in the next ten years, the partially automated examination of cellular behavior at a systemic level will improve this skill more quickly.

Although bioinformatics is a relatively new discipline, it has significantly contributed to the advancement of fundamental microbiology and biotechnology. This has been achieved through the creation of algorithms, tools, and discoveries that have improved our understanding of the abstract model of microbial cell activity. The primary influence of bioinformatics has been in the automation of microbial genome sequencing, the creation of integrated databases accessible through the internet, and the analysis of genomes to gain insights into gene and genome function.

16.3.4 Revisiting the Evolution of Microorganisms

Bioinformatics experts have conducted thorough comparisons of many genomes to establish correlations, classify the genomes into different families, and investigate evolutionary patterns. Multiple researchers have confirmed that the process of evolution involves a combination of point-based mutations that lead to the formation of new species, as well as the reorganization of genomes through gene duplications, insertions, deletions, fusions/fissions, horizontal gene transfer, and restructuring at the domain level (64, 65).

The study of evolution can be categorized into three approaches: (1) the utilization of point-based mutation approaches to construct traditional evolutionary trees by employing multiple sequence alignment of 16S rRNA, (2) the examination of genome restructuring through gene-level inversions and transpositions, and (3) the investigation based on whole genome comparisons using gene identity of orthologous genes across multiple microbial genomes (66, 67).

16.3.5 Quantum Microbiology

Quantum microbiology, similar to quantum biology, is postulated to rely on a resilient heat engine. This engine is driven by specific endothermic and exothermic biochemical events, which are dictated by precise organic genetic instructions found in DNA and RNA (68). Contrary to this perspective, it is argued that only a small number of processes in biological systems necessitate knowledge of quantum theory, even though all biochemicals have quantum beginnings (69). To what extent are quantum principles integrated into biological systems at the molecular level, such as DNA replication and photosynthesis, as well as at higher levels, including human senses and subconscious thought? We believe that significant progress in knowledge often occurs when different disciplines intersect. Therefore, it is important to thoroughly investigate, contemplate, and research all connections, as well as any gaps, plausible hypotheses, and incomplete experiments, whenever feasible. Quantum theory is fundamental to the physical world and plays a central role in all biological systems. These systems consist of physical organisms made up of specific and often interacting biochemicals, which rely on specific genetic instructions and enzyme catalysts. These organisms undergo natural selection at the organism level while striving to reproduce for the survival of their species. Nevertheless, the impact of quantum events on the emergence of non-living pre-biotic organisms, as well as the initial microbial cells on Earth, remains incomprehensible (70). From this standpoint, quantum theory is utilized to examine the impact of microbial life on Earth and its subsequent development over 3.6–4 billion years. Given that the cell serves as the fundamental building block of life, any quantum-level data obtained from microbial cells can reasonably be assumed to be relevant to eukaryotic cells.

Scientists have recently been studying photosynthesis in green sulfur bacteria that are capable of surviving in environments with low light intensities. These bacteria can efficiently transmit photonic energy at an impressive efficiency rate of about 100%. In terms of comparison, solar panels typically attain an average efficiency of approximately 20%, while specialized laboratory research models can reach up to 42% efficiency. These organisms utilize a light-harvesting chlorosome antenna to turn sunlight into energy. This energy is then transferred to a photosynthetic reaction center through a pigment-protein complex called the FMO complex, which connects the two. This pigment process complex consists of a trimer, where each monomer is made of bacteriochlorophyll molecules encased in a protein backbone. This design allows for multiple potential routes through which the absorbed energy can get from the chlorosome to the reaction center (71).

It is noteworthy that bacterial light-sensing phytochromes play a similar role in enhancing photosynthetic quantum processes. In the absence of light, most phytochromes are in a condition where they absorb red light, but upon absorbing light, they transition to a state where they absorb far red light. Recently, it has been shown that cyanobacteria have a plethora of distinct photoreceptors that exhibit a high level of sensitivity to colors that are not red. Due to the varying depths of fresh and saltwater where these creatures reside, the absorption of light by water transforms with increasing depth. It transitions from a red hue near the surface to green-blue and eventually blue as the depth increases. Bacteria can undergo chromatic adaptation by utilizing various color-absorbing pigmented phycobiliproteins in their light-harvesting antennae. This is achieved by manipulating the proportion of these light-absorbing pigments, which in turn detect the specific type of ambient light, enhancing the efficiency of photon absorption (72).

Evolution and the organic, genetic code have been consistent factors during nearly 4 billion years of life on Earth, with quantum mechanics serving as the foundation of the physical world and resulting in the existence of numerous diverse living species. According to Pattee (1967), quantum mechanics is the most suitable framework for comprehending energy and matter, whereas the theory of evolution is the optimal technique for understanding natural selection and inherited features (73, 74). Perhaps a synthesis of these two methodologies will eventually yield a comprehensive explanation regarding the genesis of microbial life, substantiated by empirical evidence. In the realm of scientific inquiry, it is customary to tackle the challenging investigations first, while the seemingly unattainable ones demand additional time and exertion. Potential factors that could contribute to a deeper comprehension of the origins of microbial life and biology include quantum mechanics, the laws of thermodynamics, time, the dual nature of light as both a particle and a wave, and the role of organic, genetic instructions. Additionally, a more thorough understanding of how quantum events can manifest as classical events, and vice versa, is crucial. It is necessary to enhance our understanding of these domains and establish correlations among them in living beings. As humanity progresses, we may gain a deeper comprehension of the human intellect and its origins as a rudimentary microorganism.

16.3.6 Forensic Microbiology

Forensic microbiology, often known as microbial forensics, is a specialized field that applies microbiological knowledge to aid in forensic investigations. Forensic microbiology can be applied in several scenarios, such as the unintentional release of biological substances, fraudulent activities, crimes, and bioterrorism (55). The significance of this research increased in 2001 with the resolution of the *Bacillus anthracis* bioterror attack on the United States postal services. This resolution was achieved through the utilization of high-resolution whole genome sequencing and comparative genomics (56). Forensic microbiology focuses on detecting microbial sources to determine the time since death (postmortem interval or PMI), identifying individuals, identifying tissues/fluids, determining criminal locations, and eventually determining the cause of death. Forensic microbiology surpasses the testing conducted by clinical microbiology and public health laboratories. This discipline is based on traditional microbiology methods, including cultures and biochemical identification, but also integrates molecular biology assays, genetics, and phylogenetics. Furthermore, it is crucial to consider forensic principles, evidence-gathering, maintaining the chain of custody, and delivering court presentations (75).

Forensic microbiology is constantly advancing, providing innovative methods for solving crimes and addressing public health issues. The combination of metagenomics, developments in DNA sequencing, and an enhanced comprehension of microbial ecology are contributing to the increasing importance of the area. To establish strong standards and maintain the credibility of forensic microbiology in legal settings, researchers, forensic practitioners, and legal authorities must collaborate.

16.3.6.1 Artificial Intelligence in Forensic Microbiology

The science of forensic microbiology is already being transformed by artificial intelligence (AI) and machine learning (ML). Microbial succession during postmortem deterioration has the potential to be a valuable method for estimating the postmortem interval (PMI) (76). Johnson et al. have devised an artificial intelligence method to efficiently examine PMI by collecting samples of skin microbiota from the nasal and ear canals (77). The study of the microbiome has the potential to serve as a valuable tool for the crucial task of individual identification. Theoretically, each person possesses a distinct collection of microbial communities that remain generally consistent over extended periods (76). Yang et al. achieved a 90% accuracy in identifying individuals by combining the *Cutibacterium acnes* 16S rRNA genotype with skin microbiome profile data using random forest machine learning (78). Efforts have also been undertaken to discern different tissues and body fluids by analyzing the composition of microorganisms, as this could significantly aid in crime reconstruction. Díez López et al. achieved a high level of accuracy in classifying body sites such as oral, cutaneous, and vaginal secretions using deep learning networks that are not dependent on taxonomy (79). The MetaSUB effort, founded in 2015, aims to improve the utilization and development of cities by detecting, analyzing, and creating urban environmental metagenomics. It focuses on the subways and urban biomes, employing an interdisciplinary approach. Huang et al. utilized data from MetaSUB and demonstrated a high accuracy of 86% in determining the geographical source of metagenomic samples by the application of machine learning, namely logistic regression (80).

Diatom analyses have traditionally been employed in forensic investigations related to drowning to ascertain the cause of death. Zhou et al. utilized an AI technique to automatically detect diatoms. They accomplished this by training convolutional neural network (CNN) models utilizing transfer learning and data augmentation techniques (81). Forensic microbiology is now in its early stages of development, but it is expected to advance significantly through the use of extensive datasets obtained from gene sequencing, which will enhance the capabilities of this area. Recent research has shown that CNN models can effectively distinguish between bacteria and algae by analyzing microbiome photos. The primary emphasis of microbial computer image analysis lies in the segmentation, grouping, classification, and quantification of microorganisms (82).

Forensic microbiology is now in its early stages of development. The utilization of advanced technologies, such as next-generation sequencing (NGS) and whole genome sequencing (WGS), has generated a substantial dataset that was previously inconceivable. Continued technological improvements will further enhance the ability to conduct forensic examinations involving microorganisms. Furthermore, artificial intelligence has significantly advanced the technological progress of forensic microbiology over the past few decades (83).

16.3.7 Microbial Biotechnology

Biotechnology employs microorganisms such as fungi, bacteria, and viruses in the domains of agriculture, food, and pharmaceuticals. Farmers and academics are collaborating to harness the potential of microorganisms as a biocontrol agent to combat pests that pose a threat to their crops. Soil microorganisms, such as bacteria and fungi, play a crucial role in breaking down organic matter and reusing deceased plant material (84). In the field of biotechnology and biomanufacturing, microorganisms and living cells serve as small-scale chemical factories that generate essential products such as amino acids, enzymes, medications, and food additives (85). Microorganisms have been utilized by humans for millennia to facilitate fermentation processes. Yeasts, molds, and bacteria have been employed to produce a wide range of food products, including bread, vinegar, beer, wine, cheese, and yogurt, as well as fermented fish, meat, and vegetables (86). Microorganisms are utilized in the process of fermenting various types of food to create a diverse range of oriental culinary goods. Large quantities of microorganisms have been utilized to create numerous biological preparations that hold significant importance in the fields of medicine and pharmacy (87).

16.3.8 Agriculture

The incorporation of microorganisms as natural fertilizers is emerging as the most popular area of research, alongside the use of hybrid seeds, high-yielding types, chemical fertilizers, and regular watering. The detrimental effects and exorbitant expenses associated with chemical fertilizers render them economically unfeasible for agricultural purposes. By 2020, an estimated 28.8 million tons of nutrients will be needed to produce 321 million tons of food grains. However, the quantity of nutrients accessible amounts to 21.6 million tons. The difference between nutrient supply and nutrient loss is 7.2 million tons (88). Natural fertilizers consist mostly of living microbial inoculants, such as algae, fungi, and bacteria, either individually or in combination. These inoculants can improve the availability of nutrients in the soil for plants (89). Through their microbial activities, they possess the capacity to transform the nutritionally significant component found in the soil from an unusable state to a usable state. These activities include phosphate solubilization, nitrogen fixation, excretion of plant growth hormones, and biodegradation in the soil. Biofertilizers have ecological benefits, enhance productivity, are readily available to small-scale farmers, and demonstrate higher efficiency (90). So, microbial use in the agriculture sector has increased since the introduction of microbial biotechnology.

Moreover, the soil harbors plant pathogenic bacteria within the rhizosphere zone, which can give rise to various plant diseases. Scientists have utilized microorganisms as a biological tool to manage and control undesirable weeds and pests. These microorganisms have genes that are capable of invading and attacking weeds, ultimately leading to their demise, and also show resistance to drought and other unfavorable changes in soil (91). Similarly, the utilization of bioinsecticides is increasing in comparison to chemical insecticides owing to their reduced level of risk (92).

16.3.9 Microbial Proteins Revolutionizing Nutritional Sciences

Microbiology education often explores the formation of microbial protein (MP) or single-cell protein (SCP) as an interesting case study. Researchers explore the use of plentiful and affordable hydrocarbon substrates to develop high-quality protein additions. This not only caters to the nutritional requirements of cattle and aquaculture, but also offers educators a chance to delve into interdisciplinary subjects by combining microbiology, nutrition, and sustainable methods (93). These developments are recognized by the FAO/WHO standards for essential amino acid scoring patterns in human nutrition, providing educators with a modern perspective to teach fundamental ideas (94).

16.3.10 Environmental Microbiology

The study of environmental microbiology concerns the microscopic domain that has shaped the development of all living forms on Earth for three billion years. Despite being invisible to human sight, these microscopic unicellular organisms have played a significant role in altering the original atmosphere, which lacked oxygen, to the current atmospheric composition of 21% (95). The bacterial ability to perform oxygenic photosynthesis was instrumental in this remarkable achievement. In the event of a catastrophic occurrence caused by nature or humans, all non-microbial life forms would be eliminated. However, certain microorganisms that are resistant and capable of surviving in

Earth's seas and crust could endure. These microorganisms may act as catalysts for a novel evolutionary mechanism following a catastrophe, adjusting to newly arising environmental circumstances as long as there is access to liquid water.

Microbial ecology, a scientific field pioneered by German scientist Ernst Haeckel in the 1800s, focuses on studying the interactions between microorganisms and their living and non-living surroundings. This field has been widely utilized since the early 1960s. It has significantly advanced our understanding of the role of living creatures, particularly microbes, in preserving the ecological balance of ecosystems and the larger environment. It is important to integrate these concepts into future microbiology instruction. Educators can utilize the historical significance of microbes in shaping Earth's evolution, highlighting their influential contribution to the composition of the atmosphere and their mutually beneficial interactions with more advanced life forms. Incorporating the study of microbial ecology into the curriculum would provide students with a comprehensive understanding of the complex interactions between microbes and their surroundings, as initiated by Ernst Haeckel in the 1800s. This method aligns with the changing nature of microbiology education, equipping students to tackle the difficulties and possibilities brought about by growing research areas in comprehending and utilizing the potential of microbes for ecological equilibrium.

Microorganisms play a vital role in various climate-related processes, including the cycling of carbon, emission of greenhouse gases, and soil health (96). Therefore, future microbiology teaching should emphasize these issues, emphasizing the interdependence of microbial activities and climate dynamics. By integrating climate change issues into microbiology courses, students can expand their horizons and prepare to participate actively in current efforts to address climate-related challenges. This comprehensive approach is aligned with the evolving field of microbiology education, equipping students with the tools to tackle urgent global challenges from a microbiological perspective.

Microbial biostimulants have the potential to be a sustainable and effective approach for mitigating the increased abiotic and biotic pressures caused by climate change (97). Furthermore, the utilization of microbial biostimulants has the potential to uphold the ecological equilibrium of agroecosystems, hence reducing the reliance on pesticides and the presence of toxic heavy metals in agricultural practices. However, some factors must be considered when striving for improved product effectiveness and broader utilization, including regulatory considerations and research and development processes (98).

Understanding these dynamics requires an interdisciplinary approach, incorporating insights from host–microbe interactions, microbial distribution alterations, and terrestrial, marine, and urban microbiology changes. Such a holistic perspective aligns with the broader goals of future microbiology education, preparing students to address multifaceted challenges at the intersection of microbiology, climate change, and sustainable practices (99).

16.4 IMPLEMENTATION OF EMERGING FIELDS IN MICROBIOLOGY EDUCATION

The implementation of emerging fields in microbiology education is essential to keep pace with the latest developments in the field. Nowadays, microbiology is constantly dealing with complex systems, emphasizing the importance of meta-analyses of publication (100). Additionally, the implementation of microbiology with emerging fields has provided new opportunities for microbiology educators. This implementation is crucial for exploring the pathogenesis of neoplasms driven by the microbiome, which is an important aspect of modern microbiology (101). The implementation of advancements and learning technologies in microbiology education is essential to develop interest in the field of microbiology (102). The implementation of some of the emerging fields such as microbiome studies, microbial ecology, synthetic microbiology, bioinformatics, biotechnology, quantum microbiology, and microbial forensics in microbiology education are discussed below.

16.4.1 Implementation of Microbiome Studies in Microbiology Education

Microbiome studies have attained significant attention in microbiology education due to their capability to improve learning outcomes and enhance the engagement of students. The implementation of microbiome projects in microbiology undergraduate research is effective in teaching the concepts of microbiology (103, 104). The American Society for Microbiology (ASM) curriculum guidelines for undergraduate students of microbiology have helped provide a comprehensive curriculum for introductory programs or courses in microbiology, further supporting the implementation of microbiome studies in microbiology education (105).

Incorporating microbiome studies in educational platforms can make microbiome research using 16S rRNA approachable for novice researchers in educational settings for microbiology research (106). Microbiology education has been revolutionized by microbiome studies, which have the potential to provide practical research experiences that improve learning outcomes and enhance student engagement. The implementation of microbiome research experiences and projects in undergraduate microbiology education aligns with the emerging microbiology landscape and the increasing demand for skills and knowledge in microbiome studies and analysis of data (107).

16.4.2 Implementation of Environmental Microbiology and Microbial Ecology in Microbiology Education

To effectively implement environmental microbiology and microbial ecology into microbiology education, it is essential to incorporate advanced tools and techniques that are presently shaping the field of environmental microbiology. For example, the use of high-resolution secondary ion mass spectrometry (NanoSIMS) and scanning transmission X-ray microscopy (STXM) has made researchers capable of linking environmental processes to the in situ functioning of microorganisms (108). Additionally, the use of Quantitative Insights into Microbial Ecology (QIIME) software has allowed students to perform taxonomic, phylogenetic, and statistical analyses on bacterial communities, providing a practical approach to understanding microbial ecology (109). Furthermore, the integration of advanced mathematical approaches with microbiological experiments can be introduced to microbial ecological studies, offering a multidisciplinary perspective that can enrich microbiology education (110).

16.4.3 Implementation of Synthetic Microbiology and Genetic Engineering in Microbiology Education

Over the last decade, synthetic microbiology has emerged as a field for the engineering of biological systems. Working with microorganisms poses unique challenges in engineering due to an incomplete understanding of natural biological systems and the tools available for manipulating them. Synthetic microbiology focuses on fundamental platforms for efficient engineering of biological systems, including DNA construction, computational tools for designing, parts libraries, and interfaces for manipulating synthetic circuits. Developing these enabling technologies requires applying an engineering mindset to microbiology education, emphasizing the implementation of generalizable techniques of synthetic microbiology and genetic engineering in microbiology education (111).

The projects of synthetic microbiology have been implemented in a wide range of training and educational practices through the iGEM competition (International Genetically Engineered Machines) (112). Many undergraduate students in Canada were introduced to synthetic microbiology through their participation in the annual iGEM competition . Transdisciplinary Synthetic Biology Education (TSBE) provides a framework for implementing best educational practices in synthetic microbiology for undergraduate microbiology students through qualitative analysis of data obtained from the iGEM competition (113).

Synthetic microbiology provides an ideal opportunity for microbiology educators to promote laboratory courses with project-based teaching styles. Implementing synthetic microbiology engages students in inquiry-based learning programs to enhance their experience in the scientific process (114).

16.4.4 Implementation of Bioinformatics and Computational Microbiology in Microbiology Education

Since the completion of the first genome sequencing in the 1990s, the field of bioinformatics has rapidly evolved, indicating the significance of bioinformatics in education (115). Microbiology depends upon bioinformatics for understanding complex interactions of microorganisms. Undergraduate microbiology students often lack the basic computer and quantitative skills required for bioinformatics analysis. To address this challenge, the course module "A Town on Fire! 16S rRNA Gene Amplicon for Analysis of Microbial Communities" was developed for an undergraduate microbiology course, which provides a resource for implementing bioinformatics in other undergraduate microbiology teaching (109).

Microbiology has utilized bioinformatics to identify pathogens, discover virulence factors, analyze resistance and type strains, among other applications. Although specialist bioinformatic skills, such as software development, are not essential for most life science researchers, it is becoming increasingly important for all researchers in this field to possess basic skills in using common bioinformatics software and interpreting results effectively (116). It is imperative to prioritize the development of this competence in the professional growth of contemporary life scientists. The learning process should begin with introductory courses at the undergraduate level to introduce and cultivate the essential abilities at an early stage. Ultimately, bioinformatics should be fully incorporated into degree programs in the field of life science. Introductory bioinformatics primarily focuses on the various bioinformatics tools, their resources (databases), and methods of application. Its purpose is to equip life scientists who are not directly involved in a bioinformatics program with a basic understanding of bioinformatics concepts. This knowledge is important to prevent the misuse and incorrect interpretation of biological data (117).

There are currently multiple research publications that have shown the effective use of active learning methodologies in teaching bioinformatics and computational biology courses (118–120). The cooperative learning approach in bioinformatics, as suggested by Wright et al., entails students collaborating in groups or pairs. The class exercises are deliberated in groups or using the think-pair-share method (121). Kruchten proposed the integration of the bioinformatics module into an existing course like general microbiology or as a separate course in the curriculum (122). Similarly, Goodman and Dekhtyar proposed an effective instructional technique that involves teaching interdisciplinary communication and collaboration skills to students studying biological science and computer science (123).

Additionally, the implementation and design of bioinformatics have been described through the Bioinformatics Life Science Curriculum (BLSC) program, emphasizing the structure and organization of the program (124).

The implementation of computational microbiology and bioinformatics in microbiology education is supported by various studies and initiatives, emphasizing the need for curriculum guidelines, student-centered approaches, and authentic research experiences. Various attempts have been made to implement bioinformatics exercises into microbiology courses. These efforts aim to enhance the understanding and application of microbiological concepts in educational settings (125).

16.4.5 Implementation of Microbial Biotechnology in Microbiology Education

Biotechnology has numerous applications in various sectors of microbiology. Microbiology is a broad field that includes many areas such as food, dairy, agriculture, industrial, environmental, marine, veterinary, and medical microbiology which require skill development and specialized training in biotechnology for individuals pursuing expertise within a specific area of microbiology. Microbiologists are in high demand in the biotechnology industry for preparing enzymes, emphasizing the need for implementing biotechnology education in microbiology courses (126).

Professional development for microbiology educators needs to cover highly technical subjects like microbial biotechnology but they experience challenges of not having enough time to practice. The Biotechnology Immersion Program (BiP) was implemented in microbiology education to address this challenge. BiP was designed to incorporate engrossing experimental systems into microbiology professional development. Microbiology educators perform hands-on activities and do assessments with the guidance of faculty expert in biotechnology. BiP focuses on three important components of professional development: personal experience, time, and connection. Through the implementation of this practice system, microbiology educators feel confident and prepared to perform new activities in the biotechnology lab in their microbiology classrooms (127).

The Department of Molecular Biology and Biochemistry of the University of Barcelona also designed a course mainly oriented to microbiology educators. The course, composed of lectures and practical classes, was designed by keeping two aims in mind: firstly, bringing educators up to date with biotechnological aspects relevant to microbiology, and secondly, providing exciting challenges for further discussion in microbiology courses (128).

16.4.6 Implementation of Quantum Microbiology in Microbiology education

To implement quantum microbiology in microbiology education, it is necessary to consider the importance of e-learning in infectious diseases and clinical microbiology (129). E-learning may give a platform for implementing concepts of quantum microbiology into the curriculum, offering an accessible and flexible approach to education. In addition, student perceptions of in-person and online laboratory experiences of microbiology can guide curriculum adaptation in the implementation of quantum microbiology concepts (130). Understanding the perceptions of students will aid in the development of effective strategies for implementing quantum microbiology into laboratory experiences.

Furthermore, the quantum-like model for the adaptive dynamics of regulation of genes in the metabolism of lactose/glucose in *E. coli* offers an opportunity to implement quantum-like formalism into genetics and microbiology education (131). This model can serve as a basis for implementing quantum principles into microbiology education, providing students with a deeper understanding of microbial behavior at the genetic level.

Moreover, a review of transdisciplinary microbiology education emphasizes integrating innovative tools like principles of quantum into biology and microbiology education (132). This supports the implementation of quantum microbiology as a transdisciplinary approach, enhancing the educational experience by bridging quantum concepts and microbiological studies.

16.4.7 Implementation of Forensic Microbiology in Microbiology education

Microbial forensics is an emerging field in microbiology that combines forensic science with microbiology (133). Microbial forensics is an integral component of microbiology which is implemented in microbiology education to teach forensic investigations for the protection of human health (134). The scientific basis, interpretations, applications, and lessons learned by forensics microbiologists need to be reported and then transferred to the next generation of scientists through microbiology education. An education program in microbial forensics is broad and provides information covering all aspects of this field (135).

The main purpose of implementing microbial forensics in microbiology education is the tracking of the source of microbial pathogens to prevent the spread and transmission of infectious diseases (136). It is essential to differentiate between naturally acquired pathogens and intentionally released microorganisms that cause infections. This information is important in the formulation of procedures to prevent transmission of microbial infections and in apprehending persons involved in biowarfare, bioterrorism, and biocrime (134).

There can be various venues and formats for the implementation of microbial forensics in microbiology education. Full programs in academic style are adopted at universities to educate microbiology learners comprehensively in this field. Law enforcement officials and microbiologists are mainly trained in relevant aspects of this new field but may require additional training to incorporate their expertise effectively to fulfill the demands of microbial forensics (108).

Implementation of broad and specific microbial forensics represents a challenge to educators. This must be met through the development of expertise and experts that we desperately need to combat biocrime and bioterrorism (133). Additional training in forensic science programs in microbiology education could serve to educate students about microbial forensics. It is essential to emphasize the implementation of microbial forensics as a specific microbiology profession (135).

16.5 REDEFINING PEDAGOGY FOR MICROBIOLOGY EDUCATION

Redefining pedagogy for microbiology education is an urgent concern as emphasized by recent research studies. The pandemic caused by COVID-19 created a crisis in society affecting many areas, including education which faced the challenge of online teaching and digital education. Online teaching made learning practical skills difficult for students. So redefining pedagogy is essential to deliver microbiology education effectively in both offline and online settings (137). The delivery of scientific information theoretically requires a redefining of pedagogy, which should include the incorporation of practical learning for effective education in clinical microbiology (138). This highlights the significance of hands-on practical experiences in microbiology education, supporting the perception that pure online microbiology education results in a lack of essential involvement with lab technology (139). The evaluation of laboratories for microbiology educators is essential for the development of a comprehensive curriculum of microbiology (140). In addition, a study underscores the need for the identification of best practices for using online education to ensure a beneficial and engaging learning environment for microbiology students (141). Thus, redefining pedagogy with a balanced plan incorporating hands-on practical and online experiences is essential for comprehensive microbiology education. Developing the role of teachers in microbiology education can improve learning outcomes and enhance the engagement of students in both offline and online settings (142).

Redefining pedagogy for microbiology education requires a deep understanding of creative methods and current trends in the teaching of microbiology. The National Science Education Standards highlight the significance of science laboratories for providing conceptual knowledge and practical skills to science students (143). This emphasizes the importance of innovative pedagogical methods in microbiology education. In addition, microbiology educators in higher education are greatly contributing to pedagogical publications that positively impact student retention, teaching quality, and course satisfaction (144).

Innovative pedagogical approaches such as active projects and problem-based learning are becoming popular in microbiology education with the aim of diverse student populations and improving training related to microbiology (145). Additionally, the shift toward student-centered methods is reflected by curricular guideline development for introductory microbiology courses.

The improvement of learning outcomes and students engagement requires a redefining of pedagogy for microbiology education (146, 147).

Furthermore, the Curriculum Guidelines by the American Society of Microbiology for undergraduate students provide outlines for introductory courses in microbiology (145). Education committees of ASM created two nationally validated and tested concepts of microbiology, for health sciences microbiology and for general microbiology, for the assessment of students' learning based on the Curriculum Guidelines of ASM. The Curriculum Guidelines of ASM for undergraduate microbiology yield 27 basic statements and 13 scientific skills for the laboratory. They provide guidance to microbiology educators in their course design. The ASM community proposes these statements to cover basic concepts that students should comprehend after a general course in microbiology. ASM provides resources and opportunities for professional development to enable more microbiology educators to implement ASM Guidelines in their classrooms, reflecting the collaborative and supportive efforts of professional societies to mold microbiology education with pedagogical reforms (145). We hope that instructors will be empowered by this work to use student-centered pedagogical approaches in introductory courses, ensuring a long-lasting legacy among microbiology educators.

In conclusion, redefining pedagogy for teaching microbiology requires a holistic approach that involves addressing challenges posed by the COVID-19 pandemic, integration of theoretical knowledge with practical experiences, adaptation of innovative approaches to pedagogy, and alignment with ASM curriculum guidelines to improve learning outcomes of students. Studies emphasize the significance of including a pedagogical framework as a specific component of the curriculum in microbiology education. The developing perspective of microbiology education demands an adaptable pedagogical framework that fulfills the diverse needs of microbiology students.

16.6 FUTURE PERSPECTIVES

Technological advancements are enhancing the efficiency of processes and products while significantly accelerating the acquisition of information. The past decade has witnessed a significant surge in health innovations, particularly in the form of internet-connected devices that facilitate the effortless sharing of information. Online learning has significantly enhanced the field of microbiology. The utilization of electronic media has been proven to enhance the performance and engagement of students in microbiology examinations (148, 149). Technology should be utilized to facilitate the continuous growth of learners, starting from their undergraduate medical education and extending into their clinical practice. The Josiah Macy Jr. Foundation has published comprehensive guidelines on the utilization of educational technologies in health professions. These encompass the utilization of technology for in-person, virtual, and simulation-driven teaching, as well as technologies for evaluation and incorporation into practical clinical settings (150).

Both learners from the current generation and instructors must be equipped to adapt to a novel education system that primarily depends on technology. Implementing cutting-edge teaching methods, along with educational technologies that encourage active participation in the learning process, can be a complex undertaking (151). Innovation in microbiology education is threatened by the increasing demands on faculty time, the expansion of class sizes and curricula, and the goal of increasing accessibility while maintaining affordability (152). In addition to these limitations, it is important to consider the perceived utility of e-learning, as well as its simplicity of use in some instances (153).

Microbiology education can be improved by integrating cutting-edge technology into teaching, learning, and assessment (154, 155). Utilizing innovative and practical technology gadgets can have an impact on the experiential learning of basic sciences such as microbiology and immunology, as educational technologies encompass materials and devices used to solve practical challenges connected to training and learning. Affordable equipment can stimulate the curiosity of microbiology students in scientific progress, particularly in the field of next-generation sequencing and its diverse uses in medicine and environmental microbiology (156). Utilizing science platforms that specifically emphasize microbiology in conjunction with educational assistance provided by university research facilities and the implementation of bioengineering techniques can enhance the acquisition of microbiology knowledge (112, 154, 157). The experiential learning of students engaged in a microbiology/interdisciplinary microbiology research project has demonstrated that their subjective attitudes change the experiential learning. These changes include an increase in science motivation, confidence, and knowledge, as well as the development of a new social niche. When experiential learning is aligned with the aims of an organization, it can contribute to enhancing quality and safety at the clinical level, as well as improving the education and competence of trainees (158).

In a period of globalization, contemporary microbiology education must create avenues for cultivating a global health workforce that comprehends the significance of culture, science and technology proficiency, and innovative strategies for addressing health and disease. With the growing globalization of health education, it will be essential to establish standardized qualifications in knowledge, competence, and professionalism. The new paradigm for educating future microbiologists will focus on teaching quality, which involves imparting knowledge, skills, and attitudes, and assessing the learning processes rather than just the content. This approach aims to provide physicians with the critical thinking abilities necessary to adapt and succeed in the evolving healthcare landscape (102).

Effective microbiology training in low- to medium-income countries can have a long-lasting beneficial effect by focusing on the key elements of planning, development, and implementation. There is a need to address the significant challenges posed by limited infrastructure, inadequate training facilities, lack of local expertise, and insufficient funding. Developing and implementing an action plan that can result in the more widespread, organized, and cogent spread of microbiology throughout the world, including developing nations, calls for the deliberate collaboration of all relevant parties, including academic institutions, governments, training centers, and researchers. Teaching microbiology in underdeveloped nations presents significant challenges due to its multidisciplinary nature, as most undergraduate students and faculty members have limited or no prior experience or expertise in the subject (27).

Davis et al. (2020), from the Canadian Society of Microbiologists (CSM) Committee on Microbiology Undergraduate Education, suggested five examples of curriculum innovation that can be integrated into microbiology education. These examples are as follows. (i) The curriculum includes the study of data analysis and critical thinking, with a particular emphasis on ethical discussions. (ii) There is a strong emphasis on experimental design in the laboratory. (iii) The curriculum also includes elements of science communication. (iv) There is a specific focus on developing competencies in bioinformatics. (v) Additionally, the curriculum incorporates principles of equity, diversity, and inclusion into the study of microbiology. To achieve a wider and more organized distribution of bioinformatics, it is necessary for all stakeholders, including research institutions, governmental agencies, teachers, trainers, and researchers, to collaborate and actively contribute to the development and implementation of an action plan (159).

REFERENCES

1. Varro, M.T., *The three books of M. Terentius Varro concerning agriculture*. 1800: At the University Press, for the author, and sold by J. Cooke, by F. and C ....
2. Aishat, A., Microbiology education in nigeria: common training deficits could be addressed with modern educational tools. *Journal of Microbiology & Biology Education*, 2019. 20(1): p. 10.1128/jmbe.v20i1.1595.
3. Kolter, R., The history of microbiology—a personal interpretation. *Annual Review of Microbiology*, 2021. 75: p. 1–17.
4. Blevins, S.M. and M.S. Bronze, Robert Koch and the 'golden age' of bacteriology. *International Journal of Infectious Diseases*, 2010. 14(9): p. e744–e751.
5. Quammen, D., *Breathless: the scientific race to defeat a deadly virus*. 2023: Simon and Schuster.

6. Peralta, A.L., et al., A social–ecological framework for "micromanaging" microbial services. *Frontiers in Ecology and the Environment*, 2014. 12(9): p. 524–531.
7. Lennon, J., et al., Microbiology and climate change: a transdisciplinary imperative. *Mbio*, 2023. 14(1): p. e03335–22.
8. Sampaio-Maia, B., et al., Wiki as a tool for microbiology teaching, learning and assessment. *European Journal of Dental Education*, 2014. 18(2): p. 91–97.
9. Pommerville, J.C., *Fundamentals of microbiology*. 2013: Jones & Bartlett Publishers.
10. Goodswen, S.J., et al., Machine learning and applications in microbiology. *FEMS Microbiology Reviews*, 2021. 45(5): p. fuab015.
11. Cuccuru, G., et al., Orione, a web-based framework for NGS analysis in microbiology. *Bioinformatics*, 2014. 30(13): p. 1928–1929.
12. Jiang, S., et al., Next-generation sequencing applications for the study of fungal pathogens. *Microorganisms*, 2022. 10(10): p. 1882.
13. Hernández-de-Diego, R., et al., The eBioKit, a stand-alone educational platform for bioinformatics. *PLoS Computational Biology*, 2017. 13(9): p. e1005616.
14. Salam, M.A., et al., Conventional methods and future trends in antimicrobial susceptibility testing. *Saudi Journal of Biological Sciences*, 2023. 30(3): p. 103582.
15. Anderson, M., M. Cecchini, and E. Mossialos, *Challenges to tackling antimicrobial resistance: economic and policy responses*. 2020: Cambridge University Press.
16. Roope, L.S., et al., The challenge of antimicrobial resistance: what economics can contribute. *Science*, 2019. 364(6435): p. eaau4679.
17. Thakral, Y., Digital monitoring for data-driven antimicrobial stewardship: a process perspective from resource-constrained contexts in India. *Frontiers in Antibiotics*, 2023. 2: p. 1214826.
18. Schemes, P., Pediatrics in a resource-constrained setting. In *Hunter's Tropical Medicine and Emerging Infectious Diseases E-Book*, ed. E.T. Ryan, D.R. Hill, T. Solomon, N. Aronson, and T.P. Endy. 2019: p. 141. Elsevier.
19. Aristovnik, A., et al., Impact of the COVID-19 pandemic on online learning in higher education: a bibliometric analysis. *Frontiers in Education*, 2023. 8: p. 225834.
20. Buchmann, C. and E. Hannum, Education and stratification in developing countries: a review of theories and research. *Annual Review of Sociology*, 2001. 27(1): p. 77–102.
21. Juma, C., et al., Global governance of technology: meeting the needs of developing countries. *International Journal of Technology Management*, 2001. 22(7–8): p. 629–655.
22. Kujur, S., et al., Constraints faced by the students in use of educational technology for learning. *Communication Approaches*, 2019. 37: p. 12.
23. Ohanu, I.B. and C.A. Chukwuone, Constraints to the use of online platform for teaching and learning technical education in developing countries. *Education and Information Technologies*, 2018. 23: p. 3029–3045.
24. Öman, C.B., K.S. Gamaniel, and M.E. Addy, Properly functioning scientific equipment in developing countries. *Analytical Chemistry*, 2006. 78(15): p. 5273–5276.
25. Dickeson, R.C., *Prioritizing academic programs and services: Reallocating resources to achieve strategic balance, revised and updated*. 2010: John Wiley & Sons.
26. Sawant, A.P., et al., Is the undergraduate microbiology curriculum preparing students for careers in their field?: an assessment of biology majors' conceptions of growth and control of microorganisms. *International Journal of STEM Education*, 2018. 5(1): p. 1–14.
27. Adenaike, O., O.E. Olabanjo, and A.A. Adedeji, Integrating computational skills in undergraduate Microbiology curricula in developing countries. *Biology Methods and Protocols*, 2023. 8(1): p. bpad008.
28. Sender, R., S. Fuchs, and R. Milo, Revised estimates for the number of human and bacteria cells in the body. *PLoS Biology*, 2016. 14(8): p. e1002533.
29. Gilbert, J.A., et al., Current understanding of the human microbiome. *Nature Medicine*, 2018. 24(4): p. 392–400.
30. Ley, R.E., et al., Evolution of mammals and their gut microbes. *Science*, 2008. 320(5883): p. 1647–1651.
31. Schwabe, R.F. and C. Jobin, The microbiome and cancer. *Nature Reviews Cancer*, 2013. 13(11): p. 800–812.
32. Xavier, J.B., et al., The cancer microbiome: distinguishing direct and indirect effects requires a systemic view. *Trends in Cancer*, 2020. 6(3): p. 192–204.

33. Cho, I. and M.J. Blaser, The human microbiome: at the interface of health and disease. *Nature Reviews Genetics*, 2012. 13(4): p. 260–270.
34. Poore, G.D., et al., Microbiome analyses of blood and tissues suggest cancer diagnostic approach. *Nature*, 2020. 579(7800): p. 567–574.
35. Song, C., et al., Expression profile analysis of microRNAs in prostate cancer by next-generation sequencing. *The Prostate*, 2015. 75(5): p. 500–516.
36. LeBlanc, V.G. and M.A. Marra, Next-generation sequencing approaches in cancer: where have they brought us and where will they take us? *Cancers*, 2015. 7(3): p. 1925–1958.
37. Wong, A.C. and M. Levy, New approaches to microbiome-based therapies. *MSystems*, 2019. 4 (3): p. 10.1128/msystems.00122-19.
38. Schlaberg, R., Microbiome diagnostics. *Clinical Chemistry*, 2020. 66(1): p. 68–76.
39. Mirzayi, C., et al., Reporting guidelines for human microbiome research: the STORMS checklist. *Nature Medicine*, 2021. 27(11): p. 1885–1892.
40. Gevers, D., et al., The Human Microbiome Project: a community resource for the healthy human microbiome. PloS Biology, 2012. 10(8): e1001377.
41. Qin, J., et al., A human gut microbial gene catalogue established by metagenomic sequencing. *Nature*, 2010. 464(7285): p. 59–65.
42. McDonald, D., et al., American gut: an open platform for citizen science microbiome research. *Msystems*, 2018. 3(3): p. 10.1128/msystems.00031-18.
43. Kashyap, P.C., et al. Microbiome at the frontier of personalized medicine. in *Mayo Clinic Proceedings*. 2017. Elsevier.
44. Moe-Behrens, G.H., R. Davis, and K.A. Haynes, Preparing synthetic biology for the world. *Frontiers in Microbiology*, 2013. 4: p. 5.
45. Heinemann, M. and S. Panke, Synthetic biology—putting engineering into biology. *Bioinformatics*, 2006. 22(22): p. 2790–2799.
46. Chubukov, V., et al., Synthetic and systems biology for microbial production of commodity chemicals. *NPJ Systems Biology and Applications*, 2016. 2(1): p. 1–11.
47. Adkins, J., et al., Engineering microbial chemical factories to produce renewable "biomonomers". *Frontiers in Microbiology*, 2012. 3: p. 313.
48. Lamsen, E.N. and S. Atsumi, Recent progress in synthetic biology for microbial production of C3–C10 alcohols. *Frontiers in Microbiology*, 2012. 3: p. 196.
49. Lee, S.J., S.-J. Lee, and D.-W. Lee, Design and development of synthetic microbial platform cells for bioenergy. *Frontiers in Microbiology*, 2013. 4: p. 92.
50. Immethun, C.M., et al., Microbial production of isoprenoids enabled by synthetic biology. *Frontiers in Microbiology*, 2013. 4: p. 75.
51. Misra, A., et al., Metabolic analyses elucidate non-trivial gene targets for amplifying dihydroartemisinic acid production in yeast. *Frontiers in Microbiology*, 2013. 4: p. 200.
52. Conway, M.F., et al., Metabolic analyses elucidate nontrivial gene targets for amplifying dihydroartemisinic acid production in yeast. 2013.
53. Fraser, T.H., The future of recombinant DNA technology in medicine. *Perspectives in Biology and Medicine*, 1980. 23(4): p. 499–512.
54. Teuber, M., Genetic engineering techniques in food microbiology and enzymology. *Food Reviews International*, 1993. 9(3): p. 389–409.
55. Robinson, W.H., et al., Protein microarrays guide tolerizing DNA vaccine treatment of autoimmune encephalomyelitis. *Nature Biotechnology*, 2003. 21(9): p. 1033–1039.
56. Zagursky, R.J., et al., Bioinformatics: how it is being used to identify bacterial vaccine candidates. *Expert Review of Vaccines*, 2003. 2(3): p. 417–436.
57. Rao, V.S., et al., Recent developments in life sciences research: role of bioinformatics. *African Journal of Biotechnology*, 2008. 7(5): p. 495–503.
58. Loman, N.J. and M.J. Pallen, Twenty years of bacterial genome sequencing. *Nature Reviews Microbiology*, 2015. 13(12): p. 787–794.
59. Gottweis, H., Governing genomics in the 21st century: between risk and uncertainty. *New Genetics and Society*, 2005. 24(2): p. 175–194.
60. Liu, J., et al., Antimicrobial drug discovery through bacteriophage genomics. *Nature Biotechnology*, 2004. 22(2): p. 185–191.

61. Nakashima, N., Y. Mitani, and T. Tamura, Actinomycetes as host cells for production of recombinant proteins. *Microbial Cell Factories*, 2005. 4(1): p. 1–5.
62. Thorne, B.C., A.M. Bailey, and S.M. Peirce, Combining experiments with multi-cell agent-based modeling to study biological tissue patterning. *Briefings in Bioinformatics*, 2007. 8(4): p. 245–257.
63. Bock, C. and T. Lengauer, Computational epigenetics. *Bioinformatics*, 2008. 24(1): p. 1–10.
64. Bansal, A.K., Bioinformatics in microbial biotechnology–a mini review. *Microbial Cell Factories*, 2005. 4: p. 1–11.
65. Suárez-Díaz, E., Making room for new faces: evolution, genomics and the growth of bioinformatics. *History and Philosophy of the Life Sciences*, 2010. 32(1): p. 65–89.
66. Bansal, A.K. and T.E. Meyer, Evolutionary analysis by whole-genome comparisons. *Journal of Bacteriology*, 2002. 184(8): p. 2260–2272.
67. Woese, C., The universal ancestor. *Proceedings of the National Academy of Sciences*, 1998. 95(12): p. 6854–6859.
68. Matsuno, K., Forming and maintaining a heat engine for quantum biology. *BioSystems*, 2006. 85(1): p. 23–29.
69. Bittner, E.R. and A. Czader, Quantum mechanics in biology: photoexcitations in DNA. In *Energy Transfer Dynamics in Biomaterial Systems*, ed. I Burghardt, V. May, D.A. Micha, and E.R. Bittner. 2009, Springer. p. 103–126.
70. Davies, P.C., Does quantum mechanics play a non-trivial role in life? *BioSystems*, 2004. 78(1–3): p. 69–79.
71. Ishizaki, A., et al., Quantum coherence and its interplay with protein environments in photosynthetic electronic energy transfer. *Physical Chemistry Chemical Physics*, 2010. 12(27): p. 7319–7337.
72. Bezy, R.P. and D.M. Kehoe, Functional characterization of a cyanobacterial OmpR/PhoB class transcription factor binding site controlling light color responses. *Journal of Bacteriology*, 2010. 192(22): p. 5923–5933.
73. Trevors, J. and L. Masson, Quantum microbiology. *Current Issues in Molecular Biology*, 2011. 13(2): p. 43–50.
74. Pattee, H.H., Quantum mechanics, heredity and the origin of life. *Journal of Theoretical Biology*, 1967. 17(3): p. 410–420.
75. Lehman, D.C., Forensic microbiology. *Clinical Laboratory Science*, 2012. 25(2): p. 114.
76. He, Q., et al., Advances in microbial metagenomics and artificial intelligence analysis in forensic identification. *Frontiers in Microbiology*, 2022. 13: p. 1046733.
77. Johnson, H.R., et al., A machine learning approach for using the postmortem skin microbiome to estimate the postmortem interval. *PloS One*, 2016. 11(12): p. e0167370.
78. Yang, J., et al., Cutibacterium acnes (Propionibacterium acnes) 16S rRNA genotyping of microbial samples from possessions contributes to owner identification. *Msystems*, 2019. 4(6): p. 10.1128/msystems.00594-19.
79. López, C.D., et al., Novel taxonomy-independent deep learning microbiome approach allows for accurate classification of different forensically relevant human epithelial materials. *Forensic Science International: Genetics*, 2019. 41: p. 72–82.
80. Huang, L., et al., A machine learning framework to determine geolocations from metagenomic profiling. *Biology Direct*, 2020. 15(1): p. 1–12.
81. Panicker, R.O., et al., Automatic detection of tuberculosis bacilli from microscopic sputum smear images using deep learning methods. *Biocybernetics and Biomedical Engineering*, 2018. 38(3): p. 691–699.
82. Ma, P., et al., A state-of-the-art survey of object detection techniques in microorganism image analysis: from classical methods to deep learning approaches. *Artificial Intelligence Review*, 2023. 56(2): p. 1627–1698.
83. Oliveira, M. and A. Amorim, Microbial forensics: new breakthroughs and future prospects. *Applied Microbiology and Biotechnology*, 2018. 102: p. 10377–10391.
84. Shin, S.G., et al., A comprehensive microbial insight into two-stage anaerobic digestion of food waste-recycling wastewater. *Water Research*, 2010. 44(17): p. 4838–4849.
85. Bhalla, T.C., N.N. Sharma, and M. Sharma, Production of metabolites, industrial enzymes, amino acid, organic acids, antibiotics, vitamins and single cell proteins. *Journal of Environmental Issues*, 2007. 6: p. 34–78.
86. Ciani, M., et al., Microbes: food for the future. *Foods*, 2021. 10(5): p. 971.

87. Patra, J.K., G. Das, and H.-S. Shin, *Microbial Biotechnology: Volume 2. Application in Food and Pharmacology*. 2018: Springer.
88. Kapoor, D., et al., Microbes in pharmaceutical industry. In *Microbial Diversity, Interventions and Scope*, ed. S.G. Sharma, N.J. Sharma, and M Sharma. 2020: Springer. p. 259–299.
89. Subba, R., *Biofertilizers in agriculture and forestry*. 1993: International Science Publisher.
90. Mohammadi, K. and Y. Sohrabi, Bacterial biofertilizers for sustainable crop production: a review. *ARPN Journal of Agricultural and Biological Science*, 2012. 7(5): p. 307–316.
91. Hoagland, R.E., et al., Bioherbicides: research and risks. *Toxin Reviews*, 2007. 26(4): p. 313–342.
92. Carlini, C.R. and M.F. Grossi-de-Sá, Plant toxic proteins with insecticidal properties. A review on their potentialities as bioinsecticides. *Toxicon*, 2002. 40(11): p. 1515–1539.
93. Rettenbacher, L.A., et al., Microbial protein cell factories fight back? *Trends in Biotechnology*, 2022. 40(5): p. 576–590.
94. Matassa, S., et al., Microbial protein: future sustainable food supply route with low environmental footprint. *Microbial Biotechnology*, 2016. 9(5): p. 568–575.
95. Mills, D.B. and D.E. Canfield, Oxygen and animal evolution: did a rise of atmospheric oxygen "trigger" the origin of animals? *BioEssays*, 2014. 36(12): p. 1145–1155.
96. Dastagir, M.R., Role of microorganisms in managing climate change impacts. In *Microbial Interventions in Agriculture and Environment: Volume 3: Soil and Crop Health Management*, ed. D.P. Singh and R. Prabha. 2019, Springer. p. 1–16.
97. Fadiji, A.E., et al., The potential role of microbial biostimulants in the amelioration of climate change-associated abiotic stresses on crops. *Frontiers in Microbiology*, 2022. 12: p. 829099.
98. Tiedje, J.M., et al., Microbes and climate change: a research prospectus for the future. *Mbio*, 2022. 13(3): p. e00800–22.
99. Sangiorgio, D., et al., Facing climate change: application of microbial biostimulants to mitigate stress in horticultural crops. *Agronomy*, 2020. 10(6): p. 794.
100. Stres, B. and L. Kronegger, Shift in the paradigm towards next-generation microbiology. *FEMS Microbiology Letters*, 2019. 366(15): p. fnz159.
101. Hamada, T., et al., Integration of microbiology, molecular pathology, and epidemiology: a new paradigm to explore the pathogenesis of microbiome-driven neoplasms. *The Journal of Pathology*, 2019. 247(5): p. 615–628.
102. Cervantes, J., The future of infectious diseases education. *Medical Science Educator*, 2020. 30(4): p. 1783–1785.
103. Zelaya, A.J., et al., The bean beetle microbiome project: a course-based undergraduate research experience in microbiology. *Frontiers in Microbiology*, 2020. 11: p. 577621.
104. Fuller, K.S. and C. Torres Rivera, A culturally responsive curricular revision to improve engagement and learning in an undergraduate microbiology lab course. *Frontiers in Microbiology*, 2021. 11: p. 577852.
105. Horak, R.E., S. Merkel, and A. Chang, The ASM curriculum guidelines for undergraduate microbiology: a case study of the advocacy role of societies in reform efforts. *Journal of Microbiology & Biology Education*, 2015. 16(1): p. 100–104.
106. Mitchell, K., et al., PUMA: a tool for processing 16S rRNA taxonomy data for analysis and visualization. bioRxiv, 2018: p. 482380.
107. Ras, V., et al., Using a multiple-delivery-mode training approach to develop local capacity and infrastructure for advanced bioinformatics in Africa. *PLoS Computational Biology*, 2021. 17(2): p. e1008640.
108. Behrens, S., A. Kappler, and M. Obst, Linking environmental processes to the in situ functioning of microorganisms by high-resolution secondary ion mass spectrometry (NanoSIMS) and scanning transmission X-ray microscopy (STXM). *Environmental Microbiology*, 2012. 14(11): p. 2851–2869.
109. Tobin, T.C. and A. Shade, A town on fire! Integrating 16S rRNA gene amplicon analyses into an undergraduate microbiology lecture class. *FEMS Microbiology Letters*, 2018. 365(10): p. fny104.
110. Haruta, S., et al., Challenges for complex microbial ecosystems: combination of experimental approaches with mathematical modeling. *Microbes and Environments*, 2013. 28(3): p. 285–294.
111. Cheng, A.A. and T.K. Lu, Synthetic biology: an emerging engineering discipline. *Annual Review of Biomedical Engineering*, 2012. 14: p. 155–178.
112. Kelwick, R., et al., Promoting microbiology education through the iGEM synthetic biology competition. *FEMS Microbiology Letters*, 2015. 362(16): p. fnv129.

113. Diep, P., et al., Advancing undergraduate synthetic biology education: insights from a Canadian iGEM student perspective. *Canadian Journal of Microbiology*, 2021. 67(10): p. 749–770.
114. Beach, D.L. and C.J. Alvarez, Biotechnology by design: an introductory level, project-based, synthetic biology laboratory program for undergraduate students. *Journal of Microbiology & Biology Education*, 2015. 16(2): p. 237–246.
115. Mishra, S., et al., Introduction to the world of bioinformatics, in *A Guide to Applied Machine Learning for Biologists*, ed. M. Badar. 2023: Springer. p. 105–126.
116. Cattley, S. and J.W. Arthur, BioManager: the use of a bioinformatics web application as a teaching tool in undergraduate bioinformatics training. *Briefings in Bioinformatics*, 2007. 8(6): p. 457–465.
117. Gauthier, J., et al., A brief history of bioinformatics. *Briefings in Bioinformatics*, 2019. 20(6): p. 1981–1996.
118. Newell, P.D., et al., A small-group activity introducing the use and interpretation of BLAST. *Journal of Microbiology & Biology Education*, 2013. 14(2): p. 238–243.
119. Greenman, N.A., et al., Genomics of environmental Salmonella: engaging students in the microbiology and bioinformatics of foodborne pathogens. *Frontiers in Microbiology*, 2021. 12: p. 592422.
120. Audouze, K. and O. Taboureau, Emerging Bioinformatics Methods and Resources in Drug Toxicology, in *In Silico Methods for Predicting Drug Toxicity*, ed. E. Benfenati. 2022, Springer. p. 133–146.
121. Wright, A.M., et al., The why, when, and how of computing in biology classrooms. *F1000Research*, 2019. 8: p.1854.
122. Kruchten, A.E., A curricular bioinformatics approach to teaching undergraduates to analyze metagenomic datasets using R. *Frontiers in Microbiology*, 2020. 11: p. 578600.
123. Goodman, A.L. and A. Dekhtyar, Teaching bioinformatics in concert. *PLoS Computational Biology*, 2014. 10(11): p. e1003896.
124. Miskowski, J.A., et al., Design and implementation of an interdepartmental bioinformatics program across life science curricula. *Biochemistry and Molecular Biology Education*, 2007. 35(1): p. 9–15.
125. Dinsdale, E., et al., NIBLSE: a network for integrating bioinformatics into life sciences education. *CBE—Life Sciences Education*, 2015. 14(4): p. le3.
126. Pal, K., R. Dey, and K. Tilak, Microbiology Education for Biotechnology Industry, in *Evolving Corporate Education Strategies for Developing Countries: The Role of Universities*, ed. B. PanduRanga Narasimharao, S. Rangappa Kanchugarakoppai, and T.U. Fulzele. 2013, IGI Global. p. 209–227.
127. Bates, K.L., C. Hung, and J.J. Steel, Biotechnology Immersion Program: professional development where the participants do the preparation, teaching, and outreach to maximize learning gains. *FEMS Microbiology Letters*, 2022. 369(1): p. fnac111.
128. Fernández-Novell, J., et al., Biotechnology, microbiology and secondary school. In *Modern Multidisciplinary Applied Microbiology: Exploiting Microbes and Their Interactions*, ed. A. Mendez-Vilas. 2006, Wiley. p. 719–725.
129. Sayıner, A.A. and E. Ergönül, E-learning in clinical microbiology and infectious diseases. *Clinical Microbiology and Infection*, 2021. 27(11): p. 1589–1594.
130. Brockman, R.M., et al., Student perceptions of online and in-person microbiology laboratory experiences in undergraduate medical education. *Medical Education Online*, 2020. 25(1): p. 1710324.
131. Asano, M., et al., Quantum-like model for the adaptive dynamics of the genetic regulation of E. coli's metabolism of glucose/lactose. *Systems and Synthetic Biology*, 2012. 6: p. 1–7.
132. Capozzi, V., G. Spano, and D. Fiocco, Transdisciplinarity and microbiology education. *Journal of Microbiology & Biology Education*, 2012. 13(1): p. 70–73.
133. Tucker, J.B. and G.D. Koblentz, The four faces of microbial forensics. *Biosecurity and Bioterrorism: Biodefense Strategy, Practice, and Science*, 2009. 7(4): p. 389–397.
134. Castro, A.E. and M.C.A. De Ungria, Methods used in microbial forensics and epidemiological investigations for stronger health systems. *Forensic Sciences Research*, 2022. 7(4): p. 650–661.
135. Schutzer, S.E., B. Budowle, and P.S. Keim, Microbial Forensics: Educating the Workforce and the Community, in *Microbial Forensics*, 2nd edn, ed. B. Budowle et al. 2011, Elsevier. p. 667–680.
136. Khan, A.S., P.S. Amara, and S.A. Morse, Forensic Public Health: Epidemiological and Microbiological Investigations for Biosecurity, in *Microbial Forensics*, 3rd edn, ed. B. Budowle. 2020, Elsevier. p. 105–122.
137. Radu, M.C., et al., Teaching microbiology-online and offline challenges during COVID-19 pandemic. *South Asian Journal of Research in Microbiology*, 2022. 12(4): p. 34–40.

138. Joji, R.M., et al., Perception of online and face to face microbiology laboratory sessions among medical students and faculty at Arabian Gulf University: a mixed method study. *BMC Medical Education*, 2022. 22(1): p. 411.
139. Robinson, J.M., et al., Twenty important research questions in microbial exposure and social equity. *Msystems*, 2022. 7(1): p. e01240–21.
140. Lindsey, N., J. Pate, and L.A. Blankenship, A case study exploring the development of a quality open education clinical microbiology lab manual and online experiential lab course. *Journal of Open Educational Resources in Higher Education*, 2022. 1(1): p. 48–67.
141. McKenzie, G. and H. Hawk, *Best practices in online education*. Best Practices in Teaching Nursing, 2023.
142. Kim, M.K. and V.A. Pollar, A modest critical pedagogy for English as a foreign language education. *Education as Change*, 2017. 21(1): p. 50–72.
143. Hofstein, A. and V.N. Lunetta, The laboratory in science education: foundations for the twenty-first century. *Science Education*, 2004. 88(1): p. 28–54.
144. Lacey, M.M. and G. Efthimiou, A guide to pedagogical research for scientists from a biological sciences background. *Access Microbiology*, 2022. 4(8): acmi000416.
145. Amaral, H.F., et al., Active projects for teaching and learning soil microbiology and applications of inoculants to increase perceived subject matter understanding and acquisition of knowledge. *Brazilian Journal of Microbiology*, 2020. 51: p. 1825–1835.
146. Tripepi, M., Microbiology laboratory simulations: from a last-minute resource during the Covid-19 Pandemic to a valuable learning tool to retain—a semester microbiology laboratory curriculum that uses Labster as prelaboratory activity. *Journal of Microbiology & Biology Education*, 2022. 23(1): p. e00269–21.
147. Kushner, D.B. and A. Pekosz, Virology in the classroom: current approaches and challenges to undergraduate-and graduate-level virology education. *Annual Review of Virology*, 2021. 8: p. 537–558.
148. Guarner, J. and S.M. Niño, Microbiology learning and education online. *Journal of Clinical Microbiology*, 2016. 54(5): p. 1203–1208.
149. Stevens, N.T., et al., Can e-learning improve the performance of undergraduate medical students in Clinical Microbiology examinations? *BMC Medical Education*, 2019. 19: p. 1–8.
150. Stuart, G. and M. Triola. *Enhancing Health Professions Education through Technology: Building a Continuously Learning Health System: Proceedings of a Conference*. 2015, Josiah Macy Jr. Foundation.
151. Huynh, R., The role of E-learning in medical education. *Academic Medicine*, 2017. 92(4): p. 430.
152. McDonald, P., *Are we prepared for the future of health care education*. Online Learning Consortium, 2017.
153. Haidaris, C.G. and J.G. Frelinger, Inoculating a new generation: immunology in medical education. *Frontiers in Immunology*, 2019. 10: p. 2548.
154. Fahnert, B., Edging into the future: education in microbiology and beyond. *FEMS Microbiology Letters*, 2016. 363(7): p. fnw048.
155. Fahnert, B., Be prepared–learning for the future. *FEMS Microbiology Letters*, 2019. 366(16): p. fnz200.
156. Hu, T., et al., Next-generation sequencing technologies: an overview. *Human Immunology*, 2021. 82(11): p. 801–811.
157. Caine, M., et al., BiOutils: an interface to connect university laboratories with microbiology classes in schools. *FEMS Microbiology Letters*, 2015. 362(20): p. fnv171.
158. Gonzalo, J.D., et al., Educating patient-centered, systems-aware physicians: a qualitative analysis of medical student perceptions of value-added clinical systems learning roles. *BMC Medical Education*, 2018. 18: p. 1–7.
159. Davis, M.C., et al., *Finding the silver lining during a global pandemic: opportunities for curriculum innovation in microbiology education*. 2020, NRC Research Press 1840 Woodward Drive, Suite 1, Ottawa, ON K2C 0P7. p. 600–602.

Index